Volcanic Hazards, Risks, and Disasters

Hazards and Disasters Series

Volcanic Hazards, Risks, and Disasters

Series Editor

John F. Shroder
Emeritus Professor of Geography and Geology
Department of Geography and Geology
University of Nebraska at Omaha
Omaha, NE 68182

Volume Editor

Paolo Papale
Istituto Nazionale di Geofisica e Vulcanologia
Via di Vigna Murata, 605 00143 Rome, Italy

ELSEVIER

AMSTERDAM BOSTON HEIDELBERG LONDON NEW YORK OXFORD
PARIS SAN DIEGO SAN FRANCISCO SINGAPORE SYDNEY TOKYO

Elsevier
Radarweg 29, PO Box 211, 1000 AE Amsterdam, Netherlands
The Boulevard, Langford Lane, Kidlington, Oxford OX5 1GB, UK
225 Wyman Street, Waltham, MA 02451, USA

Library of Congress Cataloging-in-Publication Data
Volcanic hazards, risks and disasters / volume editor, Paolo Papale.
pages cm. — (Hazards and disasters series)
ISBN 978-0-12-396453-3 (hardback)
1. Volcanic hazard analysis. I. Papale, Paolo, editor of compilation.
QE527.6.V65 2014
363.34'95–dc23
2014035686

British Library Cataloguing in Publication Data
A catalogue record for this book is available from the British Library

ISBN: 978-0-12-396453-3

Contents

15. Volcanic Gas and Aerosol Hazards from a Future Laki-Type Eruption in Iceland

Anja Schmidt

16. Explosive Super-Eruptions and Potential Global Impacts

Stephen Self

17. Integration of European Volcano Infrastructures

Giuseppe Puglisi, Kristín S. Vogfjörd, Patrick Bachelery and Teresa Ferreira

18. Integrated Monitoring of Japanese Volcanoes

Toshitsugu Fujii and Hitoshi Yamasato

Co-editors

John C. Eichelberger, University of Alaska Fairbanks, 1 907 888 0204 Fairbanks, Alaska, U.S.

Setsuya Nakada, Volcano Research Center, Earthquake Research Institute, University of Tokyo 1-1-1 Yayoi, Bunkyo-ku, Tokyo, 113-0032 Japan

Sue Loughlin, British Geological Survey, Murchison House, West Mains Road, Edinburgh EH9 3LA, UK

Hugo Yepes, Instituto Geofìsico, Escuela Politecnica Nacional, Calle Ladròn de Guevara e Isabel La Catòlica, Quito, Ecuador

Warner Marzocchi, Istituto Nazionale di Geofisica e Vulcanologia, Sezione Roma 1, via di Vigna Murata 605, 00143 Roma, Italy

Patricia A. Mothes, Instituto Geofísico, Escuela Politécnica Nacional, Quito, Ecuador

Setsuya Nakada, Volcano Research Center, Earthquake Research Institute, University of Tokyo, Bunkyo-ku, Tokyo, Japan

Augusto Neri, Istituto Nazionale di Geofisica e Vulcanologia e Sezione di Pisa, via U. Della Faggiola 32, Pisa, Italy

Chris G. Newhall, Earth Observatory of Singapore, Nanyang Technical University, Singapore

John S. Pallister, Volcano Disaster Assistance Program, US Geological Survey, Vancouver, WA USA

Josè L. Palma, Departamento de Ciencias de la Tierra, Universidad de Concepción, Concepción, Chile

Paolo Papale, Istituto Nazionale di Geofisica e Vulcanologia, Rome, Italy

Giuseppe Puglisi, Istituto Nazionale di Geofisica e Vulcanologia, Sezione di Catania – Osservatorio Etneo, Catania, Italy

Gilberto Saccorotti, Istituto Nazionale di Geofisica e Vulcanologia e Sezione di Pisa, via U. Della Faggiola 32, Pisa, Italy

Laura Sandri, Istituto Nazionale di Geofisica e Vulcanologia, Sezione di Bologna, via D. Creti 12, 40128 Bologna, Italy

Anja Schmidt, School of Earth and Environment, University of Leeds, Leeds, United Kingdom

Stephen Self, Volcano Dynamics Group, Department of Environment, Earth and Ecosystems, The Open University, Walton Hall, Milton Keynes, UK; Department of Earth and Planetary Science, University of California, Berkeley, CA, USA

Jacopo Selva, Istituto Nazionale di Geofisica e Vulcanologia, Sezione di Bologna, via D. Creti 12, 40128 Bologna, Italy

Anselm Smolka, Munich Reinsurance Company, Munich, Germany; now: GEM Foundation, Pavia, Italy

Carol Stewart, Joint Centre for Disaster Research, GNS Science/Massey University, Wellington, New Zealand

Andrew Tupper, Australian Bureau of Meteorology, Hydrology Unit, Melbourne, Victoria 3001, Australia

James W. Vallance, U.S. Geological Survey, Cascades Volcano Observatory, Vancouver, WA, USA

José G. Viramonte, Instituto Geonorte – INENCO, Universidad Nacional de Salta, and CONICET, Salta, Argentina

Kristín S. Vogfjörd, Icelandic Meteorological Office, Reykjavík, Iceland

Barry Voight, Penn State University, Geosciences Dept., University Park PA 16802, U.S.

Christina Widiwijayanti, Earth Observatory of Singapore, Nanyang Technological University, Singapore

Thomas M. Wilson, Volcanic Ash Testing Lab, Department of Geological Sciences, University of Canterbury, Christchurch, New Zealand

Gordon Woo, Risk Management Solutions, London

Takahiro Yamamoto, Geological Survey of Japan, AIST, Tsukuba, Ibaraki, Japan

Hitoshi Yamasato, Penn State University, Geosciences Dept., University Park PA 16802, U.S.

Hugo Yepes, Instituto Geofìsico, Escuela Politecnica Nacional, Calle Ladròn de Guevara e Isabel La Catòlica, Quito, Ecuador

Giulio Zuccaro, Dipartimento di Strutture per l'Ingegneria e l'Architettura, Università di Napoli Federico II, Napoli, Italy

Editorial Foreword

General hazards, risks, and disasters: Hazards are processes that produce danger to human life and infrastructure. Risks are the potential or possibilities that something bad will happen because of the hazards. Disasters are that quite unpleasant result of the hazard occurrence that caused destruction of lives and infrastructure. Hazards, risks, and disasters have been coming under increasing strong scientific scrutiny in recent decades as a result of a combination of numerous unfortunate factors, many of which are quite out of control as a result of human actions. At the top of the list of exacerbating factors to any hazard, of course, is the tragic exponential population growth that is clearly not possible to maintain indefinitely on a finite Earth. As our planet is covered ever more with humans, any natural or human-caused (unnatural) hazardous process is increasingly likely to adversely impact life and construction systems. The volumes on hazards, risks, and disasters that we present here are thus an attempt to increase understandings about how to best deal with these problems, even while we all recognize the inherent difficulties of even slowing down the rates of such processes as other compounding situations spiral on out of control, such as exploding population growth and rampant environmental degradation.

Some natural hazardous processes such as volcanoes and earthquakes that emanate from deep within the Earth's interior are in no way affected by human actions, but a number of others are closely related to factors affected or controlled by humanity, even if however unwitting. Chief among these, of course, are climate-controlling factors, and no small measure of these can be exacerbated by the now obvious ongoing climate change at hand (Hay, 2013). Pervasive range fires and forest fires caused by human-enhanced or -induced droughts and fuel loadings, megaflooding into sprawling urban complexes on floodplains and coastal cities, biological threats from locust plagues, and other ecological disasters gone awry; all of these and many others are but a small part of the potentials for catastrophic risk that loom at many different scales, from the local to planet girdling.

In fact, the denial of possible planetwide catastrophic risk (Rees, 2013) as exaggerated jeremiads in media landscapes saturated with sensational science stories and end-of-the-world, Hollywood productions is perhaps quite understandable, even if simplistically shortsighted. The "end-of-days" tropes promoted by the shaggy-minded prophets of doom have been with us for centuries, mainly because of Biblical verse written in the early Iron Age during

Introduction

Paolo Papale [1], John C. Eichelberger [2], Sue Loughlin [3], Setsuya Nakada [4] and Hugo Yepes [5]

[1] *Istituto Nazionale di Geofisica e Vulcanologia, Rome, Italy,* [2] *University of Alaska Fairbanks, Fairbanks, Alaska, U.S.,* [3] *British Geological Survey, Murchison House, Edinburgh, UK,* [4] *Volcano Research Center, Earthquake Research Institute, University of Tokyo, Bunkyo-ku, Tokyo, Japan,* [5] *Instituto Geofìsico, Escuela Politecnica Nacional, Calle Ladròn de Guevara e Isabel La Catòlica, Quito, Ecuador*

Apart from impact by large celestial bodies, a large volcanic eruption represents the single event with the highest destructive potential on Earth. This simple consideration should be sufficient to provide a motivation to this book, as well as to the increasing flow of papers published yearly in the international scientific literature on volcanic hazards, risks, and disasters. Although a rare event never witnessed during historical times, a new Yellowstone-size super-eruption somewhere in the world has a probability not very dissimilar to that of a large nuclear accident, and is up to one order of magnitude more probable than the impact with a 200-m large meteorite. The consequences of such a gigantic eruption would extend all over the world: the rapid injection into the stratosphere of the order of 10^{12} tons of finely fragmented magma and 10^{10} tons of sulfur dioxide would cause a several years-long dramatic decrease in the Sun's radiation reaching the Earth's surface, forcing the world into a volcanic winter. Besides such global effects, any form of life up to several tens or hundreds of kilometers around the vent would be lost. Much smaller, but still large, explosive eruptions with frequencies of a few every centuries can cause great devastation, affect the global climate, and destroy the social and economic fabric over regional areas. And even relatively small eruptions from volcanoes close to heavily urbanized areas, like Naples, Auckland, or Mexico City, can result in enormous tragedies and economic breakdown sufficient to bring down an entire country.

Volcanic hazards originate from a variety of processes: volcanic ash injected into the atmosphere may cause disruption to air traffic routing; damage airframes, critical components, and engines; and even cause engine shutdown; the fallout from umbrella clouds can mantle vast areas with layers of pumice and ash, causing roof collapses, shutting down road traffic and lifelines, destroying crops, damaging infrastructure, and affecting critical systems and human and animal health; pyroclastic flows and surges with up to more than 100 km/h speed, originating from the collapse of explosive eruption columns or of viscous lava domes, can destroy any form of life in their path; lahars forming from the mobilization of rapidly accumulated ash by rain, or from the quick melting of volcanic glaciers, can be as much devastating as pyroclastic flows; partial or total collapse of the edifice of submarine volcanoes, volcanic islands, or nearshore volcanoes, can cause large tsunamis and comparably large devastation; and so on. Such a variety of hazardous phenomena, the potentially global impact of volcanic eruptions, and the consideration that about 800 million people in all continents live close enough to active volcanoes to be substantially affected by their activity, put volcanic risks among the most relevant natural risks on Earth.

During recent decades volcanology has rapidly evolved from a specialist branch of the natural sciences based on observations and descriptions, to become a quantitative multidisciplinary field of study in its own right. The original observational character is complemented in modern volcanology by precise measurements, both remote and in situ, involving sophisticated networks of very precise instruments and extensive use of air- and space-borne technologies; by ambitious laboratory experiments aimed at reproducing physical and chemical processes under a range of conditions, including subcrustal pressure and temperature conditions from magma genesis to storage regions to volcanic conduits, as well as subaerial conditions dominated in explosive eruptions by particulate flows traveling up to supersonic velocities; by the development of increasingly sophisticated physical and mathematical models solved with the aid of supercomputing facilities; by the development of appropriate methods based on probabilities and statistics in order to deal with the large uncertainties dominating volcanic hazard forecasts; and last but not least, by an increasing use of social science methodologies in order to address volcanic risk assessment and volcanic risk reduction.

Anticipating volcanic eruptions is in several respects a less prohibitive task than anticipating the occurrence of earthquakes. Unlike earthquakes, the location of many volcanoes and volcanic fields is known and therefore they can be monitored more efficiently; moreover, the transfer of magma toward the Earth's surface is a process that translates into geophysical and geochemical signals allowing anticipation of a volcanic eruption if sufficient monitoring is in place. Nevertheless, there is complexity that makes the anticipation of eruptions challenging. For example, frequently erupting, open-system volcanoes may in some cases enter new eruption styles with few precursory signals detected; closed system volcanoes may display unrest phenomena for years then have large accelerations over a short time period leading to eruption, leaving little time for civil protection measures. The most intriguing volcanic systems are represented by calderas, that is, depressions left as a consequence of structural collapses following the evacuation of huge masses of magma in a short time during large-magnitude eruptions. Calderas may persist in unrest conditions for decades, periodically showing variations with amplitudes such that they would almost certainly lead to an eruption if observed at more typical stratovolcanoes. On the contrary, some observations suggest that calderas can originate a new eruption following a phase characterized by signals much less relevant than those observed in other periods not followed by any eruption.

What is more difficult to deal with, is the fact that irrespective of the nature of the volcano, still no clear general relationships exist between the intensity and duration of observed preeruptive variations, and the intensity and magnitude of the subsequent volcanic eruption. This uncertainty makes effective and timely communication between scientists and civil protection essential, and has significant implications for civil-protection operations requiring short-term preparedness by all. Much of our capability to forecast the eruption size is based on statistical analysis of previous eruptions, with only limited or exceedingly uncertain links with preeruptive observations. This uncertainty around short-term forecasting is one of the current major limits of volcanology, and one of the major reasons for concern in terms of civil protection operations and planning.

Besides the challenges presented above, getting prepared for volcanic eruptions, or anticipating and mitigating volcanic risks, is a complex issue that involves many more experts than just volcanologists, and many more disciplines than just the science of volcanoes. Risk reduction decisions may have substantial consequences on society, like closing schools, shutting down productive activities, evacuating urbanized areas, or also allowing people to get back home after a crisis, so this all requires careful consideration of costs and benefits; especially when considering that forecasts of volcanic activities are dominated by uncertainties, implying additional risks related to perceived "false alarms." As for other aspects of human life, no way exists to escape from uncertainties. We can only reduce uncertainties; they cannot be eliminated—even in principle—when dealing with highly complex systems dominated by nonlinear processes as for volcano dynamics. Governmental representatives, public officials and more generally decision-makers, as well as the societies themselves, should know that living in the proximity of active volcanoes implies acceptance of a risk that, once again, can be reduced but not eliminated. Driving a car in the traffic or on a freeway equally implies acceptance of a risk; whereas, the single person may have the illusion that such a risk is under control as he or she holds the wheel, public officials do not—they know that every time a person gets in a car and turns the engine on, a percentage likelihood exists of having an accident, and a nonzero percentage likelihood of dying in that accident.

Volcanic risks, as well as the risks associated to other natural phenomena, cannot be eliminated. Risks increase as societal vulnerabilities grow. On the lower extreme, poverty, illiteracy, marginalization, and lack of infrastructure are some of the structural factors that lie behind risks; but even developed countries often fail to pursue effective risk reduction policies. Although the

Chapter 1

Global Distribution of Active Volcanoes

Elizabeth Cottrell
National Museum of Natural History, Smithsonian Institution, 10th and Constitution Avenue NW, Washington, D.C., USA

ABSTRACT

Approximately 1,550 subaerial volcanoes in the world today are thought to have erupted in the last 10,000 years, and thousands more volcanoes ring the seafloor. These active volcanoes mainly occur in curvilinear belts that define tectonic plate boundaries. Hundreds of millions of people live on the flanks of active volcanoes and could suffer the acute affects of even a moderate-sized eruption. Island arc nations such as Indonesia, the Philippines, and Japan host the largest populations within 100 km of an active volcano; Indonesia prompts further distinction as having the most explosive eruptions on record as well as the greatest number of eruption-related fatalities. The Volcanoes of the World database (version 4.1) maintained by Smithsonian's Global Volcanism Program, documents more than 10,000 Holocene eruptions, yet only six eruptions account for more than half of the total quantified fatalities. The assembled data bring into relief large gaps in our understanding of the risks posed by volcanoes and the need for more research into volcanic hazards, risks, and timescales.

1.1 INTRODUCTION

The global distribution of volcanoes and humans are inextricably linked by the simple fact that volcanoes create the land upon which humans live. The creation of land may be one of the greatest benefits afforded to humans by volcanic activity, but it also lies at the heart of volcanic hazards and risks. It is no coincidence that Indonesia, Japan, and the islands of Southeast Asia/Philippines owe their existence to volcanic activity while also having the largest populations lying within 100 km of historically active volcanoes (Siebert et al., 2010, Table 10). These islands have also suffered more than half of all documented eruption-related fatalities (Table 1.1). Volcanic activity brings nutrients to the surface to create fertile landscapes, heat to the surface to

Volcanic Hazards, Risks, and Disasters. http://dx.doi.org/10.1016/B978-0-12-396453-3.00001-0
2015 Published by Elsevier Inc.

TABLE 1.1 History's Deadliest Eruptions from VOTW 4.1, Which Account for >50% of All Documented Volcano-Related Fatalities

Volcano	Country	R&D Spending as % GDP[a]	Human Development Index[b]	VEI	Year of Eruption	Number of Fatalities	Agent of Death
Tambora	Indonesia	0.08	0.629	7	1812	60,000	Famine, pyroclastic flows, tsunami
Krakatau	Indonesia	0.08	0.629	6	1883	36,417	Tsunami, pyroclastic flows
Ilopango[c]	El Salvador	0.08	0.68	6	450	30,000	Pyroclastic flows
Pelèe	Martinique	Not available	0.904	4	1902	28,000	Pyroclastic flows
Nevado del Ruiz	Columbia	0.19	0.719	3	1985	23,080	Lahars
Unzendake	Japan	3.36	0.912	2	1792	14,524	Tsunami, debris avalanche

[a]*2009 data from data.worldbank.org "Research and Development Expenditure."*
[b]*2012 data from The United Nations Development Program. data.undp.org.*
[c]*Not included in the analysis of Witham (2005) or Auker et al. (2013) but documented in VOTW 4.1 based on Smith (1979).*

create geothermal energy, and circulates fluids to create the mineral and gem resources sought by humans. For these reasons, humans will always live in proximity to volcanoes.

In this chapter I describe the global distribution of volcanoes, focusing on Holocene volcanoes—those with confirmed or suspected eruptions occurring in the past 10,000 years. The Global Volcanism Program defines a "volcano" as vents, edifices, and cones within a volcanic field that are believed to share a single magmatic plumbing system. Thus the Michoacán–Guanajuato volcanic field in Mexico, with 5.7 million people in close proximity, counts as a single volcano as do volcanic systems in Iceland that incorporate a central volcano and extensive linear fissure swarms aligned with the rift zone. The Volcanoes of the World (VOTW) database, compiled and hosted by Smithsonian's Global Volcanism Program (GVP) at www.volcano.si.edu, summarizes what is known of the eruptive histories and impacts of these "active," or Holocene, volcanoes. Siebert et al. (2010) gives an excellent overview and analysis of this resource. All data presented here derive from version VOTW 4.1 (not statistically different from Siebert et al. (2010)) unless otherwise attributed, downloaded in October, 2013.

It is important to note that our knowledge of the geological record, and the volcanic eruption record itself, is sparse—particularly when it comes to smaller magnitude eruptions (Deligne et al., 2010; Brown et al., 2014). The VOTW likely records <2 percent of Holocene eruptions should humble us in analysis (Siebert et al., 2010, in press). Today, eyewitness accounts and satellite imagery bring 50–70 eruptions to GVP's attention annually. The >1,500 Holocene volcanoes considered here represent only half of the 3,000 named volcanoes in the Smithsonian database with suspected activity in the last 1.8 million years. However, because historical records do not extend beyond a few thousand years and because the geological record is incomplete, these older eruptive centers only add an additional 1,514 documented eruptions (14 percent) to the 10,746 recorded for the Holocene. At the same time, it is worth pointing out that >85 percent of the largest documented eruptions—those with a Volcano Explosivity Index (VEI, Newhall and Self, 1982) >6—occurred more than 10,000 years ago (LaMEVE database, version1, VOGRIPA, Crosweller et al., 2012). Magnitude 7 and 8 eruptions will surely generate hazards on a global scale in the future (Self, 2006; Self and Blake, 2008) and are considered in detail by Self later in this volume (Self, in this volume).

1.2 PATTERNS IN GLOBAL VOLCANISM AND THEIR ASSOCIATED HAZARDS

The distribution of volcanoes on Earth can be understood through the plate tectonic paradigm. The vast majority of volcanoes occur in curvilinear belts that define tectonic plate boundaries. Volcanic belts can be broadly divided into two major tectonic settings: mid-ocean ridges extend for 75,000 km and

TABLE 1.2 Number of Volcanoes with Humans Living in Proximity and the Total Number of People Living within a Given Radius of an Active Volcano

Radius (km)	Number of Volcanoes	Population within Radius (Millions)
5	1,011	47
10	1,111	58
30	1,263	200
100	1,378	>750

Source: Data from VOTW 4.1 and Seibert et al. (2010).

volcanoes (Siebert et al., 2010). Indonesia has the largest population within 100 km of a Holocene volcano (179 million), followed by the Philippines and Japan (approximately 100 million each) based on 2010 population data.

Ewert and Harpel (2004) defined the Volcano Population Index (VPI) as the number of people living within a specified radius of a volcano, or volcanoes, using the LandScan population data set produced by Oakridge National Laboratory. The regional comparisons of Siebert et al. (2008, 2010) account for overlapping radii and prevent double-counting persons living within a given radii of more than one volcano (Table 1.2). For example, the VPI_{100} of Indonesia would be the number of persons living within a 100 km radius of a Holocene volcano in Indonesia (179 million in 2010). Table 1.3 lists volcanoes with the largest proximal populations, as documented in VOTW 4.1. Interesting from the perspective of hazards and risk, the 17 volcanoes that have the largest populations living within a 5 km radius are the same volcanoes as those with the largest populations living within a 10 km radius. Further analysis reveals that these are not the flanks of stratovolcanoes but are more "inhabitable" volcanic fields, and caldera floors (Table 1.3). Volcanic fields and calderas pose different risks than stratovolcanoes and this suggests that more needs to be done to establish if these densely populated volcanoes create a high risk. Finally, some of the volcanoes on this list do not even have documented Holocene eruptions in the VOTW database but are nonetheless considered potentially active. These volcanoes show geologic evidence for having erupted in the last several thousand years (lack of erosion, lack of vegetation, relative stratigraphy of deposits) but lack dated deposits or historical accounts (Siebert et al., 2010). Experience urges us not to dismiss these volcanoes, or volcanoes that appear to have been quiescent during the Holocene, as the potential for a large eruption could still be high, as discussed below.

Left untabulated are the *risks* posed by these volcanoes. The United Nations Office for Disaster Risk Reduction defines risk as the combination of the probability of an event and its negative consequences. Many of the volcanoes with large proximal populations are located in countries with large

TABLE 1.3 The Ten Volcanoes with the Largest Proximal Populations in a Given Radius (in Order from Smallest to Largest Population within Each Category), the Year and VEI of their Most Recent Activity, and their Primary Volcano Type

Radius (km)	Volcano Name, Country	Year of Last Documented Eruption, VEI	Primary Volcano Type
5 or 10 (Each volcano has >1 million inhabitants within the radius)	San Pablo volcanic field, Philippines	1350 +/− 100 yrs	Stratovolcano
	Druze, Jabal ad, Syria	No documented Holocene activity	Volcanic field
	Hainan Dao, China	1933	Pyroclastic cone
	Ilopango, El Salvador	1879, 3	Caldera
	Campi Flegrei, Italy	1538, 3	Caldera
	Malang Plain, Indonesia	No documented Holocene activity	Maar
	Kars Plateau, Turkey	1959 (uncertain), 2	Volcanic field
	Leizhou Bandao, China	No documented Holocene activity	Volcano field
	Tatun Group, Taiwan	4100 BCE, 1	Lava dome
	Michoacan–Guanajuato, Mexico	1943, 4 (Paricutin)	Cinder cone
30 (Each volcano has >4 million inhabitants within the radius)	Chichinautzin, Mexico	400 BCE, 3 (Xitle)	Volcanic field
	Arjuno–Welirang, Indonesia	1950 (1952, 2)	Stratovolcano
	Merapi, Indonesia	Ongoing, 2010, 4	Stratovolcano
	Ungaran, Indonesia	No documented Holocene activity	Stratovolcano
	Penanggungan, Indonesia	200 (uncertain)	Stratovolcano
	Malabar, Indonesia	No documented Holocene activity	Stratovolcano

Continued

TABLE 1.3 The Ten Volcanoes with the Largest Proximal Populations in a Given Radius (in Order from Smallest to Largest Population within Each Category), the Year and VEI of their Most Recent Activity, and their Primary Volcano Type—cont'd

Radius (km)	Volcano Name, Country	Year of Last Documented Eruption, VEI	Primary Volcano Type
	Tangkuban Perahu, Indonesia	2013, 2	Stratovolcano
	Michoacan–Guanajuato, Mexico	1943, 4 (Paricutin)	Cinder cone
	Tatun Group, Taiwan	4100 BCE, 1	Lava dome
	Laguna Caldera, Philippines	No documented Holocene activity	Caldera
100 (Each volcano has >26 million inhabitants within the radius)	Popocatepetl, Mexico	Ongoing, 2	Stratovolcano
	Arayat, Philippines	No documented Holocene activity	Stratovolcano
	Iztaccihuatl, Mexico	No documented Holocene activity	Stratovolcano
	Chichinautzin, Mexico	400 BCE, 3 (Xitle)	Volcanic field
	Papayo, Mexico	No documented Holocene activity	Lava dome
	Hakone, Japan	1170 (1200, 2)	Complex
	Tangkuban Perahu, Indonesia	2013, 2	Stratovolcano
	Perbakti-Gagak, Indonesia	1939, 1	Stratovolcano
	Salak, Indonesia	1938, 2	Stratovolcano
	Gede, Indonesia	1957, 2	Stratovolcano

The list for VPI_5 and VPI_{10} are identical.
Source: Data from VOTW 4.1.

agrarian populations near the equator (Figure 1.3 and Table 1.3). Because volcanoes create mountainous terrains (58 percent of volcanoes have elevations between 1,000 and 3,000 m above sea level), the settlement and

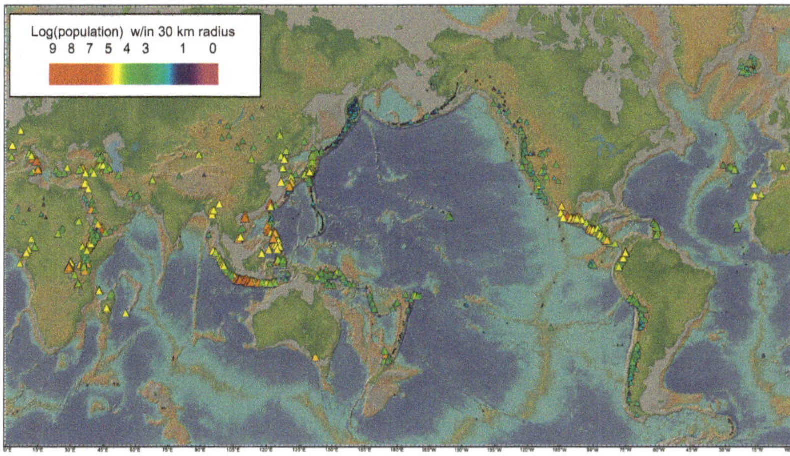

FIGURE 1.3 **Distribution of Holocene volcanoes cataloged in VOTW 4.1.** The color and size of the symbols corresponds to the size of the population living within 30 km (VPI_{30}). Volcanoes without proximal populations are shown as black triangles. *Created in GeoMapApp.org using the base map of Ryan et al. (2009).*

agriculture that take place on the flanks of volcanoes near the equator are often rendered impractical in more northern climates (Small and Naumann, 2001). Agriculture creates a link between socioeconomically vulnerable populations and those living in the shadows of the world's deadliest volcanoes—a link between hazard and risk. Moreover, while scientific infrastructure and monitoring capabilities in these nations are growing, resources for research and development are often stretched thin (Table 1.1), affecting the capacity to respond effectively to volcanic unrest and eruptions.

The global distribution of populations living in proximity to volcanoes reveals striking zones of risk in the Mediterranean, Africa, Indonesia, Southeast Asia, and Mesoamerica (Figure 1.3). Comparison with Figure 1.1 reveals that all of these regions have the potential to host eruptions with VEI 5 and greater. With Figures 1.1 and 1.3 in hand, it is tempting to analyze the proximity of large populations to centers that have experienced large magnitude eruptions in the past and to infer future risk; however, history cautions against this. In general, the highest magnitude eruptions come on the heels of the longest repose intervals. A low VEI eruption is more likely to be followed by another eruption in 1–10 years while caldera forming eruptions are likely to have repose intervals on the order of hundreds to thousands of years (Figure 1.4; Siebert et al., in press). It is ill advised to speculate beyond these generalities because some of the deadliest eruptions on record (e.g., several in Table 1.1) have emanated from volcanoes with *no* previous activity documented in VOTW. Thus the list of volcanoes and volcanic fields without any Holocene activity presented in Table 1.3 could nonetheless be sources of great risk. More prudent would be to say that all of the regions supporting large

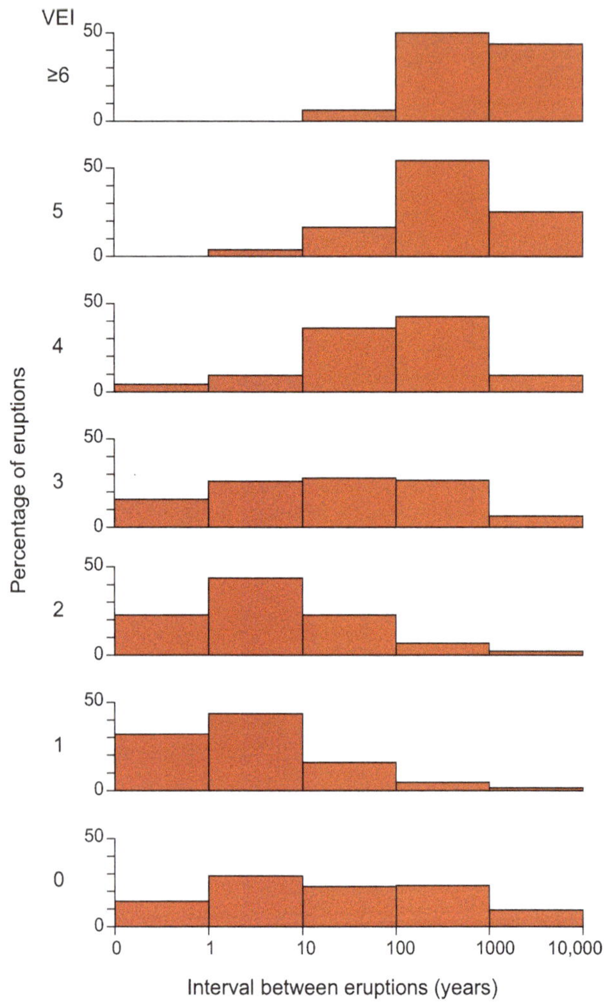

FIGURE 1.4 **Explosivity and time intervals between eruptions.** For each VEI unit, eruptions are grouped by time interval from start of previous eruption. *Modified from Siebert et al. (in press).*

populations proximal to volcanoes have also shown the potential to produce VEI 6 eruptions in the past.

1.4 PATTERNS IN VOLCANO-RELATED FATALITIES

Of the >10,000 eruptions in VOTW 4.1, only 573 are associated with fatality information. Of those 573 eruptions, only 343 are numerically quantified (e.g., "46") with an additional 151 associated with nonnumerical estimates (e.g., "many" or "several") and 79 with no additional information. Of the six Holocene eruptions with VEI = 7, only three (Tambora, Indonesia, in 1812,

Santorini, Greece, in 1610 BC, and Kikai in Japan in 4350 BC) have documented fatalities, and the number of fatalities is only quantified for Tambora. Of course Santorini's Minoan eruption and the other VEI 7's, such as the AD 1000 eruption of North Korea/China's Changbaishan, almost certainly resulted in deaths, but these remain undocumented within the GVP catalog.

Only six eruptions account for more than half of the total quantified fatalities in VOTW 4.1—each eruption with >10,000 associated deaths (Table 1.3, true whether or not Ilopango is included). Tsunamis and pyroclastic flows were agents of death in these eruptions and are responsible for more than 50 percent of volcano-related fatalities overall. The recent analysis by Auker et al. (2013), which considered the VOTW and additional data sources, found that 90 percent of volcano-related fatalities are the result of four primary causes: pyroclastic flows, tsunamis, lahars, and indirect consequences, such as famine. This is consistent with analysis of the data in VOTW 4.1 and an earlier analysis of the GVP database by Simkin et al. (2001). Layering fatality data on a map of projected world population density in 2015 we reveal new patterns (Figure 1.5 created with GeoMapApp.org with the base map from CIESIN (2005)). While Figure 1.3 gives us a snapshot of populations living in proximity to volcanoes today, Figure 1.5 shows us modern population density juxtaposed against past fatalities. Despite major shifts in global population in the last millennia, the populations at risk today are also generally the sites of past tragedies. Large fatality concentrations can be found in Indonesia, SE Asia, and Mesoamerica in particular. When considering Figures 1.1, 1.3, and 1.5, there is no mistaking the inauspicious status of the East Sunda Arc, Indonesia; it has the greatest confluence of volcanic explosivity, population density, and documented fatalities in the world (Figures 1.6 and 1.7).

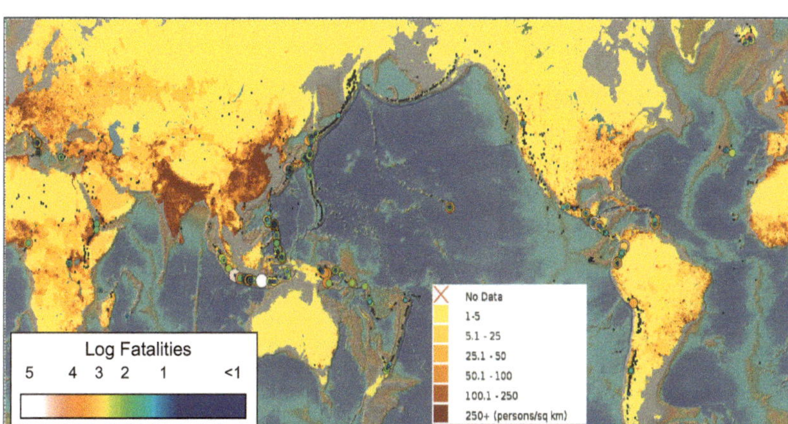

FIGURE 1.5 Projected population density in 2015 overlain with fatality data from VOTW 4.1. The color and size of symbols corresponds to the log of the number of fatalities from an eruption. Volcanoes without fatality information are shown as black circles. *Created using GeoMapApp.org with the base map from CIESIN (2005).*

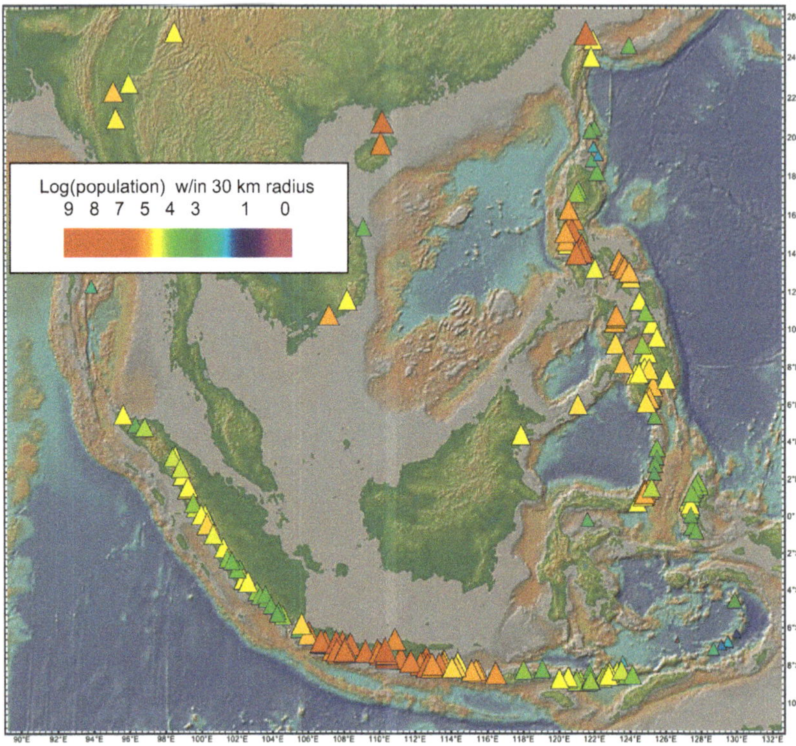

FIGURE 1.6 Distribution of Holocene volcanoes for Indonesia and SE Asia. The color and size of the symbol corresponds to the size of population living within 30 km (VPI_{30}). Volcanoes without proximal populations are shown as black triangles. *Created in GeoMapApp.org using the base map of Ryan et al. (2009).*

Despite the general similarities between Figures 1.3 and 1.5, closer inspection reveals important differences. Huge populations in Taiwan and outside of Hong Kong live in close proximity to Holocene volcanoes but have no documented fatalities. At the same time, many parts of the Philippines and Indonesia (e.g., the subregions of Sulawesi and Halmahera) seem to have experienced volcano-related fatalities in outsized proportions relative to the present size of populations living there or the size of typical documented eruptions in that region (Figure 1.1).

The disparity between populations that we might perceive to be at risk due to their proximity to volcanoes versus those that have fatalities on record comes into sharper focus when we look at Mesoamerica and the Lesser Antilles. Records suggest that the maximum VEI of explosive eruptions in the Lesser Antilles has been moderate when compared with other island arcs (Figure 1.1), and the modern population living in proximity to volcanoes is modest (Figure 1.8) albeit dense (Figure 1.9). When we look at the fatality

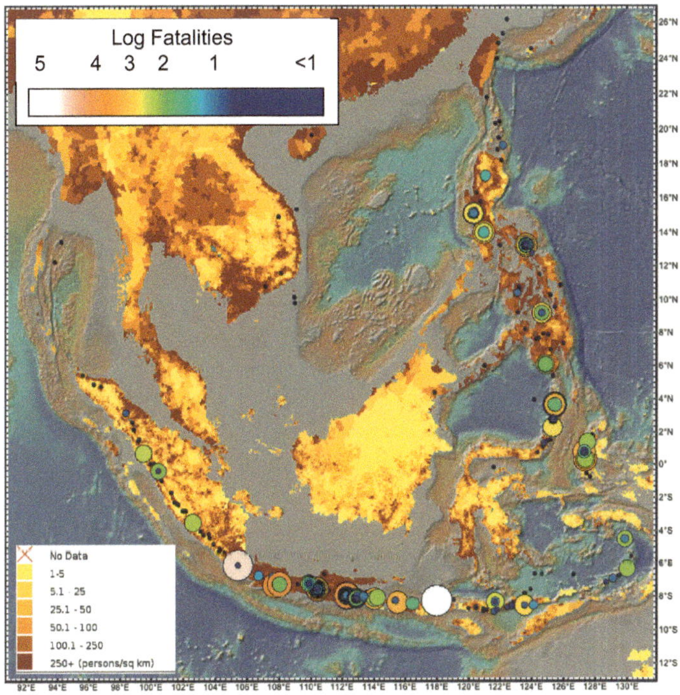

FIGURE 1.7 Projected population density in 2015 overlain with fatality data from VOTW 4.1 for Indonesia. The color and size of symbols corresponds to the log of the number of fatalities from an eruption. Volcanoes without fatality information are shown as black circles. *Created using GeoMapApp.org with the base map from CIESIN (2005).*

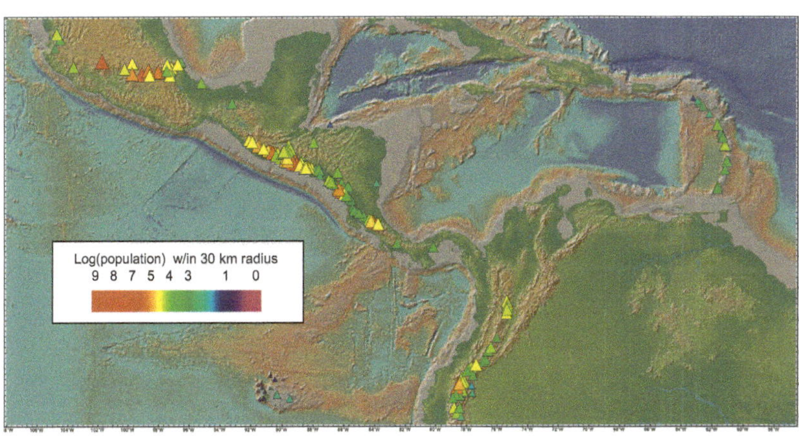

FIGURE 1.8 Distribution of Holocene volcanoes for Mesoamerica and West Indies. The color and size of the symbol corresponds to the size of population living within 30 km (VPI_{30}). Volcanoes without proximal populations shown as black triangles. *Created in GeoMapApp.org using the base map of Ryan et al. (2009).*

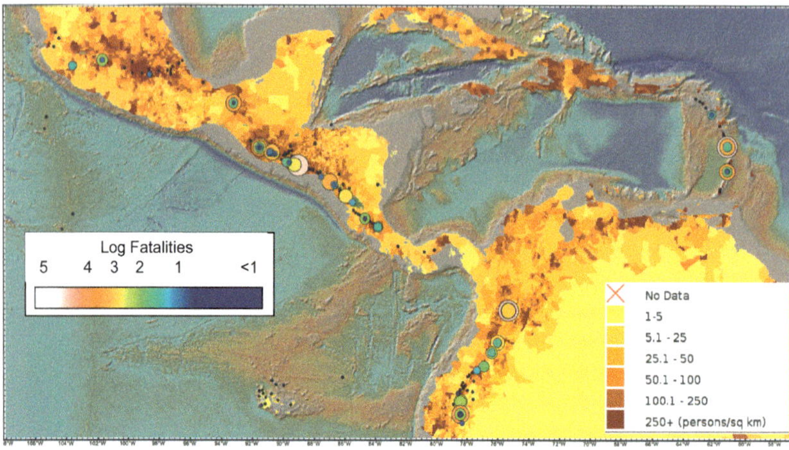

FIGURE 1.9 Projected population density in 2015 overlain with fatality data from VOTW 4.1 for Mesoamerica and the West Indies. The color and size of symbols corresponds to the log of the number of fatalities from an eruption. Volcanoes without fatality information are shown as black circles. *Created using GeoMapApp.org with the base map from CIESIN (2005).*

record, however, the Lesser Antilles appears to have borne more than its share of tragedy (also see analysis by Auker et al. (2013)). The high fatality count is not well explained by an anomalously low Human Development Index or better scientific study, reporting, or preservation (Table 1.1). It is important to recognize that fatalities in the Lesser Antilles are associated with only a handful of eruptions from an even smaller number of volcanoes (notably the eruption of Pelèe in 1902) and that our knowledge is highly incomplete. Continued study of this region (e.g., Lindsay, 2005) and others like it is warranted in order to get a much better understanding of the risks present in the modern world. The chapters that follow in this volume (Papale, in this volume) seek to explore in greater detail many of these nuances.

ACKNOWLEDGMENTS

I extend my gratitude to Paolo Papale for his invitation to write this chapter and for his editorial handling. I thank Ed Venzke for providing technical support and critical feedback and Sue Loughlin for her thorough review. Her comments greatly improved the manuscript. I am grateful for support from the Smithsonian and NSF award EAR-0841006.

REFERENCES

Auker, M.R., Sparks, R.S.J., Siebert, L., Crosweller, H.S., Ewert, J., 2013. A statistical analysis of the global historical volcanic fatalities record. J. Appl. Volcanol. 2 (2).

Brown, S.K., Crosweller, H.S., Sparks, R.S.J., Cottrell, E., Deligne, N.I., Ortiz Guerrero, N., Hobbs, L., Kiyosugi, K., Loughlin, S.C., Siebert, L., Takarada, S., 2014. Characteristics of global data on large magnitude explosive volcanic eruptions (LaMEVE). J. Appl. Volcanol. 3.

Carmichael, I.S., 2002. The andesite aqueduct: perspectives on the evolution of intermediate magmatism in west-central (105–99°W) Mexico. Contrib. Mineral. Petrol. 143, 641–663.

Center for International Earth Science Information Network – CIESIN – Columbia University, a. C. I. d. A. T.-C, 2005. Gridded Population of the World, Version 3 (GPWv3): Population Density Grid, Future Estimates. NASA Socioeconomic Data and Applications Center (SEDAC), Palisades, NY (accessed December, 2013).

Crisp, J., 1984. Rates of magma emplacement and volcanic output. J. Volcanol. Geotherm. Res. 20, 177–211.

Crosweller, S., Cottrell, E., Deligne, N.I., Guerrero, N.O., Hobbs, L., Kiyosugi, K., Loughlin, S.C., Siebert, L., Sparks, S., Takrada, S., 2012. Global database on large magnitude explosive volcanic eruptions (LaMEVE). J. Appl. Volcanol. 1 (1), 4.

Deligne, N.I., Coles, S.G., Sparks, R.S.J., 2010. Recurrence rates of large explosive volcanic eruptions. J. Geophys. Res. 115 (B6).

Ewert, J.W., Harpel, C.J., 2004. In harm's way: population and volanic risk. Geotimes 49 (4), 14–17.

Guffanti, M., Tupper. Volcanic ash and aviation risk, in this volume.

Harris, A. Lava Flows., in this volume.

Johnson, M.C., Anderson, A.T., Rutherford, M.J., 1994. Pre-eruptive volatile contents of magmas. In: Volatiles in Magmas, pp. 281–330.

Lindsay, J.M., 2005. Volcanic Hazard Atlas of the Lesser Antilles (Seismic Research).

Marty, B., Tolstikhin, I.N., 1998. CO_2 fluxes from mid-ocean ridges, arcs and plumes. Chem. Geol. 145, 233–248.

Mothes, P., Vallance, J. Lahars, in this volume.

Newhall, C.G., Self, S., 1982. The volcanic explosivity index (VEI): an estimate of explosive magnitude for historical volcanism. J. Geophys. Res. Oceans Atmos. 87, 1231–1238.

Neri, A., Ongaro, T.E. Pyroclastic Flows, in this volume.

Papale, P. Volcanic Hazards, Risks and Disasters, in this volume.

Ryan, W.B.F., Carbotte, S.M., Coplan, J.O., O'Hara, S., Melkonian, A., Arko, R., Weissel, R.A., Ferini, V., Goodwille, A., Nitsche, F., Bonczkowski, J., Zemsky, R., 2009. Global multi-resolution topography synthesis. Geochem. Geophys. Geosyst. 10 (3).

Self, S., 2006. The effects and consequences of very large explosive volcanic eruptions. Philos. Trans. R. Soc. A 364, 2073–2097.

Self, S. Giant volcanic eruptions and global impacts, in this volume.

Self, S., Blake, S., 2008. Consequences of explosive supereruptions. Elements 4, 41–46.

Siebert, L., 1984. Large volcanic debris avalanches: characteristics of source areas, deposits, and associated eruptions. J. Volcanol. Geotherm. Res. 22, 163–197.

Siebert, L., Cottrell, E., Venzke, E., Andrews, B., 2014. Earth volcanoes and their eruptions: an overview. In: Sigurdsson, H., Houghton, B., Rymer, H., Stix, J., McNutt, S. (Eds.), Encyclopedia of Volcanoes, 2nd ed. New York, Elsevier, in press.

Siebert, L., Ewert, J.W., Kimberly, P., Schilling, S.P., 2008. Population in proximity to volcanoes: a global perspective. IAVCEI 2008 General Assembly. Reykjavik, Iceland, Aug 17–22, Abs. 2008.

Siebert, L., Simkin, T., Kimberly, P., 2010. Volcanoes of the World, third ed. University of California Press, Berkeley.

Simkin, T., 1993. Terrestrial volcanism in space and time. Annu. Rev. Earth Planet. Sci. 21, 427–452.

Simkin, T., Seibert, L., 1994. Volcanoes of the World. Geoscience Press, Tuscon.

Simkin, T., Siebert, L., Blong, R., 2001. Policy forum: disasters – volcano fatalities – lessons from the historical record. Science 291 (5502), 255.

Small, C., Naumann, T., 2001. The global distribution of human population and recent volcanism. Global Environ. Change Part B: Environ. Hazards 3 (3–4), 93–109.

Smith, P.D., 1979. Environmental and cultural effects of the Ilopango eruption. In: Smith, P.D., Grayson, D.K. (Eds.), Volcanic Activity and Human Ecology. Academic Press, New York, pp. 525–564.

Smith, M.S., Shepherd, J.B., 1996. Tsunami Waves Generated by Volcanic Landslides: An Assessment of the Hazard Associated with Kick'em Jenny. Geological Society, London. Special Publications 110, 115–123.

Wallace, P.J., 2005. Volatiles in subduction zone magmas: concentrations and fluxes based on melt inclusion and volcanic gas data. J. Volcanol. Geotherm. Res. 140, 217–240.

Wilson, T.M. Impacts from Volcanic Ash Accumulation, in this volume.

Chapter 2

Basaltic Lava Flow Hazard

Andrew J.L. Harris
Laboratoire Magmas et Volcans, Clermont Ferrand, France

ABSTRACT
By virtue of their extreme heat, and their ability to surround and bury, as well as to force structural failure, lava flows present an extreme hazard to communities in their path. Here the focus is on hazards associated with the ingress of basaltic lava into vulnerable areas. The likelihood of damage is increased with increasing flow front velocity, volumetric output rate, eruption duration, poor viewing conditions, vent location, and topography (e.g., vents opening close to, or at the head of steep slopes above, population centers). Because lavas are usually slow moving and localized hazards, fatalities are rare, but under certain circumstances they do occur. However, buildings, infrastructure, communications, utilities and machinery, plant communities, and agriculture will be destroyed, requiring replacement. All of these effects, including impacts on business and commerce, incur economic losses, and may have profound impacts on community social fabric and the mental health of individuals who are part of that fabric. Loss can be reduced by preparation and planning, modifying the loss potential (i.e., effective land-use planning and/or control), or implementation of education and warning systems. In some cases, the lava flow may be diverted, but never stopped. Most successful interventions have involved appropriately sited deflection barriers.

2.1 INTRODUCTION

Lava flow hazard has been covered in many texts, the main ones being collated in Table 2.1. However, there remains no complete compilation for both lava flow hazard, response, or mitigation efforts. While Blong (1984) provided the best review of lava flow hazard, Chester (1993) detailed response issues, protocols, and mitigation. Many other excellent studies remain spread across a multitude of texts, as collated in the bibliography to this chapter and its appendix. Here, my material draws heavily on Blong (1984), bringing his compilation up-to-date, while bringing in new data which elaborates on what

Volcanic Hazards, Risks, and Disasters. http://dx.doi.org/10.1016/B978-0-12-396453-3.00002-2

TABLE 2.1 Text Books (in Chronological Order of Publication) Containing Information about Lava Flow Hazard

Author	Year	Title	Chapter	Pages
Macdonald, G.A	1972	Volcanoes	16: Volcanoes and man	419–423 (L)
Chester, D.K., et al.	1985	Mount Etna: An Anatomy of a Volcano	8: Volcanic hazard on Etna	295–336
Chester, D.K., et al.	1985	Mount Etna: An Anatomy of a Volcano	9: Volcanic hazards: human response and adjustment	337–388
Chester, D.	1993	Volcanoes and Society	8: Human responses and adjustments to volcanic hazards: a Framework for analysis	228–248
Chester, D.	1993	Volcanoes and Society	9: Responses and adjustments to volcanic hazards in different societies	249–309
Scarth, A.	1994	Volcanoes	15: Predicting volcanic eruptions and their effects	248–250 (L)
Fisher, R.V., et al.	1997	Volcanoes: Crucibles of Change	7: Lava flows	133–146
Scandone, R., Giacomel i, L.	1998	Volcanologia	14: Il rischio vulcanico	561–608
Bardintzeff, J.-M., McBirney, A.R.	2000	Volcanology	14: Volcanoes and the human environment	228–230 (L)
Bardintzeff, J.-M.	2008	Volcanologie	14: Risques volcaniques et prévisions	223–226 (L)
Parfitt, E.A., Wilson, L.	2008	Fundamentals of Physical Volcanology	11: Volcanic hazards and volcano monitoring	165–166 (L); 176–177 (L)
Lockwood, J.P., Hazlett, R.W.	2010	Volcanoes: A Global Perspective	14: Hazards and risks—monitoring and mitigation	438–442 (L)

If marked "L" page range is fcr the subsection specifically devoted to lavas, otherwise the page range of for the entire volcanic hazard Chapter within which lava hazard is included is given.

makes a lava flow hazardous and reviewing currently available mitigation measures. The chapter is thus divided into three parts:

1. What makes a lava flow hazardous?
2. What are the impacts of lava flow ingress into vulnerable populations?
3. What mitigation measures are available?

Much supporting evidence has been moved to the appendix. This evidence is flagged, sequentially, from the appropriate section. The final part of this chapter draws heavily on the works of Thomas A. Jaggar and Gordon A. Macdonald, who both did so much to understand how lava needs to be, and may be, delayed and diverted through their hands-on experiences during effusive crises on Mauna Loa and Kilauea between 1930 and 1970. Their ideas, plans, and interventions still underpin most modern actions.

2.2 WHAT MAKES A LAVA FLOW HAZARDOUS?

Lava flows are hazardous due to their extreme heat, and their ability to surround, bury, and force structural failure. The degree of hazard posed depends on five main factors, each of which is here considered in terms of the hazard presented by the actively advancing lava.

2.2.1 Flow Velocity

Faster moving flows give less time for a community to prepare and react. Velocity will depend largely on flow viscosity and terrain slope. For lava flowing in a channel, although higher viscosities serve to reduce velocity, steeper slopes promote higher velocities (Note 1a). Lavas emplaced at Ol Doinyo Lengai (Tanzania) are particularly fluid (viscosities are typically less than 100 Pa s) and demonstrate well the hazard posed by rapidly emplaced lavas. At Ol Doinyo Lengai, if lava extends onto the steep outer flanks of the cone, potential exists for rapid flow front advance, with the lava having a rheology that correlates with advance rates of several meters per second (Note 1b). Such velocities require rapid responses by downflow communities. In fact, reports by the Smithsonian Institution's Global Volcanism Program indicate that lava overflowing from Ol Doinyo Lengai's summit crater during 2006 and 2007 led to spontaneous evacuation of several villages, or atleast encouraged such a mindset (Note 1c). Such a spontaneous reaction indicates an ingrained understanding of the hazard. That is, likely very little time will exist to respond due to the high velocity of the flow front advance, and thus the need to evacuate quickly and spontaneously.

The effect of slope on increasing flow front velocity is demonstrated by the 1950 eruption of Mauna Loa (Hawai'i) as described by Finch and Macdonald (1950) and Macdonald and Finch (1950) (Note 1d). The eruption was located on Mauna Loa's SW rift zone at a site above relatively steep (up to 15°) slopes, so that flow fronts advanced at between 1 and 10 km/h. Although there were no casualties, little time existed for escape, with the main Kona-Hilo highway being cut within 2 h of the eruption, destroying Hookena post office, several houses, and a gasoline station (Figure 2.1).

Emplacement of Mauna Loa's 1950 flow field raises two important hazard issues that apply to such rapid-emplacement cases. First, there may be no time to issue an official evacuation notice, so that the population needs to be educated as to the nature of the hazard and the appropriate response. The population thus needs to be trained to recognize the signs that an eruption is underway so that they evacuate spontaneously (without having to be told to do so); with evacuation kits, escape routes and reception centers being put in place, and well-known, well in advance. The 1950 case poses an interesting additional problem in that multiple flows can rapidly cut escape routes in all directions, isolating lava-free islands where people can become trapped (Figure 2.1). Such

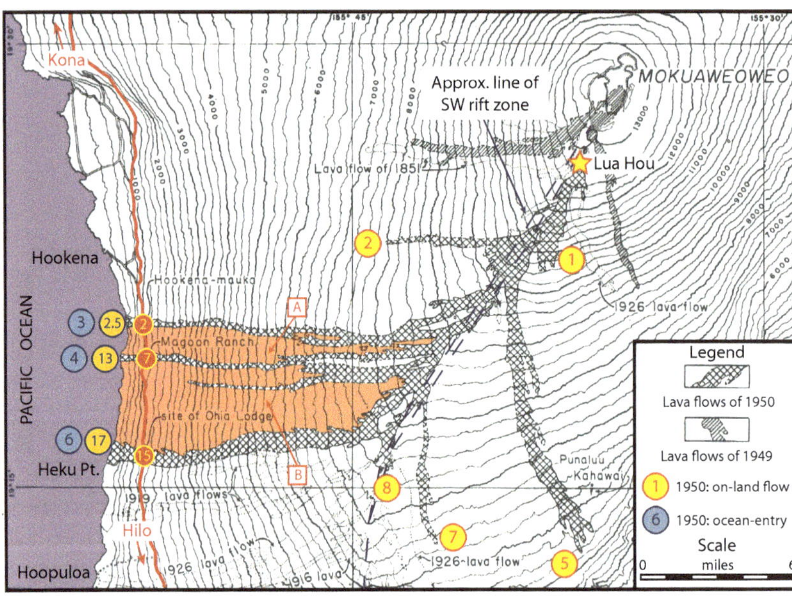

FIGURE 2.1 Macdonald and Finch's (1950) map of the June 1950 lava flow field emplaced from the Lua Hou vent on Mauna Loa's southwest rift zone. The sequence of flow unit emplacement is given by the numbers in yellow and blue circles. The time, in hours since the eruption began, that the main highway was cut is given as yellow numbers in red circles, and the time of each ocean entry is entered as blue numbers in yellow circles. The eruption quickly isolated a first zone (zone A), completely cutting road communication north to Kona after 2 h, and then south to Hilo after 7 h; the zone became completely isolated after the onset of the second ocean entry by 13 h. A second zone (zone B) became completely isolated after 17 h. These two syn-eruptive zones of extreme hazard (marked in red) represent zones: (1) within which populations can become trapped; (2) from which there is no means of over-land escape; and (3) within which there is a high risk of lava inundation by new flows descending from rift and into the risk-zone.

cases will require advance plans for prompt airborne rescue from predesignated evacuation points, and facilities to allow aircraft and/or helicopters to land, with sufficient aircraft capacity for the likely number of trapped individuals.

2.2.2 Effusion Rate

Increased effusion rate can feed longer (simple) flows of greater length than their lower effusion rate counterparts (Harris and Rowland, 2009), thus posing a hazard for more distant communities. Probably the most well-known example of high effusion rates feeding rapidly advancing lava flows is that of Nyiragongo in 1977 (Note 2). Effusion rates were likely thousands of m^3/s, so that "the front of the flow reached the outskirts of Goma in a matter of 20 min, giving an average speed of 30 km/h" (Tazieff, 1977). Such velocities mean that individuals may have no time to escape, or even outrun the lava, as was the case at Nyiragongo.

2.2.3 Duration

Increased eruption duration results in the construction of extensive compound lava flow fields of greater volume and area than flow fields produced by short-lived events. With time, tubes will develop, allowing lava to reach distant areas even at low effusion rates (Note 3a). This can build wide flow fields that steadily fill topographic lows over periods of several months to years (Note 3b). A good example of such a case is Kilauea's Aila'au flow field (Hawai'i). Emplaced around AD 1445, the flow field was fed by vents located at Kilauea's summit and flows extended to the east during a period of continuous effusion that lasted ~50 years (Clague et al., 1999). The eruption resulted in a tube-fed flow field fed at a mean output rate of 3.3 m^3/s to attain a volume of 5.2 ± 0.8 km^3 covering an area of 430 km^2 (Clague et al., 1999). The flow field extended 40 km from Kilauea's caldera to Hawai'i's Puna coast, the flow field being ~12 km wide at the coast. Such an event, if occurring in a heavily populated area, has the potential to cause widespread damage and significant population displacement. If occurring today, the Aila'au event would impact more than 23,000 people and 6,600 housing units (Table 2.2). In addition, resettlement of a devastated area during an on-going, but long-lived, event is problematic due to the likelihood of new inundation once the new settlement has been re-established (Note 3c).

2.2.4 Vent Location

Lava inundation is a high risk for locations that are on, close to, or directly downslope from active vent zones, or for communities constructed within, or next to, natural channels, such as river valleys and grabens. Below are three examples of towns and facilities at risk from, or impacted by, lava inundation

TABLE 2.2 Population and Number of Houses Built in Subdivisions that Overlie Kilauea's Aila'au Lava Flow Field

No.	Subdivision	Population	Housing Units
1	Mauna Loa Estates	No data	No data
2	Royal Hawaiian Estates	No data	No data
3	Fern Forrest Vacation Estates	No data	No data
4	Eden Rock Estates	No data	No data
5	Crescent Acres	No data	No data
6	Hawaiian Acres	1,776	843
7	Orchid Land Estates	1,731	668
8	Ainaloa	1,910	722
9	Hawaiian Paradise Park	7,051	2,671
10	Hawaiian Beaches	3,709	1,383
11	Pahoa (partial inundation)	962	352
Total		23,136	6,639

Source: Based on data from the 2000 census.

due to topographic location in relation to active vents. Two other examples (Ocean View Estates and Kalapana, both Hawai'i) are given in Note 4.

Example 1: Kapoho (Hawaii) 1959

During Kilauea's east rift zone eruption of 1960 a fissure opened within ~400 m of the center of Kapoho village. Initially, the flow direction was controlled by a shallow graben, which protected the village (which was located on the rim of the graben) by containing the lava and directing lava toward the ocean. However, once the graben had been filled, burial of Kapoho was inevitable (Figure 2.2).

Example 2: Hilo (Hawaii)

Hilo is located at the end of Mauna Loa's northeast rift zone in valley formed by the construction of Mauna Kea and Mauna Loa. As of the census of 2000, Hilo had a population of 40,759, comprising 14,577 housing units and major infrastructures including Hilo harbor and airport. Topography funnels any flow erupting from Mauna Loa's northeast rift zone toward Hilo (Figure 2.3). Eight northeast rift zone eruptions have occurred since 1843 (Lockwood and Lipman, 1987), where Note 4c describes some near-misses for Hilo.

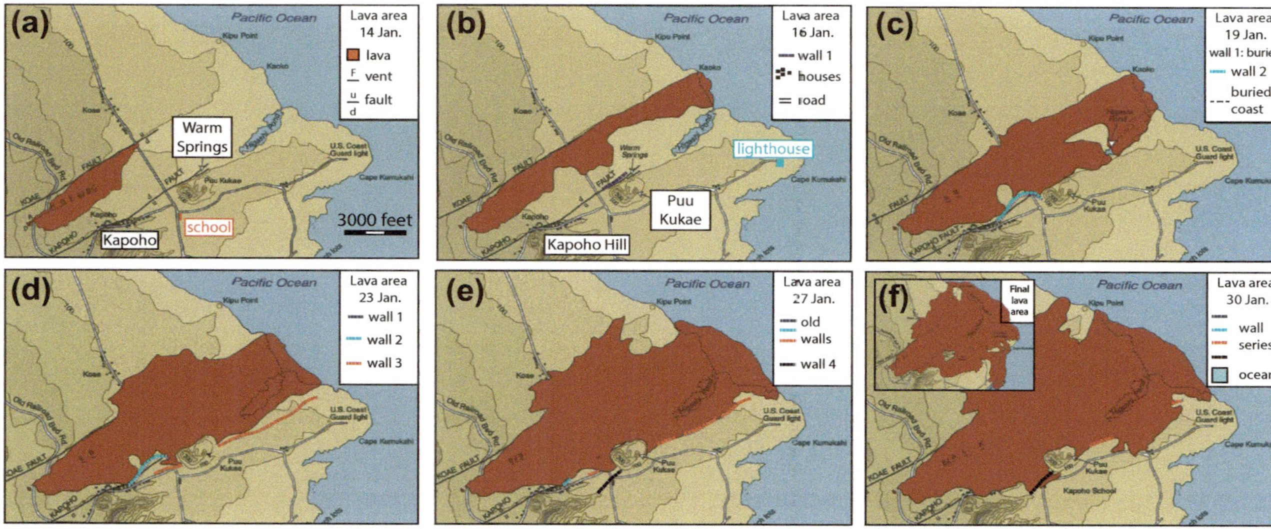

FIGURE 2.2 U.S. Geological Survey maps showing growth of the lava flow field erupted during Kilauea's January 1960 at Kapoho, and the sequence of wall construction (http://hvo.wr.usgs.gov/kilauea/history/1960Jan13/kakugawa.html). The eruption began in a shallow graben, bounded by a fault scarp that was 1.5–3 m high to the south, and on which the village of Kapoho was located. Lava effusion began at 19:30 (Hawaiian Standard Time) on January 13, 1960 from a vent opening in a sugar cane field within the graben. Initially the topography directed flows NE to the ocean (a). However, the near flat topography and shallowness of the graben, meant that Kapoho (which had been evacuated on 13 January) was at threat. Walls aimed at damming, rather than diverting, the lava were thus constructed (Macdonald, 1962). Wall 1 was planned and constructed during 16–17 January on the south rim of the graben to protect the Warm Spring resort area (b). Work on wall 2 began on 18 January; being built to fill a gap between Kapoho Hill and the Puu Kukii cinder cone and to primarily protect Kapoho school (c). Construction of wall 3 began on 21 January, and was built along the top of a ridge running from Puu Kukae to a point near Kumukahi lighthouse (d). Wall 4 was erected on 23 January (e), and was designed as a "*delaying barrier*" (Macdonald, 1962). Due to the very shallow slopes and limited capacity of the graben, eventually all walls were overwhelmed, with all of Kapoho's structures but the lighthouse being buried (f).

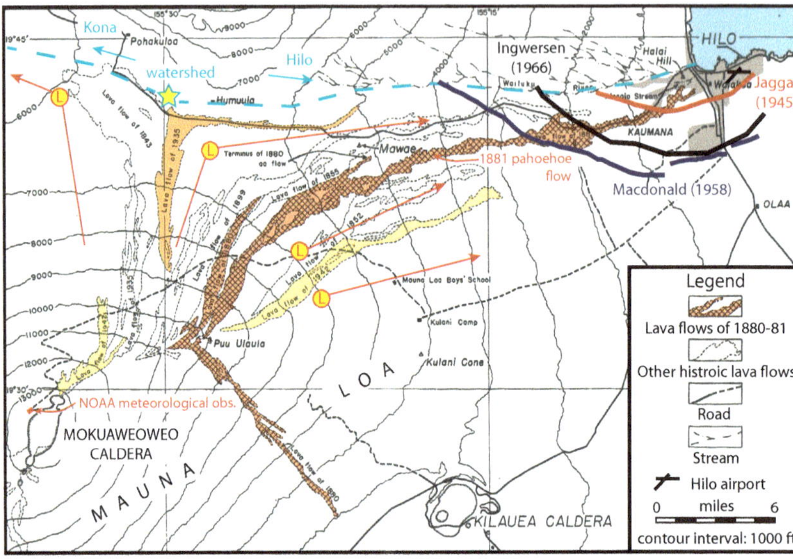

FIGURE 2.3 Location of Hilo (gray zone, from Wentworth et al.'s (1961) map), and the NOAA meteorological observatory on Mauna Loa, in relation to Mauna Loa's north-east rift zone and the lava flows of 1880–1881 (red), 1935 (orange) and 1942 (yellow); base map from Baldwin (1953). The valley between Mauna Kea and Mauna Loa (blue dashed line) funnels lavas into Hilo, as does the topography of the rift zone itself. Red lines marked "L" indicate the main directions for lava flows moving off of Mauna Loa's northeast rift zone. Note that lavas arriving to west of the Wailuku river watershed (yellow star) turn toward the Kona coast; those arriving to the east turn toward Hilo. Also marked are the lines of the deflection barriers proposed by Jaggar (1945, thick red line), Macdonald (1958, thick blue line) and Ingwersen (1966, thick black line) to protect Hilo from lava invasion.

Example 3: Piano Provenzana (Etna, Italy)

This is a tourist facility set up to serve Etna's winter ski industry and summer volcano visitors. It is located on Etna's NE flank between 1,780 and 1,830 m. The facility is located directly down slope from Etna's NE rift zone, and steep (10–20°) slopes extend from potential vent locations to the infrastructure. The Provenzana was completely destroyed and buried during Etna's 2002–2003 eruption when vents opened within a few 100 m, and directly above, its facilities (Figure 2.4).

The sighting of a population and its infrastructure is a decision that can be controlled and planned. The topographic element of lava-inundation risk is thus one that can be reduced by human decisions. For example, choice of settlement location may be made poorly in relation to active features (e.g., down slope of an active rift zone, within a graben or in a valley that has its head in a vent zone) or thoughtfully (e.g., in topographically protected zones).

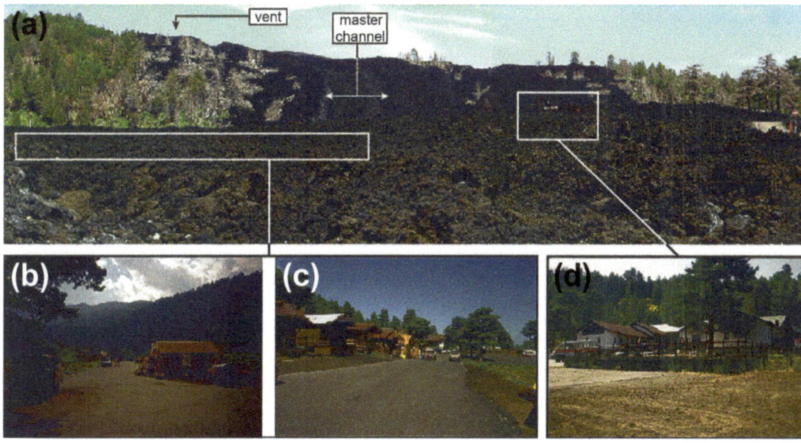

FIGURE 2.4 Piano Provenzana before and after passage of lava flows associated with Etna's 2002–2003 NE flank eruption (see also: Behncke et al., 2005). The facility was located around 800 m NE of, and 180 m below, the main effusive vent. The master channel was centered on the facility, and was 40–45 m wide and 3–4 m deep. Given a lava viscosity of 3000 Pa s, and average slope of 12.7° between the vent and the facility, lava could potentially have reached the structures within 3 min of eruption (Note 1a). Photos of mosaic given in (a) were taken in 2005 and show the master channel, the source vent and the swath cut through the preexisting forest by the 2002–2003 'a'a, as well as the approximate locations of views in (b)–(d) (white boxes). Views SW (b) and NE (c) along the main road of the facility show mobile, wooden structures (mostly souvenir shops) in 1999, and the permanent masonry structure (d) that housed a hotel, bar, restaurant, post office, and tourist office. *Source: "Before" photos by B. Behncke.*

2.2.5 Viewing Conditions

If visibility is poor, then it may be more difficult to perceive, or even see, the hazard and therefore escape. Poor visibility includes meteorological mist and fog, but also fog created by the active lava itself. In heavy rain, steam over an active flow can cause complete white-out conditions, with visibility declining to <1 m. In such conditions it is easy to become lost, disorientated, or unwittingly trapped. Night time may actually improve viewing conditions, because active zones of a flow field can be located by their glow. However, eruptions beginning during the night will give a sleeping population less time to respond, or may catch them by surprise (see Note 4d).

2.3 IMPACTS

The ability for active lava to impact a community results from its ability to burn, crush, and bury. Reynolds (1996), in his volcanic hazards assessment of Sierra Negra (Galapagos), summed this up by noting:

"lava flows are the greatest hazard...*although (they) rarely threaten human life, they crush, bury and burn everything in their path. Lava flows can also* ignite surrounding vegetation, causing wildfires."

Following Blong (1984), the main categories of impact are on:

1. Humans and animals (deaths and injuries),
2. Buildings,
3. Communications, utilities, and machinery,
4. Agriculture, habitat, and natural vegetation,
5. Economic activity, and
6. Social fabric.

Deaths and injuries caused by lava are rare because the hazard can usually be avoided, i.e., the potential victim has the time and capability to move out of the way. However, categories two through six cannot be moved; hence the damage/impact is hard felt.

2.3.1 Effects on Humans: Deaths and Injuries

Correcting and updating the listings of Blong (1984) and Witham (2005), see Note 5a, we have 342 deaths due to lava during the period 1800–2007 (Table 2.3). This is a death rate of 1.65 per year. Out of all volcanic hazards, lava flows are the seventh-to-eighth most deadly, accounting for two orders of magnitude fewer fatalities than any of the top six hazards (Table 2.4). Compared with other natural (and man-made) hazards, the likelihood of death by lava flow is utterly trivial, even among visitors to active lava flows. Perkins (2006) in his book "*Surviving Paradise*" sums this up. His book reviews the hazards likely encountered by visitors to Hawaii, where "more people die of vacation-related accidents every year than are killed by terrorists in Israel" (Perkins, 2006). Lava is ranked the fourth most deadly hazard behind road traffic accidents, swimming, surfing, and helicopters, being equal with shark attacks (Table 2.5).

As detailed in Note 5a, deaths and injuries due to lava ingress have three main causes:

1. Unexpected (or rapid) arrival of lava in a populated zone;
2. Tourism;
3. Lava-water and lava-fuel interaction.

In cases where flows were particularly fast moving, and/or have begun at night from vents near to centers of population, fatalities have occurred. The fast moving flows of Nyiragongo erupted on January 10, 1977, for example, caused many deaths, with estimates varying between 50 and 300, and probably being around 70 (Blong, 1984). For the second two cases on this list, the loss can be deemed unavoidable (Blong, 1984), because the victims approach the hazard voluntarily; the hazard does not come to them (e.g., Figure 2.5). For example, during Nyiragongo's 2002 eruption flows ignited gasoline tanks killing ~50 (maybe as many as 60) people who were trying to take fuel from an abandoned petrol station (BBC News, January 21, 2002). The source of ignition was spilled fuel that flowed to a nearby active flow; on reaching the flow the fuel ignited, and ignition spread back to the tanks (Figure 2.6).

TABLE 2.3 Deaths by Lava Flows

Volcano	Eruption Date	No. Killed	Reference
Vesuvius	1631[a]	(700)[b]	Blong (1984)
Aoba	1670	Many	Blong (1984)
San Jorge	1808	60	Blong (1984)
Kilauea	1823	Some	Blong (1984)
Etna	1832	Several	Blong (1984)
Etna[c]	1843	56	Blong (1984)
Niuafo'ou	1853	25	Blong (1984)
Vesuvius	1872	22	Blong (1984)
Kathala	1904	1	Witham (2005)
Etna	1928	0[c]	Duncan et al. (1996)
Tori-shima	1939	2	Witham (2005)
Vesuvius	1944	1	Chester et al. (2007)
Sakura-jima	1946	1	Witham (2005)
Hekla	1947	1	Blong (1984)
La Palma	1971	not known	Witham (2005)
Api Siau	1976	1	Blong (1984)
Kathala	1977	not known	Witham (2005)
Nyiragongo	1977	77	Blong (1984)
Karkar	1979	2	Witham (2005)
Nyiragongo	2002	50[d]	BBC, 21 January 2002
Rinjani	1994	31[d]	Witham (2005)
Pacaya	1995	1[d]	Witham (2005)
Kilauea	1991–2007	11[d]	Honolulu star Bulletin
Total[e]	1631–2007	342	
Total	1900–2007	179	

[a]*Date corrected from 1630 given in Blong (1984).*
[b]*Deaths due to pyroclastic density currents, not lava.*
[c]*Five casualties caused by lava invasion of Mascali are given by Blong (1984), but these are considered false reports by Duncan et al. (1996).*
[d]*Recorded since Blong (1984).*
[e]*Total is 1010−700=305, this being Blong's total minus Vesuvius 1631, plus new entries (given in Note 5a).*
Source: Modified from Blong (1984) and Witham (2005) as updated following Note 5a.

TABLE 2.4 Deaths by Volcanic Hazard, 1600–1982

No.	Hazard	1600–1900		1900–1982		Rank
1	Disease, starvation, etc.	95,313	39.9%	3,163	6.1%	3
2	Pyroclastic flows and debris Avalanches[a]	55,695	23.3%	36,787	70.4%	1
3	Tsunami	44,356	18.6%	407	0.8%	6
4	Lahars and Jokulhlaups	14,746	6.2%	6,438	12.3%	2
5	Tephra fall and ballistic projectiles	10,953	4.6%	3,019	5.8%	4
6	Unknown causes	17,182	7.2%	2,133	4.1%	5
7	Lava flows[a]	304	0.2%	664[b]	0.2%	8
8	Gases and acid rain	185	0.1%	183	0.4%	7
9	Seismic activity	89	0.04%	32	0.06%	9
10	Atmospheric effects	63	0.03%	3	0.01%	10
Total		239,000		52,000		

[a]*The 700 deaths due to lava during Vesuvius 1631 eruption have been moved from lava to pyroclastic flows.*
[b]*Value from the data base of Witham (2005).*
Source: Modified from Blong (1984) and Witham (2005).

TABLE 2.5 Perkins (2006) Book Contains the Following Sections as Separate Chapters, Which are here Listed in Order of Chapter No.

Chapter Number	Scope	Annual Rank	Notes
1	Drowning	=2	(a)
2	Volcanoes (lava)	=4	(b)
3	Surfing	=2	
4	Helicopters	3	(c)
5	Flash floods	–	One-off
6	Sharks	=4	(d)
7	Tsunami	–	One-off
8	Road traffic	1	(e)

Annual ranks are given to persistent hazards, the two assigned "One-off" happen less than annually, but upon impact can account for many fatalities in one event. Ranks are given depending on the average annual fatality rate calculated from data given in Perkins (2006) or the Honolulu Star Bulletin. (a) 24 drownings occurred on Oahu in 2002, 30 drownings at Waikiki alone between 1993 and 2003, 60–70 drowning's per year statewide; (b) based on data from Honolulu Star Bulletin (see Table 2.6); (c) In March 2007, two fatal crashes in 3 days occurred mon Kauai (Honolulu Star Bulletin), and 50 helicopter accidents occurred during 1985–2005 (~2.5 per year); (d) between 1990 and 2002 44 shark attacks occurred, but they were not always deadly (~2 year); (e) for pedestrians alone, nine fatalities occurred on just Oahu during the first 2.5 months of 2007 (Honolulu Star Bulletin), and there were 54 fatal crashes in Oahu in 2003.

FIGURE 2.5 Close approach to lava by tourists and sightseers—a hazard that can be avoided by simply keeping the observer at a safe distance from dangerous points, such as oversteepened, unstable flow fronts or fast moving, braided flows and ocean entries. Here, U.S. Army military policemen toast bread on an 'a'a flow front active at Vesuvius during March 1944 (note that the man second in line still has three large loaves to go). However, to put things in perspective, this military pass time was infinitely safer than call up to the Battle of Monte Cassino, which was raging at the time 90 km to the NE, having begun in January 1944. The four battles of Monte Cassino culminated in surrender by the German occupiers on May 18, 1944, having caused over 350,000 casualties on all sides (Parker, 2005). *Source: Black and white plate given in Atkinson (2007), as taken from the U.S. Army Signal Corps archives.*

FIGURE 2.6 Explosion due to lava–fuel interaction. Associated Press photograph, appearing in a BBC website report "*In pictures: Eruption aftermath*" (posted on January 21, 2002, 13:26 GMT), of the fire ball associated with a filling station explosion in Goma. The explosion was caused by active lava igniting fuel split by looters during Nyiragongo's 2002 eruption. The same report claimed that the explosion killed "about 50 people attempting to loot fuel" and "threw fuel drums 50 m into the air", this being the same height as the fire ball (saved from url=(0051) http://news.bbc.co.uk/1/hi/world/africa/1772326.stm). Note, lava-filled street and partially buried buildings in the foreground.

In addition, the lava may not be directly responsible for the fatality or injury. A content analysis of the *Honolulu Star Bulletin* for deaths and injuries in Hawaii Volcanoes National Park between 1991 and 2007 shows that 11 fatalities occurred during 1991–2002 (~1 fatality per year), but none were due to direct contact with lava (Table 2.6). Fatalities were mostly due to poor

TABLE 2.6 Injuries and Deaths Reported in the Hawaii Volcanoes National Park Reported in the Honolulu Star Bulletin (1991–2007)

Date	Location	Death/Injuries	Cause	Sources
1991	Caldera	1 death	Fell from caldera rim	Star Bulletin–August 6, 2001
1993	Lava bench	1 death	On bench during collapse	Perkins (2006)
1994	Halemaumau	1 death	Fell while trying to climb into pit	Star Bulletin–August 6, 2001
1996	Trail	1 death	Hyperthemia	Star Bulletin–November 11, 1998
1997	Coastal trail	1 death	Hyperthemia	Star Bulletin–October 8, 1997
1998	Ocean entry	1 death	Fell over cliff above ocean entry	Star Bulletin–April 20, 1998
2000	Flow field	1 death	Heart attack while returning from flows	Star Bulletin–October 6, 2000
2000	Flow field	2 deaths	Burnt and found inland from bench (inhaled hot steam while on bench)	Star Bulletin–November 6, 2000
2001	Caldera	1 death	Fell into caldera	Star Bulletin–October 21, 2001
2001	Caldera	1 injured	Drunk student fell ~30 m into crack—suffered cuts	Star Bulletin–April 2, 2001
2001	Caldera	1 injured	Man fell into caldera while trying to retrieve baseball cap	Star Bulletin–August 6, 2001
2002	Flow field	1 death	Woman found dead within 50 yards of active flow—cause of death unknown	Star Bulletin–October 16, 2002

preparedness, or poor decision making, by those visiting the volcano, and hiking (usually in hot, humid, and sunny conditions) up to 10 km over rough terrain to view the lava. Table 2.6 confirms Blong's idea, and extends it. Not only does tourism bring the victims to the hazard, but also their actions may cause injury and death. These victims are being exposed to an unfamiliar terrain, climate and/or situation, so that their preparations and reactions may be inappropriate. Take the example of baseball caps. We have one injury due to a tragically failed attempt to retrieve a baseball cap (Table 2.6). On August 10, 2013 Italy's *Il Tirreno* newspaper (Year 137, No. 218, Viareggio edition) reported the death of a tourist who dived from a small boat off of the coast of Elba to retrieve his cap, and drowned. In both cases, the common denominator is not the natural hazard, but the decisions and actions of the visitor.

2.3.2 Effects on Animals: Deaths and Injuries

Animal deaths are usually rare due to ease of escape. However, wild and domestic animals, as well as insects, birds, and fish, can become trapped and then burnt or boiled. Wildlife may also unwittingly enter a hazardous zone. Some documented examples of wildlife impacts do exist. Lockwood and Hazlett (2010), for example, give a picture (taken by Katia Krafft) of "a bone-filled (pigmy) elephant cast in 1977 pahoehoe near the base of Nyriagongo volcano." There are also several documented examples of wildlife deaths and injuries due at contact with active lava (and associated phenomena, such as heated water) in the Galapagos. Here, several effusive eruptions have impacted the rare species resident on the islands, causing causalities among fish, bird, iguana, and Galapagos tortoise populations (see Note 5b).

2.3.3 Effects on Buildings

Following Blong (1984), buildings will be impacted by:

1. Wall collapse due to loading and pressure exerted by the flow;
2. Burial flow inundation and/or inflation;
3. Ignition to cause fire;
4. Flooding (i.e., entry of lava into rooms);
5. Uplift and transport;
6. Foundation failure and deformation;
7. Impact from clasts tumbling from the flow front or margin.

The result is always partial or complete destruction of communities (Table 2.7), with Witham (2005) giving 21,490 made homeless due to lava between 1900 and 1999. The degree of damage will depend on flow and building type. Thin, fluid flows will likely flow around structures, and inflation will then bury the structure (Figure 2.7(a)). Thicker, less-fluid flows will have a greater ability to bulldoze and push to cause wall collapse (Figure 2.7(b)), and

TABLE 2.7 Towns and Villages Completely or Partially Inundated by Lava, 1900–2005

Town	Volcano	Eruption
Casa Bianca/Boscotrese	Vesuvius	1906
Catena (nr. Linguaglossa)	Etna	1923
Hoopuloa	Mauna Loa	1926
Mascali	Etna	1928
Paricutin and San Juan[a]	Paricutin	1943–1952
San Sebastiano	Vesuvius	1944
Hookena	Mauna Loa	1950
Kapoho	Kilauea	1960
Vestmannaejar	Eldfell	1974
Goma	Nyiragongo	1977 (January)
Piton Sainte-Rose	Piton de la Fournaisse	1977 (March)
Kalapana (and Royal Gardens)	Kilauea	1990–1991[b]
Goma	Nyiragongo	2002

[a]*Damage to three other villages: Zirosto, Angahuan, and Zacan.*
[b]*In Kalapana, during 1990–1991, 175 residences were destroyed, 21 uninsured homes were removed and 24 homes rendered uninhabitable by lava ingress (Weisel and Stapleton, 1992). The last house in Royal Gardens subdivision was destroyed in 2010.*

braided flows may branch around and surround property (Figure 2.7(c)). In terms of resistance to the impact, building design (especially building material and wall orientation) is key. For example, combustible materials will smolder and flame when contacted by lava (Figure 2.7(d)), and walls orientated at right-angles to the flow front will be more amenable to push-over than those with a parallel orientation (Blong, 1984).

2.3.4 Effects on Communications, Utilities, and Machinery

Communications (roads, railways, canals, rivers, sea ports, and airports) will be buried and cut. Examples of railways being invaded by lava include the Circum-Vesuviana line in 1906 (Vesuvius), the main Catania-Messina line in 1928 (Etna), and the Circum-Etna line in 1923 and 1981. Goma airport was invaded by lava in 2002, and the productive fishing port of Heimay (Iceland) was nearly blocked in 1973. Roads are also frequently buried on Etna, Kilauea, and Piton de la Fournaise (Rénion), as detailed in Note 6. Utilities (water, telecommunications, gas, and electricity) and machinery, plus all unrecovered household items and vehicles, will all be subject to ignition and burial

FIGURE 2.7 Building impacts by wall collapse, burial, and ignition. (a) Time series showing burial by inflation of Walter Yamaguchi's "*Kalapana Store and Drive Inn*" (Kalapana, Kilauea) during June 1990. *(Photos by J.D. Griggs in Takahashi et al. (2003), photo dates: 6/3/90; 6/6/90; 6/13/90.)* (b) Burial, push-over and deformation of masonry buildings in Mascali (Etna) by 'a'a on November 7, 1928. Note, at the right, a completely collapsed building adjacent to a white plaster building which is undergoing deformation (suffering cracking and plaster loss); the church has lost its roof and far wall. *(Photo by G. Ponte in Bonaccorso and Branca (2010).)* (c) Burial of Boscotrecase (Vesuvius) by 1906 lava *(photo by A.L. Day in Perret (1924))*. (d) Wood-framed house in Kalapana Gardens (Kilauea) ignited by pahoehoe in 1990. *(Photo by J.D. Griggs in Takahashi et al. (2003).)*

(Figure 2.8). Gas and water supplies are of particular concern (as are filling stations), as they have the potential to generate explosions (Figure 2.6).

2.3.5 Effects on Agriculture

All crops will be ignited or buried, as will habitat and natural vegetation (Figure 2.8(d)). Any livestock remaining in gated fields, being unable to

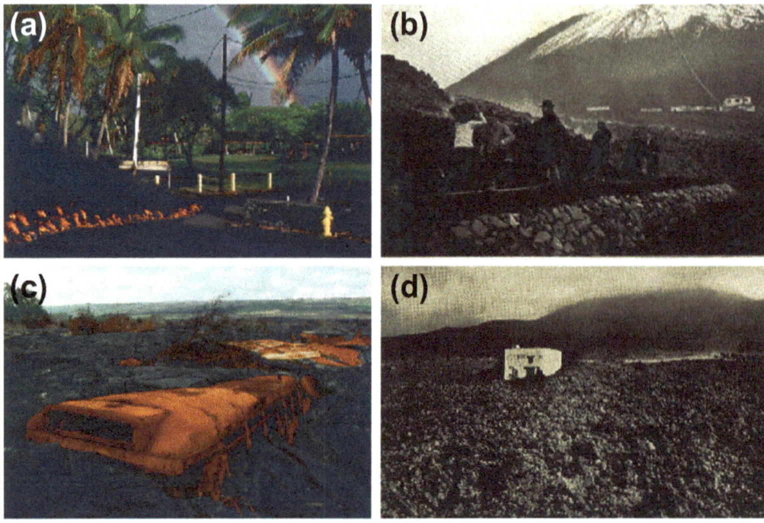

FIGURE 2.8 Destruction of communications, hardware, and agriculture. (a) Pahoehoe burning up road, and entering the Harry K. Brown Park (Kalapana, Kilauea) during 1990. *(Photo by D. Weisel in Takahashi et al. (2003).)* (b) Burial of the Vesuvian railway by 'a'a on February 7, 1906. Note: clean-up operations are already underway. *(Photo by F.A. Perret in Perret (1924).)* (c) Bus embedded in lava after the invasion of Kalapana during 1990. *(Photo by J.D. Griggs in Takahashi et al. (2003).)* (d) 'A'a from Vesuvius' 1906 eruption covering agricultural land above Boscotrecase. *(Photo by A.L. Day in Perret (1924).)*

escape, will be killed. During Etna's 1981 eruption, flows extended into cultivated areas, cut three major roads, buried a section of the circum-Etna railway, and destroyed buildings to cause ~$US 10 million of damage (Smithsonian Institution, 1989). The problem with burial of agricultural land, as well as natural vegetation, is that previously fertile land will be replaced by barren, sterile fields of lava (Figure 2.8(d)). This means that the farmer either goes out of business, must place significant investment in reclaiming buried land, or must relocate (e.g., Luhr and Simkin, 1993).

2.3.6 Effects on Economic Activity

Economic losses will result from the destruction of business and property, as well as the associated commercial, industrial, agricultural, and tourism losses (e.g., see Note 6). There will also be financial costs for: (1) evacuation, diversion efforts, relocation, and clean-up, (2) infrastructure, machinery, and possession replacement, and (3) rebuilding. By way of example, the cost of Kalapana's destruction on Kilauea between 1990 and 1991 was: $50 million damage to private property, $15 million to public utilities, $32 million to public and private roads, and $3.5 million to rebuild water systems (Weisel and Stapleton, 1992), for a total financial impact of $100 million.

2.3.7 Social Effects

Blong (1984) lists three social effects: (1) destruction of community; (2) break-down of society, including lawlessness, disorder, and looting (see Note 7b); and (3) breakdown of individual mental health. Temporary social breakdown may occur during the eruption. For example disruption of daily community routine will occur during an evacuation period, but this will reestablish if the community is untouched and the population returns. Permanent effects will result from the total destruction of community and failure to reestablish a new one (Note 7).

The degree to which social damage is done depends to a large extent on the reactions, and leadership skills, of community leaders, as well as the nature and extent of state, national, and international assistance (Blong, 1984). Examples of effective and poor leadership, and the results, during lava invasion of communities at Paricutin (Mexico) during 1943–1944, as taken from Luhr and Simkin (1993), are given in Note 8 and range from a Bishop leading the population out of town, to breakdown of new colonies due to external (hostility from locals) and internal (poor quality of land) pressures. Long-term mental scars will occur due to the traumatic experience. These scars can be increased by insensitive actions by media crews and sightseers, as well as those of other third parties, such as government agencies and insurance companies. Weisel and Stapleton (1992) noted that in Kalapana, "the insensitivity of outsiders coming to witness the destruction of their community, the legal clauses in insurance policies that required buildings to burn before their owners could get any compensation, and the demands of living in the face of lingering chaos" added to the frustration facing homeowners. The resulting mental health problems included cases of post-traumatic stress disorder (Note 7).

2.4 MITIGATION

Mitigation is the act of reducing the severity, or moderating the impact of, the natural hazard, while also alleviating the loss incurred. Chester (1993) lists three potential mitigation responses that are here considered in relation to lava flow hazard.

2.4.1 Preparation and Planning for Losses

Planning will range from individual preparations, to preparations at the municipal, regional, and national levels. Individual preparations may include obtaining adequate insurance coverage, having a family evacuation kit and plan. Such planning may be aided by the official posting of evacuation routes, meeting points, and evacuation centers. Personal planning may also involve preparing the family for the potential loss of home and community, and preparing for issues such as what the family will do should the community be destroyed.

In cases where a population is familiar with volcanic activity, the related hazard and correct response (i.e., evacuation) may be spontaneous. That is, evacuation occurs without the need for official intervention. Such spontaneous responses will likely be developed in isolated communities with no formal monitoring or response structures (an example is given in Note 9a), or in locations where activity is frequent, so that experience of the hazard is endemic (see Note 9b).

Community- and individual-based responses will likely occur at volcanoes where the population is particularly well-educated regarding the nature of the hazard and response protocols. An informed, self-motivated response can thus be aided by education programs aimed at increasing awareness to the type and nature of hazard, and correct responses, so as to aid in the formulation of appropriate plans and actions (examples of such educative measures aimed at residents and visitors are given in Note 9a). However, such preparation is more difficult than it may appear. Dibben (2008) found, especially for those who had moved to the town of Trecastagni on the flanks of Etna, that residents appeared to "cognitively diminish their perceptions of volcanic threat" to justify the positive socioeconomic reasons for moving there. Clearly, such attitudes need to be taken into account when preparing education and response plans, where residents may deny or ignore the hazard, preferring to focus on the positive reasons for living where they do.

At the municipal–regional–national level, evacuation and resettlement plans, as well as guidance for the impacted population, need to be drafted, and made well-known to the population at risk (Note 10). With effusive eruptions, the advantage is that there is normally plenty of time (hours, even days, weeks, or months) for the population to prepare and evacuate once the eruption has begun. The disadvantage is that the burial of agricultural, business, and residential land under meters of sterile lava may render quick repopulation problematic. The key is to prepare the population for the potential destruction of home and community. Inundation may, or may not happen, but potential replacement sites need to be identified and resettlement plans drafted, discussed with the community at risk and then put in place. Discussion with the community allows the population to participate in the process that determines their future. It thus allows the location of a socially acceptable displacement colony and design, thus maximizing the chance that the community stays intact. By preparing individuals for the eventuality, reality, and nature of the loss, appropriate planning, warning, and preparedness may also help reduce the loss and grief resulting from burial of the community by lava.

2.4.1.1 Jaggar (1931)

Jaggar (1931) provided a five point blueprint of "organization for preparedness" for a lava invasion which still holds today. The ideal response of Jaggar (1931) required:

1. Organization of discipline and communication, this being the burden of Civil Protection in liaison with the emergency services, the municipality, and possibly the military.
2. Organization of rescue, which requires mobilization of "large forces of laborers," trucks, and buses.
3. Organization of relief and refugee camps, including prompt installation of camp grounds. This means that tents, beds, bedding, surgical supplies, doctors, nurses, washing facilities, food supplies, water, utensils, fuel, field kitchen and cooks need to be ready, and clothes, means of communication, education, and amusement (phones, televisions, temporary schools, social centers) will need to be available.
4. Organization of relief funds, if loss has occurred.
5. Organization of rebuilding.

An example of a particularly well-planned evacuation and relocation comes from Mascali, a town with a population of ~2,000 on the east flank of Mount Etna. Mascali and ~1.6 km^2 of highly productive agricultural land was destroyed by lava during November 1928. Municipal guards, soldiers, and firemen were mobilized to aid with the evacuation, and each family was assisted in turn, so that many possessions (even roof tiles) were carried out by trucks (Chester et al., 1999). Families were billeted with friends and family, or housed in second homes and hired apartments in nearby towns (Duncan et al., 1996). The town was rebuilt 1.5 km to the east of the old town (i.e., as close as possible to the original settlement) on flat ground and in a topographically sheltered position. The new town was laid out on a grid pattern, with a large church, railway station and schools, at a cost of 25 million lira or US $1.3 million (Duncan et al., 1996; Chester et al., 1999). Each family was given 10,000 to 20,000 lira (US $525–1,050) depending on family size, this being ~60% of the cost of a new home (Duncan et al., 1996). Whereas old Mascali had narrow streets, overcrowded dwellings, high population density, poor sanitation, and a communal water supply, amenities of the new town were excellent, with running water, piped drainage, and electricity. A measure of the success of this relocation was that a visit of Mussolini in 1937, whose prewar fascist government had organized the response, was greeted with "*shear rapture*" (Chester et al., 1999). Other examples of successful response and evacuation measures are given in Note 10.

2.4.2 Modify the Loss Potential

In the case of lava inundation, loss can be avoided or modified by land-planning, land use control and/or guidance, as well as implementation of effective warning systems. In effect, communities are either prevented from, or warned against, developing in particularly hazardous zones. To aid in land use and development decisions by individuals, businesses and agencies (private and government), hazard assessment reports and hazard maps can be drafted (see Note 11). Such initiatives can only be used to guide land use planning policy and law. Without the imposition of legally binding land-use restrictions, such documents can only serve as guides to educate and inform. Developers will remain at liberty to continue building and selling in hazardous areas. Other actions that can be implemented are given in Note 12 and include the relocation of particularly vulnerable facilities.

2.4.3 Modify the Hazard

In terms of lava this means attempting to stop or divert the flow, so that it spreads in a direction that does not threaten vulnerable communities. It is not possible to stop effusion of lava from an erupting vent, but it may be possible to stall, or divert, advancing lava. One of the best classifications of ways in which we can modify the hazard posed by an advancing lava flow is given by Ganeri (2000), who lists five means of modification, a listing which is followed here.

2.4.3.1 Construction of a Barrier

A full review of the history of barrier building is given in Note 13a. A great deal of work exploring the correct design and orientation of lava diversion barriers was carried out between 1930 and 1960 by Jaggar (1937, 1945a; 1949), Macdonald (1958) and Wentworth et al. (1961), as summarized in Figure 2.3. Subsequently barriers were constructed:

- To protect the village of Nomashi (Japan) during Mihara Yama's 1951 eruption (Mason and Foster, 1953);
- To protect plantations during Kilauea's 1955 eruption (Macdonald and Eaton, 1955; Macdonald, 1958);
- To delay lava ingress into the village of Kapoho during Kilauea's 1960 eruption (Macdonald, 1962), see Figure 2.2;
- At Heimaey (Vestmannaeyjar, Iceland) during Eldfell's 1973 eruption (Williams, 1997; Jónsson and Matthiasson, 1997);
- In 1986 to protect the Mauna Loa observatory from lava inundation (Moore, 1982); and
- At Etna in 1983 (Colombrita, 1984; Lockwood and Romano, 1985), 1992 (Barberi et al., 1993; Di Palma et al., 1994), and 2001 (Barberi et al., 2003), where other attempts to divert flows on Etna occurred in 1669, 1832, 1879 and 1923 (Chester et al., 2012).

As a result, standard barrier design is well documented (Note 13b) and, following recent barrier construction projects on Etna, required resources and costs for barrier construction are well defined (see Note 14). From these examples, we can define two types of barrier: deflection barriers and damming barriers (Figure 2.9). Deflection barriers are designed to direct lava around or away from vulnerable infrastructure. They therefore need to be designed to use their angle and topography to force the flow in the desired direction (Figure 2.9(a)). Damming barriers, instead, are built to impede or hold up flow (e.g., Note 15). Once lava has reached the barrier crest, the lava will overtop the barrier (Figure 2.9(b)). Such barriers thus have a limited lifetime and can only buy time, during which the eruption may end, output rates may decline or other diversion initiatives may be successful.

2.4.3.2 Bombing

The aim of bombing is, by using either aircraft6launched or artillery6fired missiles, to explosively disrupt tube-walls and roofs, breach channel levees or remove cone segments so as to cause the lava to flow in a different direction (Figure 2.10). The use of explosives to change the course of a lava flow was attempted at Mauna Loa during its 1935 (Jaggar, 1936a,b) and 1942 (Finch, 1942; Macdonald, 1943) eruptions (see Note 16a), as well as on Etna during 1983 (Aberstein, 1983) and 1992 (Vassale, 1994). Further experiments to evaluate the use of bombs for lava diversion were carried out over pre-historic flows on Mauna Loa in 1975 and 1976 by Lockwood and Torgerson (1980). Based on the Hawaiian experiences, Finch and Macdonald (1949) noted that the diversion of lava flows by bombing depends "upon a combination of several favorable circumstances", including:

1. Lava flowing in a channel, or lava held in a source cone, at a level higher than the adjacent land surface;
2. Narrow levees, thin cone walls, or thin tube roofs that can be broken by bombs;
3. Favorable topography.

They added that time is required to select targets, prepare planes, load appropriate bomb loads, and put the plan into operation (see Note 16b).

Jaggar (1936c) pointed out the potential resistance to bombing of volcanic structures due to community beliefs which respect the volcano and the deity residing there. In the case of Hawaii, Pele and the cultural beliefs of the resident population must be taken into consideration. This can be done through appropriately designed social surveys. The key question is, does the resident population want intervention? Interviews of evacuees from Kapoho in 1960 by Lachman and Bonk (1960), and reexamined by Gregg et al. (2008), showed that 72% of the 638 respondents favored the construction of barriers at Kapoho, versus 16 percent who favored the military use of bombs.

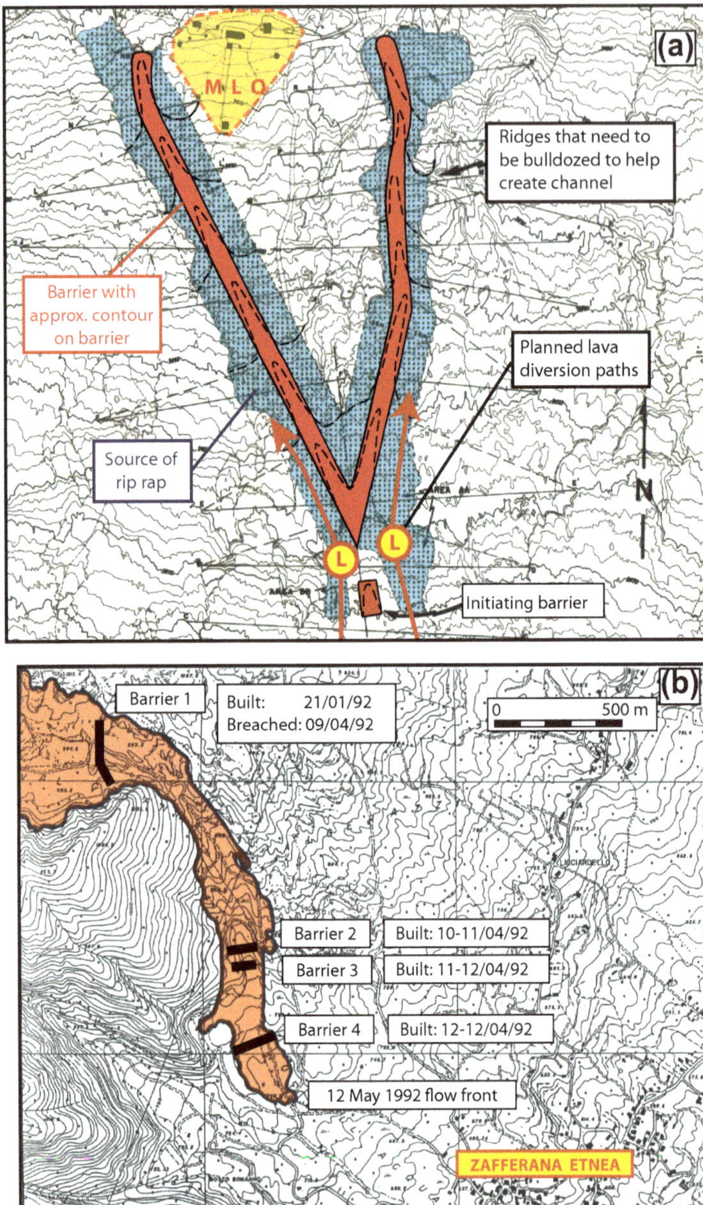

FIGURE 2.9 (a) Moore's (1982) plan for barriers built to deflect lava around the NOAA Mauna Loa (meteorological) Observatory. The barriers are made from "rip rap", this being a coarse (>1.5 inch) aggregate used in construction, and accounting for 33% of the coarse aggregate produced in the United States (Willett, 2007). It is important to note that the tip of the V-shaped barrier is where the "MLO ridge meets the valley above it" and that "rip rap for the construction of

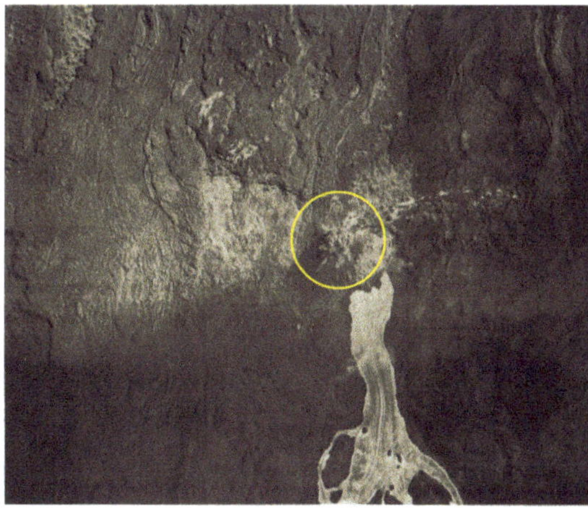

FIGURE 2.10 Six-hundred pound bomb exploding beside lower source of Mauna Loa's 1942 lava channel (plume circled in yellow). The aim was to collapse the source roof. The channel-exiting source is 15–20 m wide. *Source: Photo from Base Photo Lab., given in Finch (1942).*

2.4.3.3 Use of Water

The approach here is to spray water on the active flow to develop a quenched retaining crust, or better still to solidify the flow completely. The latter is impossible because, as long as the source vent is active, molten lava will always be supplied to the flow front. As a result, eventually, the quenched barrier will fail, or lava ponding behind the barrier will flow over, through or around it. Water may also be sprayed on buildings to retard combustion. Such operations have been attempted in Hawaii and Iceland (Note 17a) but, like the construction of damming barriers, they are time-buying measures to delay the onset of lava inundation until another diversion measure has been executed or the eruption ends. Issues and recommendations regarding the use of water, based on the Heimaey experience are given in Note 17b.

2.4.3.4 Diversion

Diversion involves deflecting the lava onto a new course or into a new area where the lava can flow or pond without doing damage. Alternatively, lava can

the barriers should be taken from the expected flow-ward side of the barrier" (Moore, 1982). This, along with the geometry of the deflection and initiating barriers, sets up conditions favorable to diverting flows arriving at the apex of the "V" into channels created by the rip rap excavation. (b) Barberi et al.'s (1993) map of damming barriers built to hold up the advance of Etna's 1992 flow toward Zafferana Etnea. Four barriers where built: one at the Portella Calanna—Barrier 1 (L = 234 m, h = 21 m, base width = 70 m, summit width = 8 m), and three in the ravine above Zafferana—Barrier 2 (L = 110 m, h = 12 m), Barrier 3 (L = 90 m, h = 6 m), and Barrier 4 (L = 160 m, h = 12 m) (Di Palma et al., 1994).

be diverted into an artificial channel that takes the flow on a different course away from, or around, vulnerable communities. Diversion attempts have been carried out in Japan and Italy (Note 18), with the most famous being diversion of lava during Etna's 1991–1993 eruption. A number of attempts were made before the final, successful, intervention, with each building on the lessons learnt from the previous experience. The final intervention required (Barberi et al., 1993; Vassale, 1994):

1. Construction of an artificial channel next to a skylight in the master tube to create a thin, but solid, septum between the natural and artificial channels;
2. Laying of explosive charges in, and blasting of, the septum to achieve partial diversion into the artificial channel, and
3. Dumping of steel hedgehogs and lava boulders into the natural channel to obstruct it and achieve maximum feed to the artificial channel.

These efforts, and required resources, are detailed and illustrated in Note 19.

2.4.3.5 Offer a Sacrifice (Turn to Faith)

This option involves turning to a form of belief to deviate the flow, stop the eruption or ameliorate the impact of the disaster. Examples of such actions are given in Note 20. Whether or not we believe such actions can stop lava flows and eruptions, turning to a belief or embracing a religious reason for the events impacting the individual and community may help in individual and community acceptance of the event. Such a turn to faith, and actions to encourage the miraculous to take place during effusive events, remain common place on both Etna and Vesuvius (Chester et al., 2008). Without a doubt, such actions can help bind a community in the face of tragedy and reduce psychological damage, where solace and reason is found in a strong belief form. Such an attitude, that it is the will of a deity or nature (or just one of those things that life throws at us), may help in the process of bearing the loss, restarting, and healing the mental scars resulting from loss.

2.5 CONCLUSIONS

The main, direct hazards posed by actively advancing basaltic lavas, have been focused upon, as well as both institutional and individual responses to the hazard. Not considered have been the multitude of secondary hazards, including:

1. Wild-fires set off by lava advancing into dry vegetation;
2. Lahars caused by lava advancing over steep slopes covered by snow and ice;
3. Explosions resulting from buildup of methane beneath lava flows;
4. Inversion of relief to cause new areas to become vulnerable;
5. Blocking of valleys to dam rivers that can build lakes, to also cause avulsion and flooding.

Also not considered have been hazards special to silicic lava flows, mainly:

6. Collapse of flow fronts to generate pyroclastic (block and ash) flows.

Secondary effects have however, been considered that were due to:

7. Water and fuel interactions to cause explosions;
8. The effects of lava-heated water (and water chemistry changes), plus lava-generated steam, on the health of humans and animals (see Note 5b); plus
9. The hazard due to collapse of unstable lava benches built during ocean entry of lava (see Note 5a), and
10. Problems due to close approach by tourists.

All are factors which, in the words of Blong (1984), need to be included in the plans and responses of those responsible for managing the hazard.

The main conclusion is that lava is a localized, but highly destructive, natural hazard that humans can usually escape. However, anchored and unretrieved structures, cannot avoid combustion and burial, and fertile land will be rendered sterile. Under opportune conditions, lava can be diverted, but can only be stopped for a finite amount of time. Those living with the daily potential of lava inundation need to be educated to the fact that their community and social infrastructure may be brutally destroyed at any time; to the extent that they know how to respond and react so that, ideally, they become hardened to the eventuality of loss (see Note 21).

The development of a hardened, realistic, resilient, and reactive mentality among the resident population is required. That is, the population becomes familiar with the nature of the hazard, ready to evacuate, prepared for the loss, and ready to re-establish or relocate. The development of such a mentality will be a function of previous experiences, familiarity with the activity of the host volcano, knowledge of the products and their related hazards and, hence, education. Such exposure will promote development of appropriate community beliefs and reactions. Likewise, acceptance that loss is an eventuality needs to become an accepted part, and way, of life.

With the recent growth of tourism (including volcano tourism), tourist facilities and large numbers of tourists have increasingly become located in volcanically hazardous areas. As a result education, training and planning needs to extend to tourist operators, tour representatives, and tourists, so that these alien groups are also in a position to respond correctly if faced with a natural disaster while visiting a location. The problem with this group is that they will be facing situations and conditions with which they are unlikely familiar (see Bird et al. (2010) and references therein).

REFERENCES

Aberstein, L., 1983. Diversion of a lava flow from its natural bed to an artificial channel with the aid of explosives: Etna, 1983. Bull. Volcanol. 47 (2), 1163–1177.

Atkinson, R., 2007. The Day of Battle: The War in Sicily and Italy, 1943–1944. Henry Holt, New York, 791 p.

Baldwin, E.D., 1953. Notes on the 1880–1881 lava flow from Mauna Loa. Volcano Lett. 520, 1–3.

Barberi, F., Carapezza, M.L., Valenza, M., Villari, L., 1993. The control of lava flow during the 1991–1992 eruption of Mt. Etna. J. Volcanol. Geotherm. Res. 56, 1–34.

Barberi, F., Brondi, F., Carapezza, M.L., Cavarra, L., Murgia, C., 2003. Earthen barriers to control lava flows in the 2001 eruption of Mt. Etna. J. Volcanol. Geotherm. Res. 123, 231–243.

Behncke, B., Neri, M., Nagay, A., 2005. Lava flow hazard at Mount Etna (Italy): new data from a GIS-based study. Geol. Soc. Am. Spec. Paper 396, 189–208.

Bird, D.K., Gisladottir, G., Dominey-Howes, D., 2010. Volcanic risk and tourism in Iceland: implications for hazard, risk, and emergency response education and training. J. Volcanol. Geotherm. Res. 189, 33–48.

Blong, R.J., 1984. Volcanic Hazards. Academic Press, Sydney, 424 p.

Bonaccorso, A., Branca, S., 2010. Fotografia storica alla luce del Vulcano: Fondo fotografico Gaetano Ponte (1876–1955). Le Nove Muse Editrice, Catania, 191 p.

Chester, D., 1993. Volcanoes and Society. Edward Arnold, London, 351 p.

Chester, D.K., Duncan, A.M., Dibben, C., Guest, J.E., Lister, P.H., 1999. Mascali. Mount Etna region Sicily: an example of fascist planning during the 1928 eruption and its continuing legacy. Nat. Hazards 19, 29–46.

Chester, D.K., Duncan, A.M., Wetton, P., Wetton, R., 2007. Responses of the Anglo-American military authorities to the eruption of Vesuvius, March 1944. J. Hist. Geography 33, 168–196.

Chester, D.K., Duncan, A.M., Dibben, C.J.L., 2008. The importance of religion in shaping volcanic risk perception in Italy, with special reference to Vesuvius and Etna. J. Volcanol. Geotherm. Res. 172, 216–228.

Chester, D.K., Duncan, A.M., Sangster, H., 2012. Human responses to eruptions of Etna (Sicily) during the late-pre-industrial era and their implications for present-day disaster planning. J. Volcanol. Geotherm. Res. 225–226, 65–89.

Clague, D.A., Hagstrum, J.T., Champion, D.E., Beeson, M.H., 1999. Kilauea summit overflows: their ages and distribution in the Puna district, Hawai'i. Bull. Volcanol. 61, 363–381.

Colombrita, R., 1984. Methodology for the construction of earth barriers to divert lava flows: the Mt Etna 1983 eruption. Bull. Volcanol. 47–4, 1009–1038.

Di Palma, S., Drago, F., Galanti, E., Pennisi, V., 1994. Earthen barriers and explosion tests to delay the lava advance: the 1992 Mt. Etna experience. Acta Vulcanol. 4, 167–171.

Dibben, C.J.L., 2008. Leaving the city for the suburbs – the dominance of 'ordinary' decision making over volcanic risk perception in the production of volcanic risk on Mt Etna, Sicily. J. Volcanol. Geotherm. Res. 172, 288–299.

Duncan, A.M., Dibben, C.D., Chester, D.K., Guest, J.E., 1996. The 1928 eruption of Mount Etna Volcano, Sicily, and the destruction of the town of Mescali. Disasters 20 (1), 1–45.

Finch, R.H., 1942. The 1942 eruption of Mauna Loa. Volcano Lett. 476, 2–5.

Finch, R.H., Macdonald, G.A., 1949. Bombing to divert lava flows. Volcano Lett. 506, 1–3.

Finch, R.H., Macdonald, G.A., 1950. The June 1950 eruption of Mauna Loa part I: narrative of the eruption. Volcano Lett. 508, 1–12.

Ganeri, A., 2000. Violenti Vulcani. Salani, Varese, Italy, 128 p.

Gregg, C.E., Houghton, B.F., Paton, D., Swanson, D.A., Lachman, R., Bonk, W.J., 2008. Hawaiian cultural influences on support for lava flow hazard mitigation measures during the January 1960 eruption of Kilauea volcano, Kapoho, Hawaii. J. Volcanol. Geotherm. Res. 172, 300–307.

Harris, A.J.L., Rowland, S.K., 2009. Effusion rate controls on lava flow length and the role of heat loss: a review. In: Thordarson, T., Self, S., Larsen, G., Rowland, S.K., Hoskuldsson, A. (Eds.), Studies in Volcanology: The Legacy of George Walker, 2. Special Publications of IAVCEI, pp. 33–51.

Ingwersen, G.P. (Lt. Col.), 1966. Review Report on Survey for Lava Flow Control, Island of Hawaii, State of Hawaii, Department of the Army, Honolulu District, Corps of Engineers, Fort Armstong. Hawaii, Honolulu, 12 p.

Jaggar, T.A., 1931. Preparedness for disaster. Volcano Lett. 338, 2–4.

Jaggar, T.A., 1936a. The bombing operation at Mauna Loa. Volcano Lett. 431, 4–6.

Jaggar, T.A., 1936b. The bombing of Mauna Loa, 1935. Volcano Lett. 442, 1–7.

Jaggar, T.A., 1936c. Native superstition. Volcano Lett. 431, 6–7.

Jaggar, T.A., 1937. Protection of Hilo from coming lava flows. Volcano Lett. 443, 1–8.

Jaggar, T.A., 1945a. Protection of harbors from lava flow. Am. J. Sci. 243A, 333–351.

Jaggar, T.A., 1949. Threat of Lava Flow to Hilo (Abstract). Hawaii Academy of Science 24th Annual Meeting, Proceedings.

Jónsson, V., Matthiasson, M., 1997. Lava Cooling on Heimaey – Methods and Procedures. U.S. Geological Survey Open File Report 97–724.

Lachman, R., Bonk, W.J., 1960. Behavior and beliefs during the recent volcanic eruption at Kapoho Hawaii. Science 31, 1095–1096.

Lockwood, J.P., Romano, R., 1985. Diversion of lava during the 1983 eruption of Mount Etna. Earthquake Inf. Bull. 17 (4), 124–133.

Lockwood, J.P., Lipman, P.W., 1987. Holocene eruptive history of Mauna Loa Volcano. In: Decker, R.W., Wright, T.L., Stauffer, P.H. (Eds.), Volcanism in Hawaii, vol. 1. U.S. Geological Survey Professional Paper 1350, p. 518.

Lockwood, J.P., Torgerson, F.A., 1980. Diversion of lava flows by aerial bombing–lessons from Mauna Loa Volcano, Hawaii. Bull. Volcanol. 43 (4), 727–741.

Lockwood, J.P., Hazlett, R.W., 2010. Volcanoes: Global Perspectives. Wiley & Blackwell, Chichester, 539 p.

Luhr, J.F., Simkin, T., 1993. Paricutin. Geoscience Press, Pheonix, 427 p.

Macdonald, G.A., 1943. The 1942 eruption of Mauna Loa, Hawaii. Am. J. Sci. 241 (4), 241–256.

Macdonald, G.A., 1958. Barriers to protect Hilo from lava flows. Pac. Sci. 12, 258–277.

Macdonald, G.A., 1962. The 1959 and 1960 eruptions of Kilauea volcano, Hawaii, and the construction of walls to restrict the spread of the lava flows. Bull. Volcanol. 24, 249–294.

Macdonald, G.A., Eaton, J.P., 1955. The 1955 eruption of Kilauea Volcano. Volcano Lett. 529, 1–10.

Macdonald, G.A., Finch, R.H., 1950. The June 1950 eruption of Mauna Loa part II: special features. Volcano Lett. 509, 1–7.

Mason, A.C., Foster, H.L., 1953. Diversion of lava flows at O Shima, Japan. Am. J. Sci. 251, 249–258.

Moore, H.J., 1982. A Geologic Evaluation of Proposed Lava Diversion Barriers for the Mauna Moa Observatory, Mauna Loa Volcano. U.S. Geological Survey Open File Report, 82–314, Hawaii, 17 p.

Parker, M., 2005. Monte Cassino: The Hardest-fought Battle of World War II. First Anchor Books, Random House, New York, 414 p.

Perkins, M.C., 2006. Surviving Paradise. Quidnunc Press, USA, 338 p.

Perret, F.A., 1924. The Vesuvius Eruption of 1906: Study of a Volcanic Cycle. Carnegie Institution, Washington D.C, 151 p.

Reynolds, R.W., 1996. Volcanic hazards at Sierra Negra. Not. Galapagos 57, 13–19.

Takahashi, T.J., Heliker, C., Diggles, M.F., 2003. Selected Images of the Pu'u 'O'o – Kupaianaha Eruption, 1983–1997. U.S. Geological Survey. Digital Data Series DDS-80.

Tazieff, H., 1977. An exceptional eruption: Mt. Niragongo, jan. 10th, 1977. Bull. Volcanol. 40–3, 189–200.

Vassale, R., 1994. The use of explosive for the diversion of the 1992 Mt. Etna lava flow. Acta Vulcanol. 4, 173–177.

Weisel, D., Stapleton, F., 1992. Aloha O Kalapana. Bishop Museum Press, Honolulu, 153 p.

Wentworth, C.K., Powers, H.A., Eaton, J.P., 1961. Feasibility of a lava-diverting barrier at Hilo, Hawaii. Pac. Sci. 15, 352–357.

Willett, J.C., 2007. Stone, crushed, United States geological survey 2005 minerals year book. U.S. Department of the Interior, 71.7–71.35.

Williams, R.S., 1997. Lava-Cooling Operations during the 1973 Eruption of Eldfell Volcano. U.S. Geological Survey Open-File Report 97–724, Heimaey, Vestmannaeyjar, Iceland.

Witham, C.S., 2005. Volcanic disasters and incidents: a new data base. J. Volcanol. Geotherm. Res. 148, 191–233.

Chapter 3

Impacts from Volcanic Ash Fall

Thomas M. Wilson [1], Susanna Jenkins [2] and Carol Stewart [3]
[1] *Volcanic Ash Testing Lab, Department of Geological Sciences, University of Canterbury, Christchurch, New Zealand,* [2] *Department of Earth Sciences, University of Bristol, Wills Memorial Building, Queens Road, Bristol, UK,* [3] *Joint Centre for Disaster Research, GNS Science/Massey University, Wellington, New Zealand*

ABSTRACT
All explosive eruptions produce volcanic ash, fragments of volcanic rock generated when magma or vent material is explosively disintegrated during eruption. Volcanic ash is convected upwards within the eruption column and carried downwind, falling out of suspension and potentially affecting communities across hundreds of square kilometers. Although ash falls rarely endanger human life directly, threats to public health and disruption to critical infrastructure services, aviation, and primary production (e.g. agriculture) can lead to significant societal impacts. Even relatively small eruptions such as the Eyjafjallajökull eruption in Iceland in 2010 (Volcanic Explosivity Index of 4) can cause widespread disruption, damage, and economic loss. Knowledge of the likely impacts can support mitigation actions, crisis planning, and emergency management activities. This chapter presents an overview of ash fall impacts for sectors of society including buildings, critical infrastructure, and agriculture; and discuss associated socioeconomic factors. We also discuss the likely response (vulnerability) of these key sectors to ash fall impacts. Broad relationships between volcanic ash thickness and levels of damage and disruption have been outlined. Understanding these vulnerabilities is an essential step towards building resilience for communities.

DEFINITIONS USED IN THIS CHAPTER (*modified from UN, 2009*):

- *Hazard*: potential threat arising from the physical phenomenon.
- *Vulnerability*: The characteristics and circumstances of a community, system or asset that makes it susceptible to the damaging effects of a hazard.
- *Impact*: function of the hazard and vulnerability of the exposed assets.
- *Exposure*: People, property, systems, or other elements present in hazard zones that are thereby subject to potential losses.

Volcanic Hazards, Risks, and Disasters. http://dx.doi.org/10.1016/B978-0-12-396453-3.00003-4

- *Risk*: function of both the probability of a hazardous event and the consequences to the exposed assets.
- *Hazard intensity*: a measure of the severity of a particular hazard at a particular location (e.g., ash fall depth or loading).
- *Vulnerability function*: relates the intensity of a hazard to percentage of loss to an exposed asset or system.
- *Societal assets*: those assets that support the normal functioning of human society.

3.1 INTRODUCTION

Explosive volcanic eruptions typically produce fragments of volcanic rock that range in size from blocks and bombs (>64 mm) to lapilli (2–64 mm) and ash (<2 mm) (Figure 3.1). These ejecta are collectively known as tephra. Blocks and bombs follow ballistic trajectories and are commonly confined to within 5 km of the vent (Blong, 1984). Lapilli and ash are convected upwards within the eruption column and dispersed by winds, with larger, denser fragments falling

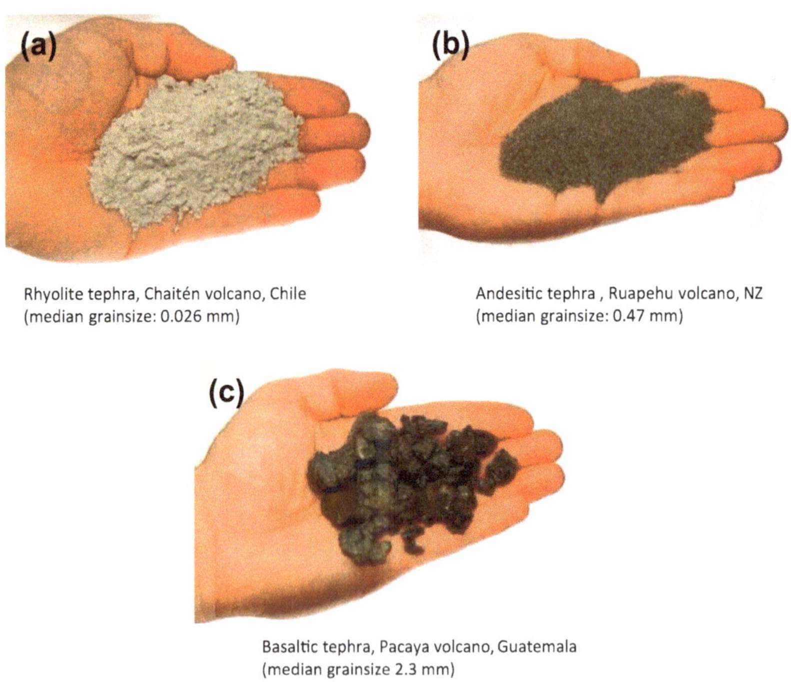

FIGURE 3.1 (a) Rhyolite ash, Chaitén volcano, Chile (median grain size of sample: 0.026 mm); (b) andesitic volcanic ash, Ruapehu volcano, NZ (median grain size of sample: 0.47 mm); (c) basaltic tephra, Pacaya volcano, Guatemala (median grain size of sample: 2.3 mm). *Photo credits: G. Wilson.*

FIGURE 3.2 a) A farm approximately 8 km from the vent of Eyjafjallajökull volcano in Iceland during the 2010 eruption. b) Three hours and a change in wind direction separate the two photos and highlight the issue of reduced visibility during ash falls. Directly under the volcanic ash plume visibility was approximately 10 m or less. *Photo credits: S. Jenkins.*

out (deposited) closer to source and smaller, lighter particles dispersed farther downwind (Figure 3.2 and Figure 3.3). The resulting deposits (referred to as "ash falls") thus typically thin rapidly with increasing distance from the volcano, in line with a reduction in grain size (e.g. Pyle, 1989), although aggregation of particles can lead to secondary maxima in fall depths and grain sizes at distance from the volcano. Volcanic ash can be transported for hundreds or even thousands of kilometers from the source volcano covering large areas. The final area of deposition depends on wind conditions, volcanic ash grain size, and density, and eruption style and magnitude (Figure 3.3). Volcanic ash deposit thicknesses and load can be measured, plotted on a map, and contoured. Lines of equal volcanic ash thickness are termed isopachs. It is rare for volcanic ash to accumulate in a symmetric, radial pattern of isopachs around a volcano, instead isopachs generally define distributions elongated in the direction of the prevailing wind during the eruption. Multiple isopach lobes may form during an eruption (or eruptive episode) due to changing winds (Figure 3.3).

Ash fall is the most frequently produced volcanic hazard, occurring in 90% of all eruptions (Newhall and Hoblitt, 2002). However, the assessment of ash fall impacts is a comparatively underdeveloped field compared to, for example, seismic impact assessment. Ash falls rarely endanger human life directly, except where very thick falls cause structural damage (e.g., roof collapse) that threatens occupants. However, ash falls can affect public health, particularly through ash inhalation. Ash falls can also cause extensive and severe damage and/or disruption of critical infrastructure (such as electricity supplies, water supplies, and transport networks), aviation, the built environment, and primary agricultural production (Wilson et al., 2012b). The loss of functionality of these societal assets can have greater consequences for affected communities than the physical impacts of volcanic ash itself, commonly when ash fall deposits are relatively thin (i.e., relatively low ash hazard intensity).

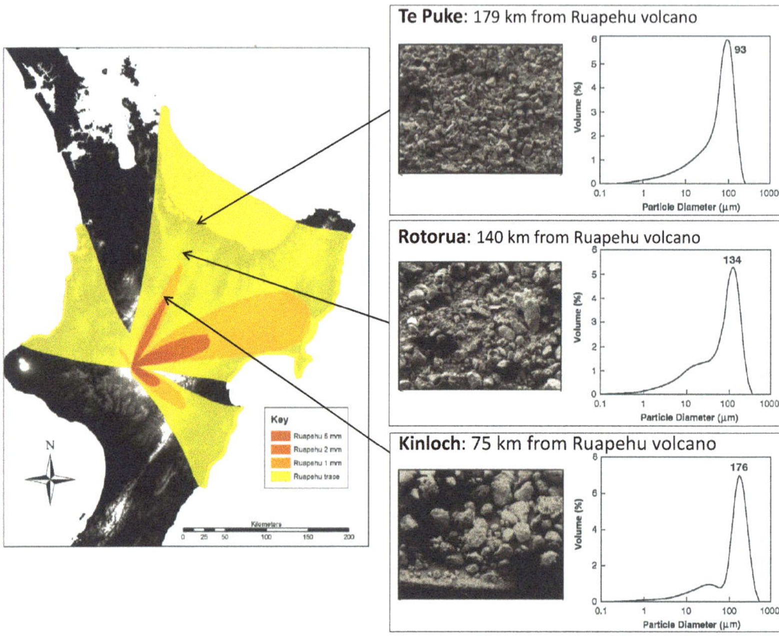

FIGURE 3.3 Isopach map of ash falls from the 1995–1996 eruption episode from Ruapehu volcano, New Zealand. Note the dispersal of volcanic ash in four major lobes, primarily driven by wind conditions at the time of the eruptions. Grain size distribution of volcanic ash deposits are presented for three different locations progressively further from the erupting volcano for the 17 June 1996 eruption. Note the changes in grain size distribution with distance from the volcano. *Isopach data is taken from Cronin et al. (1998); grain size distributions are from David Johnston.*

This chapter focuses on how volcanic ash can impact communities, and at what approximate hazard intensities. In particular, we highlight characteristics of volcanic ash that influence impacts to different societal assets. We also consider temporal dimensions of ash fall, including similarities and differences in impact between large, single ash falls ("acute exposure"), and repeated smaller ash falls resulting from long-duration eruptions or remobilization of deposits ("chronic exposure"). Drawing from data collected following past eruptions and through experimental and theoretical studies, we focus on identifying notable ash fall impacts to critical infrastructure, the built environment, and primary production (e.g., agriculture). In quantifying likely ash fall impacts, the hazard and the vulnerability of the asset in question are important. Understanding the breadth and severity of impacts from ash falls of different amounts and characteristics, and the vulnerability of major sectors of society, is key to reducing its impact on livelihoods and socioeconomic activities. In the final section of the chapter we identify sectors of society that are

FIGURE 3.4 Processes occurring within volcanic plumes (refer to Section 3.2). *Figure used with permission of GNS Science.*

most vulnerable to ash fall impacts and outline quantitative relationships between the hazard intensity (in this case ash fall thickness) and impact levels of damage or disruption. We only briefly summarize the potential health impacts from volcanic ash in this chapter, as these are addressed more comprehensively in Chapter 12 (Figure 3.4).

3.2 ASH FALL CHARACTERISTICS AND HOW THEY INFLUENCE IMPACTS

The impact of volcanic ash fall on society can be viewed as a complex relationship between the quantity and physical and chemical properties of the ash fall, the characteristics of the exposed assets and their environment as well as the level of preparation and mitigation (e.g., planning), and adaptive capacity of all levels of society.

Volcanic ash has a wide range of possible physical and chemical characteristics, largely controlled by magma chemistry and eruption conditions.

Deposits are made up of various proportions of vitric (glassy, noncrystalline) and crystalline or lithic (nonmagmatic) particles. Such particles are typically very hard and angular, making them abrasive. Vitric particles often contain small voids, known as vesicles, which are created when magmatic gas expands before the enclosing magma solidifies. Vesicular particles can have very high surface area-to-volume ratios; during eruption and transport exsolved magmatic gases condense onto these surfaces (Figure 3.4). The presence of strong mineral acids H_2SO_4 (formed from the oxidation of SO_2), HCl, and HF can make these surface coatings on fresh volcanic ash highly acidic. More than 55 different soluble components have been measured in volcanic ash leachates, with some components, such as fluoride, being potentially toxic to humans and livestock (Witham et al., 2005; Stewart et al., 2006; Cronin et al., 2003). Freshly deposited volcanic ash is, therefore, potentially acidic, corrosive, toxic, and electrically conductive (Wilson and Stewart, 2013).

Climatic and environmental conditions will also control volcanic ash dispersion and influence its physical and chemical properties from site to site. This variability both between different eruptions and during the same eruption makes it challenging to develop generic estimates of expected volcanic ash impacts. Historically, the thickness of ash fall has been the most common parameter used to assess hazard intensity in relation to damage because it is relatively easy to measure in the field or interpret from eyewitness accounts (Wilson et al., 2012b). More recently, volcanic ash loading (in mass per unit area) has become a preferred measure as it is more informative when considering impacts to structures and agriculture, and is the common output of numerical ash fall models. Measurements or estimates of load also avoid complications due to volcanic ash compaction following deposition. However, physical loading is not the only mechanism by which volcanic ash can cause damage or disruption; for instance simply the anticipation of an ash fall or the mere presence of volcanic ash can be damaging or disrupt societal activities. Further, volcanic ash grain size and density, chemical surface coating composition, abrasiveness, and friability can all influence, or even control, the level of damage or disruption incurred (Tables 3.1 and see Tables 3.2–3.6). Below we provide more detail on how each characteristic influences potential impact:

- *Thickness (common unit: mm)*: Fall thickness has been widely used to describe the physical presence of volcanic ash, and most published data use this intensity measure; it is quick and intuitively obvious for observers, although it is commonly inaccurately measured or reported (e.g., Engwell et al., 2013). We note that postevent compaction of volcanic ash deposits may be up to 50% (Blong, 2003). However, we consider this metric to be the most generally useful in relating hazard intensity to against exposed assets.

TABLE 3.1 Representative images of ash fall characteristics that are influential in determining the nature and severity of impacts during or following an ash fall.

Ash Characteristic	Caption	Image
Thickness	*~100mm of rhyolitic ash fall smothering pasture on farmland, Amarillo valley, Chile, following the 2008 Chaiten eruption. Credit: T. Wilson*	
Loading	*Ash cleaned off a hospital roof in Heimaey following 1973 Eldfell eruption, Iceland (tractor for scale). Credit: G. Oskarsson*	
Airborne Ash Concentration	*Windy conditions in Jacobacci, Argentina on 9 Sept 2011 leading to high levels of fine airborne ash due to remobilisation of fall deposits from June 2011 eruption of Cordon Caulle. Credit: J. Mellado*	

Continued

TABLE 3.1 Representative images of ash fall characteristics that are influential in determining the nature and severity of impacts during or following an ash fall.—cont'd

Particle Density	*Ash-water suspensions after one hour continuous mixing at 100 rpm followed by one hour settling. Key:* *1 – 1996 Ruapehu ash;* *2 – 2010 Pacaya ash <1 mm;* *3 – 2008 Chaitén ash;* *4 – 2010 Pacaya ash >1 mm* *Note – these are ashes from the same eruptions as in Figure 1.* *Credit: C. Stewart*	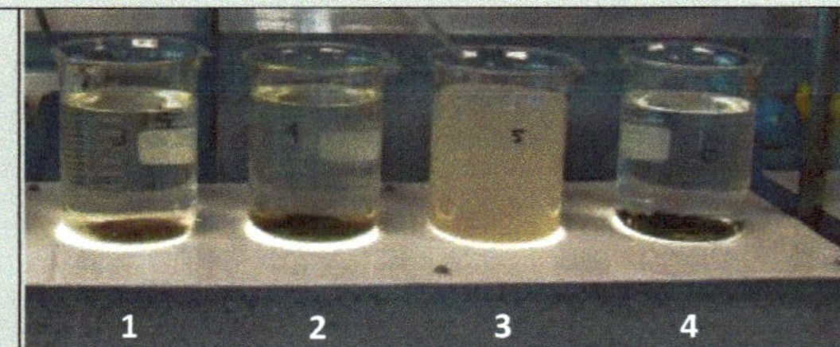
Particle Size Distribution (grain size)	*Following the June 2011 erupton of Cordon Caulle volcano, suspended ash entered the city of Bariloche water treatment plant through the water intakes as well as by direct fall out. Modes of damage depend on particle size. Fine particles caused less abrasion damage to pump impellors, but blocked the pores in the sand filter beds (pictured). Coarser particles caused more abrasion damage and were more difficult separate from the sand in filter beds. Credit: Departamento Provinvial de Agua, Bariloche*	
Surface Composition	*Chillis damaged by acidic surface coating of the adhering ash during the Merapi 2006 eruption, Indonesia. Credit:G. Kaye* *~3mm of wetted ash on electrical insulator string during flash-over. The leaching soluble salts from the ash increase the electrical conductivity and reduces electrical insulation. Credit: E. Welch*	 

Mechanical Strength	*The 2010 eruption of Pacaya volcano deposited between 20-30mm of ash on Guatemala City.Much of the coarser ash was highly friable which was easily crushed by vehicles on paved surfaces. The resulting finer particles more readily remobilised, blocking storm water drains in some instances. Credits: Director of Works, Municipality of Guatemala City & G. Wilson*		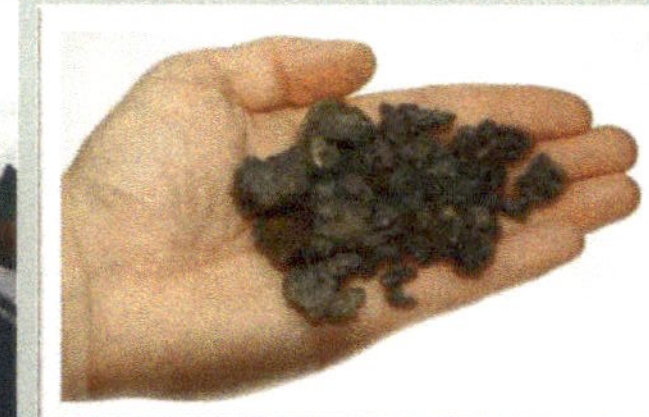
Abrasiveness	*Suspended ash in waste water cased accelerated wear to pumping station impellors in Bariloche waste-water network, Argentina, following the 2011 eruption of Cordón Caulle. Credit: C. Stewart*		

TABLE 3.2 **Potential Impacts from Trace to Light (>0.1 mm) Ash Falls, Categorized by Sector and Including the Key Volcanic Ash Characteristics That Determine the Level of Damage of Disruption**

			Potential Consequences	Relevant Volcanic Ash Characteristics T	L	Φ	C	ρ	P	SC	M	A	Examples
Sector	Buildings	S	None										
		NS	Minimal. Volcanic ash may also accumulate in gutters (where fitted to the building) depending on roof catchment area. Possible corrosion of metallic elements.	XX		X		X		X		XX	
		C&F	Volcanic ash may contamination of building interiors, leading to risk of adverse health impacts for building occupants; damage to sensitive equipment and abrasion damage to flooring.	XX			XX	X		X		XX	
	Infrastructure	GT	Transport networks possibly affected by reduced visibility. Roadway and railway may experience reduced surface traction.	X		XX			XX	X	X	X	Reduced visibility from <0.5 mm-thickness ash fall deposit during the Shinmoedake 2011 eruption, Japan © C. Magill.
		AP	Airports are generally very sensitive to ash fall. They may close, for instance,	XX		XX	XX	X	X		XX	X	

	due to reduced runway friction (braking effectiveness), airborne volcanic ash in near airspace, reduced visibility when volcanic ash on the ground is disturbed by engine exhausts during take-off and landing, ingestion of remobilized volcanic ash into jet engines during taxi, takeoff and landing, deposition of volcanic ash on hangars and parked aircraft, contaminated ground-support systems. Almost always requires clean-up.									
Agriculture	Likely to be minimal; but could include coverage of feed supplies or produce and contamination of water supplies.	X	X	X						
Clean-up	Usually no action required. Suspended ash may require removal from closed environments, particularly for people susceptible respiratory illness.	XX		X	X		X			X

Note that all impacts experienced at the lower thickness thresholds will also be experienced under greater thicknesses in addition to the impacts at that thickness. See Table 3.7 for subsector abbreviations. Each ash characteristic has been assigned a unique letter or symbol: thickness, T; loading, L; grain size distribution, Φ; airborne volcanic ash concentration, C; presence, P; particle density, ρ; surface composition (includes electrical conductivity), SC; mechanical strength, M; abrasiveness, A. Weighting is given to the importance of each characteristic's influence on sector impacts, as follows: key characteristic, XX; possible influence, X.

TABLE 3.3 Potential Impacts from Thin (>1 mm) Ash Falls, Categorized by Sector and Including the Key Volcanic Ash Characteristics That Determine the Level of Damage of Disruption

				Relevant Volcanic Ash Characteristics									
			Potential Consequences	**T**	**L**	**Φ**	**C**	**ρ**	**P**	**SC**	**M**	**A**	**Examples**
Sector	Buildings	S	Un ikely to be a problem at this volcanic ash hazard intensity.										
		NS	Volcanic ash may block gutters and downpipes, leading to localized flooding and damage, especially of roofs, drainage networks, and ceilings spaces. Internal gutters are particularly at risk and are not easily accessible for cleaning. Roofing materials may be abraded or damaged by human actions during ash removal.	XX	XX	X		XX		XX		XX	
		C&F	Heating, ventilation and air-conditioning (HVAC) systems can be disrupted by obstructed filters, condensers and air intakes. Loss of the building's essential services may also occur due to volcanic ash impacts to critical infrastructure (see section below).	XX		XX	XX	X	X	XX		X	Sheep covered in ~1 mm volcanic ash during the Ruapehu 1996 eruption, New Zealand © NZ Herald.

Infrastructure	ALL	Corrosion and/or abrasion, e.g., of paintwork, windscreens, metallic elements, some air-handling, mechanical and electrical equipment or engines. Potential clogging of air and oil filters. Control systems may be disrupted by volcanic ash ingress into HVAC systems blocking intakes leading to reduced performance, and affecting dependent systems.	XX		XX	XX	X	XX	X	X
	GT	Visibility from direct fall and remobilization; reduction in traction; most road markings obscured; increased wear of engine and brakes; possible signal failure on railway lines. Possible clogging of air-handling and filtration systems, mechanical and electrical equipment.	XX		XX	XX	X	X	X	XX
	PS	Flashover of power line and transformer insulators may occur with <3 mm of ash fall provided a significant portion of the insulator creepage distance (>50%) is covered in moist ash. Hydroelectric stations dealing with volcanic ash laden water may see abrasion damage to turbines. Thermal plants may be vulnerable to volcanic ash blockage to air intakes and cooling systems. Volcanic ash may reduce the resistivity of substation ground gravel cover, reducing tolerable step and touch voltages.	XX	X	XX			XX		X

Dampening of 1–2 mm deposited volcanic ash to reduce re-suspension during the Eyjafjallajökull 2010 eruption, Iceland © S. Jenkins.

(*Continued*)

WS	Raw water sources: probable increase in turbidity; possible increase in acidity and levels of soluble elements. Operational problems for water treatment plants include: increased turbidity of source water, overloading filtration equipment, accelerated wear on pump impellers, penetrate bearings and seals and overload motors. Ash-induced electricity outages are the most common cause of disruptions to water production after an eruption and increased maintenance can be expected. During clean-up operations following an ash fall a considerable increase in demand for water can occur. Demand may remain high for months afterwards if water is needed to dampen downwind remobilized ash.	XX	X	XX	XX	XX	X	XX
WW	Volcanic ash may enter sewer lines where there is inflow or infiltration (through illegal connections, cross connections, gully traps, manhole covers, cracks in sewer pipe work etc.). Cities with combined storm water and sewer lines are particularly at risk. Volcanic ash may form unpumpable masses in sewer lines and catch-pits which may cause blockages and overflows. It may also cause	XX	X	XX	XX	X	X	XX

			T	L	Φ	C	P	ρ	SC	M	A
		accelerated damage to pump impellers (pitting and thinning of metal). If ash-laden sewage reaches a wastewater treatment plant, mechanical pretreatment and treatment equipment is highly vulnerable to damage from ash-laden raw sewage.									
	TC	Possible damage to external telecommunication components and power cables.	XX	XX		X			X		
	AP	As above, with higher likelihood of more severe impacts.	XX			XX		X			
Agriculture		Volcanic ash coverage of crops, may lead to aesthetic discoloration of leaves and fruits—may lead to scratching during clean-up; coverage of livestock feed; possible leachable element toxicity hazards (volcanic ash specific). Generally minor health hazards for livestock.	XX	XX	XX	X	X		XX		XX
Clean-up		Minor clean-up required: sweeping of roads, paved areas and roofs/gutters usually sufficient. Ash falls of only a few mm depth will generate large volumes of volcanic ash for collection and disposal. Clean-up of urban areas and critical sites, such as airports, after an ash fall is a time-consuming, costly and resource intensive operation.	XX	X	XX	X	XX		X	X	X

Note that overloading of a system through high-demand or damage can lead to closure, exhaustion or restrictions at critical times, e.g., the exhaustion of water supplies during clean-up activities. See Table 3.7 for subsector abbreviations. Each ash characteristic has been assigned a unique letter or symbol: thickness, T; loading, L; grain size distribution, Φ; airborne volcanic ash concentration, C; presence, P; particle density, ρ; surface composition (includes electrical conductivity), SC; mechanical strength, M; abrasiveness, A. Weighting is given to the importance of each characteristic's influence on sector impacts, as follows: key characteristic, XX; possible influence, X.

TABLE 3.4 Potential Impacts from Moderate (>10 mm) Ash Falls, Categorized by Sector and Including the Key Volcanic Ash Characteristics That Determine the Level of Damage of Disruption

				Relevant Volcanic Ash Characteristics									
			Potential Consequences	T	L	Φ	C	ρ	P	SC	M	A	Examples
Sector	Buildings	S	In rare instances, nonengineered and long-span roofs may be vulnerable to damage, particularly when ash falls wet or is subsequently wetted by rainfall.	X	XX	X							
		NS	As above, with higher likelihood of more severe impacts.	XX	XX	X		XX		X		X	
		C&F	As above, with higher likelihood of more severe impacts.	X	X	XX	XX	XX	X	XX		XX	
	Infrastructure	GT	Ground transport networks disrupted by visibility and traction problems.	XX		XX	XX	X		XX	XX	XX	
		PS	Insulator flashover is the most likely impact. Volcanic ash accumulation may also overload lines, weak poles and light structures, and cause additional tree-fall onto lines.	XX	X	XX				XX		X	Chillis damaged by acidic surface coating during the Merapi 2006 eruption, Indonesia © G. Kaye.
		WS & WW	Possible damage to water and wastewater networks if volcanic ash enters systems, such as pumping	XX	X	XX		XX		XX			

		equipment, hydraulic seals, water treatment plants, waste-water treatment plants, etc. Increase in turbidity and chemical contamination of surface water storage. Possible blockage of storm drains, ditches and/or sewers, leading to surface flooding (depending on catchment size and system design).									
	TC	As above	XX	XX		X			X		
	AP	Airports closed for clean-up.	XX			XX		X			
Agriculture		Volcanic ash coverage of crops may lead to structural damage and more severe UV blockage and leachable element damage. Coverage of livestock feed more intense and longer duration, possible requiring supplementary feed. Higher chance of leachable element toxicity hazard.	XX	XX	XX	X	X		XX		XX
Clean-up		All roads and paved areas on public and private properties require cleaning; private properties usually require assistance with clean-up/disposal. Need for coordination of clean-up. Volcanic ash dump(s) established.	XX		XX	X	XX		X	X	X

20–30 mm of volcanic ash covering aeroplanes during the 2011 Puyehue-Cordón Caulle eruption, Chile. © Bariloche Airport.

Clean-up in Jacobacci, Argentina, following ~50 mm of ash fall from Puyehue-Cordón Caulle on 9 June 2011. Credit: Ailen Rodriguez.

See Table 3.7 for subsector abbreviations. Each ash characteristic has been assigned a unique letter or symbol: thickness, T; loading, L; grain size distribution, Φ; airborne volcanic ash Concentration, C; presence, P; particle density, ρ; surface composition (includes electrical conductivity), SC; mechanical strength, M; abrasiveness, A. Weighting is given to the importance of each characteristic's influence on sector impacts, as follows: key characteristic, XX; possible influence, X.

TABLE 3.5 Potential Impacts from Thick (>100 mm) Ash Falls, Categorized by Sector and Including the Key Volcanic Ash Characteristics That Determine the Level of Damage of Disruption

Sector			Potential Consequences	Relevant Volcanic Ash Characteristics											Examples
				T	L	Φ	C	ρ	P	S	C	M	A	H	
Sector	Buildings	S	Structural damage due to excessive volcanic ash loading. Very thick volcanic ash deposits (>100 mm) may cause roof collapse, although this volcanic ash thickness is rare. Nonengineered, long-span and low-pitched roofs are typically the most vulnerable.	X	XX	X									
		NS	Nonstructural elements, such as gutters, are more vulnerable to failure. Gutters may accumulate volcanic ash from the roof, reducing the drainage capacity and further increasing loading.	XX	XX	X		XX		X			X		Collapse of a weak metal sheet roof under 150–300 mm of volcanic ash during the 1991 Pinatubo eruption, Philippines © NOAA/NGDC, T.J. Casadevall, USGS.
		C&F	As above, with higher likelihood of more severe impacts.	X	X	XX	XX	XX	X	XX			XX	X	
	Infrastructure	GT	Transport networks severely affected by reduced traction; very thick ash	XX		XX	XX	X		XX		XX	XX		

		falls may create extra loading on bridges, especially when wet.					
	PS	Structural damage to lightly engineered components and buildings possible. Substantial clean-up of sites required. Most generation plants likely to be disrupted.	XXX	XX		XX	X
	WS & WW	Significant contamination of exposed surface waters. Probable blockage or sedimentation of water supply pipes or canals, storm drains and/or sewers, leading to surface flooding. High risk of severe damage to water- and wastewater- treatment plants if volcanic ash enters systems.	XXX	XX	XX	XXX	XX
	TC	Possible damage to communication dishes and microwave towers due to excess volcanic ash loading.	XXXX	X		X	
	AP	As above, with higher likelihood of more severe impacts.	XX	XX	X		
Agriculture	P & H	Burial of pasture and most crops. Loading damage to tree crops likely. The buried soil's fertility may wane over time as key soil processes are broken by volcanic ash blanket (i.e., water, oxygen and nitrogen cycles). Damage/burial of farm assets,	XXX	XX	X	XX	XX

Forest killed by heavy ash falls on the flanks of Bromo volcano, Indonesia, during an eruption in 2011. © R. Roscoe.

(Continued)

TABLE 3.5 Potential Impacts from Thick (>100 mm) Ash Falls, Categorized by Sector and Including the Key Volcanic Ash Characteristics That Determine the Level of Damage of Disruption—cont'd

		Potential Consequences	Relevant Volcanic Ash Characteristics										Examples
			T	L	Φ	C	ρ	P	SC	M	A	H	
		including irrigation systems, access-ways, etc.											
	F	Tree damage possible in production forests.		XX			X				X		
Clean-up		As above, but with much larger volumes which will require greater resources and/or cleaning time; vegetated areas (e.g., parks and gardens) may require cleaning too.	XX	X	XX	X	XX		X	X	X		

See Table 3.7 for subsector abbreviations. Each ash characteristic has been assigned a unique letter or symbol: thickness, T; loading, L; grain size distribution, Φ; airborne volcanic ash concentration, C; presence, P; particle density, ρ; surface composition (includes electrical conductivity), SC; mechanical strength, M; abrasiveness, A. Weighting is given to the importance of Each characteristic's influence on sector impacts, as follows: key characteristic, XX; possible influence, X.

TABLE 3.6 **Potential Impacts from Very Thick (>1000 mm) Ash Falls, Categorized by Sector and Including the Key Volcanic Ash Characteristics That Determine the Level of Damage of Disruption**

			Relevant Volcanic Ash Characteristics										
		Potential Consequences	T	L	Φ	C	ρ	P	SC	M	A	H	Examples
Sector	Buildings	Structural damage to most buildings likely.	X	XX			XX						 Volcanic ash cleaned off a hospital roof in Heimaey following 1973 Eldfell eruption, Iceland (tractor for scale). Credit: G. Oskarsson.
	Infrastructure	Structural damage to all structures possible. Expect most systems to be disrupted for extended periods. Ground transport networks likely to be impassable.	XX	XX	X		XX		XX				
	Agriculture	Total burial of crops/pastures. The buried soil's fertility will decrease over time as key soil processes are sealed off. Tree damage likely in production forests.	XX	XX	X		X		XX				
	Clean-up	As above, but with much larger volumes which will require greater resources and/or cleaning time.	XX	X	XX	X	XX		X	X	X		

(Continued)

TABLE 3.6 Potential Impacts from Very Thick (>1000 mm) Ash Falls, Categorized by Sector and Including the Key Volcanic Ash Characteristics That Determine the Level of Damage of Disruption—cont'd

	Relevant Volcanic Ash Characteristics										
Potential Consequences	T	L	Φ	C	ρ	P	SC	M	A	H	Examples
											 Abandoned farmland (buried fence line) following 1991 Hudson eruption, Chile. Photo taken in 2008. Credit: C. Stewart.

See Table 3.7 for subsector abbreviations. Each ash characteristic has been assigned a unique letter or symbol: thickness, T; loading, L; grain size distribution, Φ; airborne volcanic ash concentration, C; presence, P; particle density, ρ; surface composition (includes electrical conductivity), SC; mechanical strength, M; abrasiveness, A. Weighting is given to the importance of each characteristic's influence on sector impacts, as follows: key characteristic, XX; possible influence, X.

TABLE 3.7 Subsector Abbreviation Used in Tables 3.2-3.6

Sector	Code	Sub-
Buildings	S	Structural
	NS	Nonstructural
	C&F	Contents and fittings
Infrastructure	GT	Ground transport
	PS	Power system
	WS	Water supply
	WW	Waste water
	AP	Airport
	TC	Telecommunications
Agriculture	P	Pastoral farming
	H	Horticulture farming
	F	Forestry (silviculture)
Applicable to all	ALL	Applicable to all

- *Loading (common units: kg/m^2, kPa)*: Under large enough loads, the pressure exerted by accumulated ash fall can cause structural damage to buildings and other infrastructure, as well as crush crops and vegetation. Additionally, water saturation can increase deposit load by up to 100% (Macedonio and Costa, 2012). This metric allows the accurate estimation of masses over exposed areas. Thickness can be calculated from loading if bulk density (kg/m^3) is known, and vice versa. Note that bulk density will change if the ash deposit compacts.
- *Airborne ash concentrations (common unit: μg/m^3)*: The presence of suspended fine ash particles in air is an important consideration for public health (see Chapter 12). Of particular relevance for assessing human population exposure to airborne respirable ash is PM_{10} (the concentration in μg/m^3 of particles less than 10 μm diameter sampled with a size selective inlet). The metric $PM_{2.5}$ is also of increasing importance due to increasing awareness of the role of very fine particles in causing disease (World Health Organisation, 2013). Airborne ash is also important for aviation safety (see Chapter 5) and may cause impacts to equipment with air intakes such as electric power generators, and air handling systems for large buildings, such as hospitals.
- *Presence of volcanic ash*: In the event of an explosive eruption or volcanic unrest episode, communities within range of the volcano may experience

increased anxiety of the impacts of the volcanic ashfall on health, pets, livestock and property, regardless of whether these communities receive ashfall. There is likely to be a heightened demand for information about ashfall impacts, which should be anticipated by emergency and risk managers. Temporary and sometimes unnecessary socioeconomic disruption may result from measures such as evacuation, school closures, and flight cancellations. Losses may also result from precautionary risk management measures, e.g., if a large number of dust masks are purchased unnecessarily, covering open water supplies, precautionary evacuations, or loss of business activity (Johnston et al., 2000).

- *Particle size (grain size) distribution*: Smaller particles are carried farther downwind from the vent, are more easily eroded and remobilized by wind, and can penetrate smaller openings than larger grains. The penetration of ash particles into the respiratory tract is dependent primarily on particle size, with particles >10 μm lodging in the upper respiratory tract while those <4 μm may penetrate into the terminal bronchi and alveoli (Horwell and Baxter, 2006; Weinstein et al., 2013). Small particles can adhere more readily to vegetation and infrastructure components.
- *Particle density (common unit: kg/m^3)*: The density of individual particles influences their downwind transport. This may in turn influence the susceptibility of deposits to resuspension by wind or rain, and also influence difficulty of removal in cleanup operations, for example, if volcanic ash enters drainage networks. Sometimes, there can be a range of particle densities within a single ash fall deposit (Figure 3.3).
- *Surface composition (common unit mg/kg dry weight for individual elements)*: Fresh volcanic ash commonly has a surface coating of readily soluble salts as a consequence of condensation of strong mineral acids onto volcanic ash surfaces in a cooling plume (Ayris and Delmelle, 2012). This surface coating is highly variable in composition and concentration both between and within eruptions. These salts may be released upon contact with water or physiological fluids. Typical effects of ash fall on drinking water supplies are the elevation of the metals Fe, Mn, and Al which produce unpleasant tastes but little direct health impact (Stewart et al., 2009). In addition, acidic coatings can accelerate corrosion of metal components such as roofing, fencing, and other exposed metal surfaces (Oze et al., 2014). High concentrations of readily water-soluble salts on volcanic ash can result in a conductive coating unless the particles are dry. Such increases in conductivity have led to documented cases of flashover (the unintended electrical discharge around or over the surface of an insulator) on line and station insulators in electricity transmission networks (Wardman et al., 2012a). The cementing of volcanic ash deposits can form an impermeable, durable horizon which complicates clean up, disrupts soil nutrient and gas cycles, and reduces infiltration of water which can in turn increase surface runoff and accelerate fluvial erosion.

- *Mechanical strength: The mechanical strength of volcanic ash particles depends on particle shape (i.e. vesicularity) and composition (i.e., volcanic glass, crystal shard or lithic)* and will affect the ease with which a fall deposit can be crushed into smaller particles, which may then be remobilized and infiltrate, for example, into engines. Less-friable deposits may have more potential for beneficial re-use, such as construction aggregate.
- *Abrasiveness*: Volcanic ash particles are typically highly abrasive. Abrasiveness is a function of both hardness and particle shape (Gislason et al., 2011). Abrasive damage from volcanic ash can be widespread and include health impacts (e.g., scratching of corneas) to abrasion of aircraft leading edges to accelerated wear and tear on equipment such as pump impellers (Gislason et al., 2011; Wilson et al., 2012b) and to surfaces such as bituminous coatings (i.e., paved roads and runways) during cleanup operations (Wardman et al., 2012b).

3.3 VOLCANIC ASH IMPACT: SPATIAL AND TEMPORAL DIMENSIONS

The scope of damage and disruption due to ash falls is arguably one of the most complex and wide-ranging of all volcanic hazards. The impact depends upon the amount of volcanic ash and its physical and chemical characteristics, which are in turn a function of the eruption magnitude and style and also the vulnerability of the exposed community, system, or asset.

3.3.1 Impacts by Sector

The 1980 Mount St Helens eruption in the United States triggered a more quantitative approach to volcanic impact assessment. The study of this eruption and its effects was one of the first of its kind and results provided information for more robust estimation of the likely impact of ash falls on people, buildings, infrastructure, agriculture, and livelihoods. Since 1980, a number of eruption impact studies have been done that cover a variety of different styles of explosive eruption, climates, built environments, and socioeconomic conditions. Drawing on the available literature, we have broadly categorized volcanic ash impacts across four sectors of society. Specific impacts will depend strongly on asset, network or system design, typology, ash fall volume and characteristics, and the effectiveness of any mitigation strategies. Key eruption case studies and review articles are suggested at the beginning of each sector discussion.

- *Buildings* (e.g., Vulcan and Tavurvur VEI 4, 1994, Papua New Guinea: see Blong, 2003; Pinatubo VEI 6, 1991, Philippines: see Spence et al., 1996): Under large enough thicknesses or loads (>100 mm), roofs and load-bearing walls can suffer structural damage or even collapse; however, this is strongly dependent upon building style and materials, which can vary greatly. Nonstructural damage, e.g., collapse of gutters, blocking of drains, which can then create flooding and water damage, and contamination of interiors can affect the habitability of a building. However, attributing a generically applicable hazard intensity to nonstructural damage can be difficult because damage depends on, for example, the roof catchment area, pitch, span, and fixings, as well as the characteristics of the ash fall deposit.

- *Critical infrastructure* (e.g., Chaiten VEI 4, 2008: Wilson et al., 2012b): This category includes those assets, frequently taken for granted, which are essential for the functioning of a society and economy. This includes such systems as power systems (generation, transmission, and distribution), water treatment and distribution, waste-water network and treatment, transportation (ground and aviation), fuel supply, telecommunications, and beyond. Ash fall is typically disruptive for critical infrastructure systems and networks but the degree of disruption in highly dependent upon system design, level of preparedness, and mitigation strategies (Wilson et al., 2012b).
- *Agriculture, forestry, and fishing* (e.g., Ruapehu, 1995–1996: Cronin et al., 1998; Mount St. Helens, 1980: Cook et al., 1981): Ash falls can directly damage vegetation, soils, livestock, and fisheries; production facilities and assets; interrupt harvest and processing for extended periods of time; and can be strongly season and climate dependent (e.g., Cook et al., 1981; Cronin et al., 1998; Wilson et al., 2011). However, in some cases, ash falls may be beneficial, improving soil fertility and thus agricultural yields (Cronin et al., 1998).
- *Clean-up operations* (Shinmoedake, 2011: Magill et al., 2013; Mount St Helens, 1980: Blong, 1984): Clean-up during and following an ash fall is usually essential to restore functionality in urban (and sometimes rural) environments. It is one of the most common impacts communities experience, and can be surprisingly time- and resource-intensive (Wilson et al., 2012). Repeated ash falls can lead to communities developing effective mitigation strategies e.g., Kagoshima city receives regular ash falls from Sakurijima volcano, Japan, and has developed an efficient clean up system and mitigation measures for buildings and critical infrastructure. Conversely, ongoing ash contamination of inhabited areas and the burden of repeated clean-up can also have considerable direct and indirect socioeconomic impacts (see Section 3.3.2).

Here we provide a qualitative overview of volcanic ash impacts as a function of primary volcanic ash thickness for each of the key sectors listed above (Tables 3.2–3.6). We discuss impacts with reference to volcanic ash thickness because it is the parameter easiest to measure and most commonly used as metric of ash fall magnitude. However, as discussed earlier other volcanic ash characteristics also exert important influence or control on resulting impacts, and these have been highlighted. Impacts associated with thinner ash falls will also exist at greater thicknesses, thus impacts can generally be considered accumulative. In addition to the physical impacts defined in Tables 3.2–3.6, we touch lightly on the important aspects of health and psycho-social impacts from ash fall below. The reader is referred to Chapter 12 for a more thorough overview.

- *Human health*: While ash fall is generally not a major cause of mortality in eruptions, it has the potential to cause both short and long-term morbidity and may also cause indirect health hazards, for example, if injuries are sustained by people cleaning ash from roofs or if ash deposition on roads creates dangerous driving conditions such as loss of traction (Magill et al., 2013). The area that has received the most sustained attention from the research community is the respiratory health hazards of volcanic ash (see Horwell and Baxter, 2006 for a comprehensive review article). A recent, comprehensive review of evidence (Kar-Purkayastha et al., 2012) for the potential health impacts of volcanic ash on the population of the United Kingdom and Republic of Ireland suggests that important information streams for the effective management of health risks include: surveillance of clinical symptoms and health service usage; air quality data (PM_{10} and $PM_{2.5}$) from regional monitoring networks; and physicochemical and toxicological profiling of the ash. Characteristics of volcanic ash that are relevant to a respiratory health hazard assessment include grain size, bulk composition, particle shape, crystalline silica content, surface area, and surface reactivity (Horwell et al., 2013; Weinstein et al., 2013).
- *Psychosocial and community impacts*: In general, ash falls do not cause large numbers of casualties unless widespread roof collapse is experienced due to severe ash loading, although areas where heavy ash fall impacts are sufficiently severe may become uninhabitable due to contamination or critical infrastructure service disruption (Wilson et al., 2012b). More commonly, ash falls can affect psychological functioning due to factors such as disruption of livelihoods and consequent anxiety. For example, although the 1995–1996 eruptions of Mt Ruapehu were small (VEI 3), the economic consequences were considerable, particularly to the ski industry (three ski-fields were located on the volcano) and related services, and communities at the base of the mountain experienced considerable uncertainty, anxiety, and economic hardship (Johnston et al., 2000). One manifestation of this was increased levels of posttraumatic stress in school-aged children (Ronan, 1997). These effects are likely to be most pronounced for long-duration crises with continuous or intermittent ash fall (Sword-Daniels et al., 2014), or extended remobilization episodes (e.g., Wilson et al., 2011). This chapter does not focus on this element, but we note that it is an important area which needs consideration when managing ash fall impacts (Gaillard, 2006; Tobin & Whiteford, 2002).

3.3.2 Spatial Dimensions

The variation of ash fall impact across the landscape (its spatial dimension) is strongly controlled by ash dispersal with distance from the volcano. Dispersal is in turn controlled by eruption style, duration, and environmental conditions (e.g., wind, rain and snow flushing, etc.) at different altitudes and distances from the vent. Broadly speaking, potentially three main zones of impact exist on the ground during a large magnitude eruption; each requiring a different approach to management and planning (Figure 3.5): (1) destructive and immediately life-threatening (proximal); (2) damaging/disruptive (medial); (3) mildly disruptive/nuisance (distal). In smaller magnitude eruptions with very thin ash falls, zones 1 and 2 may not be present. In larger eruptions, volcanic ash deposits thick enough to collapse buildings and cause significant clean-up generally occur within tens of km of the vent. For instance, during the 1991 eruption of Mt Pinatubo, Philippines, areas relatively close to the volcano such as Olongapo City received thick ash fall (>100 mm) and experienced structural damage to buildings and complete burial of agricultural areas (Spence et al., 1996). An estimated 320 people died and 279 were injured during the eruption, and most of these casualties were believed to have been due to the collapse of roofs under the weight of wet volcanic ash on June 15 (Spence et al., 1996). In comparison, for similarly large eruptions, relatively thin (<few mm) falls will occur over very large areas (up to many hundreds of kilometers downwind) and will cause a different range of problems. For instance the 2011 eruption of Cordon-Caulle volcano in Chile (VEI 4), dispersed volcanic ash over 150,000 km^2 of southern Chile and Argentina, impacting several towns with moderate to thin ash falls, including San Carlos de Bariloche (pop. ~108,000), which received ~30 mm of ash fall. This caused damage and disruption to power, water, and sewage and ground transportation networks. In areas several hundreds of kilometers away from the Cordon-Caulle volcano, ash falls of several millimeters were sufficient to disrupt road networks for days to weeks (Wilson et al., 2012a). Wind and water remobilization of those ash deposits led to visibility and sedimentation problems and isolated debris flow hazards disrupted road networks for over nine months after the ash fall. The eruption also resulted in considerable livestock losses due to ash covering and/or destroying sparse available feed. Large, coordinated cleaning operations were required in affected urban centers. The severity of impacts typically decreased with distance from the volcano, however, several elongate zones of ash fall to the east-southeast and northeast resulted in expanded zones of disruption downwind (Wilson et al., 2012a).

3.3.3 Temporal Dimensions

Volcanic eruptions, extend temporally, on average, around seven weeks, with 10% lasting less than 1 day and approximately 10% lasting longer than 2 years (Simkin and Siebert, 2000); over half are characterized by more than one explosive ash-producing event (Jenkins et al., 2007). Therefore, over the course of one eruption, a number of different wind conditions and ash emissions can

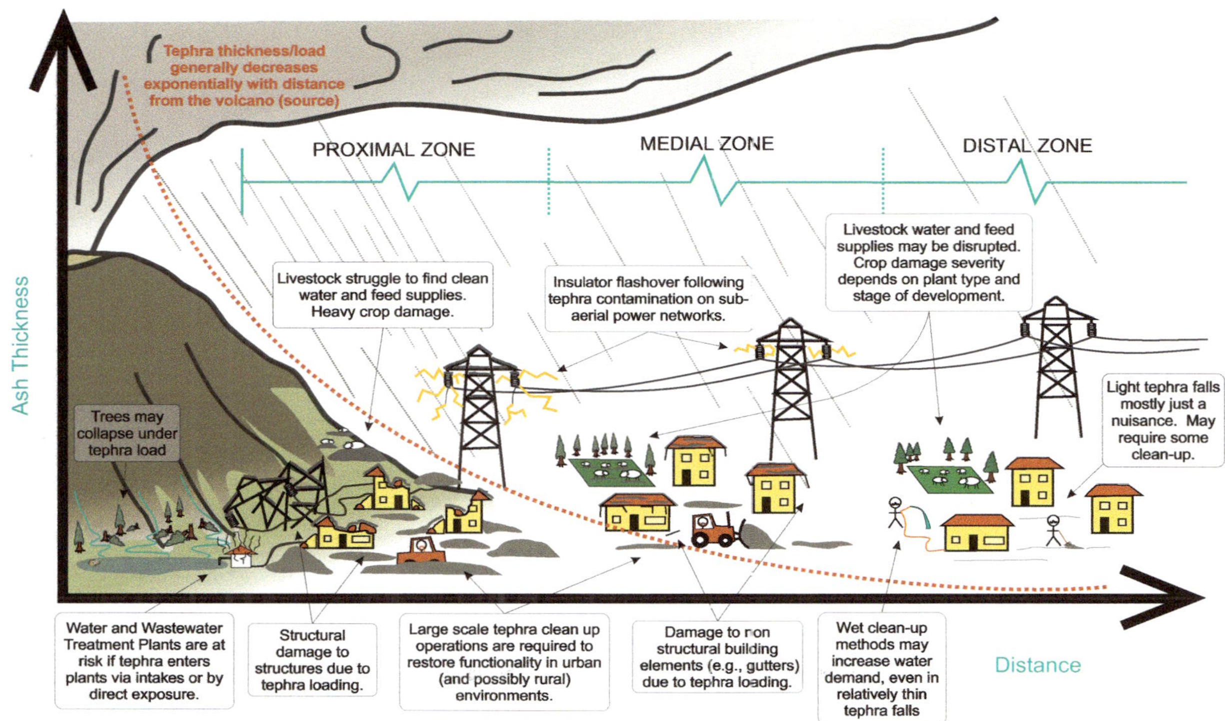

FIGURE 3.5 Three main zones of ash fall impacts. (1) destructive and immediately life-threatening (proximal); (2) damaging/disruptive (medial); and (3) mildly disruptive/nuisance (distal).

mean that any one site downwind of an erupting volcano may experience a number of ash falls of varying intensity and duration. The duration and timing of ash falls within an eruption strongly determines what impacts occur, and what their impacts may be, particularly for the agriculture sector and clean-up of urban environments (Cook et al., 1981; Wilson et al., 2011, 2012b). One or few large ash falls within a short period of time can lead to acute impacts of significant magnitude but brief duration. Affected communities may have limited experience with ash fall events and few tools or resources which can exacerbate the original impacts and a comparatively poor capacity to cope. Although assessing and preparing for acute eruption impacts is important, it is also necessary for hazard and risk assessments, and any associated emergency planning, to consider the different set of challenges created by repeated or lengthy ash falls. Extended volcanic ash emissions can produce a more widely dispersed ash fall than a discrete event because of changing wind conditions, likely affecting a greater number of people and assets. Some communities downwind of the volcano may be subject to recurrent ash falls over years or even decades, with repeated disruptions to lifeline utilities, business continuity, and social infrastructure. This situation can lead to successful adaptations but can also result in cumulative losses which are difficult to sustain in the long-term (Sword-Daniels et al., 2014) (Table 3.7).

In addition to the eruption duration and timing with respect to season, environmental factors such as wind and rain on volcanic ash dispersal and remobilization, will dictate the level of damage or disruption a community experiences. For example, dry, windy environments facilitate remobilization of volcanic ash deposits near the ground, which can "ash-blast" vegetation and cause challenging living conditions for decades after the eruption (e.g., Hudson 1991 eruption: Wilson et al., 2011). Conversely, large amounts of rain can saturate volcanic ash deposits and increase loading causing greater structural and agricultural damage as well as large, potentially damaging mudflows (e.g., Pinatubo 1991: Newhall and Punongbayan, 1996).

Repeated exposure to volcanic ash over months or years (either the result of long-duration eruptions or repeated remobilization of existing deposits) *can* significantly disrupt socioeconomic activities and lead to health hazard anxiety across a wide area. The way in which society adapts to and manages these impacts can profoundly influence how disruptive the volcanic ash events are. The success of adaptation may be related in part to how frequently the population is exposed to ash fall impacts. For example, over the last 50 years, Kagoshima city in Japan has been frequently affected by ash fall from Sakura-jima volcano but mitigation measures have successfully reduced the disruption of ongoing ash falls (Durand et al., 2001). For instance, the city has modified critical services (such as covering waste-water treatment plants and jacketing electrical insulators to reduce the likelihood of flashover), changed building designs to prevent volcanic ash ingress, and invested in volcanic ash clean-up equipment and development of plans for clean-up operations. As such, the city and its people cope well with the frequent ash falls. Disruption for typical eruptive events is minimal (Durand et al., 2001). Conversely, repeated episodes of wind remobilization of volcanic ash deposits from the 1991 eruption of Mt Hudson in Chile over a period of 10–15 years led to the abandonment of lands and the collapse of agricultural communities. Farm systems were stressed by the relatively low-intensity but long-duration impacts, but ultimately led to many farming families relocating from the area. For example "ash-storms" affecting Puerto Ibañez (located at the eastern extent of the Ibañez river valley) meant agricultural production waned and ultimately the population declined 30% from ~4,000 in 1991 to ~2,800 by 2008 (Wilson et al., 2011).

3.4 QUANTIFYING VULNERABILITY TO ASH FALL

Reducing risks posed by ash fall events requires first an understanding of the potential range and severity of impacts from ash falls of different depths and characteristics. Understanding the vulnerability of key community assets to ash fall is also fundamental in assessing potential impacts and developing appropriate mitigation strategies. A commonly used approach is to categorize impacts by sector (see Section 3.3) and consider the relationship between hazard intensity and damage/disruption state (i.e., including loss of functionality) for each sector. In Tables 3.8–3.10, we present some broad guidelines that relate volcanic ash thickness to key sectors. A number of assumptions have been made in these impact estimates:

- One discrete ash fall event of 1 week duration or less;
- No removal (either by wind/water or anthropogenic actions/processes);
- No mitigation actions taken either before or after the ash fall event (i.e., internal strengthening of the building or cleaning);

Accounting for, and quantifying where possible, the range of possible outcomes and associated uncertainties and limitations, is imperative when communicating potential ash fall impacts. The impact estimates in Tables 3.8–3.10 are subject to a number of uncertainties and limitations:

- Variations in volcanic ash volume and characteristics at the site/region.
- The vulnerability of buildings, primary industries and infrastructure systems to ash fall impacts is strongly dependent on condition, construction, design, age, environmental setting, as well as the effectiveness of the preparedness and mitigation strategies which are applied.
- The vulnerability estimates shown here define very broad typologies and systems that in reality will vary in condition, age, design, and construction. We urge caution in using the impact estimates in Tables 3.8–3.10 for risk or impact assessments; wherever possible, local studies of the vulnerability should inform a specific risk assessment.
- Documentation of impacts of volcanic eruptions is relatively limited. This is in part because large explosive eruptions impacting populated areas are relatively infrequent and because few comprehensive quantitative damage

TABLE 3.8 Approximate Median Hazard Intensities (Using Thickness as a Proxy and an Assumed Load Density of 1,000 kg/m^2) That Relate to Key Damage and Functionality States for a Range of Generic Roof Types

<table>
<tr><td colspan="2">Code:</td><td>D0</td><td>D1</td><td>D2</td><td>D3</td><td>D4</td><td>D5</td></tr>
<tr><td colspan="2">Description:</td><td>No damage</td><td>Minor/basic repair required</td><td>Moderate repair required</td><td colspan="2">Major/specialist repair required</td><td>Beyond economic repair</td></tr>
<tr><td colspan="2">Function:</td><td>Functional</td><td>Repeated clean-up required; Loss of functionality for some contents and fittings</td><td colspan="3">Ash infiltration or threat of roof collapse may prohibit habitation</td><td>Retired</td></tr>
<tr><td colspan="2">Structural damage:</td><td>No damage</td><td>No damage</td><td>No damage to principal roofing supports</td><td colspan="2">Partial or complete failure of the supporting structure, e.g. battens or trusses; Partial or moderate damage to the vertical structure</td><td>Collapse of roof and supporting structure over 50% of roof area; External walls destabilised</td></tr>
<tr><td colspan="2">Non-structural damage:</td><td>No damage</td><td>Minor damage to roof coverings, e.g. abrasion and corrosion of metal roofs</td><td>Potential damage to gutters, roof overhangs and coverings, e.g. excessive bending</td><td colspan="2">Severe damage or partial collapse of roof overhangs; Collapse or partial collapse of roof covering</td><td>Partition wall/s destroyed</td></tr>
<tr><td colspan="2">Contents and fittings:</td><td>Some infiltration of ash possible</td><td>Volcanic ash infiltration and potential damage to fittings, e.g. air-con, and appliances</td><td colspan="3">Variable levels of contamination and damage</td><td>Irreversible damage to most contents and fittings</td></tr>
<tr><td rowspan="4">Approximate median volcanic ash thicknesses (mm) for each damage state and generic roof types:</td><td>Timber board on timber supports</td><td rowspan="4">1?</td><td rowspan="4">10?</td><td colspan="4">200</td></tr>
<tr><td>Tiles on timber supports</td><td colspan="4">300</td></tr>
<tr><td>Sheeting on timber supports</td><td colspan="4">300</td></tr>
<tr><td>Reinforced concrete</td><td colspan="4">700</td></tr>
</table>

Brief descriptions of likely disruption (impact on functionality) and damage are outlined for each state (D0 through D5). These hazard-damage/disruption relationships are based on very few data and subject to uncertainty, assuming generic building types of average condition and construction. These estimates are thus intended as broad guidelines based on the evidence available and should be considered with appropriate error bars, and refined for individual cases and conditions. Typologies are ordered by increasing dry ash fall thicknesses for severe damage or disruption (D4).

TABLE 3.9 Approximate Median Hazard Intensities (Using Thickness as a Proxy) That Relate to Key Damage and Functionality States for a Range of Critical Infrastructure

Damage and functional states and approximate median volcanic ash thicknesses for each sector:	Code:	D0	D1	D2	D3	D4	D5
	Description:	**No damage**	**Cleaning required**		**Repair required**		**Beyond economic repair**
Airports	**Function**	Fully functional	Closure of runway				Indefinite closure
	Damage	No damage		Possible runway surface degradation	Runway surface degradation		Complete burial
	Thickness (mm)	**0**	**>0**				**> 500**
Power	**Function**	Fully functional	Temporary disruption, e.g. flashover of insulators		Disruption requiring repair		Permanent disruption
	Damage	No damage	No damage to components		Damage to component		Structural damage
	Thickness (mm)	**0**	**5**		**20**		**>100**
Railways	**Function**	Fully functional	Reduced visibility	Signal disrupted	Loss of traction making operation unsafe; Possible derailing through ash accumulation		Impassable
	Damage	No damage		Possible abrasion and/or corrosion of signal components and track			Complete burial
	Thickness (mm)	**0**	**0.5**	**1**	**30**		**>200**
Roads	**Function**	Fully functional	Reduced visibility and traction	Road coverings obscured	2WD vehicles obstructed	4WD vehicles obstructed	Impassable
	Damage	No damage		Possible road surface and marking abrasion	Road surface and marking abrasion		Complete burial
	Thickness (mm)	**0**	**0.5**	**2**	**50**	**150**	**>300**

Brief descriptions of likely disruption (impact on functionality) and damage are outlined for each state (D0 through D5). These hazard-damage/disruption relationships are based on few data and subject to uncertainty. The response of a system to ash fall thicknesses will depend upon the system design and type, its components and the characteristics of the ash fall. These estimates are thus intended as broad guidelines based on the evidence available and should be considered with appropriate error bars, and refined for individual cases and conditions. Sectors are ordered by increasing ash fall thicknesses for severe damage or disruption (D4).

TABLE 3.10 Approximate Median Hazard Intensities (Using Thickness as a Proxy) That Relate to Key Levels for Loss of production for a Range of Agriculture Types

	Code:	D0	D1	D2	D3	D4	D5
	Description:	No damage	Disruption to harvest operations and livestock grazing of exposed feed	Minor productivity loss: less than 50 %/crop	Major productivity loss: more than 50 %/crop; Remediation required	Total crop loss; Substantial remediation required	Retirement of land
Approximate median volcanic ash thicknesses (mm) appropriate to each loss of production state:	Horticulture/Arable	0	1	5	50	100	300
	Pastoral	0	3	50	100	150	300
	Paddies	0	1	30	100	200	300
	Forestry	0	5	200	1000	1500	?

These hazard-loss of productivity relationships are based on expert judgment and very few empirical and experimental data. Impacts assume that the crops are in the growing stage (a worst-case impact) and season modifiers will allow impacts during other periods of the growing cycle to be accounted for. These estimates are thus intended as broad guidelines and should be considered with appropriate error bars, and refined for individual cases and conditions. Sectors are ordered by increasing ash fall thicknesses for severe damage or disruption (D4).

studies have been carried out (e.g., exceptions include Blong, 2003; Spence et al., 1996; Wilson et al., 2011, 2012b). Impacted zones can be dangerous and/or inaccessible for long periods of time. Experimental data (e.g., Oze et al., 2014) are also limited, and, where available, refer exclusively to materials, strengths, building codes and standards exhibited by the tested building stock. In the absence of empirical data, theoretical calculations of component, system or network response to hazard parameters can be prepared (e.g., Spence et al., 2005), but these are also subject to various uncertainties. As a result, vulnerability estimates for volcanic impact are therefore typically less robust than those for earthquake or wind, for example, which are rapid-onset, near instantaneous events that therefore are not subject to the same spatial and temporal variations as ash fall.

This is not an exhaustive list of societal assets or economic activities that may be impacted by ash fall. Some impacts are strictly system or context dependent and therefore have no established intensity–impact relationship, as for example, telecommunication interference by volcanic cloud lightning and signal attenuation during ash fall. In many parts of the world the failure, disruption or reduced functionality of infrastructure is likely to have a larger impact on livelihoods and the local economy than direct damage to buildings. In some cases it can exacerbate existing economic, social, or agronomic decline (e.g., Wilson et al., 2011) because of a high systemic vulnerability (interdependencies between physical, economic, and social systems). We therefore recommend that comprehensive physical vulnerability and risk assessments for volcanic areas should always include surveys of exposed buildings *and* infrastructure, and consider direct and indirect socioeconomic impacts.

3.5 MITIGATING ASH FALL IMPACTS

Core components of disaster-risk reduction include: (1) providing guidance on the nature and extent of the hazard, likely impacts, and best-practice mitigation strategies; and (2) encouraging communities or organizations to adopt preparedness measures that increase their ability to manage hazard consequences, and thus, increase their capacity to manage risk (UNISDR, 2011). Severe ash falls (>100 mm) are typically rare events in most places over the time span of a generation and so commonly there is a relatively poor awareness exists of the hazard and what the likely impacts will be (Johnston et al., 2000). During an ash fall emergency, various studies have reported that organizations, communities, and individuals have exhibited strong abilities to respond to, withstand, and recover from any negative impacts, producing novel and unique approaches to managing and mitigating the impacts (e.g., Johnston et al., 2000; Durand et al., 2001; Wilson et al., 2012b). However, the lack of preparation actions was costly in terms of the time of restoration, delay of key decisions by authorities, and ultimately the damage caused. A trial and error approach delays an effective response and heightens the risk of negative public perceptions of the quality of the emergency management. Therefore, sustained preparedness activities are valuable and should include providing stakeholders with access to specialized, sector-specific impact, preparedness, and post-event response/recovery information. Such preparedness activities need to be undertaken in a context of foster stakeholders (e.g., communities and organizations) to actively participate as partners in the management of ash fall risk (Covello and Allen, 1988; Paton et al., 2005). Communication regarding the hazard and recommended mitigation steps should also be transparent, repeated, and from multiple sources which are authoritative, credible, and trusted (Berlo, 1960).

The availability of sophisticated models of ashfall and volcanic ash dispersion increases the possibility for providing exposed communities with effective and timely warnings. Ash fall warnings are now standard procedures in many countries, such as in the United States, Japan, and New Zealand. Although there has been limited, if any research, specifically on what makes an effective ash fall warning, such warnings can help facilitate appropriate protective actions e.g., seal buildings, shutdown vulnerable systems, minimize exposure, etc. It may also allow development of thresholds that trigger-specific risk management actions, tailored to end-user requirements (e.g., forecast of >1 mm of ash fall may trigger insulator cleaning programs on vulnerable power lines).

Pre-event planning to manage ash fall impacts in urban communities can avoid costly clean-up and recovery operations. The development of clean-up plans that prioritize critical areas or lines of communication and identify volcanic ash disposal sites and procedures is highly recommended. The establishment of mutual support or continuity agreements between municipal authorities, critical infrastructure organizations, and businesses can facilitate greater access to resources to deal with ash fall events (e.g., human resources, equipment, etc.).

Finally, a major concern to both the affected populace and authorities following ash falls are the potential health impacts. As a relatively rare public and agricultural health hazard, it can be difficult for health and agricultural agencies to know which collection and analysis methods are appropriate and can be used to compare against robust benchmark studies. This can be resolved by application of a standardized physical and chemical volcanic ash characterization protocols. Recent international collaboration through the International Volcanic Health Hazard Network (IVHHN) has developed and regularly refined several such protocols, which we recommend for use. They can be found at www.ivhhn.org. The use of such standardized protocols is essential for effective communication about the extent of the risk posed by ash. These issues are dealt with more extensively in Chapter 12.

3.6 MOVING FORWARD

Recent empirical, experimental, and theoretical studies have greatly improved our understanding of the response of society to ash fall. However, systematic and comprehensive documentation for a wide range of eruptions and volcanic ash conditions, including investigation of cascading impacts, is lacking. For example, studies have typically not focused on impacts associated with thin (<2 mm) ash falls, but on very large eruptions and impacts associated with ash falls >10 mm. Thin ash falls of a few mm or less are therefore a source of significant uncertainty for emergency management planning and loss assessment models, despite being the most common volcanic hazard. The information requirements of end-users and the rapidly evolving field of volcanic risk assessment will require the research community to provide robust estimates of probable impacts from ash falls before, during, and after an eruption crisis. Given the wide range of physical and chemical properties of volcanic ash (as discussed above), and increasing vulnerability (e.g., the rapid growth and evolution) of society, this poses a significant challenge. In particular, fast-evolving technology, increasingly interdependent networks and systems, and the low frequency (and therefore human experience) of explosive eruptions in populated areas make it difficult to determine whether existing mitigation recommendations are appropriate for modern or future equipment (Wilson et al., 2012b). A mechanism to continually re-evaluate guidance is therefore needed.

Accordingly, for future damaging ash fall events, quantitative impact studies will be critical to improving the sparse existing empirical data set and for calibrating existing vulnerability estimates and results of experimental and theoretical assessments (Jenkins et al., 2014). To compare multiple damage studies, standard data collection guidelines should be followed; for example the volcanic ash collection and analysis protocols promoted by the International Volcanic Health Hazard Network (www.ivhhn.org) and the building survey guidelines of Jenkins et al. (2014).

Although posteruption impact assessments will continue to strengthen and diversify the understanding of volcanic ash impacts, this reactive research model rarely allows for detailed analysis to determine with any certainty how and why observed impacts occur. There is a lack of empirical knowledge of how the broad range of volcanic ash characteristics can impact the built environment, infrastructure, agriculture, and human. Acquiring a better understanding of the physical and chemical properties of volcanic ash should be a focus of future research in this area. However, it is difficult to explore impact thresholds in the field due to limited time, access difficulties, and incomplete ranges of volcanic ash properties such as thickness, grain size, and surface coating composition. The empirical performance testing of critical components using varying intensities and types of volcanic ash under laboratory conditions can begin to address these gaps and assess performance thresholds for different configurations of equipment. Crucially, it also provides a setting for the proactive experimentation of potential mitigation strategies.

ACKNOWLEDGMENTS

We thank Paul Ashwell and Grant Wilson for assistance with figures. David Johnston assisted with data and conceptual design of several figures. Much of the content for this work has been funded from the New Zealand Government MBIE Natural Hazard Research Platform Subcontract: C05X0804 (TW, CS), AXA Research Fellowship, and VOLDIES ERC contract 228064 (SJ) and is supported by the Global Volcanism Model (www.globalvolcanomodel.org). We thank Tina Neal for a very thorough review which greatly improved the quality of the chapter. Finally, we thank editor Paolo Papale.

REFERENCES

Ayris, P.M., Delmelle, P., 2012. The immediate environmental effects of volcanic ash emission. Bull. Volcanol. 74 (9), 1905–1936.

Berlo, D.K., 1960. The Process of Communication. Holt, Rinehart & Winston, New York.

Blong, R.J., 1984. Volcanic Hazards: A Sourcebook on the Effects of Eruptions. Academic Press Australia, 424 pp.

Blong, R.J., 2003. Building damage in Rabaul, Papua New Guinea, 1994. Bull. Volcanol. 65 (1), 43–54.

Cook, R.J., Barron, J.C., Papendick, R.I., Williams, G.J., 1981. Impact of agriculture of the Mount St. Helens eruptions. Science 211, 16–22.

Covello, V.T., Allen, F., 1988. Seven Cardinal Rules of Risk Communication. U.S. Environmental Protection Agency, Washington, D.C.. Policy Document OPA-87–020.

Cronin, S.J., Hedley, M.J., Neall, V.E., Smith, R.G., 1998. Agronomic impact of ashfall out from the 1995 and 1996 Ruapehu Volcano eruptions, New Zealand. Environ. Geol. 34 (1), 21–30.

Cronin, S.J., Neall, V.E., Lecointre, J.A., Hedley, M.J., Loganathan, P., 2003. Environmental hazards of fluoride in volcanic ash: a case study from Ruapehu volcano, New Zealand. J. Volcanol. Geotherm. Res. 121, 271–291.

Durand, M., et al., 2001. Impacts of, and Responses to Ashfall in Kagoshima from Sakurajima Volcano – Lessons for New Zealand. Institute of Geological & Nuclear Sciences Science Report 2001/30, 53 pp.

Engwell, S.L., Sparks, R.S.J., Aspinall, W.P., 2013. Quantifying uncertainties in the measurement of ashfall thickness. J. Appl. Volcanol. 2, 5. http://dx.doi.org/10.1186/2191-5040-2-5.

Gaillard, J.C., 2006. Traditional societies in the face of natural hazards: the 1991 Mt. Pinatubo eruption and the Aetas of the Philippines. Int. J. Mass Emerg. Disaster 24 (1), 5–43.

Gislason, S.R., Hassenkam, T., Nedel, S., Bovet, N., Eiriksdottir, E.S., Alfredsson, H.A., Hem, C.P., Balogh, Z.I., Dideriksen, K., Oskarsson, N., Sigfusson, B., Larsen, G., Stipp, S.L.S., 2011. Characterization of Eyjafjallajökull volcanic ash particles and a protocol for rapid risk assessment. Proc. Natl. Acad. Sci. 108 (18), 7307–7312.

Horwell, C.J., Baxter, P.J., 2006. The respiratory health hazards of volcanic ash: a review for volcanic risk mitigation. Bull. Volcanol. 69, 1–24.

Horwell, C.J., Baxter, P.J., Hillman, S.E., Calkins, J.A., Damby, D.E., Delmelle, P., Donaldson, K., Dunster, C., Fubini, B., Kelly, F.J., Le Blond, J.S., Livi, K.J.T., Murphy, F., Nattrass, C.,

Sweeney, S., Tetly, T.D., Thordarson, T., Tomatis, M., 2013. Physicochemical and toxicological profiling of ash from the 2010 and 2011 eruptions of Eyjafjallajokull and Grimsvotn volcanoes, Iceland using a rapid respiratory hazard assessment protocol. Environ. Res. 127, 63–73.

Jenkins, S.F., Magill, C.R., McAneney, K.J., 2007. Multi-stage volcanic event: a statistical investigation. J. Volcanol. Geotherm. Res. 161, 275–288.

Jenkins, S.F., Spence, R.J.S., Fonseca, J.F.B.D., Solidum, R.U., Wilson, T.M., 2014. Volcanic risk assessment: quantifying physical vulnerability in the built environment. J. Volcanol. Geotherm. Res. 276, 105–120.

Johnston, D.M., Houghton, B.F., Neall, V.E., Ronan, K.R., Paton, D., 2000. Impacts of the 1945 and 1995–1996 Ruapehu eruptions, New Zealand: an example of increasing societal vulnerability. Geol. Soc. Am. Bull. 112 (5), 720–726.

Kar-Purkayastha, I., Horwell, C., Murray, V., 2012. Review of evidence on the potential health impacts of volanci ash on the population of the UK and ROI. Centre for Radiation, Chemical and Environmental Hazards, Health Protection Agency, United Kingdom. Available online at: http://www.hpa.org.uk/webc/HPAwebFile/HPAweb_C/1317136779478.

Macedonio, G., Costa, A., 2012. Rain effect on the load of volcanic ash deposits. Nat. Hazards Earth Syst. Sci. Brief Commun. 12, 1229–1233.

Magill, C., Wilson, T.M., Okada, T., 2013. Observations of ashfall impacts from the 2011 Shinmoedake eruption, Japan. Earth Planets Space 65 (6), 677–698. http://dx.doi.org/10.5047/eps.2013.05.010.

Newhall, C.G., Hoblitt, R.P., 2002. Constructing event trees for volcanic crises. Bull. Volcanol. 64, 3–20. http://dx.doi.org/10.1007/s004450100173.

Newhall, C.G., Punongbayan, R.S., 1996. Fire and Mud. Philippine Institute of Volcanology and Seismology, Quezon City and University of Washington Press, Seattle and London.

Oze, C., Cole, J.W., Scott, A., Wilson, T.M., Wilson, G., Gaw, S., Hampton, S.J., Doyle, C., Li, Z., 2014. Corrosion of metal roof materials related to volcanic ash interactions. Nat. Hazards 71 (1), 785–802. http://dx.doi.org/10.1007/s11069-013-0943-0.

Paton, D., Smith, L., Johnston, D.M., 2005. When good intentions turn bad: promoting natural hazard preparedness. Aust. J. Emerg. Manage. 20 (1), 25–30.

Pyle, D.M., 1989. The thickness, volume and grain size of ashfall deposits. Bull. Volcanol. 51, 1–15.

Ronan, K.R., 1997. The effects of a "benign" disaster: symptoms of post-traumatic stress in children following a series of volcanic eruptions. Aust. J. Disaster Trauma Stud. 1997 (1).

Simkin, T., Siebert, L., 2000. Earth's volcanoes and eruptions: an overview. In: Sigurdsson, H. (Ed.), Encyclopedia of Volcanoes. Academic Press, San Diego, pp. 249–261.

Spence, R., Pomonis, A., Baxter, P., Coburn, A., White, M., Dayrit, M., Field Epidemiology Training Program Team, 1996. Building damage caused by the Mount Pinatubo eruption of June 15, 1991. In: Newhall, C.G., Punongbayan, R.S. (Eds.), Fire and Mud: eruptions and lahars of Mount Pinatubo, Philippines, Philippines Institute of Volcanology and Seismology. University of Washington Press, Seattle, pp. 1055–1061.

Spence, R., Kelman, I., Baxter, P., Zuccaro, G., Petrazzuoli, S., 2005. Residential building and occupant vulnerability to ashfall. Nat. Hazards Earth Syst. Sci. 5, 477–494.

Stewart, C., Johnston, D.M., Leonard, G.S., Horwell, C.J., Thordarson, T., Cronin, S.J., 2006. Contamination of water supplies by volcanic ash fall: a literature review and simple impact modelling. J. Volcanol. Geotherm. Res. 158, 296–306.

Stewart, C., Pizzolon, L., Wilson, T., Leonard, G., Dewar, D., Johnston, D., Cronin, S., 2009. Can volcanic ash poison water supplies? Integr. Environ. Assess. Manage. 5 (3), 713–716.

Sword-Daniels, V., Wilson, T.M., Sargeant, S., Rossetto, T., Twigg, J., Johnston, D.M., Loughlin, S.C., Cole, P.D., 2014. Consequences of long-term volcanic activity for essential services in Montserrat: challenges, adaptations and resilience. In: Memoir of the Geological Society of London, Special volume. The Eruption of Soufriere Hills, Montserrat, pp. 471–488. http://dx.doi.org/10.1144/M39.26.

Tobin, G.A., Whiteford, L.M., 2002. Community resilience and volcano hazard: the eruption of Tungurahua and evacuation of the faldas in Ecuador. Disasters 26 (1), 28–48.

UN, 2009. United Nations Office for Disaster Risk Reduction Terminology on Disaster Risk Reduction. Available from: http://www.unisdr.org/we/inform/terminology (accessed 06.03.14).

UNISDR, 2011. HFA Progress in Asia-Pacific: Regional Synthesis Report 2009–2011.

Wardman, J.B., Wilson, T.M., Bodger, P.S., Cole, J.W., Stewart, C., 2012a. Potential impacts from tephra fall to electric power systems: a review and mitigation strategies. Bull. Volcanol. 74 (10), 2221–2241. http://dx.doi.org/10.1007/s00445-012-0664-3.

Wardman, J., Sword-Daniels, V., Stewart, C., Wilson, T., 2012b. Impact Assessment of the May 2010 Eruption of Pacaya Volcano, Guatemala. GNS Science Report 2012/09. 90 pp.

Weinstein, P., Horwell, C.J., Cook, A., 2013. Volcanic emissions and health. In: Selinus, O., Alloway, B., Centeno, J.A., Finkelman, R.B., Fuge, R., Lindh, U., Smedley, P. (Eds.), Essentials of Medical Geology. Springer, p. 805.

WHO, 2013. Review of Evidence on Health Aspects of Air Pollution – REVIHAAP Project. Technical Report. Available from: http://www.euro.who.int/__data/assets/pdf_file/0004/193108/REVIHAAP-Final-technical-report-final-version.pdf.

Wilson, T.M., Stewart, C., 2013. Volcanic ash. In: Bobrowsky, P.T. (Ed.), Encyclopedia of Natural Hazards. Springer, New York, p. 1000.

Wilson, T., Cole, J., Stewart, C., Cronin, S., Johnston, D., 2011. Ash storms: impacts of wind-remobilised volcanic ash on rural communities and agriculture following the 1991 Hudson eruption, southern Patagonia, Chile. Bull. Volcanol. 73 (3), 223–239.

Wilson, T.M., Stewart, C., Bickerton, H., Baxter, P.J., Outes, V., Villarosa, G., Rovere, E., 2012a. Impacts of the June 2011 Puyehue Cordon-caulle Volcanic Complex Eruption on Urban Infrastructure, Agriculture and Public Health. GNS Science. 88 pp.

Wilson, T.M., Stewart, C., Sword-Daniels, V., Leonard, G.S., Johnston, D.M., Cole, J.W., Wardman, J., Wilson, G., Barnard, S.T., 2012b. Volcanic ash impacts on critical infrastructure. Phys. Chem. Earth Part A/B/C 45-46, 5–23.

Witham, C.S., Oppenheimer, C., Horwell, C.J., 2005. Volcanic ash leachates: a review and recommendations for sampling methods. J. Volcanol. Geotherm. Res. 141, 299–326.

Chapter 4

Volcanic Ash Hazards and Aviation Risk

Marianne Guffanti[1] and Andrew Tupper[2]

[1] *U.S. Geological Survey, Reston, VA, U.S.,* [2] *Australian Bureau of Meteorology, Hydrology Unit, Melbourne, Victoria 3001, Australia*

ABSTRACT

The risks to safe and efficient air travel from volcanic-ash hazards are well documented and widely recognized. Under the aegis of the International Civil Aviation Organization, globally coordinated mitigation procedures are in place to report explosive eruptions, detect airborne ash clouds and forecast their expected movement, and issue specialized messages to warn aircraft away from hazardous airspace. This mitigation framework is based on the integration of scientific and technical capabilities worldwide in volcanology, meteorology, and atmospheric physics and chemistry. The 2010 eruption of Eyjafjallajökull volcano in Iceland, which led to a nearly week-long shutdown of air travel into and out of Europe, has prompted the aviation industry, regulators, and scientists to work more closely together to improve how hazardous airspace is defined and communicated. Volcanic ash will continue to threaten aviation and scientific research will continue to influence the risk-mitigation framework.

4.1 INTRODUCTION

Airborne volcanic ash has known deleterious effects on jet aircraft and poses a frequent hazard to the safety and efficiency of air travel on a global scale. Volcanic ash is present in the atmosphere at flight altitudes of jet aircraft virtually every day somewhere around the world because multiple long-lived explosive eruptions produce recurring ash emissions. The exposure of aviation to ash hazards is substantial, as thousands of commercial flights occur daily along the many major air routes that overlie the world's volcanically active regions. Numerous encounters of aircraft with ash clouds over the decades of modern jet travel have confirmed the severe damage that can result from the ingestion of ash particles into a jet engine, even to the point of engine shutdown during flight. Major economic and transportation disruptions also

Volcanic Hazards, Risks, and Disasters. http://dx.doi.org/10.1016/B978-0-12-396453-3.00004-6

can occur when flights are canceled or rerouted to avoid ash-contaminated airspace, as a large part of the world experienced during the 2010 eruption of Iceland's Eyjafjallajökull volcano.

The risks to aviation from volcanic ash are exceptionally dynamic in the short term. Explosive eruption columns can reach cruise altitudes in a few minutes; winds aloft then disperse ash as clouds at speeds that can reach 150 km/h. Aircraft flying in the same airspace as dispersing ash may approach an ash cloud at ~800 km/h. Airborne volcanic ash is a nonlocalized hazard. Unlike an established community in the path of a ground hazard from a particular volcano, the at-risk aviation "community" (air travelers) continually changes location relative to multiple hazard sources (active volcanoes underlying an air route) and shifting hazard zones (dispersing ash clouds).

Mitigating actions accordingly are dynamic and fast paced. Specialized warning messages that describe the expected movement of ash clouds are written and disseminated within tens of minutes of detection of explosive eruptions and ash in the atmosphere; updated messages are issued a few minutes to several hours later as the ash cloud moves. Decisions to divert en-route aircraft to avoid ash are made in real time by flight crews communicating with ground-based dispatch and air-traffic control centers; planning for future flights is set into motion, taking into account the forecast location of ash clouds.

The mitigation framework of ash avoidance is based on globally coordinated eruption reporting, ash-cloud detection and forecasting, and rapid dissemination of warnings and is under continual improvement by aviation regulators, the volcanological and meteorological scientific communities, and airline operators. The extent of aviation risk also undergoes evaluation. For example, the possibility that aircraft can safely tolerate transit through dilute, well-dispersed ash clouds without imminent threat to safety of flight was raised during the 2010 Eyjafjallajökull eruption. Should that be possible—the question is not yet settled—the need remains for the aviation sector to be rapidly informed of, and adjust to, the presence of ash in airspace.

4.2 A VOLCANOLOGICAL AND METEOROLOGICAL HAZARD

Volcanic ash clouds—as their name indicates—are both a volcanological and meteorological phenomenon. Explosive eruptions provide the hazardous volcanic material which then interacts over time with the Earth's dynamic atmosphere.

About 40–50 of the world's ~500 historically active volcanoes erupt each year with enough explosivity (VEI 2 or greater; Siebert et al., 2010) to inject significant amounts of ash into airspace traversed by aircraft. Typically, episodes of eruptive activity are separated by noneruptive intervals of hours to months. The duration of a single episode usually ranges from a few minutes to tens of hours. The entire eruptive period of a volcano can last for months to years. Thus, one eruption can be responsible for repeatedly generating many ash clouds. For example, Tungurahua Volcano in Ecuador, which has been erupting episodically since 1999, produced ash emissions on at least 128 days in 2012 alone.

In an explosive volcanic event, gas-charged magma abruptly depressurizes as it nears the Earth's surface, jetting out of a constrained vent. The ejected magma fragments into glassy shards and mineral particles that—along with gases released from the decompressing magma and air entrained from the surroundings—rise upward in a convective plume (Sparks et al., 1997). Eruption columns rise at tens of meters per second and can quickly reach cruise altitudes of jet aircraft and beyond, to nearly 50 km (Self and Walker, 1994). Plumes are generally affected by the atmosphere, entraining moisture to increase their buoyancy or bending under the effect of local wind fields (Carey and Bursik, 2000). In stronger eruptions, the plume punches through the tropopause (the level below which most meteorological clouds are confined) to form a mushroom-shaped cap similar to an intense thunderstorm (Figure 4.1). Volcanic fragments larger than several tens of microns fall out of a plume within hours and are generally deposited within a few hundred kilometers of the volcano, whereas some smaller particles (termed fine ash) remain suspended in the atmosphere for days.

Once ejected into the atmosphere, the ash and gas are dispersed by prevailing winds as a cloud of small particles and aerosol droplets (Figure 4.2); if wind direction varies with height (wind shear), the eruption plume may disperse in multiple directions. Ash clouds can extend over large areas and travel great distances from the source volcano. For example, the June 2011 eruption of Cordon Caulle in Chile produced an ash cloud that was tracked by satellite-based instruments as it circled the Southern Hemisphere twice over the period of weeks, with "old" ash mingling with "new" ash as the plume

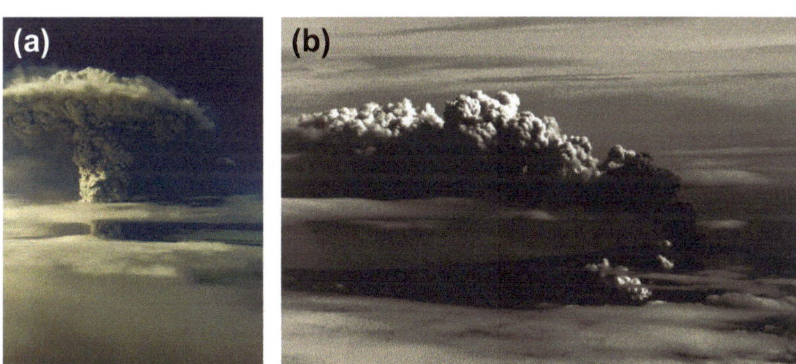

FIGURE 4.1 (a) A strong eruptive plume with an umbrella cap at Puyehue-Cordon Caulle volcanic chain, Chile, on June 5, 2011. *Photo by Chilean Air Force.* (b) A weak, wind-bent eruptive plume at Eyjafjallajökull, Iceland, on May 4, 2010. *Image provided courtesy of T. Jonsson, Icelandic Meteorological Office.*

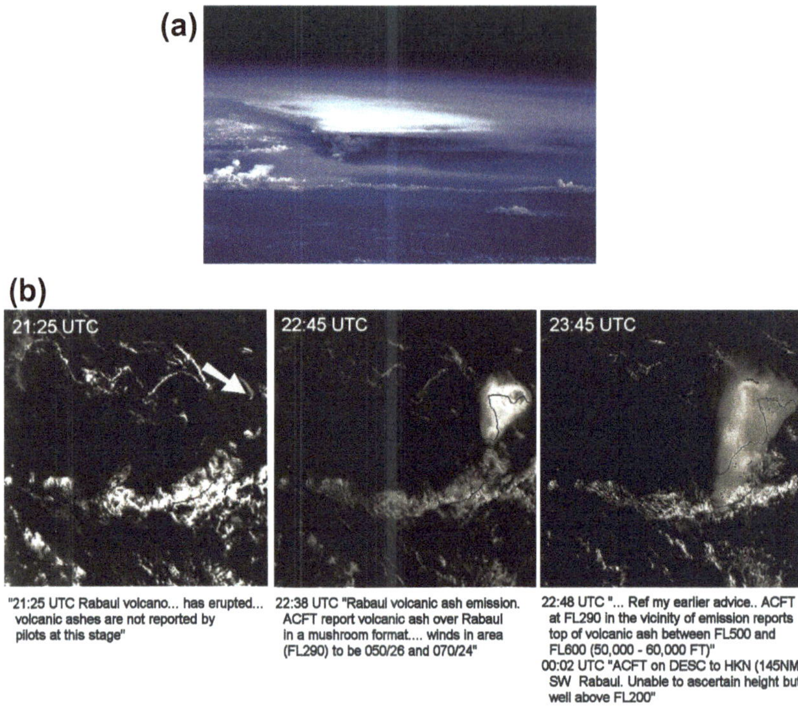

FIGURE 4.2 (a) Space Shuttle (STS-64) photo of the volcanic cloud produced by the eruption of Rabaul Volcano, taken on September 19, 1994. *Image provided courtesy of NASA.* (b) Excerpts from the air-traffic control and pilot reports during the early parts of the September 19, 1994 eruption of Rabaul Volcano, Papua New Guinea, together with the matching satellite images (after Tupper and Kinoshita (2003)). GMS-4 visible images are shown for (1) 21:25, (2) 22:45, and (3) 23:45 UTC on September 18, 1994. As this eruption occurred next to an airport and during daylight working hours, the flow of pilot-sourced information was excellent. However, the potential for confusion still existed—in particular, note the quick clarification between 22:38 and 22:48 UTC that the height mentioned (FL290, or approximately 9 km) was the height of the aircraft, not of the eruption. This clear communication is absolutely essential during the fast-moving early stages of a major eruption, when much of the information can be conflicting.

completed its circumnavigation over the still-erupting volcano (Vernier et al., 2013). Ash clouds tend to become stratified into thinner layers as they age, with the best data on such stratification coming from lidar (light detection and ranging, or "light radar") measurements of the especially well-studied cloud from the 2010 Eyjafjallajökull eruption (Winker et al., 2012; Marenco et al., 2011; Schumann et al., 2011). Results of those studies indicate that the days-old ash was distributed nonuniformly in the troposphere as discrete, inclined, or subhorizontal layers hundreds of meters to a few kilometers in vertical extent.

An ash cloud eventually dissipates in the atmosphere as the particles progressively disperse, fall out by gravitational settling, and (or) are removed by precipitation. Ash concentrations decrease until the cloud can no longer be seen by the human eye or detected by in situ or remote-sensing instruments. However, as yet no single, well-documented threshold concentration exists at which ash is known to pose no or minimal harm to aircraft. Work on that issue is hampered by the scarcity of data on engine tolerances to ash ingestion, as well as difficulties in accurately determining the constituents (solid particles, gases, aerosols), density, and three-dimensional shape of a volcanic cloud at particular times and locations.

Numerous inadvertent encounters of aircraft with ash clouds (Guffanti et al., 2010), as well as some engine-test data (Dunn and Wade, 1994; Dunn, 2012; Shinozaki et al., 2013), provide evidence of the aircraft damage that can occur. Ash particles can cause abrasion damage to forward-facing surfaces, including windscreens, fuselage surfaces, and compressor fan blades. Moreover, the temperature at which glassy silicate material in an ash cloud can soften and stick to engine surfaces is lower than combustion temperatures in jet engines; consequently, ingested ash particles can accumulate as resolidified deposits in cooler parts of an engine (Przedpelski and Casadevall, 1994). The overall result of an aircraft's flying into an ash cloud can be degraded engine performance—even to the point of in-flight loss of thrust power—diminished windscreen visibility, and the failure of critical navigational and operational instruments.

The compilation of Guffanti et al. (2010) documents 79 damaging ash/aircraft encounters that were reported through 2009. Twenty-six of the damaging encounters involved significant to very severe aircraft damage, including nine encounters in which engine shutdown occurred during flight. Fortunately in each of those cases, at least one engine was able to be restarted or did not shutdown, and there have been no crashes resulting from ash ingestion. Aircraft have been damaged by eruptions ranging from small, recurring ash emissions (such as at Soufriere Hills Volcano in Montserrat) to very large, infrequent events (such as the 1991 eruption of Mt. Pinatubo). Most, but not all, of the damaging encounters occurred within 24 h of the onset of ash production at distances less than 1000 km from the source volcanoes.

Volcanic activity also has caused significant adverse effects to numerous airports worldwide, with local to far-ranging effects on travelers and commerce (Guffanti et al., 2009). The most common effect is temporary operational disruption, ranging from flight cancellations to airport closures for periods of hours to weeks. The main hazard is ash fall. The accumulation of only a few millimeters of ash on runways is sufficient to force temporary closure of an airport, although disruptions also have been caused by airborne ash in the vicinity of airports without the deposition of ash on the ground. The accumulation of more than a trace amount requires removal of the ash in a manner that prevents it from being remobilized by wind and aircraft and during the clean-up process itself. Recommended clean-up and other mitigation procedures for airports are specified in a manual of the International Civil Aviation Organization (2007).

4.3 DEVELOPMENT OF A GLOBAL FRAMEWORK TO AVOID ASH CLOUDS

The decade of the 1980s brought the risk to aviation from the volcanic-ash hazard into dramatic focus. Damaging encounters of jet aircraft with ash clouds had occurred in the 1970s and during the eruption of Mt. St. Helens in 1980. However, it was not until eruptions of Mt. Galunggung (Indonesia) in 1982 and Redoubt Volcano (USA) in 1989 caused in-flight engine failures in passenger jets that the ash hazard to aviation became widely recognized by aviation regulators, the airline industry, and the scientific community (Casadevall, 1994; Miller and Casadevall, 2000). In the Redoubt incident, all four of the aircraft's jet engines lost thrust power, and the aircraft descended nearly 4.5 km (~14,600 ft) before the engines could be restarted, narrowly averting a crash into mountainous terrain. The aircraft landed safely, but it sustained more than $80 million in damages (~$150 million in 2013 dollars) (Tuck and Huskey, 1992).

Motivated by the potential severity of the emerging hazard, the International Civil Aviation Organization (ICAO)—the United Nations entity charged with regulating international air navigation—established the International Airways Volcano Watch (IAVW) in 1987 with a requirement for the international dissemination of information about airborne volcanic ash. The IAVW requirement set the stage for the development of globally coordinated procedures that are now in place to: (1) report eruptions that are imminent or underway, (2) detect ash clouds in the atmosphere and forecast their expected movement hours into the future, and (3) issue special warning messages to aviation meteorological offices, airline operation centers, and air-traffic management centers (International Civil Aviation Organization, 2007). Together, these actions are aimed at reducing risk by giving the parties that plan and execute flights the information needed to ensure that flights avoid potentially hazardous airspace.

A major achievement of the IAVW was to define the role of Volcanic Ash Advisory Centers (VAACs) to detect ash in the atmosphere and forecast movement of ash clouds. Through the efforts of ICAO, various countries agreed to incorporate ash-cloud detection and forecasting functions into their existing national weather organizations, thus establishing by 1997 a worldwide system of nine VAACs (Figure 4.3). VAACs issue Volcanic Ash Advisories that describe current and expected locations of ash clouds. VAACs rely primarily on data from civilian meteorological satellites, Volcano Observatory notifications, and pilot reports to detect and track ash clouds, and they run atmospheric dispersion models to forecast cloud movement hours and

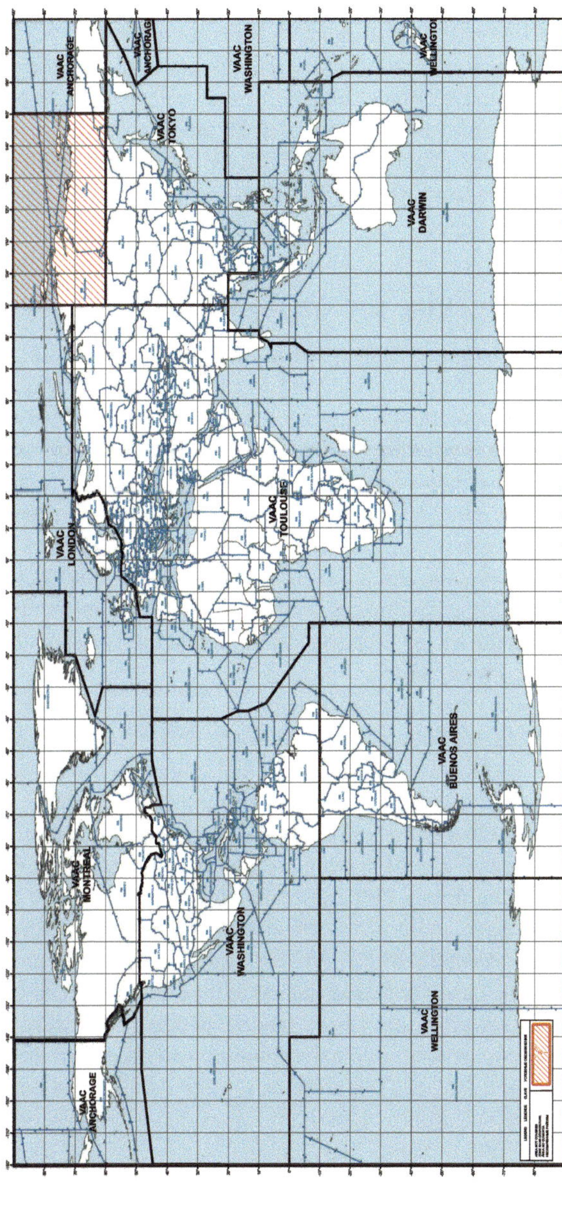

FIGURE 4.3 Areas of responsibility of the world's nine Volcanic Ash Advisory Centers (VAACs). *Image provided courtesy of ICAO.*

sometimes days into the future. VAAC Advisories, in turn, form the basis of formal hazard warnings (called SIGMETs for SIGnificantMETeteorological information) that are issued by the worldwide system of aviation Meteorological Watch Offices to inform airline dispatchers, pilots, and air-traffic managers of weather-related hazards including ash clouds.

Effective ash-hazard advisories and warnings are based on the integrated use of the scientific and technical capabilities in volcanology, meteorology, and atmospheric physics and chemistry that are available worldwide to monitor volcanic activity, characterize ash clouds, and forecast ash dispersion. As these capabilities improve, the ash hazard is more accurately described and aviation risk better managed.

4.3.1 Volcano Monitoring and Eruption Reporting

Timely reporting of volcanic activity, beginning with information about the precursory build-up phase and carrying through to eruption cessation, supports a variety of risk-management activities by the aviation sector. Forewarning of an impending eruption is used by airlines and air-traffic managers to optimize flight planning over periods of hours to days and by VAACs to focus their attention on likely ash sources. Prompt reports of the onset of eruptive activity and observations of ash emissions are needed by airlines and air-traffic managers to carry out safe and efficient en-route flight diversions and by VAACs to quickly issue advisories and initiate ash dispersion models. Accurate reports of the end of an eruption allow all parties to step down their responses and hazard messaging.

Scientific agencies that monitor volcanoes (Volcano Observatories) are the primary source of eruption reports, although en-route pilots and VAACs may be among the first to detect the actual onset of eruptive activity, particularly at volcanoes lacking real-time ground-based sensors. Approximately 60 Volcano Observatories worldwide are engaged in monitoring many—but by no means all—of the world's potentially hazardous volcanoes. Only about 60% of the nearly 500 explosive volcanoes that have erupted in the past 2000 years have some form of continuous ground-based monitoring in place (Ewert and Newhall, 2004). The majority of these have only seismic monitoring, in many cases a single sensor. Since 2004, improvements in volcano-monitoring coverage have been made, but there are still roughly 200 recently active volcanoes with explosive potential that remain unmonitored with real-time sensors. In cloudy conditions remote sensing can be inhibited, and ground-based information is essential; for example, a major eruption of Manam Volcano in 2005 in Papua New Guinea, to a height of 23 km, was not remotely sensed in real time due to atmospheric conditions (Tupper et al., 2007). The potential for sudden, unreported ash-producing eruptions at undermonitored volcanoes constitutes a significant weakness in aviation-risk mitigation.

Historically, priority has been given to installing monitoring networks on volcanoes that threaten nearby ground-based populations, and most Volcano Observatory operations have been designed to mitigate impacts to people on the ground rather than in the air (Ewert and Newhall, 2004). But places exist where high aviation risks have helped to justify significant volcano-monitoring investments by governments. In the United States, seismic networks were installed from 1996 to 2007 at nearly two dozen sparsely populated Alaska volcanoes primarily to help mitigate aviation risk that had been dramatically highlighted in 1989 by the dangerous engine-failure incident during the eruption of Redoubt Volcano. At Mt. Etna in Italy, in part as the result of repeated airport and air-traffic disruptions from its recurring explosive activity, a concerted effort has been made over the past decade to improve monitoring networks, alerting systems, and ash-dispersion forecasting capabilities (Alparone et al., 2007; Scollo et al., 2009). In 2008 in Chile, the unexpected eruption of Chaitén volcano—unmonitored at the time—generated lahars that devastated the town of Chaitén and ash clouds that caused extensive airport and air-traffic disruptions in Chile and Argentina; within a few months, the Chilean government had approved a multiyear program to significantly expand volcano monitoring and hazard assessment studies (Ewert et al., 2008). Most recently, the nearly week-long shutdown of air travel in Europe caused by the 2010 eruption of Eyjafjallajökull in Iceland was a substantial part of the rationale for a European Commission project called FUTUREVOLC that selected Iceland as a site for integration of ground- and space-based observations for improved volcano monitoring and hazard assessment (Sigmundsson et al., 2013). Thus, although ash hazards to aviation are seldom the sole hazard focus of volcano-monitoring programs, these examples attest to the contribution aviation risk can make in setting national hazard-monitoring priorities.

An alert-level system for communicating information about volcanic activity has been developed specifically for aviation (Guffanti and Miller, 2013). The system was devised in 1990 by the Alaska Volcano Observatory during the 1989–1990 eruption of Redoubt Volcano, and a modified version (Table 4.1) subsequently has been adopted by ICAO as the recommended guidance for Volcano Observatories worldwide. The system uses a simple, color-based ranking that focuses on conditions at a volcano (not on the ash cloud as it moves downwind) and works for both increasing and decreasing activity: Green–normal background; Yellow—signs of unrest; Orange—heightened, possibly precursory unrest or minor ash eruption; Red—major ash eruption imminent or underway (http://www.icao.int/icaonet/dcs/9766/9766_cons_en.pdf/). The color-code definitions purposely exclude much volcanological detail and instead distill highly technical data into information that readily can be understood by users who are not volcanologists, such as pilots, airline dispatchers, and meteorologists. To convey the color-code rankings and other operationally critical information, a specialized message template—called a

TABLE 4.1 Aviation Color Codes

GREEN	Volcano is in typical background, non-eruptive state, *or, after a change from a higher level,* volcanic activity has ceased and volcano has returned to non-eruptive background state.
YELLOW	Volcano is exhibiting signs of elevated unrest above known background level, *or, after a change from a higher level,* volcanic activity has decreased significantly but continues to be closely monitored for possible renewed increase.
ORANGE	Volcano is exhibiting heightened or escalating unrest with increased potential of eruption, timeframe uncertain, *or,* Eruption is underway with no or minor volcanic-ash emissions.
RED	Eruption is imminent with significant emission of volcanic ash into the atmosphere likely, *or,* eruption is underway or suspected with significant emission of volcanic ash into the atmosphere.

Volcano Observatory Notice for Aviation (VONA)—has been developed by ICAO for use by Volcano Observatories. Along with the basic volcano information (name, identifying number, location, current and previous color code), the VONA gives a brief, jargon-free summary of volcanic activity and observations about ash emissions (or lack thereof). The VONA template and guidance on its dissemination are given in the ICAO's Handbook on the IAVW (International Civil Aviation Organization, 2004).

The aviation color-code system is the only standardized international volcano-alert system in effect (Fearnley et al., 2012). As of 2013, it is used by Volcano Observatories in the United States, Russia, New Zealand, and Iceland and by the Darwin VAAC in Australia as part of its Ash Advisories for the Southwest Pacific region. Some barriers exist to implementation and further use of the aviation color code. As noted previously, ground hazards are generally the highest priority of Volcano Observatories; moreover, if a color-code system is used for ground hazards, there is the possibility of confusion with respect to aviation colors (this is why the U.S. Geological Survey uses terms such as "Watch" or "Warning" rather than colors for ground-based volcanic hazards; see Gardner and Guffanti, 2006). Furthermore, in some countries real-time seismic monitoring networks are sparse, and limited resources at some of these Observatories makes assigning color codes more difficult. Nevertheless, substantial progress already has been made under the aegis of ICAO, and with continued education and outreach to Volcano Observatories worldwide, greater use of the aviation color code is achievable.

4.3.2 Ash Cloud Detection

Once ash is present in the atmosphere, VAACs are responsible for advising the aviation sector of the detected and forecast (expected future) locations of ash clouds. VAACs issue their advisories in both text and graphical form, usually with a forecast timeframe of 6–18 h from the time an ash cloud is detected. These advisories are the focus of continued refinement and improvement. Although each VAAC operates autonomously within its national structure, they also coordinate as a group through ICAO on various issues, for example, establishing protocols for issuing advisories for ash clouds that disperse across VAAC boundaries.

The major technology used by VAACs and others for detecting ash clouds is satellite remote sensing, as it provides recurring synoptic views of the atmosphere over the spatial scales necessary for tracking ash cloud movement (hundreds to thousands of kilometers). Several countries operate satellite systems that carry sensors useful for tracking volcanic clouds, and data from these sensors are shared freely through international agreement. Such sensors were not designed specifically for detecting volcanic ash and gas, but data-processing techniques for that use have been developed over time by various atmospheric and volcanological researchers.

Prata (2009) and Zehner (2010) provided overviews of techniques for the satellite detection of volcanic clouds, in the context of aviation hazards. Satellite sensors detect volcanic ash and gas (primarily sulfur dioxide gas, SO_2) in the infrared through visible to ultraviolet wavelengths, and sensors on both geostationary and polar-orbiting satellites provide useful data. The mainstay of operational volcanic ash detection has been the so called "reverse absorption" technique, based on the different absorption properties of volcanic ash clouds in the thermal infrared (TIR) spectrum compared with water or ice clouds (Prata, 1989). The technique relies on certain conditions, for example that the ash cloud is the highest cloud layer, ash particles are not completely encased in ice, and the thermal perturbation caused by the ash cloud is large enough to be detected. This technique is typically used together with pattern recognition techniques and is considered operationally robust for clouds in the appropriate conditions (Tupper et al., 2004). Recent research advances have produced rapid, automated techniques to retrieve ash-cloud properties such as mass loading (ash mass per unit area), cloud-top height, and effective particle sizes from TIR measurements (Figure 4.4 and Pavolonis, 2010). As these techniques move from the research to operational realm, VAACs will have more quantitative tools to delineate ash clouds.

Volcanic-cloud observations are generally most difficult in the moist tropics. The Maritime Continent (comprising mainly Indonesia, Papua New Guinea, and the Philippines) with its hundreds of active volcanoes is also the most important energy source in the world for meteorology, with sustained atmospheric convection from the Tropical Warm Pool reaching altitudes of

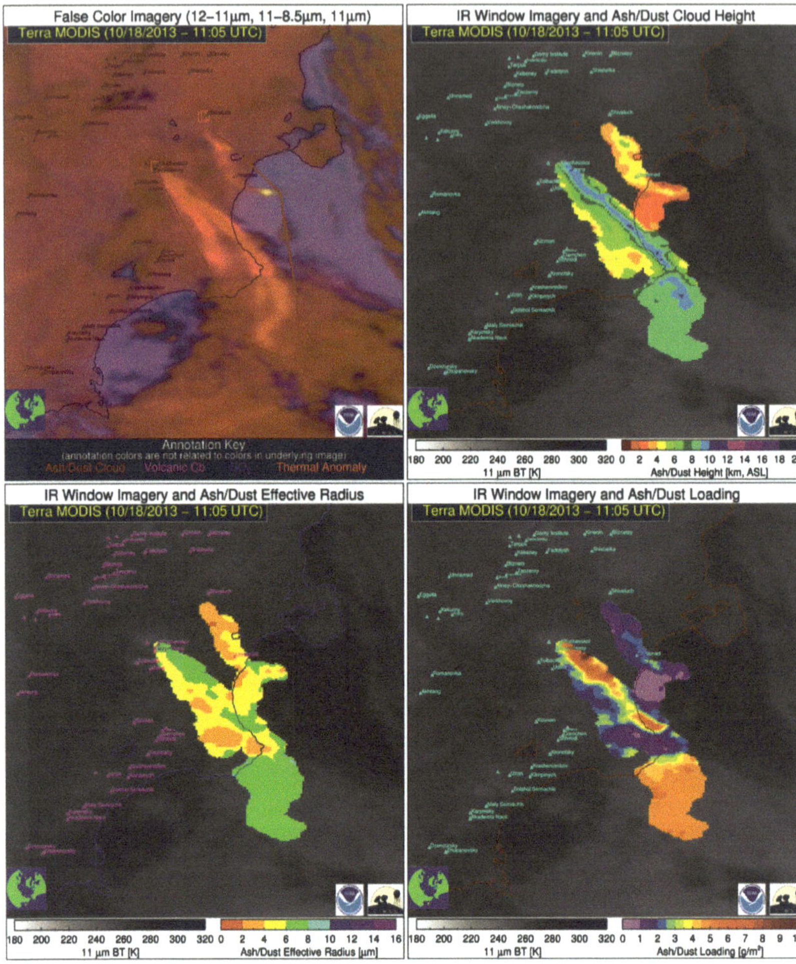

FIGURE 4.4 Results of automated technique of retrieving ash-cloud height, particle size (effective radius), and ash loading (mass per area) from satellite-based thermal infrared measurements of the Kamchatka region on October 18, 2013. *Image provided courtesy of M. Pavolonis, NOAA.*

15–20 km and forming cirrus "shields" that may obscure observation of ash clouds for many hours. Because aircraft generally cruise at altitudes of 10–12 km, volcanic ash can be encountered by aircraft in circumstances where the ash cannot be observed from space. The variation of the height of the tropopause (the typical "lid" for both meteorological and volcanic clouds) with latitude also has operational implications; at mid latitudes ash erupted to the height of the tropopause likely will be present at cruise altitudes, whereas ash erupted closer to the poles is more likely to be constrained to below cruise altitudes.

Ice formed from water that was either ejected with the magma or entrained from the atmosphere can coat ash, forming ice-ash aggregates, and thus obscuring the ash from remote sensing techniques even if the plume itself is visible, such as in the eruption of Rabaul in 1994 (Figure 4.2 and Rose et al., 1995). Moist entrainment is highest in the tropics and contributes to an increase in the plume height (Sparks et al., 1997; Tupper et al., 2009). At the smaller end of the eruption scale, eruptions can act as a trigger for moist convection that contains little ash but reaches the high altitude of tropical convection. Ash in such clouds cannot usually be detected in reverse absorption imagery, but aerosols if present in sufficient quantity can be detected through the application of algorithms designed to identify polluted clouds and through the application of SO_2 sensing techniques (Tupper et al., 2005).

Many eruptions emit SO_2 gas along with ash, and eruptions are the only source of significant quantities of SO_2 at jet cruise altitudes (Carn et al., 2009). The hazard to aircraft from exposure to SO_2 is not well investigated, although it is thought to pose a corrosion risk to key components and in sufficient quantities may pose a health risk to aircraft occupants. Ultraviolet (UV) instruments are effective at detecting coerupted volcanic SO_2 that may be present in ash clouds. SO_2 can persist longer in the atmosphere than ash particles subject to gravitational settling. Although it does not always coexist with ash, detection of SO_2 is very useful to supplement ash detection, particularly where direct sensing of volcanic ash is obscured by ice in the cloud.

With respect to the aviation hazards, a major limitation of TIR and UV satellite remote sensing is the intrinsic difficulty of determining the vertical thickness and mass concentration (particle density) of an ash cloud, both key parameters for optimal identification of what airspace to avoid. For example, the mass loading values derived from TIR data for a thin, dense ash cloud may be similar to a thicker but more dilute one. A promising technique to determine ash-cloud thickness is lidar, as demonstrated with measurements of ash clouds from the eruption of Eyjafjallajökull in 2010 and Cordon Caulle in 2011. Lidar commonly is a ground-based instrument (Pappalardo et al., 2004), but can also be airborne (e.g., Marenco et al., 2011), or space borne (Winker et al., 2010). The advantage of airborne and space borne platforms is a much greater chance of intercepting targets of opportunity.

The eruption of Eyjafjallajökull also prompted airborne in situ sampling volcanic clouds (Schumann et al., 2011; Johnson et al., 2012). Aircraft carrying various sensors flew into dispersed clouds and made measurements of particles and trace gases. Combined with ground and satellite observations, this process revealed the cloud structure and composition in unprecedented detail.

4.3.3 Forecasting Ash Cloud Movement

Having analyzed all available data, including possibly contradictory real-time data such as conflicting pilot reports and cloud-height estimates, a VAAC must delineate the expected locations of the ash cloud in a way that is easily translated into concise warnings for aircraft (Figure 4.5). Inevitably this involves assumptions and simplifications, as the clouds may in reality be extremely complex.

To forecast ash-cloud movement, each VAAC uses a particular atmospheric transport and dispersion model such as HYSPLIT (Hybrid Single Particle Lagrangian Integrated Trajectory Model; Draxler and Hess, 1998), NAME (Numerical Atmospheric dispersion Modeling Environment; Jones et al., 2007), or MLDP0 (Modèle Lagrangien de Dispersion de Particules; D'Amours et al., 2010). In addition to having different internal physics, these models also can input different numerical weather predictions and different properties of the erupting plume (eruption source parameters). In an intercomparison of these models, Witham et al. (2007) found strong similarities between model results, despite the different models and setup conditions.

However, regardless of the model used substantial inherent uncertainties exist in model output—i.e., in the forecasts of cloud location and concentration. As discussed by Bonadonna et al. (2012), forecast uncertainty is a

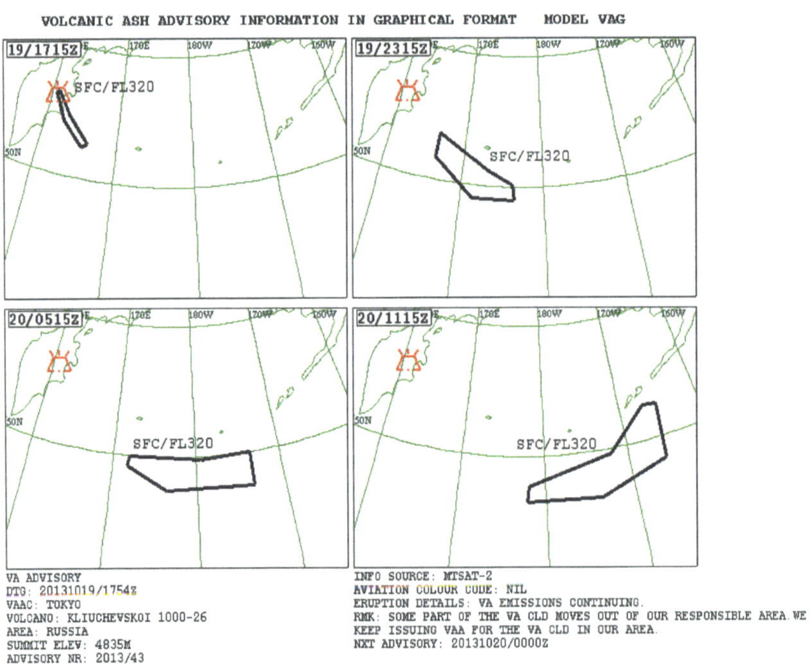

FIGURE 4.5 Example of a Volcanic Ash Advisory graphic by the Tokyo VAAC. A simple polygon depicts ash erupted from the Russian volcano Kliuchevskoi and moving toward Alaska. The ash is defined here as extending from the surface to Flight Level 320 (nominally, 32,000 ft). The bulk of the ash in this situation is likely to be concentrated into thin layers at one or more altitudes, but it is rare for detailed observations to be available in real time that delineate such layering.

combination of: (1) uncertainty in input parameters related to the eruption source (plume height, eruption start and stop time, mass eruption rate, particle size distribution, and distribution of mass with elevation in the erupting plume; Mastin et al., 2009), (2) uncertainty in meteorological parameters (wind velocity and direction, temperature, humidity), and (3) limitations in model accuracy, particularly in relation to how models simulate removal of ash from the atmosphere (e.g., by aggregation of fine particles to form larger clusters that fall out more quickly). Bonadonna et al. (2012) recommended a variety of approaches to reduce forecast uncertainties. One approach is to incorporate (assimilate) observational data, such as from remote sensing or airborne sampling, into dispersion-model output then reinitialize the model carrying the corrections forward. For example, Stohl et al. (2011) showed how using remote sensing and a reverse application of a Lagrangian dispersion model can refine the eruption source parameters. Vernier et al. (2013) considered the problem of a long-lived eruption plume and showed that Calipso lidar data can be assimilated into a Lagrangian model and projected forwards, refreshing continuously as new corrective data is received.

It is worth emphasizing that no single observational or modeling technique used by the VAACs represents the full picture of the potential hazard—each observation or forecast assists in describing operational threat areas and their future configuration. Typically, the start of an eruption is accompanied by high degrees of uncertainty, and as more information is gleaned and operational consensus becomes more and more informed, the confidence of VAAC products progressively improves.

4.3.4 Communicating the Hazard

The core of the IAVW risk-mitigation strategy is to transform complex volcanological and meteorological information into a suite of useful messages that directly reach the parties who ultimately make operational decisions about preflight route changes, flight cancellations, and in-flight diversions—i.e., airline dispatchers and pilots, air-traffic controllers, and aviation meteorologists. Table 4.2 summarizes the primary types and sources of operational ash-hazard messages for aviation. The dynamic nature of the ash hazard means that these messages are updated and re-issued over time periods of minutes to hours; for a longer-lived ash cloud, dozens of messages will be generated and disseminated over a period of days. Most of the messages are disseminated via the global telecommunication networks organized for air-traffic services under the authority of ICAO, and some are also posted on agency web sites. As described earlier, the content of the messages is not generated automatically; human minds must determine the essential information to be synthesized into a succinct warning message.

Given the importance of the hazard messages, a great deal of IAVW protocol is devoted to governing their generation and dissemination. Certainly,

TABLE 4.2 Sources and Types of Volcanic-Ash Hazard Messages for Aviation

Source	Message Type
En-route flight crew	*Special Air Report (AIREP)*: Report of observations of volcanic activity made by flight crew during flight. Also known as pilot report (PIREP). *Volcanic Activity Report (VAR)*: ICAO form for reporting observed volcanic activity or an aircraft encounter with an ash cloud.
National civil aviation authorities	*Notice to Airmen (ASHTAM or ash NOTAM)*: Message to alert pilots of any ash hazards en route or at a specific location (such as an airport).
Volcanic Ash Advisory Centers	*Volcanic Ash Advisories (VAA)*: Text and graphical description (VAG) of the location of detected ash cloud with forecast of its expected movement hours into the future.
Meteorological Watch Offices	*SIGnificant METeorological Information (SIGMET)*: Warning message about meteorological hazards to flights, including volcanic ash clouds.
Volcano Observatories	*Volcano Observatory Notice for Aviation (VONA)*: Succinct description of activity at a volcano with focus on ash emission.

there is room for improvement in ash-hazard messaging. VONA are provided by a minority of the world's Volcano Observatories. Some Meteorological Watch Offices do not issue volcanic ash SIGMET consistently, leaving users to rely on VAAC Advisories. Pilot reports do not always reach VAACs. There is considerable variation among VAACs in the graphical displays of their forecasts. But overall, the messaging element is fundamentally well organized and conducted, and the deficiencies gradually are being remedied.

4.4 EYJAFJALLAJÖKULL SHIFTS PERCEPTION OF RISKS AND GALVANIZES EFFORTS TO QUANTIFY HAZARDS

The three decades from 1980 to 2010 saw major shifts in the perception of risk to aviation from volcanic-ash hazards. From the advent of regular intercontinental jet travel in the 1960s until 1980, the risk to aviation was essentially unrecognized, although a few low-severity encounters did occur. Then in the 1980s, dramatic incidents involving in-flight engine shutdowns highlighted the life-threatening potential of ash encounters and led to widespread endorsement by regulators, airlines, and scientists of the strategy of strict ash avoidance. In 1991, an international symposium to improve the scientific knowledge base that underpins ash avoidance (Casadevall, 1994) was preceded by only one month by the eruption of Mt. Pinatubo in the Philippines, a huge eruption that caused 17 damaging ash/aircraft encounters including two in-flight engine shutdowns (Casadevall et al., 1996). The policy of strict ash avoidance to ensure the safety of flight was reinforced by the Pinatubo eruption and continued to be widely accepted until the 2010 eruption of Eyjafjallajökull in Iceland. That event disrupted air transport on such an unprecedented scale that the need to avoid even the very dilute parts of clouds was questioned by European aviation authorities, regulators, and the aviation industry. Concern about ensuring the *efficiency* of flight was added to concern about ensuring *safety* of flight.

The eruption that began from Eyjafjallajökull's summit on April 14, 2010 was sent by prevailing winds southeasterly toward Europe (Gudmundsson et al., 2012). Ash advisories derived solely from dispersion-model output were issued by the London VAAC, depicting the presence of ash over large parts of Europe and the North Atlantic. Based on those advisories, over 300 airports in about two dozen countries and a correspondingly large airspace were closed during April 15–21, 2010, resulting in massive impacts to air travel worldwide. Over 100,000 flights were canceled over that week, affecting seven million passengers and resulting in USD1.7 billion in lost revenue to airlines (Oxford Economics, 2010). To reopen airspace and get the global air transportation system moving again, European aviation authorities and associated meteorological offices created a new type of advisory product that delineated zones, again based on dispersion model output, where low ash concentrations ($<2-4\ mg/m^3$) were forecast to occur in Eyjafjallajökull's ash clouds (Chivers, 2010). Input received from aircraft-engine manufacturers indicated that such zones could be transited with expectation of no or minimal risk of immediate aircraft damage, under the condition of enhanced risk assessment by airlines including more frequent aircraft inspections. Concentration charts, as the new product came to be called, were a substantial change from existing VAAC graphics (Figure 4.5) that implied an "ash-no ash" cloud boundary.

In view of the huge impact that Eyjafjallajökull's eruption had on air travel, ICAO formed the International Volcanic Ash Task Force (IVATF) in May 2010, charging it to examine how best to define hazardous airspace and manage aviation risk. The IVATF included representatives from government and industry groups involved in aviation regulation, operations, and scientific investigations. An important element of their work was to incorporate scientific advice on best methods to detect ash in the atmosphere and forecast zones of hazardous airspace. The IVATF finished its work in June 2012, and a record of its results is available at http://www.icao.int/safety/meteorology/ivatf/Pages/default.aspx.

A major science-based outcome of the IVATF was the decision that concentration charts forecasting areas with specific ash-concentration values would not be required of the world's VAACs. The basis for that decision was the realization that such concentrations, which are based primarily on dispersion models, can have at least an order of magnitude in uncertainty (Guffanti et al., 2012) primarily because of uncertainty in estimates of mass eruption rate (Mastin et al., 2009). Accordingly, the concentration charts did not delineate hazardous airspace with the level of confidence needed for operational decisions. Moreover, the tolerance threshold of jet engines to ash ingestion is not well documented in the scientific literature, so the specific values chosen for concentration charts were open to question. Subsequent work by Webster et al. (2012) compared modeled ash concentrations against observed concentrations in the 2010 Eyjafjallajökull cloud and further documented the scientific challenges involved in forecasting concentration in an operational mode.

However, the situation did not revert to the status quo ante April 2010. The impacts of the Eyjafjallajökull eruption exposed the need for a coordinated response across numerous national boundaries, and a more unified response capability is now being developed through multinational contingency plans and practical exercises involving regulators, airlines, VAACs, and Volcano Observatories. VAACs are working together to define their common "best practices" for evaluating the best available sources of meteorological and volcanological information to derive evidence-based, globally consistent ash-cloud analyses and forecasts. On the scientific front a notable increase in volcanic-cloud research has occurred. Many scientists in various government agencies and academic institutions around the world have adjusted or accelerated their research programs to focus on topics of volcanic-plume characterization and ash-cloud detection and forecasting in ways that innovatively integrate volcanology, meteorology, and atmospheric chemistry and physics. A burst of scientific articles has been published, including in special journal issues (Hasager et al., 2014; Langmann et al., 2012; J. Geophys. Res., 2012).

4.5 CONCLUSIONS

Over the past three decades, volcanic ash has become widely recognized as posing significant risks to aviation. Risks were first recognized in the 1980s after severely damaging encounters of commercial jet aircraft with ash clouds led to in-flight engine failures that imperiled passenger safety. Through ICAO, a global, multifaceted, mitigation framework based on avoiding ash clouds and preventing encounters was developed in the 1980s and 1990s. The costly economic disruption caused by the 2010 eruption of Eyjafjallajökull forced a further realization that aviation risks include risk to efficiency of flight as well as to safety of flight and that methods to safely minimize flight disruptions were needed. The Eyjafjallajökull eruption highlighted the need for more quantitative characterization of ash clouds and more data on aircraft tolerance to ash ingestion so that more informed risk decisions can be made by aviation regulators and the aviation industry.

In the aftermath of the Eyjafjallajökull eruption, ICAO's mitigation framework of eruption reporting, ash cloud detecting and forecasting, and specialized messaging remains in place, with added impetus among all stakeholders to improve how hazardous airspace is defined and communicated. Volcanic ash will continue to pose hazards to aviation, and scientific research in volcanology, meteorology, atmospheric chemistry and physics, and aeronautics will continue to influence the risk-mitigation framework.

REFERENCES

Alparone, S., Andronico, D., Sgroi, T., Ferrari, F., Lodato, L., Reitano, D., 2007. Alert system to mitigate tephra fallout hazards at Mt. Etna Volcano, Italy. Nat. Hazards 43, 333–350.

Bonadonna, C., Folch, A., Loughlin, S., Puempel, H., 2012. Future developments in modelling and monitoring of volcanic ash clouds: outcomes from the first IAVCEI-WMO workshop on ash dispersal and civil aviation. Bull. Volcanol. 74, 1–10. http://dx.doi.org/10.1007/s00445-011-0508-6.

Carey, S., Bursik, M., 2000. Volcanic plumes. In: Sigurdsson, H. (Ed.), Encyclopedia of Volcanoes. Academic Press, San Diego, pp. 527–544.

Carn, S.A., Krueger, A.J., Krotkov, N.A., Yang, K., Evans, K., 2009. Tracking volcanic sulfur dioxide clouds for aviation hazard mitigation. Nat. Hazards 51, 325–343. http://dx.doi.org/10.1007/s11069-008-9228-4.

Casadevall, T.J. (Ed.), 1994. Proceedings of the First International Symposium on Volcanic Ash and Aviation Safety: 1991 July. US Geological Survey Bulletin 2047, Seattle, USA.

Casadevall, T.J., Delos Reyes, P.J., Schneider, D.J., 1996. The 1991 Pinatubo eruptions and their effects on aircraft operations. In: Newhall, C.G., Punongbayan, R.S. (Eds.), Fire and Mud—Eruptions and Lahars of Mount Pinatubo. University of Washington Press, and Quezon City, Philippine Institute of Volcanology and Seismology, Philippines. Seattle, pp. 1071–1085.

Chivers, H., June, 2010. VAAC and predicting the movement of volcanic ash. Meteorological Technol. Int., 62–65.

D'Amours, R., Malo, A., Servranckx, R., Bensimon, D., Trudel, S., Gauthier-Bilodeau, J.-P., 2010. Application of the atmospheric Lagrangian particle dispersion model MLDP0 to the 2008 eruptions of Okmok and Kasatochi volcanoes. J. Geophys. Res. 115, D00L11. http://dx.doi.org/10.1029/2009JD013602.

Draxler, R.R., Hess, G.D., 1998. An overview of the HYSPLIT_4 modeling system of trajectories, dispersion, and deposition. Aust. Meteor. Mag 47, 295–308.

Dunn, M.G., 2012. Operation of gas turbine engines in an environment contaminated with volcanic ash. J. Turbomach. 134, 051001/1-18.

Dunn, M.G., Wade, D.P., 1994. Influence of volcanic ash clouds on gas turbine engines. In: Casadevall, T.J. (Ed.), US Geological Survey Bulletin 2047, 107–118.

Ewert, J.W., Newhall, C.G., 2004. Status and challenges of volcano monitoring worldwide. In: Proceedings of the Second International Conference on Volcanic Ash and Aviation Safety, 2004 June 21–24, Alexandria, USA. Washington DC. Office of the Federal Coordinator for Meteorological Services and Supporting Research, pp. 9–14 session 2.

Ewert, J.W., Lara, L.E., Moreno, H., 2008. The 2008 eruption of Chaitén volcano, Chile and national volcano-monitoring programs in the U.S. and Chile. Am. Geophys. Union Fall Meet. Abstr., V43D–V2177.

Fearnley, C.J., McGuire, W.J., Davies, G., Twigg, J., 2012. Standardisation of the USGS volcano alert level system (VALS): analysis and ramifications. Bull. Volcanol. 74, 2023–2036.

Gardner, C.A., Guffanti, M.C., 2006. US geological Survey's alert notification system for volcanic activity. US Geol. Surv. Fact Sheet 2006–3139, 4. http://pubs.usgs.gov/fs/2006/3139/.

Gudmundsson, M.T., Thordarson, T., Höskuldsson, A., Larsen, G., Björnsson, H., Prata, F.J., Oddsson, B., Magnússon, E., Högnadóttir, T., Petersen, G.N., Hayward, C.L., Stevenson, J.A., Jónsdóttir, I., 2012. Ash generation and distribution from the April–May 2010, pp. 1–12. Eruption of Eyjafjallajökull, Iceland. Nature. http://dx.doi.org/10.1038/srep00572. Scientific Reports 2: number 572.

Guffanti, M., Miller, T.P., 2013. A volcanic alert-level system for aviation: review of its development and application in Alaska. Nat. Hazards 69, 1519–1533. http://dx.doi.org/10.1007/s11069-013-0761-4. http://www.springerlink.com/openurl.asp?genre=article&id=doi:10.1007/s11069-013-0761-4/.

Guffanti, M., Casadevall, T.J., Budding, K., 2010. Encounters of aircraft with volcanic ash clouds—a compilation of known incidents, 1953–2009. US Geol. Surv. Data Ser. 545. http://pubs.usgs.gov/ds/545/.

Guffanti, M., Mayberry, G.C., Casadevall, T.J., Wunderman, R., 2009. Volcanic hazards to airports. Nat. Hazards 51, 287–302. http://www.springerlink.com/content/v3236p5733245713/fulltext.pdf/.

Guffanti, M., Mastin, L., Schneider, D.J., Tupper, A., 2012. Science in support of aviation-risk management since the April 2010 eruption of Eyjafjallajökull, Iceland. Am. Geophys. Union Fall Meet. Abstr., V41E–V2325.

Hasager, C.B., Birmili, G., Pappalardo, P., Prata, F. (Eds.), 2014. Atmospheric Implications of the Volcanic Eruptions of Eyjafjallajökull, Iceland. Atmospheric Chemistry and Physics. http://www.atmos-chem-phys.net/special_issue212.html.

International Civil Aviation Organization, 2004. Handbook of the International Airways Volcano Watch. Document 9766-AN/986, second ed. Available from: http://www.icao.int/safety/meteorology/iavwopsg/Pages/default.aspx/.

International Civil Aviation Organization, 2007. Manual on Volcanic Ash, Radioactive Material and Toxic Chemical Clouds. Document 9691-AN/954. The International Civil Aviation. Available from: http://www.paris.icao.int/news/pdf/9691.pdf.

Johnson, B., Turnbull, K., Brown, P., Burgess, R., Dorsey, J., Baran, A.J., Webster, H., Haywood, J., Cotton, R., Ulanowski, Z., Hesse, E., Woolley, A., Rosenberg, P., 2012. In situ observations of volcanic ash clouds from the FAAM aircraft during the eruption of Eyjafjallajökull in 2010. J. Geophys. Res. 117, D00U24. http://dx.doi.org/10.1029/2011JD016760.

Jones, A.R., Thomson, D.J., Hort, M., Devenish, B., 2007. The U.K. Met Office's next-generation atmospheric dispersion model, NAME III. In: Borrego, C., Norman, A.-L. (Eds.), Air Pollution Modeling and its Application XVII. Proceedings of the 27th NATO/CCMS International Technical Meeting on Air Pollution Modelling and its Application. Springer, pp. 580–589.

Langmann, B., AladosArboledas, L., Folch, A., Matthias, V. (Eds.), 2012. Volcanic Ash over Europe During the Eruption of Eyjafjallajöekull on Iceland, April–May 2010, 48, pp. 1–240. Special issue Atmos Environ. http://www.sciencedirect.com/science/journal/13522310/.

Marenco, F., Johnson, B., Turnbull, K., Newman, S., Haywood, J., Webster, H., Ricketts, H., 2011. Airborne lidar observations of the 2010 Eyjafjallajökull volcanic ash plume. J. Geophys. Res. 116, D00U05. http://dx.doi.org/10.1029/2011JD016396.

Mastin, L.G., Guffanti, M., Servranckx, R., Webley, P., Barsotti, S., Dean, K., Durant, A., Ewert, J.W., Neri, A., Rose, W.I., Schneider, D., Siebert, L., Stunder, B., Swanson, G., Tupper, A., Volentik, A., Waythomas, C.F., 2009. A multidisciplinary effort to assign realistic source parameters to models of volcanic ash-cloud transport and dispersion during eruptions. J. Volcanol. Geotherm. Res. 186 (1–2), 10–21. http://dx.doi.org/10.1016/j.jvolgeores.2009.01.008.

Miller, T.P., Casadevall, T.J., 2000. Volcanic ash hazards to aviation. In: Sigurdsson, H. (Ed.), Encyclopedia of Volcanoes. Academic Press, San Diego, pp. 915–930.

Monitoring volcanic ash from space. In: Zehner, C. (Ed.), Proceedings of the ESA-EUMETSAT Workshop on the 14 April to 23 May 2010 Eruption at the Eyjafjöll Volcano, South Iceland, 26–27 May 2010, Frascati, Italy. Zehner. ESA Publication STM-280. Eur. Space Agency, Paris. http://dx.doi.org/10.5270/atmch-10-01. http://earth.eo.esa.int/workshops/Volcano/files/STM-280_ash101004_V2.pdf.

Oxford Economics, 2010. The Economic Impacts of Air Travel Restrictions Due to Volcanic Ash. A Report Prepared for Airbus. Oxford Economics. www.oxfordeconomic.com.

Pappalardo, G., Amodeo, A., Pandolfi, M., Wandinger, U., Ansmann, A., Bösenberg, J., Matthias, V., Amiridis, V., De Tomasi, F., Frioud, M., Iarlori, M., Komguem, L., Papayannis, A., Rocadenbosch, F., Wang, X., 2004. Aerosol lidar intercomparison in the framework of the EARLINET project:. 3. Raman lidar algorithm for aerosol extinction, Backscatter, and lidar Ratio. Appl. Opt. 43, 5370–5385. http://dx.doi.org/10.1364/AO.43.005370.

Pavolonis, M.J., 2010. Advances in extracting cloud composition information from spaceborne infrared radiances—a robust alternative to brightness temperatures, part I—theory. J. Appl. Meteor. Climatol. 49, 1992–2012. http://dx.doi.org/10.1175/2010JAMC2433.1.

Prata, A.J., 1989. Infrared radiative transfer calculations for volcanic ash clouds. Geophys. Res. Lett. 16, 1293–1296.

Prata, A.J., 2009. Satellite detection of hazardous volcanic clouds and the risk to global air traffic. Nat. Hazards 51, 303–324. http://dx.doi.org/10.1007/s11069-008-9273-z.

Przedpelski, Z.J., Casadevall, T.J., 1994. Impact of volcanic ash from 15 December 1989 Redoubt Volcano eruption on GE CF6-80C2 turbofan engines. In: Casadevall, T.J. (Ed.), Proceedings of the First International Symposium on Volcanic Ash and Aviation Safety: 1991 July. US Geological Survey Bulletin 2047, Seattle, USA, pp. 129–135.

Rose, W.I., Delene, D.J., Schneider, D.J., Bluth, G.J.S., Krueger, A.J., Sprod, I., McKee, C., Davies, H.L., Ernst, G.G.J., 1995. Ice in the 1994 Rabaul eruption cloud: implications for volcano hazard and atmospheric effects. Nature 375, 477–479.

Schumann, U., Weinzierl, B., Reitebuch, O., Schlager, H., Minikin, A., Forster, C., Baumann, R., Sailer, T., Graf, K., Mannstein, H., Voigt, C., Rahm, S., Simmet, R., Scheibe, M., Lichtenstern, M., Stock, P., Rüba, H., Schäuble, D., Tafferner, A., Rautenhaus, M., Gerz, T., Ziereis, H., Krautstrunk, M., Mallaun, C., Gayet, J.-F., Lieke, K., Kandle, K., Ebert, M., Weinbruch, S., Stohl, A., Gasteiger, J., Groß, S., Freudenthaler, V., Wiegner, M., Ansmann, A., Tesche, M., Olafsson, H., Sturm, K., 2011. Airborne observations of the Eyjafjalla volcano ash cloud over Europe during air space closure in April and May 2010. Atmos. Chem. Phys. 11, 2245–2279. http://dx.doi.org/10.5194/acp-11-2245-2011. www.atmos-chem-phys.net/11/2245/2011/.

Scollo, S., Prestifilippo, M., Spata, G., D'Agostino, M., Coltelli, M., 2009. Monitoring and forecasting Etna volcanic plumes. Nat. Hazards Earth Syst. Sci. 9, 1573–1585.

Self, S., Walker, G.P.L., 1994. Ash clouds—Characteristics of eruption columns. In: Casadevall, T.J. (Ed.), Proceedings of the First International Symposium on Volcanic Ash and Aviation Safety: 1991 July. USA. US Geological Survey Bulletin 2047, Seattle, pp. 65–74.

Siebert, L., Simkin, T., Kimberly, P., 2010. Volcanoes of the World, third ed. University of California Press, Berkeley.

Sigmundsson, F., Vogfjord, K., Gudmundsson, M.T., Kristinsson, I., Loughlin, S., Ilyinskaya, E., Hooper, A., Kylling, A., Witham, C., Bean, C., Braiden, A., Ripepe, M., Prata, F., Jordan, C.J., 2013. FUTUREVOLC—a European volcanological supersite in Iceland, a monitoring network for the future. Geophys. Res. Abstr. Eur. Geophys. Union 15, 9495.

Shinozaki, M., Roberts, K.A., van de Goor, B., Clyne, T.W., 2013. Deposition of ingested volcanic ash on surfaces in the turbine of a small jet engine. Adv. Eng. Mater. 15, 986–994.

Sparks, R.S.J., Bursik, M., Carey, S.N., Gilbert, J.A., Glaze, L.S., Sigurdsson, H., Woods, A.W., 1997. Volcanic Plumes. John Wiley and Sons, Chichester.

Special collection on the eyjafjallajökull volcanic eruption in 2010. J. Geophys. Res. 117 (B9), 2012, D20. http://onlinelibrary.wiley.com/doi/10.1002/jgrd.v117.D20/issuetoc.

Stohl, A., Prata, A.J., Eckhardt, S., Clarisse, L., Durant, A., Henne, S., Kristiansen, N.I., Minikin, A., Schumann, U., Seibert, P., Stebel, K., Thomas, H.E., Thorsteinsson, T., Tørseth, K., Weinzierl, B., 2011. Determination of time- and height-resolved volcanic ash emissions and their use for quantitative ash dispersion modeling: the 2010 Eyjafjallajökull eruption. Atmos. Chem. Phys. 11, 4333–4351. http://dx.doi.org/10.5194/acp-11-4333-2011. www.atmos-chem-phys.net/11/4333/2011/.

Tuck, B.H., Huskey, L., 1992. The Economic Consequences of the 1989–90 Mt. Redoubt Eruption. University of Alaska Anchorage. Report prepared for the Alaska Volcano Observatory. http://www.iser.uaa.alaska.edu/Publications/EconConseq1989MtRedoubtEruption.pdf/.

Tupper, A.C., Kinoshita, K., 2003. Satellite, air and ground observations of volcanic clouds over islands of the Southwest Pacific. S. Pac. Study 23, 21–46.

Tupper, A.C., Itikarai, I., Richards, M., Prata, A.J., Carn, S., 2007. Facing the challenges of the international airways volcano watch: the 2004/05 eruptions of manam, Papua New Guinea. Weather Forecast. 22, 175–191. http://journals.ametsoc.org/doi/pdf/10.1175/WAF974.1.

Tupper, A., Carn, S., Davey, J., Kamada, Y., Potts, R., Prata, F., Tokuno, M., 2004. An evaluation of volcanic cloud detection techniques during recent significant eruptions in the western ring of fire. Remote Sens. Environ. 9, 27–46. http://dx.doi.org/10.1016/j.rse.2004.02.004.

Tupper, A., Oswalt, J.S., Rosenfeld, D., 2005. Satellite and radar analysis of the volcanic-cumulonimbi at Mt Pinatubo, Philippines, 1991. J. Geophys. Res. Atmos. 110, D09204. http://dx.doi.org/10.1029/2004JD005499.

Tupper, A., Textor, C., Herzog, M., Graf, H., Richards, M.S., 2009. Tall clouds from small eruptions: the sensitivity of eruption height and fine ash content to tropospheric instability. Nat. Hazards 51, 375–401.

Vernier, J.P., Fairlie, T.D., Murray, J.J., Tupper, A., Trepte, C., Winker, D., Pelon, J., Garnier, A., Jumelet, J., Pavolonis, M., Omar, A.H., Powell, K.A., 2013. An advanced system to monitor the 3D structure of diffuse volcanic ash clouds. J. Appl. Meteor. Climatol. 52, 2125–2138. http://dx.doi.org/10.1175/JAMC-D-12-0279.1.

Webster, H.N., et al., 2012. Operational prediction of ash concentrations in the distal volcanic cloud from the 2010 Eyjafjallajökull eruption. J. Geophys. Res. 117, D00U08. http://dx.doi.org/10.1029/2011JD016790.

Winker, D.M., Pelon, J., Coakley, J.A., Ackerman, S.A., Charlson, R.J., Colarco, P.R., Flamant, P., Fu, Q., Hoff, R.M., Kittaka, C., Kubar, T.L., LeTreut, H., McCormick, M.P., Megie, G., Poole, L., Powell, K., Trepte, C., Vaughan, M.A., Wielicki, B.A., September 2010. The CALIPSO mission: a global 3D view of aerosols and clouds. Bull. Am. Meteorological Soc. http://dx.doi.org/10.1175/2010BAMS3009.1.

Winker, D.M., Liu, Z., Omar, A., Tackett, J., Fairlie, D., 2012. CALIOP observations of the transport of ash from the Eyjafjallajökull volcano in April 2010. J. Geophys. Res. 117, D00U15. http://dx.doi.org/10.1029/2011JD016499.

Witham, C.S., Hort, M.C., Potts, R., Servranckx, R., Husson, P., Bonnardot, F., 2007. Comparison of VAAC atmospheric dispersion models using the 1 November 2004 Grimsvötn eruption. Meteorol. Appl. 14, 27–38. http://dx.doi.org/10.1002/met.3.

Chapter 5

Pyroclastic Density Current Hazards and Risk

Augusto Neri[1], Tomaso Esposti Ongaro[1], Barry Voight[2] and Christina Widiwijayanti[3]

[1] *Istituto Nazionale di Geofisica e Vulcanologia e Sezione di Pisa, via U. Della Faggiola 32, Pisa, Italy,* [2] *Penn State University, Geosciences Dept., University Park PA 16802, U.S.,* [3] *Earth Observatory of Singapore, Nanyang Technological University, Singapore*

ABSTRACT

Pyroclastic density currents (PDCs) represent a formidable challenge for volcanology. This is true in terms of scientific understanding of their dynamics, and also in terms of assessment of their hazard and risk. These phenomena occurred in the famous, lethal 79 AD Pompeii eruption of Mt. Vesuvius, although recognition in "modern" times dates from observations made in 1902 at Montagne Pelée. Major progress in understanding has developed in the last few decades, due both to direct observations and improved modeling capability. Currently, PDC generation and propagation mechanisms are deeply investigated by field reconstructions, and also via detailed experimental and modeling studies. In this chapter we review the current status of knowledge on the dynamics of PDCs, and highlight the key processes and impacts as revealed by recent eruptive events. Some current approaches and modeling studies aimed at PDC hazard assessment are then illustrated, with specific reference to short-to-medium term hazard assessments carried out at the Soufriere Hills volcano at Montserrat (West Indies, UK), and to long-term assessments at the high-risk Vesuvius volcano and Campi Flegrei caldera (Italy). We conclude this review with remarks on the current and evolving state of knowledge of this dangerous volcanic phenomenon, and implications on the challenges for hazards mitigation. Significant limitations exist in our current ability to accurately forecast zones at risk from some PDCs, and in practice, hazard evaluations require cautious good judgment and not over-reliance on computations based on oversimplified algorithms and inputs.

5.1 INTRODUCTION

Pyroclastic density currents (PDCs) or pyroclastic flows and surges as end-member types, are probably the most destructive and complex volcanic phenomena with respect to the dynamics of flow (Sigurdsson et al., 1987;

Volcanic Hazards, Risks, and Disasters. http://dx.doi.org/10.1016/B978-0-12-396453-3.00005-8

Druitt, 1998; Branney and Kokelaar, 2002). They have also been the most difficult to observe directly, due to the extreme hazard, and to the associated ash-cloud that conceals internal processes.

Pliny the Younger was the first to describe them as *flames of fire* that destroyed Pompeii and Herculaneum in the famous eruption of Mount Vesuvius in AD 79 (Sigurdsson et al., 1982; Cioni et al., 1992). These PDCs were generated by collapse of a vertical Plinian eruption column (Neri et al., 2007; Esposti Ongaro et al., 2008a), although true understanding of the lethal phenomena was delayed to modern times. Similarly, the last sub-Plinian eruption of Vesuvius in 1631 caused the death of more than 4000 people, largely due to the impact of similar PDCs (Rosi et al., 1993; Scarth, 2009). The first direct observations and descriptions of PDCs occurred with the eruption of Montagne Pelée (Martinique) in 1902, when the town of St. Pierre was completely destroyed causing about 28,000 fatalities (Lacroix, 1904; Tanguy, 1994). A somewhat similar lethal eruption in 1902 occurred at Soufriere St. Vincent (Anderson et al., 1903), and the international reaction to these eruptions led to the founding of modern volcanology.

Some interest in PDCs and their hazards persisted through the twentieth century, with significant events recorded in Indonesia (Voight et al., 2000a; Self and Rampino, 2012) and New Guinea (Taylor, 1954, 1958), but few eruptions in populated areas. An accelerated interest required the extensive observations of a large explosive eruption at Mount St Helens in 1980. This event led to a new understanding of highly mobile and intensively destructive PDCs produced by a lateral volcanic blast (Figure 5.1), in addition to the documentation of column-collapse PDCs (Hoblitt 1986, 2000; Lipman and Mullineaux, 1981; Newhall, 2000; Belousov et al., 2007).

An even closer observation of dangerous PDCs occurred during the dome-forming but sporadically explosive eruption of the Soufriere Hills volcano, on the small island of Montserrat in the Caribbean Sea. This eruption began in 1995 and has continued through 2010, producing PDCs of variable types and scales (Druitt and Kokelaar, 2002; Wadge et al., 2014a). Several villages,

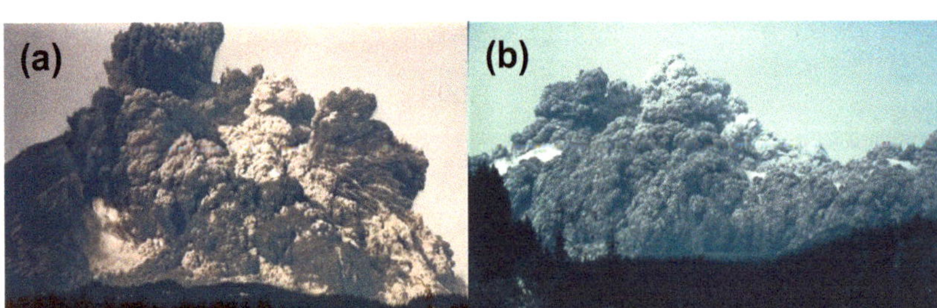

FIGURE 5.1 Initial explosion (a), and associated pyroclastic density current (b), produced by the opening blast of May 18, 1980 eruption of Mt. St. Helens. *Copyright photo courtesy of G. Rosenquist (a) and K. Ronnholm (b).*

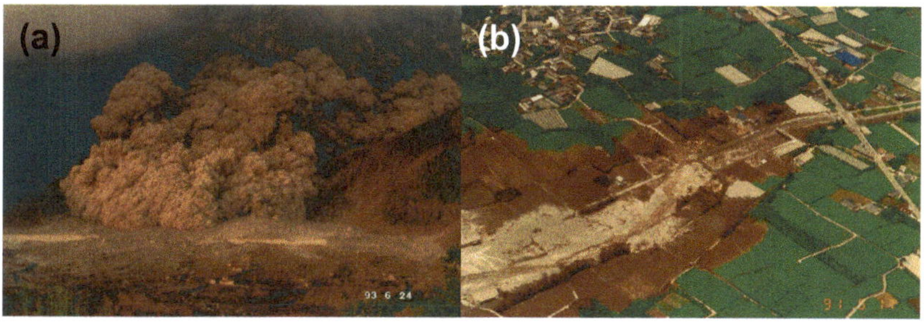

FIGURE 5.2 (a) Large pyroclastic density current (PDC) generated at Unzen volcano (Japan) affecting the Sengongi area on June 24, 1993 (note buildings for scale), and (b) deposit of PDC that occurred at Unzen volcano on June 8, 1991, showing pyroclastic flow deposit plus scorch zone from pyroclastic surge. *Photos courtesy of K. Ohta and Shimabara Earthquake and Volcano Observatory.*

including the island capital town Plymouth, were destroyed by these flows, thus allowing a first modern observation and measurement of the impacts produced by these currents on urbanized areas and people engulfed outdoors (Baxter et al., 2005).

The eruptions of El Chichon, Mexico, in 1982 (Sigurdsson et al., 1987), Unzen Volcano in 1900-1995 (Figure 5.2) (Nakada et al., 1999; Yamamoto et al., 1993) and Mt. Pinatubo in 1991 (Newhall and Punongbayan, 1996; Scott et al., 1996), as well as the recent eruptions of Merapi in the 1990s, 2006 and 2010 (Komorowski et al., 2013; Voight et al., 2000a,b,c; Charbonnier and Gertisser, 2009; Charbonnier et al., 2013), offer other well-documented examples of hazardous PDCs with remarkable impacts on nearby territory and populations.

In the following sections we briefly describe the main features of PDC processes and dynamics, as well as the key variables that determine their diverse hazards. We present also some of the methodologies presently used to assess PDC hazards and impacts. Examples include short–medium term hazards assessments at the Soufriere Hills volcano, and the present effort to quantify long-term pyroclastic flow hazards at the high-risk volcanoes Vesuvius and the Campi Flegrei caldera in Italy. The last section focuses on discussion of some open issues and present challenges in the assessment of PDC hazards.

5.2 PDC GENERATION AND DYNAMICS

PDCs are formed in several different ways (Sparks et al., 1997; Branney and Kokelaar, 2002). Typically, they are produced by: (1) the gravitational collapse of a portion of the volcanic column due to reduction of efficiency of air entrainment into the hot, ash-laden plume; (2) the sudden explosion of a gas-enriched dome or cryptodome (lateral blast); or (3) the gravitational collapse and disintegration of a gas-enriched crystallized lava dome or flow front.

Depending on the scale of the explosive events, these mechanisms can produce PDCs with volumes ranging from a few 1,000 m^3 up to tens or hundreds of cubic kilometers.

In all cases, the PDC consists of a multiphase mixture of hot volcanic gases and pyroclasts of different sizes and properties, such as ash, lapilli, and blocks, that typically moves as a gravity current. The density, velocity, and temperature of the current are governed by the complex interactions between pyroclastic particles and the highly turbulent gaseous flow (Valentine, 1987; Sigurdsson et al., 1987; Druitt, 1998; Freundt and Bursik, 1998; Wohletz, 1998; Branney and Kokelaar, 2002). The pyroclastic particles are generated by the fragmentation of the bubbly magma or hot vesicular dome rock during its decompression and by erosion, and their distribution within the current can change drastically as a function of the eruption mechanisms. During the gravity-influenced flow propagation (Figure 5.3), pyroclastic particles tend to segregate accordingly to their size, shape, and density and a steep density stratification is produced, with solid concentrations ranging from dense packing at the base (i.e., particle volume fraction >0.5) to very dilute (particle volume fraction <0.001) near the top flow boundary. Therefore PDC dynamics are controlled by the competing effects of sedimentation and erosion, and turbulent mixing within the current and with the surrounding atmosphere. Particles are mostly suspended by turbulence in the more diluted part of the current, whereas in the basal layer they are mainly supported by fluid pressure and particle—particle collisions.

This recent view of stratified PDCs includes the two end-members of the traditional field-based classification of pyroclastic deposits (e.g., Cas and

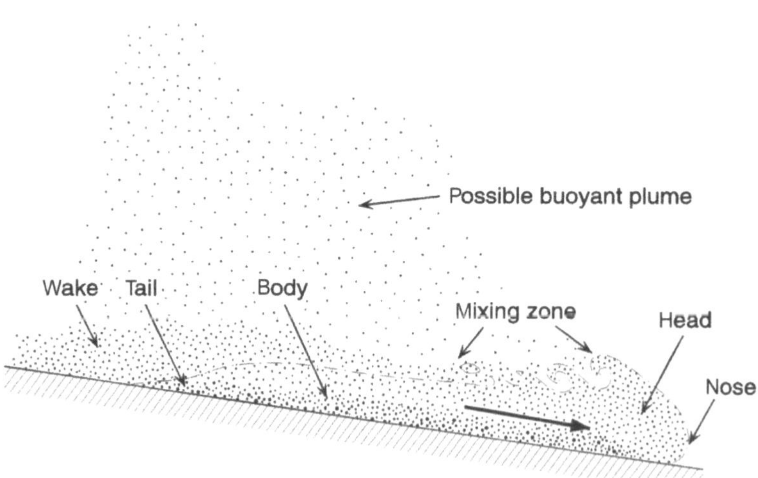

FIGURE 5.3 Sketch of pyroclastic density current illustrating its main components. *Modified from Branney and Kokelaar, 2002.*

Wright, 1987). According to it, *pyroclastic surges* are dilute, thick, and energetic currents that can decouple from a more concentrated, underlying current and are able to traverse topographical obstacles at high speed and high temperatures (Fisher, 1995; Wohletz, 1998). Ash-cloud surges typically comprise about 5−10 percent of the total flow volumes (Calder et al., 1999). *Pyroclastic flows sensu stricto*, are denser, thinner, and concentrated currents that typically move below the dilute surge and are more strongly influenced by topography.

A basic understanding of PDC dynamics is essential for assessing and remediating their hazards. In the past such an understanding has mostly depended upon studies of the field deposit sequences (Cas and Wright, 1987; Sigurdsson et al., 1987; Druitt, 1998; Branney and Kokelaar, 2002; Cassidy et al., 2009). These studies allowed inferences on the main mechanisms of generation, propagation, and mobility of the currents that provided estimates of some variables relevant for hazard assessment, such as height of fall (H) and runout (L), flow volume, deposit area, inundation area, and characteristic grain size distribution (e.g., Hayashi and Self, 1992; Yamamoto et al., 1993; Calder et al., 1999; Widiwijayanti et al., 2008).

In recent decades, PDC dynamics have been investigated through experimental and theoretical studies of a variety of types. Experimental investigations have been carried out at both laboratory scale (e.g., Roche et al., 2005) and in large-scale facilities (e.g., Dellino et al., 2007). These studies allowed investigation, for both dense and dilute regimes, of the rheological properties of flow under controlled conditions, although problems of scaling relative to the natural phenomenon remain largely unsolved.

Field and experimental observations are consistent with results of computational multiphase flow models that describe the PDC as a dense, granular coarse-grained basal layer underlying a fines-rich dilute flow (Valentine and Wohletz, 1989; Burgisser and Bergantz, 2002; Neri et al., 2003; Dartevelle, 2004; Dufek and Bergantz, 2007; Esposti Ongaro et al., 2011a). In particular the simulations show that at high concentration, the friction between particles becomes the controlling mechanism with the establishment of a granular and possibly pseudoplastic regime in the flow. In contrast, in the upper dilute current, particles are in a kinetic-collision regime of random motion, with the flow viscosity arising from the momentum transfer due to particle collisions and the complex interaction between particles and the gas turbulence. Therefore, unlike common homogeneous fluids, the flow viscosity also depends upon specific multiphase flow properties such as the nature and concentration of particles and the flow shear, among other factors.

The assessment and zoning of PDC-related hazards have been traditionally defined on the basis of the recognized extent of pyroclastic flows and surge deposits of historical age, or of prehistoric deposits (Crandell et al., 1984; Scott, 1989). The technique is limited in that deposit preservation is often poor, particularly for surges, and differences in vent location, edifice height,

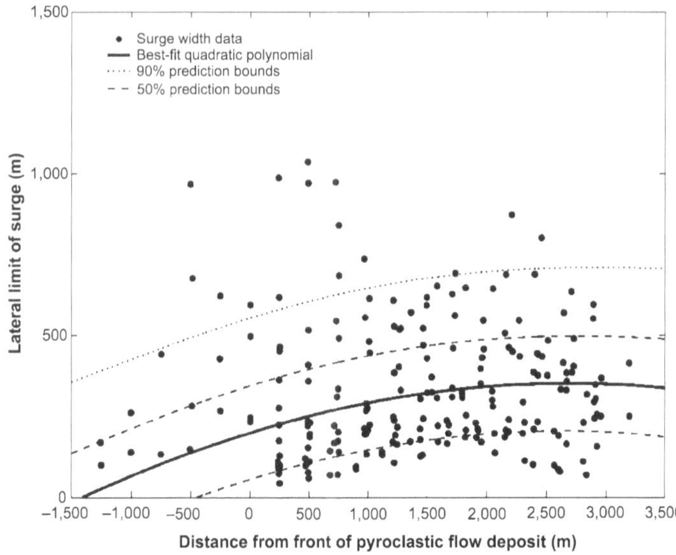

FIGURE 5.4 Lateral extent of ash-cloud surge from main channel as a function of position relative to pyroclastic flow deposit front, at Merapi and Unzen volcanoes. *After Widiwijayanti et al., 2008.*

and channel topography at the time of prior eruptions may be unknown but could greatly affect the potential extent of future PDCs. Thus computer models based on current topography were first recognized as useful to check deposit-based hazard boundaries, and with more sophisticated development became a primary technique for mapping PDC hazards around active volcanoes. Because surges can travel beyond the limits of the basal pyroclastic flow, in principle all pyroclastic flow zones should be encompassed within the envelope of a surge zone (Widiwijayanti et al., 2008). Empirical data of typical pyroclastic surges with small to moderate-sized pyroclastic flows as those shown in Figure 5.4 indicate extension of 1 km or more from the main flow channel.

5.3 HAZARDOUS BEHAVIORS OF PDCS

PDCs represent an extreme danger for the regions exposed to them, due to several distinct hazardous behaviors. Preventive evacuation of territory threatened by this hazard is generally required when the risk is high.

In addition to the obvious threat represented by the capability of voluminous PDCs to bury the areas invaded under thick layers of pyroclastic deposits, the main hazards on land are represented by their dynamic pressure, lethal temperature, fine ash-in-air concentrations, and the capacity to incorporate and transport loose rocks, trees, and construction material that contribute added

FIGURE 5.5 (a) Dynamic pressure damage produced by Boxing Day (26 December) 1997, pyroclastic density current (PDC) at Montserrat (UK). Area affected by intermediate level damage (about 1–10 kPa) with significant variation within a few hundreds meters. (b) House damaged and burnt by high temperature, low momentum PDCs on north flank of Soufriere Hills Volcano, September 1997. Fences were intact but furniture combustion produced complete destruction of the roof. (c) House with collapsed outer and internal walls, by combination of dynamic pressure and a large tree trunk acting as a missile transported by the Boxing Day PDC. (d) Concrete building destroyed by combination of dynamic pressure and impacts from individual large blocks entrained in the February 11, 2010 PDC, Montserrat. *Photos from Baxter et al., 2005; B. Voight.*

impact forces (Figure 5.5). These sources of hazard act in concert, e.g., with building damage exposing inhabitants to hot ash.

Dynamic pressure measures the average kinetic energy per unit volume of the PDC and is expressed as one-half of the product of the mixture flow density times its velocity squared ($P_{dyn} = ½\, \rho v^2$). This parameter is clearly one of the main determinants of the destructiveness observed in areas impacted by PDCs (Valentine, 1998; Clarke and Voight, 2000). Examples include the ruins of Pompeii (Gurioli et al., 2005), the devastation of St. Pierre, Martinique (Lacroix, 1904; Baxter et al., 2005), and the impressive tree blow-down observed at Mt. St. Helens in 1980 (Lipman and Mullineaux, 1981). A good recent example of the effect of dynamic pressure is provided by the housing destruction in Montserrat from PDCs generated by disintegration of the lava dome in December 1997 (Sparks et al., 2002). The PDC destroyed or severely damaged many houses along the central flow-path astride an evacuated community (Figure 5.5(a)). The building damage clearly decreased from the main axis of the flow to the periphery, as the load on the building envelope decreased with the flow dynamic pressure (Esposti Ongaro et al., 2002, 2008b; Baxter et al., 2005). Along the flow axis, the main walls of concrete structures were completely destroyed, whereas at the flow periphery the damages were modest and wood boards fastened over the windows were often able to resist the flow pressure.

An important consideration is that in areas affected by pyroclastic surges considerable damage may occur, but only thin deposits result (see Figure 5.2(b)). A rapid dilute current can generate a dynamic pressure much more powerful than deposit thickness might suggest, and with accompanying lethal temperatures. This is highly relevant since volcanologists often try to estimate hazards of an area by examining deposits (Scott, 1989). A problem is that key deposits representing potentially lethal events can be eroded or only a very thin deposit trace may remain, impeding interpretation, so in many cases it is necessary to be cautious about conclusions on hazards.

High temperatures and ash-in-air concentrations also have devastating and lethal consequences (Baxter, 1990; Baxter et al., 1998). This is particularly true for the dilute portion of PDCs (surge-like flows) which are less constrained by topographic obstacles. These effects have been clearly seen at Unzen volcano, Japan, in 1991 (Yamamoto et al., 1993; Nakada et al., 1999) and Merapi volcano, Indonesia, on frequent occasions over the last century (Voight et al., 2000a,b,c; Charbonnier and Gertisser, 2009; Charbonnier et al., 2013; Lube et al., 2011; Jousset et al., 2013). At the Soufriere Hills volcano on June 25, 1997, a dilute surge detached from the main PDC, invaded a populated region and killed 19 persons (Loughlin et al., 2002). Much of the surge destructiveness was caused by high temperature (estimated at 400 °C), which caused ignition of furnishings inside buildings wherever the current forced access through openings (Figure 5.5(b)). The effects of the high temperature were seen also on victims, with limbs showing the pugilistic pose characteristic of death under extreme heat, exemplified by body casts of victims of the Pompeii eruption (Mastrolorenzo et al., 2001).

The forceful impacts of dense solid material entrained into a PDC, referred to as "missiles," enhance the damaging effects of dynamic pressure. The maximum size of missiles transported by the flow is proportional to the dynamic pressure of the flow, whereas, for compact objects, their kinetic energy (representative of the observed damage) is proportional to the fourth power of the flow dynamic pressure (Wills et al., 1998; Esposti Ongaro et al., 2008b). Figure 5.5(c) and (d) illustrates the damage suffered by buildings due to impact of a tree and coarse blocks during PDC events at Montserrat.

The correlation of hazardous behaviors to damage, initially derived by analogy from nuclear and high-explosive blast data (Valentine, 1998), has been recently revised based on field observation of recent eruptions. Founded on Montserrat field data and analytical studies of materials and structure vulnerability (Spence et al., 2004), a five-level scale was developed to relate dynamic pressure to building damage (Baxter et al., 2005). Accordingly, structural failures on Montserrat were associated with dynamic pressures of about 8–25 kPa. The study revealed that the interiors of most modern buildings would be penetrated and partially destroyed by fire for dynamic pressures of 3–10 kPa, with the intense heat of the PDC (300–400 °C) and high concentrations of ash (>0.1 mg/m^3) causing 100 percent mortality of occupants in few minutes (Baxter et al., 1998). Data from Merapi with less resistant structures show that lesser dynamic pressures, and temperatures (~100–200 °C) also can be lethal (Voight and Davis, 2000; Shelley et al., 1995; Jousset et al., 2013).

The above phenomenological descriptions of PDC activity on buildings should not hide the complexity of the interactions between a pyroclastic current and engulfed structures. Such complexity is illustrated in Figure 5.6 by the application of a three-dimensional multiphase flow model (Esposti Ongaro et al., 2011b) and reflects: (1) the steep density and velocity stratification of the current; (2) its transient nature; and (3) the large variability of particle sizes and properties forming the mixture. Numerical simulations such as Figure 5.6(a) demonstrate nonintuitive dynamic features such as flow separation on the upstream building edge, the reattachment of the current downstream, and the formation of complex and unsteady eddy structures, with associated particle decoupling and deposition. The time-dependent action is computed by integrating the modeled pressure field along the building envelope as a function of time. As in the case of homogeneous flow, the computed action is correlated to the average dynamic pressure of the flow, but the relationship is nonlinear and strongly dependent on flow stratification and, ultimately, on grain density and size distribution (Figure 5.6(b)).

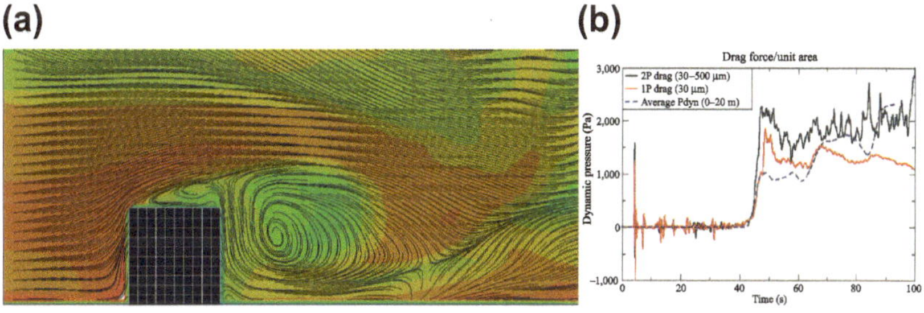

FIGURE 5.6 (a) Lateral view of modeled 3D vortex structures around a building engulfed by a pyroclastic density current (PDC). The streamlines represent gas velocity at 50 s on the longitudinal (xz) plane at half width, whereas color contours indicate the Log_{10} of total particle volume fraction on the same plane. (b) Drag force per unit surface exerted by the PDC on a building as a function of time. The red line indicates a model run using a single particle class, 30 μm, and the black line indicates a run with two particle sizes, 30 and 500 μm. The blue line shows the flow dynamic pressure over time, computed as average value over the first 20 m (height of the building). *Modified from Esposti Ongaro et al., 2011b.*

Finally, additional complex hazards are associated with PDCs generated in the proximity of the sea or a lake. Important examples include Krakatoa, Montagne Pelée, Aniakchak caldera, Alaska (Lacroix, 1904; Latter, 1981; Waythomas and Neal, 1998), and Montserrat (Edmonds and Herd, 2005; Mattioli et al., 2007). Typically there are two distinct flowage components, a dense avalanche component of the PDC that submerges and transforms to a long-runout submarine debris flow, and a surge part that moves as a hot sandy cloud over water, sometimes for great distances. On Montserrat and probably elsewhere the explosive interaction of seawater with the blocky hot flows of the PDC caused an inland-directed base surge (Edmonds and Herd, 2005). The impact of the avalanche portion of a PDC into the water body can generate a tsunami hazard affecting islands and shorelines tens to hundreds of kilometers distant (e.g., Latter, 1981; Carey et al., 1996).

5.4 HAZARD SCENARIOS AND PROBABILISTIC HAZARD ASSESSMENT

In this section we present examples of some approaches and techniques currently used to assess PDC hazards. Examples include the short-to-medium term hazard assessments during recent lava-dome collapse and Vulcanian explosive activity of the Soufriere Hills volcano, long-term assessment of sub-Plinian activity at densely-populated Vesuvius, with consideration of the important Monte (Mt.) Somma topographic influence on PDC distribution, and the assessment of long-term hazard at the high-risk caldera of Campi Flegrei (Italy), with uncertainty in future vent position and eruptive intensity. These scales of activity and techniques are representative of both the current worldwide state-of-practice, and the use of advanced techniques in special circumstances.

5.4.1 Soufriere Hills Volcano, Montserrat

From 1995 to 2010 the Soufriere Hills volcano displayed intense and diverse activity, during which the generation of PDCs represented the major source of risk for the small island. A great variety of eruptive styles were displayed (Druitt and Kokelaar, 2002; Wadge et al., 2014b): lava extrusion, cyclic behavior of conduit flow and dome extrusions, collapses of the dome triggered by lava intrusion or by intense rainfall (Calder et al., 2002; Elsworth et al., 2004), vertical explosions triggered by dome collapse, repetitive Vulcanian explosions with pumice flows generated by collapse of the eruption column (Druitt et al., 2002a; Clarke et al., 2002), formation of surge-generated secondary pyroclastic flows (Druitt et al., 2002b), structural collapse of the flank with accompanying lateral blast (Sparks et al., 2002), and the generation of lahars from remobilization of proximal deposits by intense rainfall.

The 2000–2010 period saw wholesale collapses of the dome in 2003 and 2006 (Herd et al., 2005), the occurrence of offshore-generated hydrovolcanic explosions with on-land and offshore deposits (Edmonds and Herd, 2005), Vulcanian plumes with 6.5 km PDC runout threatening a populated area near the Belham Valley, and generation of tsunamis related to major dome collapse (Mattioli et al., 2007). PDCs generated from dome collapse generally were due to retrogressive break-offs from an advancing lava shear-lobe and moved toward a certain sector (Voight, 2000). The current sometimes bifurcated, but in such cases generally it was possible to know in advance which sector and main valleys would be affected. This is true also for edifice collapse and lateral blasts (Young et al., 2002; Sparks et al., 2002; Belousov et al., 2007). With Vulcanian explosions and collapse of the eruption column, PDCs were generated in several valley drainages simultaneously (Clarke et al., 2002; Druitt et al., 2002a; Cole et al., 2014; Ogburn et al., 2014).

The space shuttle image in Figure 5.7(a) illustrates Montserrat island in 2009. About two thirds of the island has been affected by PDCs. Some flows have entered the sea such as on the east coast below the crater, and further north where the airport was overrun (Figure 5.7(b)), and the associated ash-cloud surges skimmed across the sea surface for distances as great as 2 or 3 km (Edmonds and Herd, 2005). The capital town of Plymouth was destroyed sequentially and the island population was much reduced by emigration, but some populated zones near the Belham River valley remain within potential reach of PDCs (see right hand side of the island coast). Thus hazard zonation has been required on land, and maritime hazard zones have also been established to mitigate the offshore surge hazards.

FIGURE 5.7 (a) Oblique true-color image of Montserrat from the International Space Station on October 11, 2009 (view to southwest). An eruption plume rises above the volcano and is then advected to the west, hiding the destroyed capital town of Plymouth. Lava extruding two days earlier marks the start of eruption Phase 5, with fresh (paler) block-and-ash pyroclastic density current (PDC) deposits from these first few days of activity seen extending to the south via White River to delta off south coast. Area encircled indicates location of currently populated area of Salem and Olveston, adjacent to the Belham River, a potential route for PDCs *(copyright NASA)*. (b) View southwest toward crater, looking over new deposits and shoreline extended by 650 m after major PDC events of February 11, 2010. Compare deposits with Figure 5.7(a). *Photo courtesy of R. Syers and Montserrat Volcano Observatory/Government of Montserrat.*

Hazards assessments and recommendations for hazard zone boundaries have been developed by the Montserrat Volcano Observatory and/or the "Scientific Advisory Committee on Montserrat Volcanic Activity," or SAC. As examples we present some selected outcomes of SAC hazards assessments made in 2007 and 2010, following heightened concerns of large PDCs descending the Belham Valley to the populated region (Wadge et al., 2014b; Voight, 2011a,b). The problem has two main components: the mass (volume) of the dome capable of being disintegrated to generate a collapse, and the subsequent PDC mobility and region impacted.

A crisis developed after January 2007 when rapid growth in the large dome threatened the northwest sector and Belham Valley, and SAC initiated a series of flow simulations to better assess PDC hazard. Dome growth at the volcano has tended to be directed to a given sector (such as northwest to enter the Belham valley) for weeks at a time, and this provided the source of most PDCs. Models showed that about 20 Mm^3 represented the estimated upper limit to the volume of dome mass that could enter the Belham River valley in one collapse event. The PDCs derived from dome collapse included both a dense block-and-ash flow and a less dense and more widely dispersed ash-cloud surge extending beyond the limits of the block-and-ash flow deposit. From a hazards perspective both are effectively fatal to humans and so it is the definition of the surge limit that can be taken as the key output for flow simulations.

Three numerical models were used to simulate dome-collapse pyroclastic flows: *Pyroflow* (Wadge et al., 1998; Wadge, 2009), *PFz* (Widiwijayanti et al., 2008) and *Titan2D* (Patra et al., 2005; Widiwijayanti et al., 2004, 2010). The first two models are based on empirical laws of flow propagation. *Pyroflow* uses a three-parameter correlation to predict a one-dimensional flow path, with parameters calibrated by best-fitting run-out data for PDCs of different volume. Repeated runs with different start points and parameters enable areal coverage, computing with each run the associated ash-cloud surge limits orthogonal to the PDC path (using the Bursik and Woods (1996) model). *PFz* simulates cross-sectional area and areal extent of inundation by block-and-ash pyroclastic flows using the Digital Elevation Model (DEM) and statistical relationships between areas and volume of deposit calibrated from known cases. Surges are considered independently, e.g., using the area of inundation of an independent simulation by assuming a different volume to act as a proxy for the area of enveloping surge, or by applying an envelope about the flow deposit, based on empirical data (e.g., Figure 5.4).

The third model, *Titan2D* (Patra et al., 2005), is a fluid-dynamic model based on two-dimensional, depth-averaged equations, describing a PDC as a gravity-driven dry avalanche controlled by topography and granular friction. Surge limits were estimated with model runs using a reduced basal friction parameter. Basal friction angles of 7−9° were used to simulate most block-and-ash flows, and friction angles of 5−7° were used for surges, giving reasonable results compared to observed flows and surges (Widiwijayanti et al., 2004, 2010).

The simulations effectively demonstrated the extent to which dome-collapse PDCs could reach populated areas of the Belham Valley (SAC8, 2007). The success of this modeling effort was subsequently highlighted by the fidelity to which the 2007 *Titan2D* simulation replicated the inundated area of the February 11, 2010 PDC event (Figure 5.8).

The fourth model used in the hazard assessment work was the 3D multi-phase flow model *PDAC* (Neri et al., 2003; Esposti Ongaro et al., 2007), which was used to simulate a lateral blast northward from the dome. Lateral blasts were recognized as a potential hazard for Montserrat, because one had occurred during the 1997 Boxing Day event (Sparks et al., 2002) when collapse of an older part of the edifice triggered explosion of the pressurized dome lava behind it, causing immense damage. This severe blast had been simulated a few years before 2007 by Esposti Ongaro et al. (2008b) with *PDAC*, and comparison of model outcomes with field observations demonstrated the potential use of three-dimensional models to simulate complex explosive events.

During the 2007 "northwest-threat" crisis, the *PDAC* model was applied to estimate the consequences of a volcanic blast in the Belham valley. This was the first time that such a complex model had been applied as a response to any current volcanic crisis (Voight, 2011a). The resulting blast-incursion maps and videos, of which an example is shown in Figure 5.9, indicated that for a 10 Mm^3 blast the impacted area extended well beyond the limits of the "conventional" surges from ordinary pyroclastic flows, as previously discussed. The blast started out toward the north, but then spread to both northeast and northwest sectors, guided by valley topography. Figures 5.9 (a) and (b) shows the distribution of total particle volume fraction and maximum dynamic pressure, and clearly illustrate the impact potential for a blast covering a significant part of the populated area.

A scenario similar to that of the 2007 crisis developed in 2010 when PDCs were again observed to step-wise approach the populated area, and the response was similar, with the same models being used but with new high-resolution DEM topography (Cole et al., 2010; Stinton et al., 2010; SAC15, 2010). In Figure 5.10 we show some of the 2010 simulations, starting with *Pyroflow* simulations of a northwest-directed 20 Mm^3 PDC in the Belham valley (Figure 5.10(a)). The colors represent the number of overlaid inundations. Most of the flows follow the Belham Valley down to the sea, but several jump out of the valley at the Cork Hill bend and flow southwest. The map also reveals the high gradient of hazard near the margins. *Titan2D* and *PFz* simulations (Figure 5.10(b) and (c)) show a similar result, demonstrating a potential threat posed to workers at a new geothermal site.

During the various crises of the Soufriere Hills volcano, the above methods were used to develop state-of-practice modeling results and interpretations,

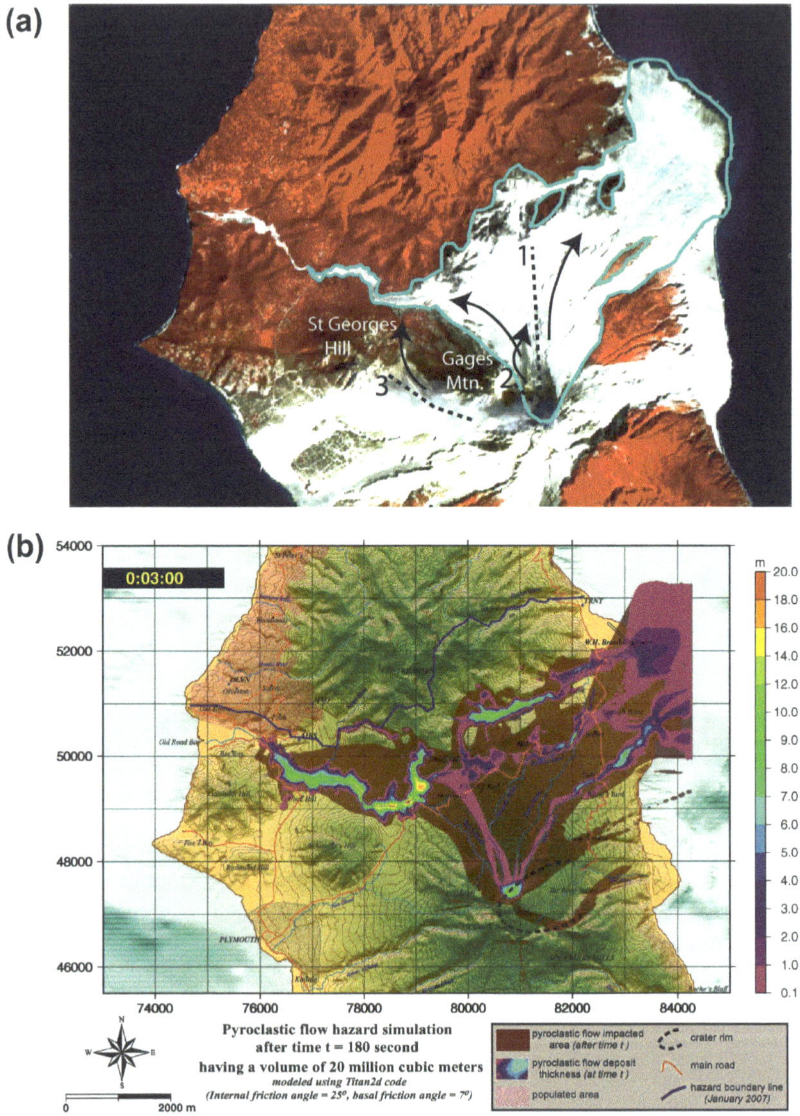

FIGURE 5.8 Comparison of the extent of the pyroclastic flow deposits of February 11, 2010 (a) and a computer simulation (*Titan2D* model with 20 Mm^3 collapsing to the north) of a pyroclastic flow and deposits produced by C. Widiwijayanti, D. Hidayat, B. Voight for the crisis response considered by SAC8 in March 2007 (b). Cf. Figure 5.7(a) and (b). In the top map, the image is from the ASTER satellite taken after the February 11, 2010 event and the pale blue lines show the outline of the deposits from that event *(courtesy G. Wadge)*. The black dashed lines (1–3) are the "watersheds" (topographic divides) that affected the directions (arrowed lines) of PDCs on the Farrell's plain, Gages Mountain and Gages Valley, respectively. They were determined from post-February 11, 2010 topography measured by radar interferometry using TerraSAR-X data.

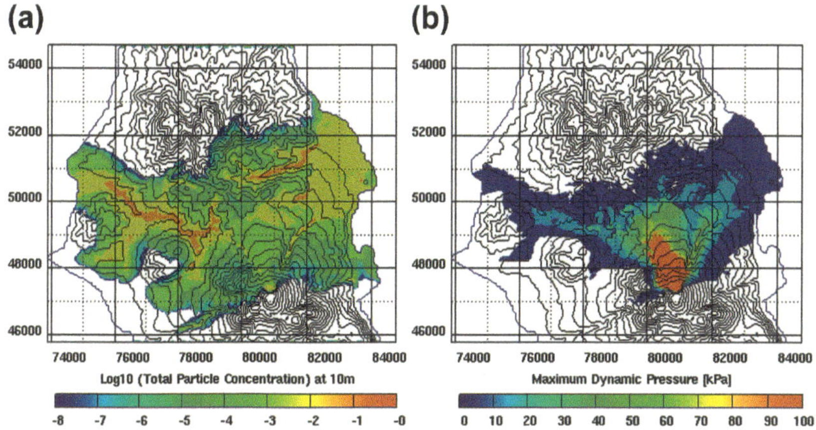

FIGURE 5.9 (a) Log10 of total volume particle fraction, and (b) maximum dynamic pressure of the pyroclastic density current for a 10 Mm^3 volcanic blast northeast and northwest of Soufriere Hills Volcano, as computed by the *PDAC* numerical model. The values reported refer to the estimates computed in the first 10-m cell above the ground and to the first 200 s from the onset of the blast. The simulation suggested that significant parts of the populated area north of the Belham Valley were potentially at risk from a blast.

and these contributed to generally useful responses on the part of the government. The government, not scientists, were charged with making decisions on zoning and responses to warnings (Aspinall et al., 2002; Voight, 2011a,b). The modeling results were presented to government by SAC in the framework of quantitative risk analyses, aimed at estimating the hazard probabilities and the average risk to the public in different areas of the island (Wadge and Aspinall, 2014). The risk levels were mainly expressed as potential loss-of-life estimates and as annualized individual risk exposures—that is, the risk of suffering a given number of casualties in the society as a whole, or the risk of a hypothetical individual losing his or her life during 1 year. Overall, the scientific assessment performed at Montserrat translated into useful and practical zoning decisions by government, although some zoning decisions were arguably imprudent. An example is that in early 2007, the hazard zoning defined by government effectively mitigated the area of highest (1-in-14) dome-collapse occurrence probability, but the region of modeled blast impact (at that time, about a 1-in-110 probability) was effectively ignored. Thus some governmental actions were inconsistently correlated with high hazard probabilities, and/or did not necessarily err on the side of caution. On balance the responses reflect difficulties of risk communication, problems of acceptance and response to risk estimates and uncertainty, inexperience of decision makers with risk-based tools, the partly effective resistance of influential local people potentially affected by zoning, news editorials disrespecting the credibility of scientists, incursions of island politics during volcanic crises, the

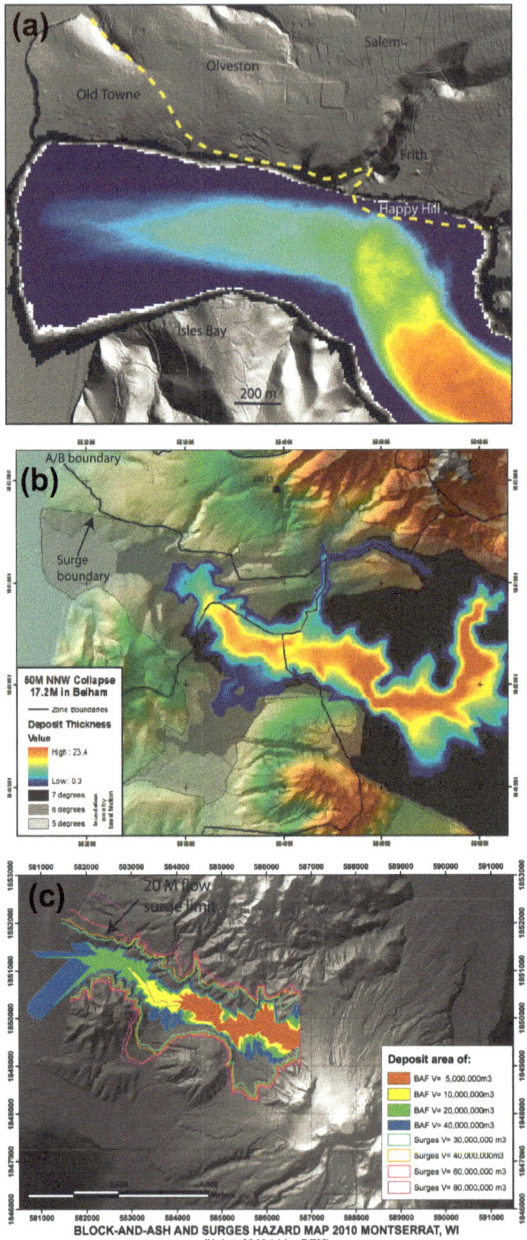

FIGURE 5.10 (a) Ensemble of 300 *Pyroflow* simulations for a pyroclastic flow and surge of about 20 Mm^3 within the lower Belham Valley with the 2010 DEM as a base. The colors within the simulated flow inundation field run from red (high frequency of inundation) to dark blue (low frequency of inundation). A second cycle of colors at the flow field margin from white to black represents the 1–5 percent (black–white) probability of inundation. A narrow 5 percent margin

frequent turnover of government officials, misperceptions regarding so-called "false alarms," and the practicalities of life and governance on a very small island with a dangerous volcano (Voight, 2011a,b).

5.4.2 Vesuvius

Vesuvius is arguably the most dangerous volcano in the world due to its explosive nature and the high urbanization of its flanks. The generation of PDCs has been always associated with Plinian and sub-Plinian eruptions, and this represents the primary source of risk for the over half million inhabitants living within a radius of about 7 km from the Gran Cono, the vent of the last eruption in 1944 (Cioni et al., 2008). PDCs were responsible for the majority of casualties during the 79 AD eruption, and during the last sub-Plinian event in 1631 (Cioni et al., 1992; Rosi et al., 1993).

Hazard assessments have included estimates of PDC dispersal based on stratigraphic analyses, and frequency-of-inundation maps have been produced as an aid to hazards response planning (Gurioli et al., 2010). Likewise, PDC deposit temperatures have been estimated for selected eruptions and PDC flow units (Cioni et al., 2004).

Despite large epistemic (system-based) uncertainties over the input parameters, during the last decades 2D and 3D numerical models have been applied at Vesuvius as a means to quantify PDC hazards in urban areas. Dynamic pressure, temperature, velocity, and density of the flow were computed by the *PDAC* numerical simulation models (Esposti Ongaro et al., 2007) for sub-Plinian eruptions to study PDC generation and dispersion. The various models considered selected 2D axisymmetric lines representative of the southern and northern slopes of Vesuvius as well as the fully 3D topography of the volcano (Esposti Ongaro et al., 2002, 2008a; Todesco et al., 2002; Neri et al., 2007).

Due to the limited resolution of the numerical grid (typically of the order of several meters), these simulations represented adequately the dynamics of the most diluted portion of the flow, but were less accurate for the underlying denser portion. Properties of the magmatic system and flow conditions at the

means that there is a higher gradient of hazard than a broader margin. The dashed yellow line shows the approximate location of the 20 Mm^3 simulation result from 2007 (SAC8, 2007) (b) *Titan2D* inundation (pyroclastic flow and proxy surge) areas (grays) for 17.2 Mm^3 pyroclastic flow within the Belham valley using three different basal friction values: 7° (dark gray), 6° (medium gray) and 5° (light gray). The resultant deposit for the 7° flow is shown in colors. (c) Block-and-ash deposit area map produced by a dome collapse of 5 up to 40 Mm^3 flow down the Belham river valley as computed by the *PFz* model. The color lines represent the extension of the surge component computed for proxy surge volumes. Simulations discussed in Cole et al. (2010) and SAC15 (2010).

vent were selected from expected or predicted values (Todesco et al., 2002). Time-dependent and peak curves of the main hazard variables were thus produced at key locations and as a function of the scenario defined (Esposti Ongaro et al., 2002). Similarly, hazard maps of PDC incursion, as well as the distribution of PDC impacts, were produced by transient 3D simulations as a function of the vent conditions assumed (Esposti Ongaro et al., 2008a).

Importantly, simulations enabled quantification of the effects of Mt. Somma, the open-sided rimmed edifice still present on the northern side of Vesuvius, as well as local topography, on the propagation of PDCs and their associated hazards. Figure 5.11 illustrates the distribution of particle volume fraction and velocity field for a numerical simulation characterized by a quasi-total collapse (i.e., boiling-over) of the column (Neri et al., 2007; Esposti Ongaro et al., 2008a). This simulation, as well as others performed by using different eruptive conditions at the vent, demonstrated the important control of Mt. Somma in effectively blocking the flow and redirecting it in preferential directions. These simulations also highlighted the concept that PDC mobility, together with their density and velocity, appears to be mainly correlated to the amount of mass that collapses to the ground, instead of collapse height as commonly assumed.

Simulation output variables were also used to make deterministic estimates of building damage in test scenarios, including the infiltration of the hot gas and particles inside buildings for estimating survival times of occupants, and for cases where the openings may have resisted the PDC due to sheltering or other reasons (Baxter et al., 2008; Zuccaro et al., 2008). Based on the

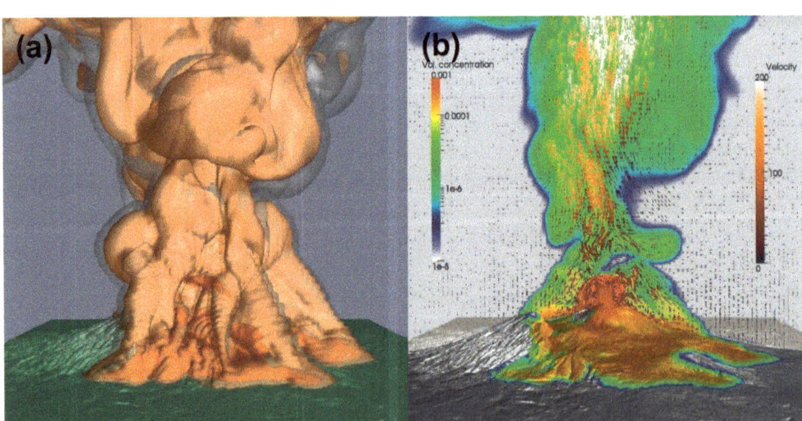

FIGURE 5.11 Simulation output of *PDAC* multiphase flow model for a quasi-total (boiling-over) collapse sub-Plinian scenario at Vesuvius. (a) isosurfaces of the total volume particle fraction corresponding to 10^{-4} (dark brown), 10^{-6} (light red) and 10^{-8} (gray). (b) distribution of the total volume particle fraction over a selected vertical plane passing through the vent (Gran Cono) and over the ground. See Esposti Ongaro et al. (2008a) for simulation input parameters and details.

Montserrat findings, dynamic pressures of few kilopascals would break most glazed windows to allow entry of hot ash, whereas at moderate and high dynamic pressures, large building structural damage would be expected.

The simulations also illustrated the additional hazards presented by the isotropic pressure of the PDC which acts over the building envelope due to its weight, the large-scale atmospheric circulations produced by the dispersal process, and the fire hazard due to high temperature of the flows (Baxter et al., 1998; Esposti Ongaro et al., 2002). For instance, Figure 5.12 illustrates the distribution of the column-collapse PDC temperature, 800 s after the same simulation represented in Figure 5.11. The simulation results indicate the significant effect of topography on the decay of the radial temperature distribution, as well as the capacity of the flow to maintain temperatures of a few hundreds degree centigrade even several kilometers away from the source (Esposti Ongaro et al., 2008a). Model results also appeared to be consistent with field investigations carried out at Pompeii, aimed at determining the effect of the urban structures on PDC dynamics (Gurioli et al., 2005; Baxter et al., 2008).

At Vesuvius, field reconstructions of PDC dispersal, and outcomes of the 3D simulations of PDCs produced by sub-Plinian events, were combined to define first-order probabilistic hazard maps for PDCs produced by a sub-Plinian eruption with a crater location coinciding with the present Gran

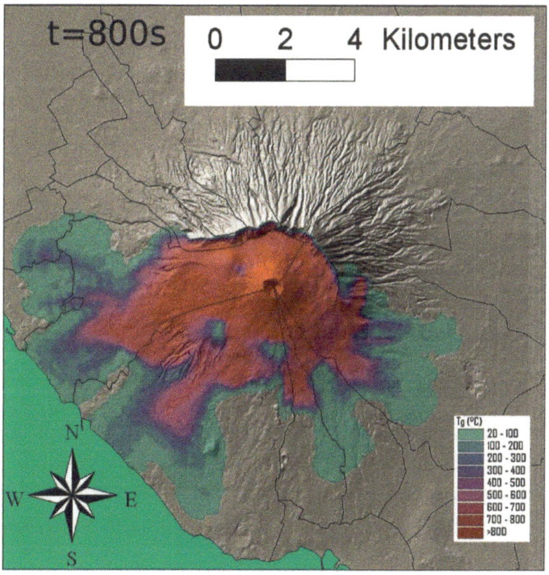

FIGURE 5.12 Distribution of PDC temperature in the first computational cell along the ground for the collapsing column represented in Figure 5.10. The distribution refers to the temperature of the gas phase and to a time interval of 800 s since the beginning of the collapse phase. *Modified from Esposti Ongaro et al., 2008.*

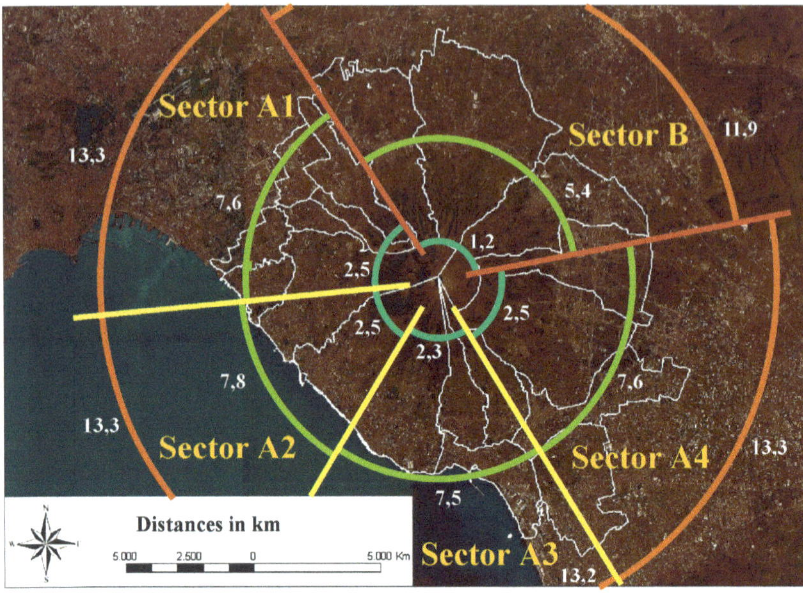

FIGURE 5.13 Probabilistic map of PDC flow invasion at Vesuvius as produced by expert elicitation techniques. Broad segmentation of area around Vesuvius recognizing the first-order effect of Mt. Somma topography in determining areas that might be invaded by PDCs as the result of a sub-Plinian eruption. The bracketed values in each sector show elicited estimates of maximum run-out distances (in km) for PDCs occurring during a sub-Plinian eruption, by sector. Inner arcs (blue) are 95 percentile confidence levels for exceeding distance shown (e.g., 2.5 km for Sector A1), central arcs (green) are expected (50th percentile) values, and outer arcs (orange) are the run-out distances assessed has having only a 5 percentile chance of being exceeded. *Modified from Neri et al., 2008.*

Cono (Figure 5.13, Neri et al., 2008; Neri, 2011). This was done in an attempt to incorporate the key evidence and outcomes from the different methodologies, as well as to account for the different sources of uncertainty. By using structured expert elicitation techniques, the various strands of evidence were summarized by subdividing the volcano into different sectors and by assigning probabilities for each sector affected by the PDCs, and their runout (Figure 5.13) (Aspinall et al., 2002; Baxter et al., 2008; Neri et al., 2008). Resulting maps, although related to a single eruptive scenario and in need of continuous updating with advancement of knowledge, enabled the consideration of uncertainty on influencing the hazard assessment process—a very large influence even for such a well-known volcano as Vesuvius.

5.4.3 Campi Flegrei

Here we discuss the case of high-risk volcanoes for a caldera setting, where long-term hazard assessments are needed for urban planning and for design of appropriate mitigation measures in case of reactivation (Newhall and Hoblitt, 2002; Marzocchi and Woo, 2009). The mapping of PDC hazards in caldera settings is particularly challenging due to the large variability of vent type and location, eruption scale and typology, in addition to the complex dynamics of the currents over the irregular caldera topography and/or beyond the caldera boundary. Nevertheless, the probabilistic mapping of possible PDC incursion is needed, aimed at accounting for the intrinsic uncertainties affecting the system, particularly for highly populated regions (Newhall and Hoblitt, 2002; Neri et al., 2008).

Campi Flegrei is an example of such an active and densely populated caldera with a very high risk associated with PDCs produced by explosive events. Its eruptive record is characterized by the occurrence of large caldera-forming events, such as the Campanian Ignimbrite and the Neapolitan Yellow Tuff, as well as eruptions of Plinian and smaller scale from isolated volcanic vents (Orsi et al., 2004). The last explosive event, named Monte Nuovo, occurred in 1538 and was characterized by the generation of a tuff-cone and impulsive explosive activity of limited scale.

Currently the Campi Flegrei caldera is populated by over 350,000 people. In the past few decades it has been affected by clear inflations and deflations reflecting volcanic processes, with ground level variations as much as several meters. With the aim to properly assess the hazards associated with a reactivation of the volcano, several field, and modeling studies have been carried out with specific focus on PDC hazards assessment. For instance, Orsi et al. (2004) reconstructed the dispersal of PDCs in the last 15,000 years of activity and sketched a first qualitative PDC hazard zonation map. In parallel, by using the *PDAC* numerical model, Todesco et al. (2006) investigated 2D PDC propagation over the eastern sector of the caldera, with specific reference to the Agnano Monte Spina Plinian event (De Vita et al., 1999). Model results highlighted the complexities of flow propagation in the caldera setting, and included evidence that in some cases PDCs could overcome the caldera rim and directly threaten Naples.

By using the distribution of relevant geophysical and geochemical variables in the caldera to quantify the spatial probability of new vent opening, and the energy-line model to determine the areas affected by PDCs, Alberico et al. (2002, 2011) developed qualitative incursion maps for selected eruptions scaled to VEI. Similarly, Rossano et al. (2004) produced first-order quantitative PDC incursion maps by assuming vent opening in the central part of the caldera, and using the simple 1D dynamic model of McEwen and Malin (1989) treating PDCs as homogeneous Bingham fluids.

More recently, Bevilacqua et al. (2012) used vent-opening probability maps, such as that proposed by Selva et al. (2011), and the deposit reconstructions of Orsi et al. (2004) to compute the first-order quantitative PDC incursion maps, considering both spatial vent location and eruptive scale. In this case, the areas affected by PDCs were assumed circular in form, based

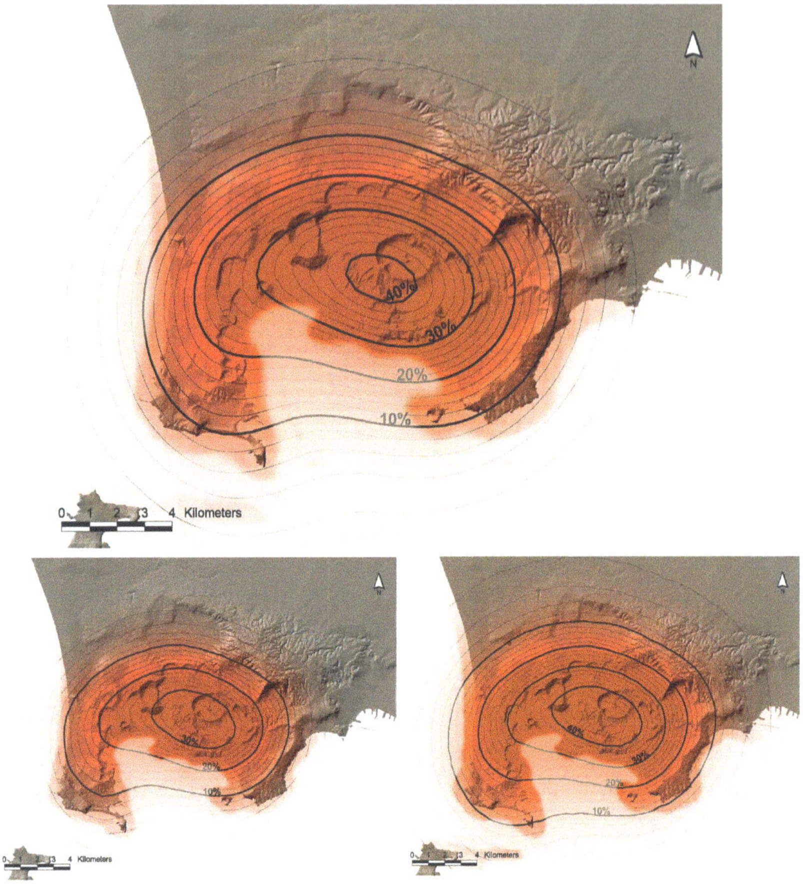

FIGURE 5.14 Average (top) and 5th percentile (bottom left) and 95th percentile (bottom right) maps of probability of PDC inundation at Campi Flegrei. The three maps consider variability of eruptive record of last 15 kyrs and probability of vent opening proposed by Selva et al. (2011). PDC invasion areas were assumed circular and based on reconstructed distribution of deposits. Reddish colors refer to probability values reported on isocontour lines. *Modified from Bevilacqua et al., 2012.*

loosely on the reconstruction of past PDC deposits. One specific objective of this analysis was to incorporate into the PDC incursion map some of the main uncertainties affecting the system, such as uncertainty on the probability of vent opening. Figure 5.14 illustrates, for instance, the average map of PDC invasion with the associated 5th and 95th percentile maps, considering the uncertainty on the vent location from Selva et al. (2011). Such a simplified approach is not able to account for the effects of existing topography, but it provides a first-order estimate of the probability of flow incursion including consideration of the variability of the two main unknowns of the volcanic system.

Further developments of the above approaches are ongoing and include the use of simplified flow models of PDC invasion calibrated by *PDAC* numerical simulations, which are able to account for the main topographic features of the caldera.

5.5 CONCLUDING REMARKS

Understanding of PDC dynamics and assessment of their hazards continue to represent controversial and challenging issues to the present-day volcanology community. This is mostly due to the complex and diverse nature of the phenomenon, as well as to the unpredictability and nonreproducibility of volcanic processes. Nevertheless, important progress has been made in the last few decades in the representation and forecasting of these phenomena, thanks to significant advances in field, experimental and theoretical investigations.

Regarding field observations, new insights have been largely due to the opportunity to closely observe, and sometimes measure, PDC behavior during recent volcanic events. The eruptions of Mt. St. Helens (1980), Unzen (1991), Pinatubo (1991), Merapi (2010 and previous decades), and Soufriere Hills volcano (Montserrat, 1995–2010), particularly, yielded well-documented images and observations of PDCs. The diverse nature of pyroclastic flows and surges, their different and sometimes independent responses to topographic effects, the occurrence of lateral blasts and secondary pyroclastic flows, the interaction and travel of PDCs with and over seawater are just some examples of new processes of PDC dynamics discovered or illuminated by these direct observations. In some cases, observations enabled quantification of key data such as H/L ratios, the areas inundated by PDCs, and their volumes (e.g., Yamamoto et al., 1993; Calder et al., 1999; Widiwijayanti et al., 2008). Such data can be applied directly for simple hazard evaluations, and also to usefully guide parameter selection for flow models. Some dynamic variables of the flow such as its velocity and temperature could be measured directly with techniques such as video/photography, seismology and acoustics, and thermal imagery, whereas flow density and dynamic pressure are still mostly inferred indirectly from impact analyses.

In parallel, theoretical and experimental investigations have made significant progress with the development of a variety of physical and numerical models of PDCs. Based on the underlying assumptions and simplifications, different models have been applied to the simulation of different types of flows and with different aims. Simplified 1D/2D flow models and statistically-based representations of the flows neglect several physical processes as well as 3D and transient effects of their dynamics. Nevertheless, once tuned with available field data sets, they enabled helpful modeling during crises. They also could be adapted to Monte Carlo simulations to investigate model sensitivity to input parameters, such as flow volume and vent location, and to produce probabilistic hazard maps. Alternatively, multidimensional and multiphase flow models are becoming increasingly effective in representing the complex behavior of PDC phenomena, and enable simulations that include remarkable details on specific scenario conditions. Such models can provide estimates of key PDC hazard variables, such as dynamic pressure and temperature, that can be used to quantify the expected impacts of the phenomenon. However, at this time, they cannot yet be readily adapted to Monte Carlo simulations due to the excessive computing time required.

Importantly, whereas in principle all pyroclastic flow hazard zones should be encompassed within the envelope of a surge zone, understanding the very real limitations of our current ability to accurately forecast zones at risk from surge inundation remains a key concern. Available models may supply some "reasonable" first-order estimations of surge dynamics, but can only approximate the effects of ash-cloud surge mass distribution as affected by topography and other factors, and ignore ambient strong wind effects. In some cases existing approaches could seriously underestimate the surge hazards in specific locales. Surges have detached from parent flows at (sometimes nonobvious) constrictions, channel cross-sections of reduced area, and channel bends, generating unexpected "lethal surprises" affecting populated regions (Yamamoto et al., 1993; Fisher, 1995; Abdurachman et al., 2000; Voight et al., 2000a,b; Loughlin et al., 2002; Charbonnier and Gertisser, 2009; Charbonnier et al., 2013; Lube et al., 2011; Ogburn et al., 2014). Thus in practice the problem requires cautious good judgment and not an over-reliance on computations based on oversimplified algorithms and inputs. We advocate use of a surge buffer zone consistent with empirical observations (such as Figure 5.4) and broad enough to override possible deficiencies in modeling. Use of a broad surge buffer zone is also helpful to remediate the limitations of the DEM used in current hazard evaluation, which may not precisely represent the topography during a future hazardous PDC.

Therefore, despite significant progress, assessments of PDC hazard are still influenced by a remarkable amount of uncertainty. The quantification and communication of the diverse uncertainty sources affecting hazard assessments appear to be a principal future challenge. Physical and numerical models of PDCs are indeed still subject to a variety of uncertainties, and necessarily so. The physical formulation of the process dynamics is not univocal. Similarly, initial and boundary conditions of the volcanic phenomenon are subject to large epistemic (systematic) and aleatoric (statistical) uncertainties that are just partially constrained by direct observations. The volumes and precise source locations for potential threatening PDCs are generally poorly determined. The current DEM representation of topography may only crudely approximate the topography at the time of a future hazardous flow. These factors imply that the absolute validation of a model is presently impossible, despite the continuous progress in the experimental validation of models at different spatial and temporal scales.

The sources of uncertainty noted above directly affect assessments and also add considerable complexity to communications conveyed to the authorities responsible for decision making. Such problems clearly emerged, for instance, during Montserrat assessments in 1995–2014 when scientists had to meld results from different PDC models, elicit risks for different sectors, and communicate findings to officials and a government decision committee inexperienced in considering risk uncertainty and influenced by parts of the population resistant to evacuation (SAC8, 2007; SAC15, 2010; Voight, 2011a,b; Wadge and Aspinall, 2014). The risk is that hazard zones may be incautiously defined, with the margin of error dangerously disproportionate to our ability to forecast reliable PDC inundation limits. Similarly, hazards mapping at high-risk volcanoes such as Vesuvius or Campi Flegrei, where decisions on preventive evacuation of hundreds of thousands of people might arise, requires challenging decisions and consideration of many sources of uncertainty.

Current efforts, including examples outlined in this chapter, aimed at estimating PDC hazard and risk on a fully probabilistic basis, taking into account the relevant uncertainties, represent an important step toward the long-term goal of developing an interdisciplinary and integrated approach to risk reduction. More accurate observations and measurements of the volcanic process, development and validation of more accurate physical models, innovative probabilistic approaches to risk assessment incorporating vulnerability and the main sources of uncertainty are current main challenges for improving effective volcanic crisis management. Added to these are the complex challenges of effective communications and education on risk and uncertainty conveyed to authorities responsible for decision making, because, accurate assessments alone are not synonymous with effective mitigation.

ACKNOWLEDGMENTS

AN and TEO thank many colleagues of previous and ongoing national and EU projects related to PDC hazard assessment for their precious contributions. BV thanks colleagues at the Montserrat Volcano Observatory, on the SAC, at the USGS Cascades Volcano Observatory, in INGV, at Merapi and Unzen volcanoes, and in Colombia and Kamchatka for ideas and support of PDC studies. This research was supported by the European Union (Exploris project), Dipartimento della Protezione Civile and Regione Campania, Italy (Speed and V1 projects), and by NSF EAR-1125069-and 0732415. We thank Steve Self for a helpful review.

REFERENCES

Abdurachman, E.K., Bourdier, J.L., Voight, B., 2000. Nuees ardentes of November 22, 1994 at Merapi Volcano, Indonesia. J. Volcanol. Geotherm. Res. 100, 345–361.

Alberico, I., Lirer, L., Petrosino, P., Scandone, R., 2002. A methodology for the evaluation of long-term volcanic risk from pyroclastic flows in Campi Flegrei (Italy). J. Volcanol. Geotherm. Res. 116, 63–78.

Alberico, I., Petrosino, P., Lirer, L., 2011. Volcanic hazard and risk assessment in a multi-source volcanic area: the example of Napoli city (Southern Italy). Nat. Hazards Earth Syst. Sci. 11, 1057–1070.

Anderson, T., Flett, J.S., McDonald, T.M., 1903. Report on the eruptions of the Soufriere, in St. Vincent, in 1902, and on a Visit to Montagne Pelée, in Martinique. Part I. R. Soc. Lond. Philos. Trans. Ser. A 200, 353–553.

Aspinall, W.P., et al., 2002. The Montserrat Volcano Observatory: its evolution, organization, role and activities. In: Druitt, T.H., Kokelaar, B.P. (Eds.), The eruptions of Soufriere Hills Volcano, Montserrat, from 1995 to 1999, 21. Geological Society, London, Memoirs, pp. 71–92.

Baxter, P.J., 1990. Medical effects of volcanic eruptions. 1. Main causes of death and injury. Bull. Volcanol. 52, 532–544.

Baxter, P.J., Neri, A., Todesco, M., 1998. Physical modeling of human survival in pyroclastic flows. Nat. Hazards 17, 163–176.

Baxter, P., Boyd, R., Cole, P., Neri, A., Spence, R., Zuccaro, G., 2005. The impacts of pyroclastic surges on buildings at the eruption of the Soufriere Hills Volcano, Montserrat. Bull. Volcanol. 67, 292–313.

Baxter, P.J., Aspinall, W., Neri, A., Zuccaro, G., Spence, R.S.J., Cioni, R., Woo, G., 2008. Emergency planning and mitigation at Vesuvius: a new evidence-based approach. J. Volcanol. Geotherm. Res. 178 (3), 454–473. http://dx.doi.org/10.1016/j.jvolgeores.2008.08.015 (1).

Belousov, A.B., Voight, B., Belousova, M., 2007. Directed blasts and blast-generated pyroclastic density currents: a comparison of the Bezymianny 1956, Mount St. Helens 1980, and Soufrière Hills, Montserrat 1997 eruptions and deposits. Bull. Volcanol. 69, 701–740. http://dx.doi.org/10.1007/s00445-006-0109-y.

Bevilacqua, A., Fourmentraux, C., Bertagnini, A., Bisson, M., Esposti Ongaro, T., Flandoli, F., Isaia, R., Neri, A., Rosi, M., 2012. Mapping Pyroclastic Flow Hazard in a Caldera Setting: Application to Campi Flegrei Caldera. EGU General Assembly 2012. Vienna, Austria.

Branney, M.J., Kokelaar, B.P., 2002. Pyroclastic density currents and the sedimentation of ignimbrites. Mem. Geol. Soc. 27, 8 pp.

Burgisser, A., Bergantz, G.W., 2002. Reconciling pyroclastic flow and surge: the multiphase physics of pyroclastic density currents. Earth Planet. Sci. Lett. 202, 405–418.

Bursik, M.I., Woods, A.W., 1996. The dynamics and thermodynamics of large ash flows. Bull. Volcanol. 58 (2), 175–193.

Calder, E.S., Luckett, R., Sparks, R.S.J., Voight, B., 2002. Mechanisms of lava dome instability and generation of rockfalls and pyroclastic flows at Soufriere Hills Volcano, Montserrat. In: Druitt, T.H., Kokelaar, B.P. (Eds.), The eruptions of Soufriere Hills Volcano, Montserrat, from 1995 to 1999, 21. Geological Society, London, Memoirs, pp. 173–190.

Calder, E.S., Cole, P.D., Dade, W.B., Druitt, T.H., Hoblitt, R.P., Huppert, H.E., Ritchie, L., Sparks, R.S.J., Young, S.R., 1999. Mobility of pyroclastic flows and surges at the Soufriere Hills volcano, Montserrat. Geophys. Res. Lett. 26 (5), 537–540.

Carey, S., Sigurdsson, H., Mandeville, C., Bronto, S., 1996. Pyroclastic flows and surges over water: an example from the 1883 Krakatau eruption. Bull. Volcanol. 57 (7), 493–511.

Cas, R.A.F., Wright, J.V., 1987. Volcanic Successions, Modern and Ancient: A Geological Approach to Processes, Products, and Successions. Allen & Unwin.

Cassidy, N.J., Calder, E.S., Pavez, A., Wooler, L., 2009. GPR-derived facies architecture: a new perspective on mapping pyroclastic flow deposits. In: Thordarsson, T., Self, S., Larsen, G., Rowlands, S.K., Hoskuldsson, A. (Eds.), Studies in Volcanology: The Legacy of George Walker, 2. Spec. Pub. IAVCEI, pp. 181–210. Geol. Soc. London.

Charbonnier, S., Gertisser, R., 2009. Numerical simulations of block-and-ash flows using the Titan2D flow model: examples from Merapi Volcano, Indonesia. Bull. Volcanol. 71, 953–959. http://dx.doi.org/10.1007/s00445-009-0299-1.

Charbonnier, S.J., et al., 2013. Evaluation of the impact of the 2010 pyroclastic density currents at Merapi volcano from high-resolution satellite imagery, field investigations and numerical simulations. J. Volcanol. Geotherm. Res. 261, 295–315. http://dx.doi.org/10.1016/j.jvolgeores.2012.12.021.

Cioni, R., Marianelli, P., Sbrana, A., 1992. Dynamics of the A.D. 79 eruption: stratigraphic, sedimentological and geochemical data on the successions from the Somma-Vesuvius southern and eastern sectors. Acta Vulcanol. 2, 109–123.

Cioni, R., Gurioli, L., Lanza, R., Zanella, E., 2004. Temperatures of the A.D. 79 pyroclastic density current deposits (Vesuvius, Italy). J. Geophys. Res. 109, B02207. http://dx.doi.org/10.1029/2002JB002251.

Cioni, R., Bertagnini, A., Santacroce, R., Andronico, D., 2008. Explosive activity and eruption scenarios at Somma-Vesuvius (Italy): towards a new classification scheme. J. Volcanol. Geotherm. Res. 178 (3), 331–346.

Clarke, A.B., Voight, B., 2000. Pyroclastic current dynamic pressure from aerodynamics of tree or pole blow-down. J. Volcanol. Geoth. Res. 100 (1), 395–412.

Clarke, A.B., Voight, B., Neri, A., Macedonio, G., 2002. Transient dynamics of Vulcanian explosions and column collapse. Nature 415, 897–901.

Cole, P., et al., 2010. Report on Activity between 28 February 2010 and 31 October 2010. MVO Report to SAC15, OFR 10–02a (report available via MVO website).

Cole, P., et al., 2014. Vulcanian explosions at Soufrière Hills Volcano, Montserrat between 2008 and 2010. In: Wadge, G., Robertson, R.E.A., Voight, B. (Eds.), The Eruption of Soufrière Hills Volcano, Montserrat from 2000 to 2010, 39. Geological Society, London, Memoirs, pp. 93–110. http://dx.doi.org/10.1144/M39.5.

Crandell, D.R., Booth, B., Kusumadinata, K., Shimozuro, D., Walker, G.P.L., Westercamp, D., 1984. Sourcebook for Volcanic-hazards Zonation. UNESCO, 97 pp.

Dartevelle, S., 2004. Numerical modelling of geophysical granular flows: 1. A comprehensive approach to granular rheologies and geophysical multiphase flows. Geochem. Geophys. Geosyst. 5, Q08003. http://dx.doi.org/10.1029/2003GC000636.

Dellino, P., Zimanowski, B., Buttner, R., La Volpe, L., Mele, D., Sulpizio, R., 2007. Large scale experiments on the mechanics of pyroclastic flows: design, engineering, and first results. J. Geophys. Res. 112, B04202. http://dx.doi.org/10.1029/2006JB004313.

De Vita, S., Orsi, G., Civetta, L., Carandente, A., D'Antonio, M., Deino, A., Di Cesare, T., Di Vito, M.A., Fisher, R.V., Isaia, R., Marotta, E., Necco, A., Ort, M., Pappalardo, L., Piochi, M., Southon, J., 1999. The Agnano–Monte Spina eruption (4100 years BP) in the restless Campi Flegrei caldera (Italy). J. Volcanol. Geotherm. Res. 91, 269–301.

Druitt, T.H., 1998. Pyroclastic density currents. In: Gilbert, J.S., Sparks, R.S.J. (Eds.), The Physics of Explosive Volcanic Eruptions, 21. Geological Society Special Publications No. 145. Geological Society, London, pp. 145–182.

Druitt, T.H., Kokelaar, B.P. (Eds.), 2002. The Eruption of Soufrière Hills Volcano, Montserrat: From 1995 to 1999, 21. Geol. Soc. Lon. Mem.

Druitt, T.H., Young, S.R., Baptie, B., Bonnadonna, C., Calder, E.S., Clarke, A.B., 6 others, 2002a. Episodes of cyclic Vulcanian explosive activity with fountain collapse at Soufriere Hills Volcano, Montserrat. In: Druitt, T.H., Kokelaar, B.P. (Eds.), The eruptions of Soufriere Hills Volcano, Montserrat, from 1995 to 1999, 21. Geological Society, London, Memoirs, pp. 281–306.

Newhall, C.G., Punongbayan, R. (Eds.), 1996. Fire and Mud: Eruptions and Lahars of Mount Pinatubo, Philippines. Philippine Institute of Volcanology and Seismology, Quezon City, p. 1126.

Newhall, C.G., 2000. Mount St Helens, Master Teacher. Science 288, 1181–1183.

Newhall, C.G., Hoblitt, R.P., 2002. Constructing event trees for volcanic crises. Bull. Volcanol. 64, 3–20.

Ogburn, S.E., Calder, E.S., Cole, P.D., Stinton, A.J., 2014. The effect of topography on ash-cloud surge generation and propagation. In: Wadge, G., Robertson, R.E.A., Voight, B. (Eds.), The Eruption of Soufrière Hills Volcano, Montserrat from 2000 to 2010, 39. Geological Society, London, Memoirs, pp. 177–192. http://dx.doi.org/10.1144/M39.10.

Orsi, G., Di Vito, M., Isaia, R., 2004. Volcanic hazard assessment at the restless Campi Flegrei caldera. Bull. Volcanol. 66, 514–530.

Patra, A.K., et al., 2005. Parallel adaptive numerical simulation of dry avalanches over natural terrain. J. Volcanol. Geotherm. Res. 139, 1–21. http://dx.doi.org/10.1016/j.jvolgeores.2004.06.014.

Roche, O., Gilbertson, M.A., Phillips, J.C., Sparks, R.S.J., 2005. Inviscid behaviour of fines-rich pyroclastic flows inferred from experiments on gas–particle mixtures. Earth Planet. Sci. Lett. 240, 401–414.

Rosi, M., Principe, C., Vecci, R., 1993. The 1631 Vesuvius eruption. A reconstruction based on historical and stratigraphical data. J. Volcanol. Geotherm. Res. 58, 151–182.

Rossano, S., Mastrolorenzo, G., De Natale, G., 2004. Numerical simulation of pyroclastic density currents on Campi Flegrei topography: a tool for statistical hazard estimation. J. Volcanol. Geotherm. Res. 132, 1–14.

SAC8, 2007. Assessment of the Hazards and Risks Associated with the Soufrière Hills Volcano, Montserrat. Eighth Report of the Scientific Advisory Committee on Montserrat Volcanic Activity, based on a meeting held between 20 and 22 March 2007 at the Montserrat Volcano Observatory, Montserrat (report available via MVO website).

SAC15, 2010. Assessment of the Hazards and Risks Associated with the Soufrière Hills Volcano, Montserrat. Fifteenth Report of the Scientific Advisory Committee on Montserrat Volcanic Activity, based on a meeting held between 20 and 22 March 2007 at the Montserrat Volcano Observatory, Montserrat (report available via MVO website).

Scarth, A., 2009. Vesuvius: A Biography. Princeton Univ. Press, 342 pp.

Scott, W.E., 1989. Volcanic-hazards zonation and long-term forecasts. In: Tilling, R.I. (Ed.), Volcanic Hazards Short Course Presented at 28th Int. Geol. Congress. American Geophysical Union, Washington, D.C, pp. 25–49.

Scott, W., Hoblitt, R., Torres, R., Self, S., et al., 1996. Pyroclastic flow deposits from the 1991 Pinatubo eruption. In: Newhall, C.G., Punongbayan, R.S. (Eds.), Fire and Mud: Eruptions and Lahars of Mount Pinatubo, Philippines. Philippine Institute of Volcanology and Seismology, Quezon City, and University of Washington Press, Seattle, pp. 545–570.

Self, S., Rampino, M.R., 2012. The 1963–1964 eruption of Agung volcano (Bali, Indonesia). Bull. Volcanol. 74, 1521–1536. http://dx.doi.org/10.1007/s00445-012-0615-z.

Selva, J., Orsi, G., Di Vito, M.A., Marzocchi, W., Sandri, L., 2011. Probability hazard map for future vent opening at the Campi Flegrei caldera (Italy). Bull. Volcanol. 74, 497–510. http://dx.doi.org/10.1007/s00445-011-0528-2.

Shelley, I., Voight, B., Nugroho, B., Tohma, Y., 1995. Medical consequences of dome-collapse nuées ardentes: November 1994 eruption at Merapi Volcano, Indonesia. EOS Trans. AGU 76 (46 Suppl.), 660.

Sigurdsson, H., Carey, S.N., Fisher, R.V., 1987. The 1982 eruption of El Chichon Volcano, Mexico (3): physical properties of pyroclastic surges. Bull. Volcanol. 49, 467–488.

Sigurdsson, H., Cashdollar, S., Sparks, R.S.J., 1982. The eruption of Vesuvius in A.D. 79: reconstruction from historical and volcanological evidence. Am. J. Archaeol. 86, 39–51.

Sparks, R.S.J., Bursik, M.I., Carey, S.N., Gilbert, J.S., Glaze, L.S., Sigurdsson, H., Woods, A.W., 1997. Volcanic Plumes. John Wiley, New York.

Sparks, R.S.J., Barclay, J., Calder, E.S., Herd, R.A., Komorwski, J.-C., Luckett, R., Norton, G.E., Ritchie, L.J., Voight, B., Woods, A.W., 2002. Generation of a debris avalanche and violent pyroclastic density current on 26 December (Boxing Day) 1997 at Soufriere Hills Volcano, Montserrat. In: Druitt, T.H., Kokelaar, B.P. (Eds.), The Eruption of Soufriere Hills Volcano, Montserrat, from 1995 to 1999, 21. Geological Society, London, Memoirs, pp. 409–434.

Spence, R.J.S., Zuccaro, G., Petrazzuoli, S., Baxter, P.J., 2004. The resistance of buildings to pyroclastic flows: analytical and experimental studies and their application to Vesuvius. Nat. Hazards Rev. 5, 48–59.

Stinton, A., Cole, P.D., Ogburn, S., 2010. Pyroclastic flow modelling in the Belham valley using Titan2D. In: Report to the SAC on volcanic activity at Soufriere Hills Volcano, Montserrat, Appendix B, MVO Open File Report 10-02b, pp. 23–35.

Tanguy, J.C., 1994. The 1902–1905 eruptions of Montagne Pelée, Martinique: anatomy and retrospection. J. Volcanol. Geotherm. Res. 60 (2), 87–107.

Taylor, G.A., 1954. Volcanological observations of Mount Lamington 29th May, 1952. Bull. Volcanol. 15 (1), 81–89.

Taylor, G.A., 1958. The 1951 eruption of Mount Lamington, Papua. Bureau of Mineral Resources, Geology and Geophysics (Commonwealth of Australia), Bull. No. 38, 117 pp.

Todesco, M., Neri, A., Esposti Ongaro, T., Papale, P., Macedonio, G., Santacroce, R., Longo, A., 2002. Pyroclastic flow hazard assessment at Vesuvius by using numerical modelling. 1. Large scale dynamics. Bull. Volcanol. 64, 155–177.

Todesco, M., Neri, A., Esposti Ongaro, T., Papale, P., Rosi, M., 2006. Pyroclastic flow dynamics and hazard in a caldera setting: application to Phlegrean Fields (Italy). Geochem. Geophys. Geosyst. 7, Q11003. http://dx.doi.org/10.1029/2006GC001314.

Valentine, G.A., 1987. Stratified flow in pyroclastic surges. Bull. Volcanol. 49 (4), 616–630.

Valentine, G.A., 1998. Damage to structures by pyroclastic flows and surges, inferred from nuclear weapons effects. J. Volcanol. Geotherm. Res. 87, 117–140.

Valentine, G.A., Wohletz, K.H., 1989. Numerical models of plinian eruption columns and pyroclastic flows. J. Geophys. Res. 94, 1867–1887.

Voight, B., 2000. Structural stability of andesite volcanoes and lava domes. Philos. Trans. R. Soc. Lond. A 358, 1663–1703.

Voight, B., Davis, M.J., 2000. Emplacement temperatures of the November 22, 1994 nuée ardente deposits, Merapi Volcano, Java. J. Volcanol. Geotherm. Res. 100 (1), 371–377.

Voight, B., Constantine, E.K., Siswowidjoyo, S., Torley, R., 2000a. Historical eruptions of Merapi volcano, central Java, Indonesia, 1768–1998. J. Volcanol. Geotherm. Res. 100 (1), 69–138.

Voight, B., Young, K.D., Hidayat, D., Subandrio, Purbawinata, M.A., Ratdomopurbo, A., Suharna, Panut, Sayudi, D.S., LaHusen, R., Marso, J., Murray, T.L., Dejean, M., Iguchi, M., Ishihara, K., 2000b. Deformation and seismic precursors to dome-collapse and fountain-collapse nuées ardentes at Merapi Volcano, Java, Indonesia 1994–1998. J. Volcanol. Geotherm. Res. 100, 261–287.

Voight, B., Sukhyar, R., Wirakusumah, A.D., 2000c. Special issue on Merapi Volcano. J. Volcanol. Geotherm. Res. 100.

Voight, B., 2011a. Modeling pyroclastic flow and surge hazards: assumptions, capabilities, limitations and integration with civil defense strategies, International workshop on Strategies of

Volcanic Disaster Mitigation 2011. In: NIED-YIES Workshop Proceedings, pp. 278–323 discussion pp. 484–489; abstract pp. 525–526.

Voight, B., 2011b. Scientific and emergency management response to volcanic risk: Lessons from Montserrat and other volcanoes. In: International Workshop on Strategies of Volcanic Disaster Mitigation 2011, NIED-YIES Workshop Proceedings, pp. 28–60 discussion pp. 61–68, 229–245; abstract pp. 515–517.

Wadge, G., Jackson, P., Bower, S.M., Woods, A.W., Calder, E., 1998. Computer simulations of pyroclastic flows from dome collapse. Geophys. Res. Lett. 25. http://dx.doi.org/10.1029/98GL00710. ISSN:0094-8276.

Wadge, G., 2009. Assessing the pyroclastic flow hazards from dome collapse at Soufriere Hills Volcano, Montserrat. In: Thordarsson, T., Self, S., Larsen, G., Rowlands, S.K., Hoskuldsson, A. (Eds.), Studies in Volcanology: The Legacy of George Walker, 2. Spec. Pub. IAVCEI, pp. 211–224. Geol. Soc. London.

Wadge, G., Aspinall, W.P., 2014. A review of volcanic hazard and risk assessments at the Soufriere Hills Volcano, Montserrat from 1997 to 2011. In: Wadge, G., Robertson, R.E., Voight, B. (Eds.), The Eruption of Soufriere Hills Volcano, Montserrat from 2000 to 2010. Memoir 39, Geological Society, London.

Wadge, G., Robertson, R.E., Voight, B. (Eds.), 2014a. The Eruption of Soufriere Hills Volcano, Montserrat from 2000 to 2010. Memoir 39, Geological Society, London.

Wadge, G., Voight, B., Sparks, R.S.J., Cole, P.D., Loughlin, S.C., Robertson, R., 2014b. An overview of the eruption of Soufriere Hills Volcano, Montserrat from 2000 to 2010. In: Wadge, G., Robertson, R.E., Voight, B. (Eds.), The Eruption of Soufriere Hills Volcano, Montserrat from 2000 to 2010. Memoir 39, Geological Society, London.

Waythomas, C.F., Neal, C.A., 1998. Tsunami generation by pyroclastic flow during the 3500-year BP caldera-forming eruption of Aniakchak Volcano, Alaska. Bull. Volcanol. 60 (2), 110–124.

Widiwijayanti, C., Voight, B., Hidayat, D., Patra, A., Pitman, E.B., 2004. Validation of TITAN2D flow model code for pyroclastic flows and debris avalanches at Soufrière Hills Volcano, Montserrat, BWI. Eos Trans. AGU 85 (47). Fall Meet Suppl Abstract V43B-1428.

Widiwijayanti, C., Voight, B., Hidayat, D., Schilling, S.P., 2008. Objective rapid delineation of areas at risk from block-and-ash pyroclastic flows and surges. Bull. Volcanol. 71, 687–703. http://dx.doi.org/10.1007/s00445-008-0254-6.

Widiwijayanti, C., Voight, B., Hidayat, D., Patra, A., Pitman, E.B., 2010. TITAN2D simulations of dome-collapse pyroclastic flows for crisis assessment on Montserrat. Eos Trans. AGU 91 (52). Fall Meet. Suppl., Abstract V13A-2342.

Wills, J., Wyatt, T., Lee, B., 1998. Warnings of High Winds in Densely Populated Areas. UK National Coord Committee for the IDNDR. Thomas Telford Publisher, London.

Wohletz, K.H., 1998. Pyroclastic surges and compressible two-phase flow. In: Freundt, A., Rosi, M. (Eds.), From Magma to Tephra: Modelling Physical Processes of Explosive Volcanic Eruptions. Elsevier, Amsterdam, pp. 247–312.

Yamamoto, T., Takarada, S., Suto, S., 1993. Pyroclastic flows from the eruption of Unzen volcano. Jpn. Bull. Volcanol. 55, 166–175.

Young, S.R., Voight, B., Barclay, J., Herd, R., Komorowski, J.-C., Miller, A.D., Sparks, R.S.J., Stewart, R.C., 2002. Hazard implications of small-scale edifice instability and sector collapse: a case history from Soufriere Hills Volcano, Montserrat. In: Druitt, T.H., Kokelaar, B.P. (Eds.), The Eruption of Soufriere Hills Volcano, Montserrat, from 1995 to 1999. Geological Society, London, Memoirs, 21, pp. 349–361.

Zuccaro, G., Cacace, F., Baxter, P.J., Spence, R., 2008. Impact of explosive eruption scenarios at Vesuvius. J. Volcanol. Geotherm. Res. 178, 416–453.

Chapter 6

Lahars at Cotopaxi and Tungurahua Volcanoes, Ecuador: Highlights from Stratigraphy and Observational Records and Related Downstream Hazards

Patricia A. Mothes [1] and James W. Vallance [2]

[1] *Instituto Geofísico, Escuela Politécnica Nacional, Quito, Ecuador,* [2] *U.S. Geological Survey, Cascades Volcano Observatory, Vancouver, WA, USA*

ABSTRACT

Lahars are volcanic debris flows that are dubbed primary when triggered by eruptive activity or secondary when triggered by other factors such as heavy rainfall after eruptive activity has waned. Variation in time and space of the proportion of sediment to water within a lahar dictates lahar flow phase and the resultant sedimentary character of deposits. Characteristics of source material and of debris eroded and incorporated during flow downstream may strongly affect the grain-size composition of flowing lahars and their deposits. Lahars borne on the flanks of two steep-sided stratocones in Ecuador exemplify two important lahar types. Glacier-clad Cotopaxi volcano has been a producer of primary lahars that flow great distances downstream. Such primary lahars include those of both clast-rich and matrix-rich composition—some of which have flowed as far as 325 km to the Pacific Ocean. Cotopaxi's last important eruption in 1877 generated formidable syneruptive lahars comparable in size to those that buried Armero, Colombia, following the 1985 eruption of Nevado del Ruiz volcano. In contrast, ash-producing eruptive activity during the past 15 years at Tungurahua volcano has generated a continual supply of fresh volcaniclastic debris that is regularly remobilized by precipitation. Between 2000 and 2011, 886 rain-generated lahars were registered at Tungurahua. These two volcanoes pose dramatically different hazards to nearby populations. At Tungurahua, the frequency and small sizes of lahars have resulted in effective mitigation measures. At Cotopaxi 137 years have passed since the last important lahar-producing eruption, and there is now a high-risk situation for more than 100,000 people living in downstream valleys.

Volcanic Hazards, Risks, and Disasters. http://dx.doi.org/10.1016/B978-0-12-396453-3.00006-X

TABLE 6.1 Triggering Mechanisms and Consequences of Primary and Secondary Lahars Events at Various Volcanoes in the Recent Past

Triggering Mechanism	Volcano/Event	Volumes	Distance Traveled (km)	Lahar Category, Comments
Pyroclastic flows and heated blasts	Cotopaxi 1877 w/~1 km^3 glacier	2×10^8 m^3; 10% of icecap melted	325	1000 deaths? *Syneruptive, primary*
Pyroclastic flow enters stream	Asama, Japan, 1783	1–4 × 10^8 m^3	10–15 km	Pyroclastic flow dammed ephemeral lake. Breakout 1 h later, 1200 deaths. *Primary*
Lava flows melted ice. Ponded water breaks out.	Villarica, Chile, 1949, 1963, 1971	Villarica lake rose 1 m in 1949	15 km	*Primary*
Avalanche generated	Rainier USA- Numerous from 7200 to 500 yBP; St Helens, USA, 1980	Each about 2×10^8 m^3	75	Most were eruption-related avalanches that liquefied. *Primary*

Crater lake expulsion	Kelut, Indonesia; numerous events between 1826 and 1951.	1919 event 4×10^7 m^3	40	5100 deaths *Primary*
Crater lake breakout due to volcanic activity	Mt Pelee, Martinique 1902	Small	4	*Primary*
Jökulhlaups eruption beneath ice cap	Katla, Iceland, 1918, 1955	8 km^3	30 and 75	*Primary*
Rain during eruption	Pinatubo, Philippines, 1991. Typhoon Yunya coeval with 15 June eruption.		25–30	*Primary*
Heavy rain, posteruption	Pinatubo, Philippines, post 1991	Lahars have remobilized >2.5 km^3 of pyroclastic flow deposits	25–30	*Secondary*
Crater lake failure	Kelut, Indonesia, 1875	38×10^6 m^3	13 km	Failure of crater wall destroyed 30 villages *Secondary*

Source: Modified from Vallance (2005). Nevado del Ruiz data from Pierson et al. (1990) and Thouret (1990).

6.3 PRIMARY LAHARS AND THEIR GENERATION AT COTOPAXI

Cotopaxi volcano (latitude 00°41'S; longitude 78°26'W) is located in the Eastern Cordillera of the Ecuadorian Andes, 60 km south of Quito and 45 km northeast of Latacunga, a regional capital. The 5,897-m-high active volcano is notable for its extreme relief, its massive size (22-km diameter), and its glacier-clad slopes (Figure 6.1). Along with other large, potentially active stratovolcanoes such as Tungurahua, Sangay, Antisana, and Cayambe, Cotopaxi lies within the eastern row of andesitic volcanoes in Ecuador, some 35 km behind the volcanic front of the Western Cordillera (Hall et al., 2008). Two main drainage systems at Cotopaxi are the northern drainage, which flows 325 km to the Pacific Ocean, and the southwestern drainage, which ends at the Atlantic after traversing the Amazon River system. The smaller *Ríó* Tambo drainage flows directly eastward to the Amazon basin (Figure 6.2).

Most lahars produced at Cotopaxi volcano have been andesitic, matrix-poor, sandy, and clast-rich. The most recent significant andesitic eruption on June 26, 1877 produced lahars that were clast-rich and matrix-poor, dominated by gravel and sand (Figure 6.3). By mixing with stream water they transformed to hyperconcentrated flows (HCFs) some 40 km downstream from the vent (Mothes et al., 2004). Recent historical eruptions (since AD 1532) of Cotopaxi have been andesitic, and most have been clast-rich.

About every 2,000 years the volcano has generated exceptionally large rhyolitic eruptions that have generated matrix-rich flows (Mothes et al., 2004; Hall and Mothes, 2008). Their deposits comprise a high percentage of ash

FIGURE 6.1 View looking south at Cotopaxi volcano, 5,897 m elevation, with its perennial glacier.

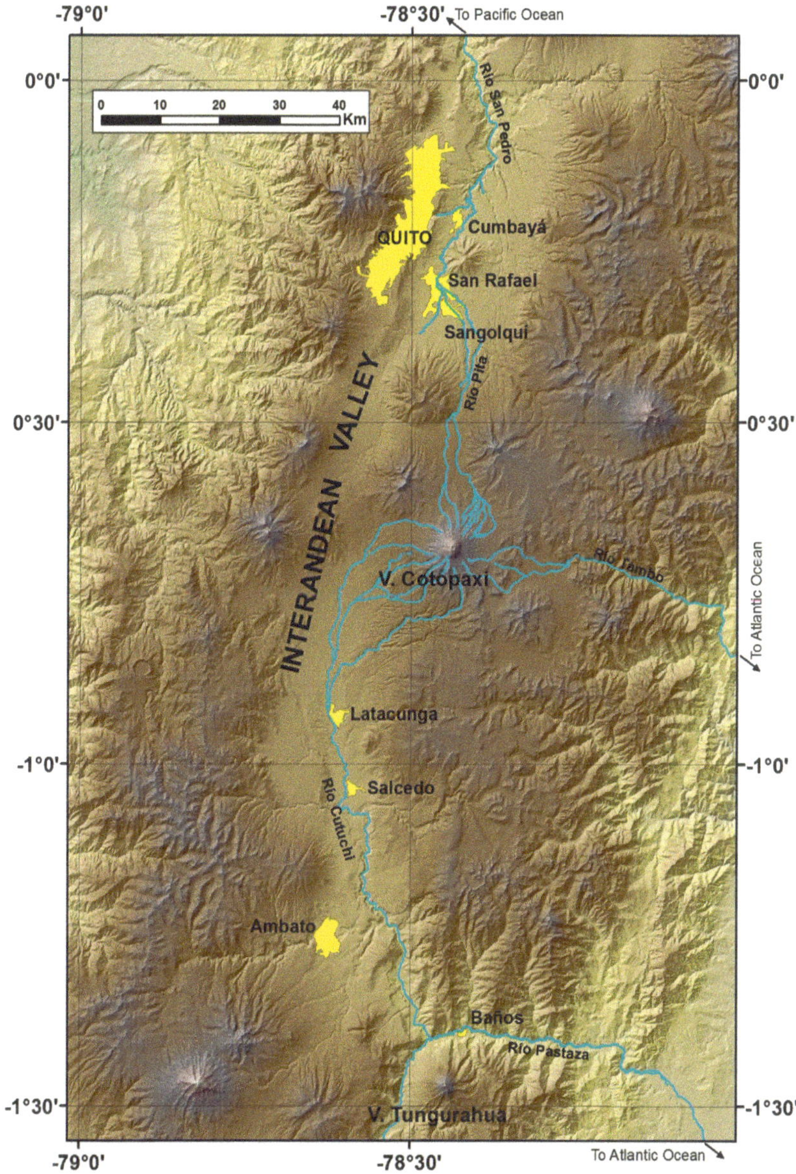

FIGURE 6.2 Map of Ecuador's Inter-Andean Valley with Cotopaxi volcano in the center and Tungurahua volcano at the southern end of the map. Shown are major rivers that drain Cotopaxi and the cities and towns that are situated within or beside lahar routes. The Chillos Valley is that part of the Inter-Andean valley east of Quito, approximately from *Río* Pita—*Río* San Pedro on the map.

FIGURE 6.3 Quarry cut of the clast- and sand-rich 1877 lahar deposit, about 20 km north from source along the east bank of the *Ríó* Pita. Note predominance of fresh gray-lithic clasts and red scoria, components typical of this deposit.

whose origin derives chiefly from rhyolitic pyroclastic flows originating at Cotopaxi (Mothes et al., 1998). Such enormous lahars do not undergo downstream dilution to form water-rich lahar phases, like HCF, owing to insufficient volume of water in the downstream river system (Vallance and Scott, 1997). Huge prehistoric examples remind us that, whereas the volcano repetitively produces similar types of "expected" events in historical time, it may also generate far bigger but less-frequent outliers, with greater risk to downstream populations.

6.3.1 Historical Clast-Rich Lahars of Cotopaxi

On June 26, 1877, erosion and melting of the ice cap by scoria-rich pyroclastic flows generated lahars that had runouts to about 10 km beyond Cotopaxi's apron (Figure 6.4). The main components in these lahars were scoria-rich sands and gravel. Preexisting blocks and debris were also eroded and

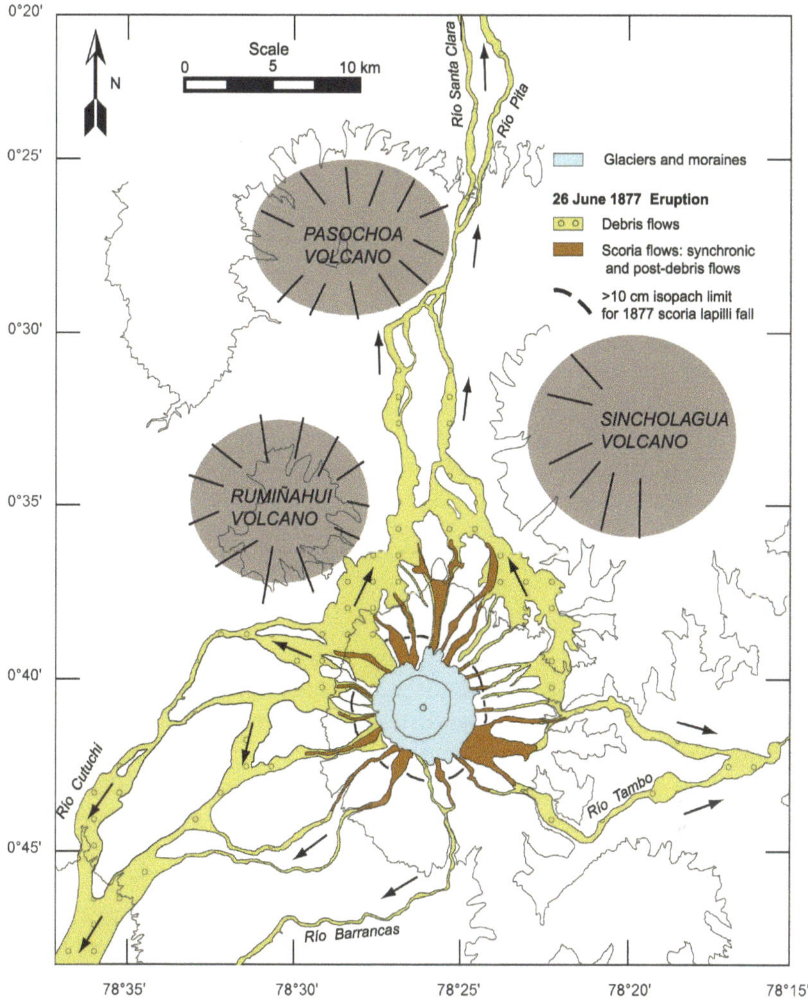

FIGURE 6.4 Map of the proximal eruptive products of the June 26, 1877, VEI 4 eruption of Cotopaxi.

incorporated by the flows. Even now one can walk the upper limit of the lahar deposits and find the conspicuous trim line of the scoria "krumpkoff" clasts, so named by the German geologist-priest Theodore Wolf because the scoria clasts reminded him of cauliflower heads (Mothes et al., 2004). After inspecting Cotopaxi's glaciers following the June 1877 eruption, Wolf (1878) wrote, "there was still enough glacier to produce ten times the amount of lahar that was formed." He also reported 10-m-deep crevasses gouged in the ice by the descending pyroclastic flows.

The 1877 lahars had velocities of 25–30 m/s near the cone and slowed to 10–15 m/s 40 km downstream on lower gradients where the restricted canyon regime opens onto a low-gradient depositional surface. In both the northern and southwestern drainages, the lahars arrived at the main population centers of Latacunga and the Chillos Valley in a little less than an hour. The valley and city of Latacunga suffered because of the broad reach of the lahars and because of lahar surges coming down several main *quebradas*, in particular the Alaquez *Quebrada* (Figure 6.5). Along the northern drainage toward the town of San Rafael, the transit of the lahar from the volcano was via narrow canyons. At only 5 km upstream of this population center, the lahar burst out of the confining canyon, spread out, and slowed down (Figure 6.6). At that time, the San Rafael community mainly comprised haciendas, most of which were buried where they were located beside the *Ríó* Pita. Heading northward from San Rafael, the lahars again became restricted in the *Ríó* San Pedro–Guayllabamba canyon until arriving at the coastal plain just south of the Pacific-coastal city of Esmeraldas. Historical accounts note that the northern-flowing lahar took 18 h to arrive at the sea, and it was still transporting fragments of houses and bodies of people and animals, all from the Sierra, more than 300 km upstream.

Along the southern *Ríó* Cutuchi route the lahar was similarly destructive and far traveled. Once it passed through Latacunga and Salcedo, it then dropped into a deep canyon and advanced to the Amazon lowland via the Pastaza River system. At the outskirts of Baños more than 85 km downstream at the foot of Tungurahua volcano, the lahar overtopped and destroyed the 60-m-high San Martin bridge where it spanned the *Ríó* Pastaza (Figure 6.7).

Volcano edifice and slope morphology control the path of medium-size to small lahars on Cotopaxi. The crater rim is lower on the west side than elsewhere and thus preferentially favors westward debouching of pyroclastic flows into the headwater of the Latacunga Valley.

Along the northwest to northeast foot of the volcano, a broad sloping plain acts as a sediment trap for lahars. Called Limpiopungo, this vast plateau is underlain by a lava package tens of meters thick that extends 40 km to the north. This plateau has served as an immense depositional zone for lahars—stranding a portion of each flow before capture in the deep and narrow *Ríó* Pita canyon (Figure 6.8). Beginning with the 1532 eruption, only four historical lahars are recognized in the northward-draining Chillos Valley downsteam of Limpiopungo (Mothes et al., 2004; Hall and Mothes, 2008).

In contrast, west and south of Cotopaxi, the Latacunga valley is not moderated by a broad lava plateau that causes lahars to spread out, slow down and deposit. Upon initiation, all lahars on the western and southern flanks immediately channel into deep canyons which carry them 25 km to the head of the Latacunga valley, where only then can significant deposition begin. As a result of the westward-oriented crater and the lack of a broad distilling plain,

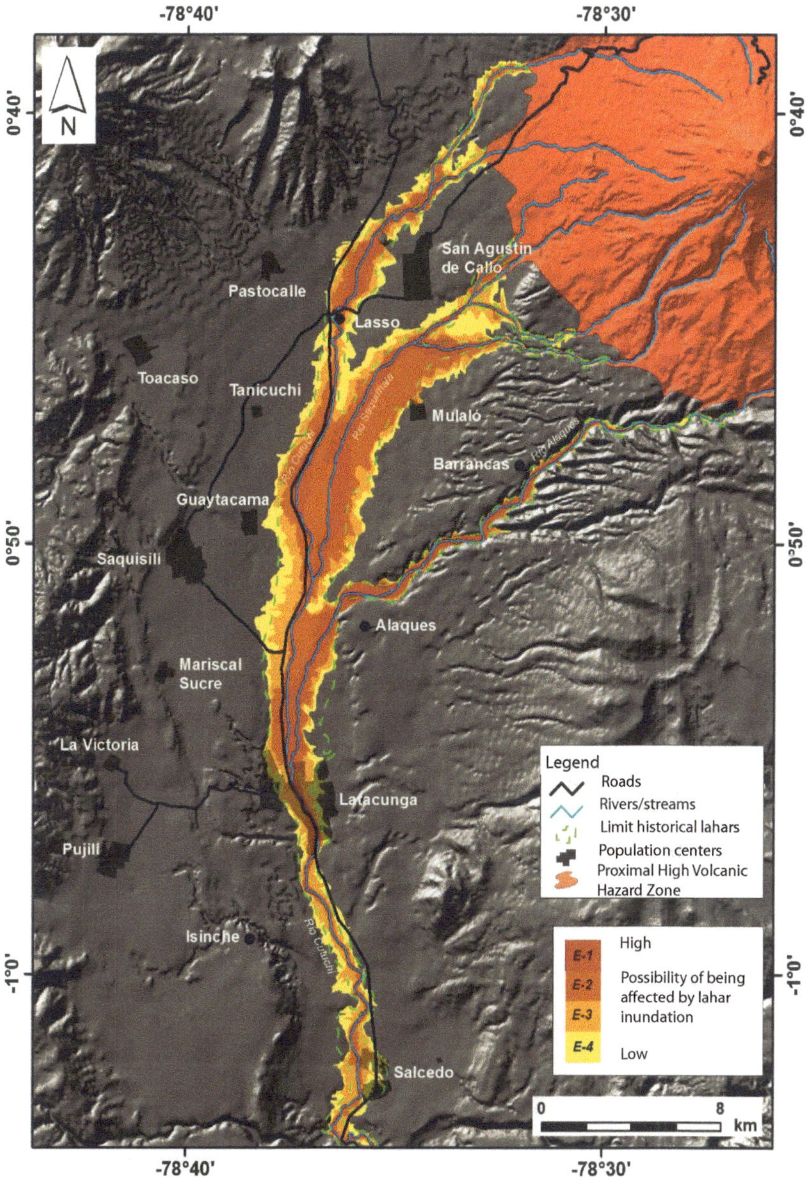

FIGURE 6.5 Maps showing lahar-inundation zones that grade from those having the greatest risk (greatest probability of inundation) and smallest area hazard zone to those having the least risk (least probability of inundation) and largest area hazard zone for the southern drainage in the Latacunga Valley. *Lahar zones are based on LAHARZ modeling and modified from an IGEPN internal report by Ordoñez et al. (2013).*

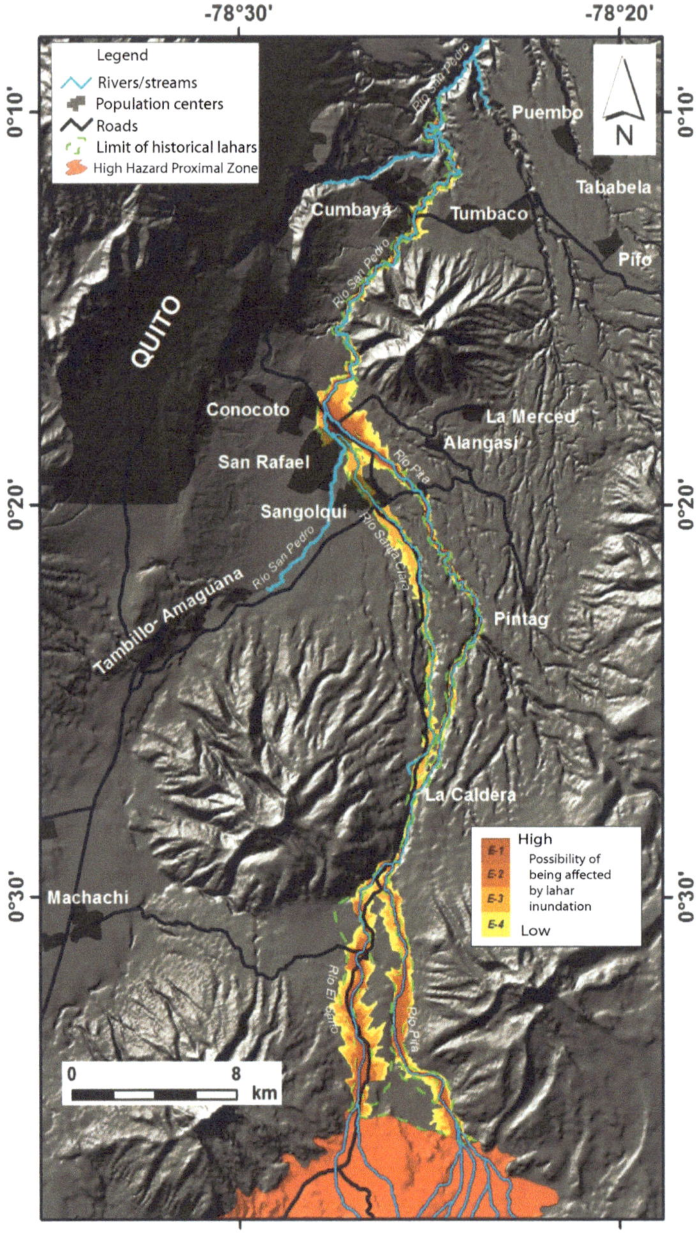

FIGURE 6.6 Maps showing lahar-inundation zones that grade from those having the greatest risk (greatest probability of inundation) and smallest area hazard zone to those having the least risk (least probability of inundation) and largest area hazard zone for the northern drainage in the Chillos Valley. *Lahar zones are based on LAHARZ modeling and modified from an IGEPN internal report by Ordoñez et al. (2013).*

FIGURE 6.7 Photograph showing the remaining bridge abutment on the *Río* Pastaza's left margin in the San Martin sector at Baños, Tungurahua, about 85 km downstream from Cotopaxi volcano. The bridge was destroyed by Cotopaxi's 1877 lahar that rose to more than 60 m deep in the narrow lava-walled canyon.

FIGURE 6.8 The vast Limpiopungo plain on the northwest to northeast foot of Cotopaxi is a great depositional trap for lahars and pyroclastic flows. View to the northwest.

the valley of Latacunga has suffered about 12 historical lahar events (Hall and Mothes, 2008; Mothes et al., 2004).

Lahar deposits of the historical eruptions of 1532, 1742, 1744, and 1768, were more voluminous, thicker, more clast-rich (possibly from bulking), and had broader inundation areas than those of the 1877 eruption, partly because of more extensive past glaciation. Pyroclastic flows eroding the glaciers, which hundreds of years ago were bigger than those of the present, caused all of these lahars. German geologist Reiss (cited in Meyer, 1907) indicated that glacier termini in 1872 averaged elevation of about 4,600 m, whereas now they average 5,000 m on all flanks. Cotopaxi glacier ice now covers about 4.7 km^2 in the northern drainage, 5.3 km^2 in the southwestern drainage, and 3.9 km^2 in the eastern drainage—a total area of 14 km^2. The total ice volume is about 0.5 km^3 (Samaniego et al., 2011).

The 1877 lahar in the *Rı́o* Pita channel had maximum stage height similar to that of the *Rı́o* Azufrado lahar generated by the November 13, 1985 eruption of Nevado del Ruiz volcano, Colombia, which destroyed the town of Armero (Pierson et al., 1990; Voight et al., 2013). The peak discharge of the 1877 lahar was probably similar to that of the Azufrado lahar, ~48,000 m^3/s. During flow of Cotopaxi's 1877 lahar through confined parts of the *Rı́o* Pita canyon, stage heights surpassed 35 m and estimated discharge was 40,000 m^3/s (Figure 6.9(a)). Lahar velocities ranged from 17 to 23 m/s in the proximal zone and dropped to 5–10 m/s (Figure 6.9(b)) in the distal depositional zone of San Rafael (Figure 6.2). Lahars of 1877 had maximum flow averaging about 2,000 m wide, but lahars of the eighteenth century tended to have wider cross sections and were more voluminous (Figure 6.9(c)) (total volume calculation of 1877 lahars is given in Table 6.2).

For the 1877 lahar and earlier lahars, gravel proportions in deposits of northern flank are greater than those in deposits of the southwestern drainage. Grain-size composition of the eighteenth century lahar deposits shows that these flows carried 50–80% gravel-size clasts (Figure 6.10). In all of the deposits, it is common to see boulders that are meters in diameter, especially in channel facies (Figure 6.11).

Relic structures buried by the Cotopaxi lahars and now uncovered by erosion or excavation provide testimony of the onslaught that previous generations faced in the Inter-Andean Valley. One such discovery in the Latacunga Valley is a partially buried stone defense wall built by the then embattled Incas at *Hacienda* San Augustine de Callo to restrain the invading Spanish *conquistadores*—just before Cotopaxi's AD 1532 eruption (D. Brown, personal communication, 2012). Investigators commonly find colonial roof tiles, bones, and molds of *agaves* within lahar deposit exposures (Mothes et al., 2004).

6.3.2 Matrix-Rich Lahars at Cotopaxi

Cotopaxi's matrix-rich lahars, although not clay-rich, are silt- and sand-enriched. They owe the abundance of these grain sizes to the syneruptive

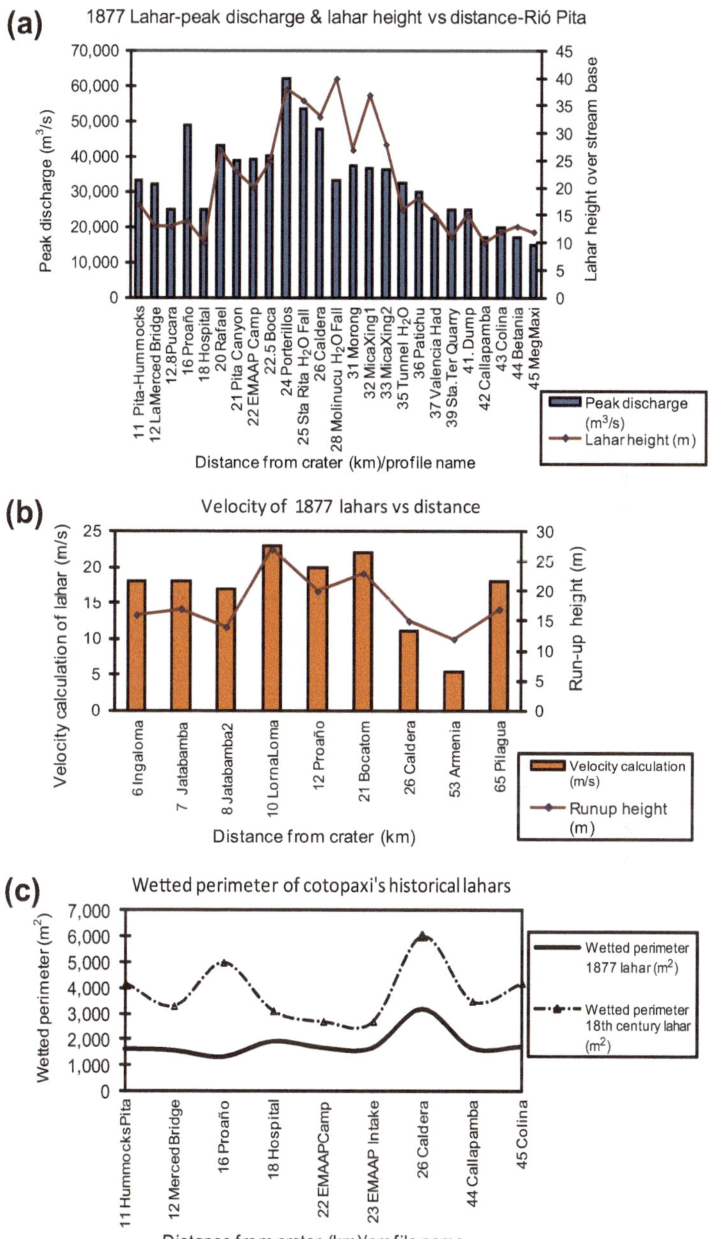

FIGURE 6.9 Graphs showing peak discharge, lahar height, velocity, and wetted perimeter of both the 1877 lahar and earlier historical lahars.

TABLE 6.2 Total Estimated Volumes for Cotopaxi's 1877 Lahars in the Southern and Northern Drainages

Drainage	Estimated Volumes for 1877 Lahar	Rivers Traversed	Comments
Northern drainage	75–100 Mm^3	Pita, Salto, Sta. Clara, San Pedro, Guayllabamba and Esmeraldas	Unknown volume was deposited on the Limpiopungo surface at the foot of the cone.
Southwest drainage	80–100 Mm^3	Cutuchi, Saquimala, Burruhuaicu, Alaquez, San Felipe, Patate, Pastaza	Minimal deposition occurred on the steep flanks and in ravines.

processes that took place during the (Volcano Explosivity Index) VEI 5 explosive eruptions that produced fine-grained rhyolitic pyroclastic flows (Mothes et al., 1998; Hall and Mothes, 2008). The water source for the lahars was glacier melt and pore water in cases where the lahars also incorporated collapsing avalanche debris. The most representative example of a matrix-rich lahar at Cotopaxi is that of the Chillos Valley Lahar (CVL). It formed 4500–5000 ^{14}C years (5700–6000 calendar years) ago in association with a series of rhyolitic eruptions and a sector collapse on the N–NE flank of Cotopaxi (Mothes et al., 1998). This lahar had a flow depth of 100 m in the Chillos Valley, a maximum flow width of 11 km and a total bulk volume of just less than 3 km^3. The matrix-rich character of the CVL shows little variation from source to its terminus on the Pacific coastal plain (Figure 6.12(a) and (b)) because the Guayllabamba–Esmeraldas river system contained insufficient water to dilute it and thus initiate downstream transformation to HCF.

Higher than normal fines (chiefly silt) contents have also been identified in lahars associated with Cotopaxi's 1742 eruption (VEI 4), which occurred after 200 years of repose (Mothes, 2010). A beige pumice lapilli layer fell over the volcano's W flank and was followed by the descent of a matrix-rich lahar in nearby stream channels. This lahar deposit directly overlies the pumice fall unit. The deposit of the matrix-rich 1742 lahar crops out in the Cutuchi quarry, 25 km from vent on the southwestern flank of Cotopaxi. While in transit the lahar carried large 2–3 m diameter boulders, which today are found down valley at Mulalo (30 km from vent). Silt concentrations in the AD 1742 lahar deposits are about 20%. At Salcedo, 60 km downstream, less silt is present and the largest size clasts are missing, suggesting downstream mixing and dilution with river water. Nonetheless, the 1742 lahar retained "matrix-rich" characteristics for more than 50 km downstream.

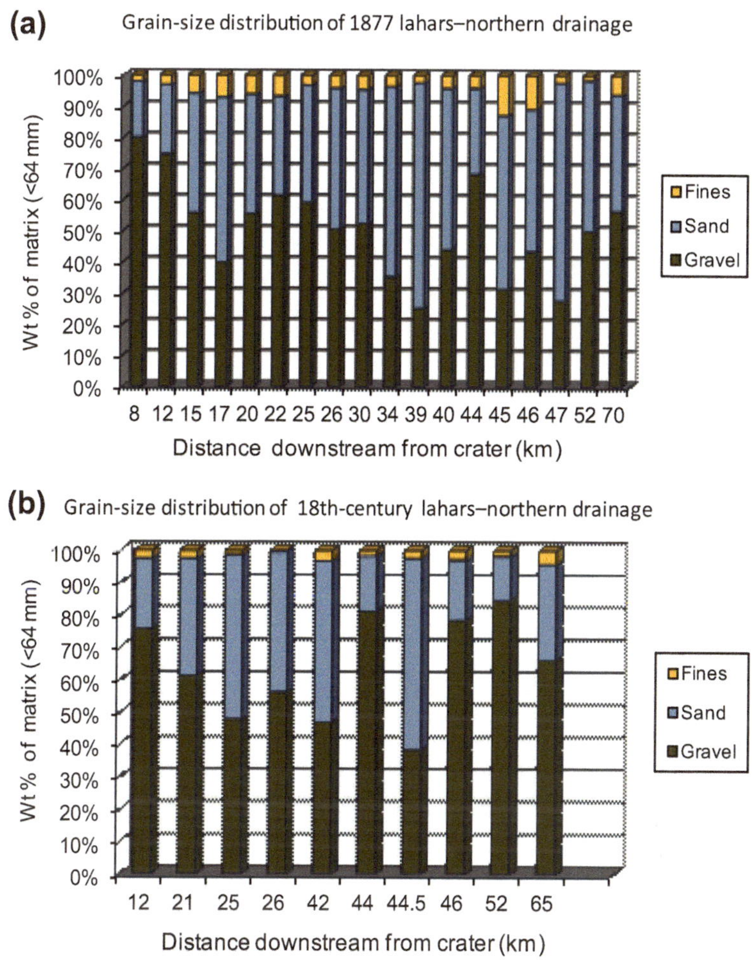

FIGURE 6.10 Grain-size distribution of 1877 and 18th century lahars along the northern, *Río* Pita drainage.

At Cotopaxi, matrix-rich lahars are more voluminous and spread over larger areas than other lahars. Because they are so voluminous and commonly do not or are slow to transform downstream to more water-diluted flows, matrix-rich lahars potentially have greater impact on essential infrastructure, property, and life. Voluminous lahars have greater downstream discharge, much broader inundation swaths, and higher flow velocities than less voluminous lahars (Mothes, 2010).

Cotopaxi now stores less water on its flanks than it did in historical times owing to receding glaciers; nonetheless, the remaining 0.5 km^3 of glacier ice provides an ample source of water for lahars when future hot turbulent pyroclastic flows sweep across the ice cap. Thus, during future

FIGURE 6.11 The giant boulder of Chilantosa on the fan of the Burruhuaicu *quebrada* (southwest flank), near San Ramon, was possibly deposited in AD 1742.

explosive eruptions, voluminous destructive lahars will threaten downslope populations.

6.3.3 Rain-Generated Secondary Lahars at Tungurahua Volcano

Tungurahua volcano (latitude 01°28′S; longitude 78°27′W) is located in the Eastern Cordillera of the Ecuadorian Andes, 120 km south of Quito, 33 km southeast of Ambato, the capital of Tungurahua Province and 85 km south of Cotopaxi Volcano. This 5,023-m-high active volcano is notable for its extreme relief (3,200 m), steep sides, and frequent eruptive cycles. Its 400-m-wide crater is situated slightly northwest of its summit with the lowest rim (4,770 m) on its northwest side, favoring spillage of pyroclastic flows onto this sector.

Tungurahua experienced important eruptive activity between AD 1300 and 1700 (Le Pennec et al., 2008). Five major eruptive episodes are recognized in historical reports: 1641–1646, 1773–1781, 1886–1888, 1916–1918, and the 1999 to present episode. Lava flows, pyroclastic flows, ash falls, and lahars characterize the volcano's activity. Tungurahua is a dangerous volcano that threatens the tourist town of Baños (population of ~20,000) as well as other small villages located around the base of the cone (Figure 6.13).

From 1999 until 2014 Tungurahua has experienced periods of low to intermittent eruptive activity, characterized mainly by Strombolian episodes, but interspersed lately by several Vulcanian eruptive episodes of varying intensity. Frequent explosions, fire fountaining, ejection of incandescent blocks,

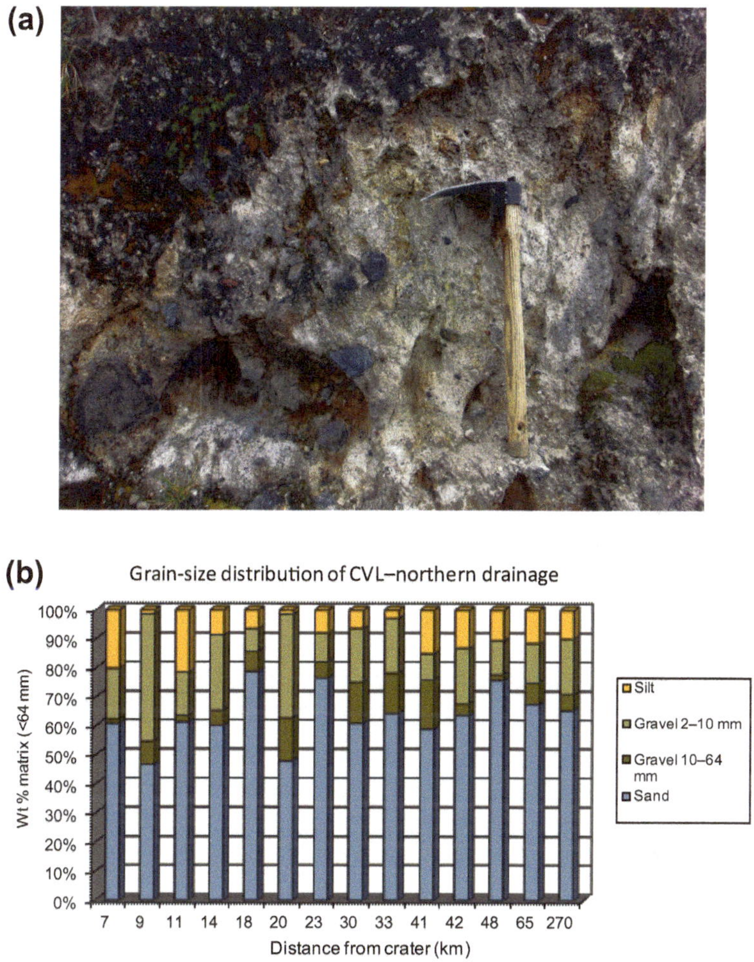

FIGURE 6.12 (a) Close-up photograph of the typical matrix of the Chillos Valley matrix-rich lahar at an outcrop on Cotopaxi's NE flank. Incorporation of rhyolitic ash gives the deposit its white coloration. (b) Grain-size distribution of the Chillos Valley Lahar, from the volcano to the Pacific coastal plain, 270 km from source. Note the relatively constant values from source to the distal outcrop.

frequent ash falls, and rain-generated lahars have characterized Strombolian activity. Eruptions on August 16–17, 2006 generated sizeable pyroclastic flows. During these eruptions, Vulcanian-like explosive events accompanied the most intense Strombolian eruptions and characteristically included very large explosions, strong infrasound signals, and powerful audible booms during paroxysmal phases of the eruption (Ruiz et al., 2006; Hall et al., 2013).

Since Tungurahua's reactivation in 1999, the Instituto Geofísico of the Escuela Politécnica Nacional (IGEPN) has maintained a permanently staffed

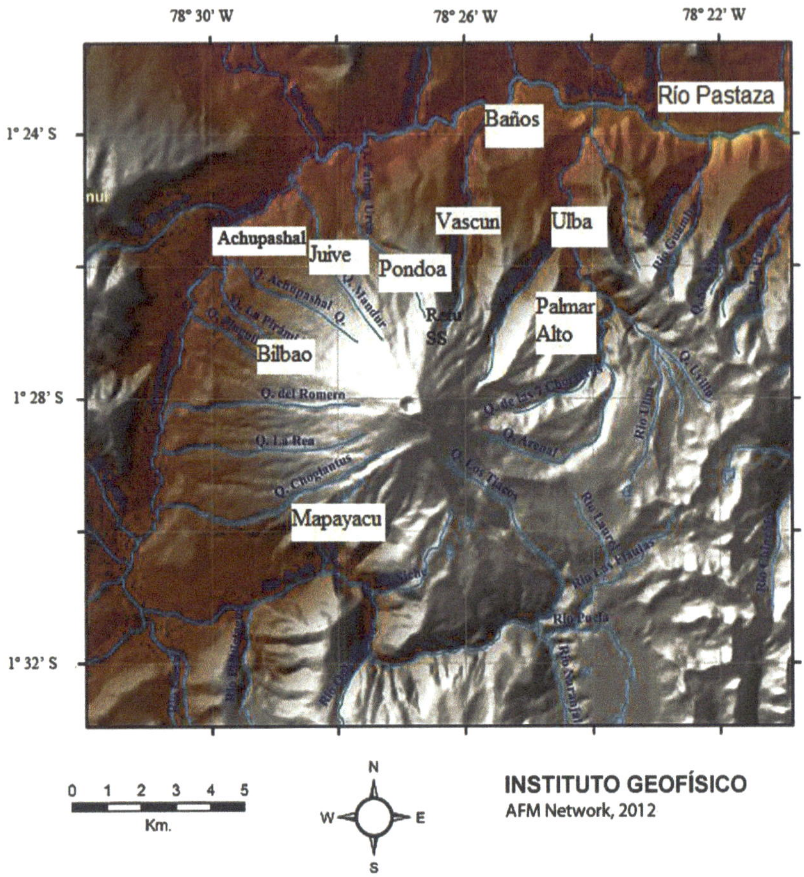

FIGURE 6.13 Digital elevation map of Tungurahua volcano showing the eight acoustic flow monitoring stations, the RETU short period seismic station, the city of Baños and major drainages, including the Vazcún *quebrada*, whose channel lies on the western side of Baños.

OVT located 13 km north-northwest of the crater and has operated its principal office in Quito 24 h a day, 7 days a week. The continual monitoring of Tungurahua employs seismic (short period and broadband instruments), SO_2 gas emission Correlation Spectrometry (COSPEC) and Differential Optical Absorption Spectrometry (DOAS), and geodetic methods (Electronic Distance Meter Electronic Distance Meter (EDM), tiltmeters, Global Positioning System Global Positioning System (GPS), and occasionally INSAR). (Correlation Spectrometry and Differential Optical Absorption Spectroscopy) In addition, thermal imagery (airborne and ground-based) during the past 10 years has proven to be invaluable for nighttime observations. IGEPN uses eight acoustic-flow-monitor (AFM) (Hadley and LaHusen, 1995) stations to detect and register secondary lahar activity. Data from these stations have been important for detecting the descent of both rain-generated lahars and pyroclastic flows (Figure 6.13).

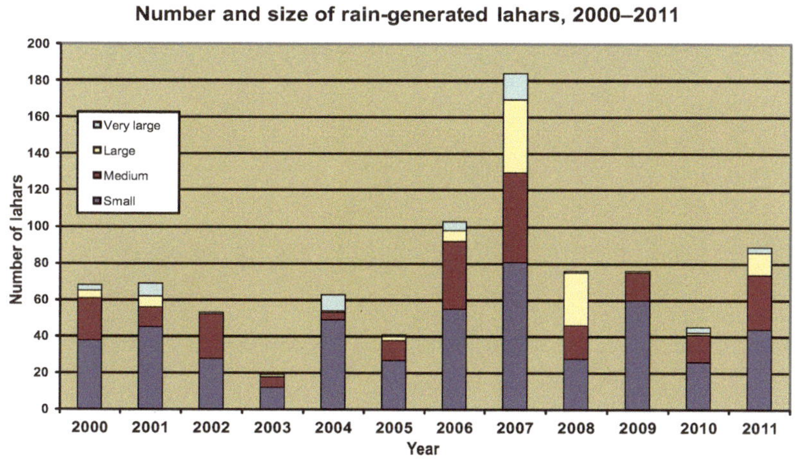

FIGURE 6.14 Graph showing the number and relative size of the 886 rain-generated lahars registered by OVT monitoring scientists and the AFM network during 2000–2011. Note that the spike in the 2007 events is owing to remobilization of the approximately 50 Mm^3 of volcaniclastic debris that was deposited on Tungurahua's flanks during the August 2006 eruptions. OVT, observatory; AFM, acoustic-flow-monitor.

Since 1999, rain-generated lahars have numbered over 880 events (J. Bustillos, written communication, 2013). Most lahars were detected by the AFM system, and others were observed by OVT personnel and also by local volunteer volcano observers, *vigías*. Not all drainage channels have AFM stations. The initial ash falls of late October 1999 changed run-off patterns on the volcano's steep flanks and lahars ensued. In 2007, a major increase occurred in the number of lahars (Figure 6.14), presumably resulting from the large volume (40 Mm^3) of new volcaniclastic deposits emplaced by the 2006 eruptions. The volcano has a minute glacier remnant on the NE summit, which is isolated from pyroclastic-flow paths. The only other source of water to generate lahars is regional rainfall and localized thunderstorms, which are commonly generated as warm air rises up the narrow Pastaza River canyon from the Amazon basin. Occasionally, these moisture-saturated air masses reach the upper slopes of the northeastern to northwestern flanks to create intense rainfall and generate the most significant lahars, which flow down the *Ríó* Vazcún (bordering the western sector of Baños) and Achupashal canyon on the western flank of the volcano. Most of the lahars remain within the canyons and drop into the *Ríó* Pastaza. To secure highway travel, bridges have had to be raised over active lahar channels, and concrete bases have had to be constructed for roads where they cross *quebradas* (ravines). A lahar alert system along the main road between Baños and Penipe gives residents and motorists early warning prior to lahar descent, particularly for lahars descending the Achupashal *Quebrada* (Figure 6.15).

FIGURE 6.15 One of the many typical rain-generated lahars produced by moderate precipitation at Tungurahua. Here at the Achupashal *quebrada* on February 11, 2011, the road was closed for several hours until the flow waned.

6.4 THE FEBRUARY 12, 2005 RAIN-GENERATED LAHAR IN THE RÍO VAZCÚN CANYON, BAÑOS

A heavy downpour on February 12, 2005 shortly before 13:00 (local time) from a localized thunderstorm on the northeastern slopes of Tungurahua produced a flood pulse down the *Río* Vazcún and, at 12:59, caused an immediate jump in seismic signal amplitudes of the short-period seismic station, RETU, located near Tungurahua's climbers refuge at 3,900 m elevation. OVT personal issued an alert at 13:12 over the emergency radio system that a big lahar was *en route*. Civil Defense sent out the order via the same radio system that all people in the El Salado hot spring bathing facility located within the Vazcún *Quebrada* should evacuate immediately. By 13:30 the AFM registry for the station (elevation 2,600 m) located on the west bank of the *Río* Vazcún had increased to more than 3,800 counts in the 10 Hz band (Figure 6.16). Personnel of the Baños Civil Defense reported that the lahar arrived at the bathing pools at 13:28 and had entirely inundated them by 13:40 (Figure 6.16(a)–(c)). Fortunately, all bathers had heeded the warning and abandoned the facility, in particular the 13 people who had been soaking in small pools near the riverside. Field measurements afterward showed that the peak discharge had been about 100 m^3/s and that flow velocities were 3–7 m/s (Williams et al., 2008). The total volume of the lahar was estimated at 54,000 m^3 (IGEPN internal report, 2005). During this particular storm, no rain was registered in Baños or at the El Salado hot spring bathing facility. The storm was localized at the north–northeast summit area (elevation 5,000 m) and clear atmospheric conditions allowed OVT monitoring scientists to see the storm buildup from the OVT. No rain gauge existed high on the volcano to record the precipitation rate or its accumulation; consequently, the triggering threshold for this rainfall-induced lahar is not known.

When heavy rains generate lahars with little warning, an AFM and seismic monitoring system provides the initial alert signal. The rapid interpretation of these signals by monitoring scientists and communication of their significance over a dependable radio network is fundamental in providing fast and reliable lahar warnings. During a lahar event, OVT personal provide near real-time interpretation of the incoming AFM and seismic data and warn of a lahar's rapid descent.

Most lahars produced by Tungurahua from 2000 to 2011 were small (Figure 6.15).. Moderate rainfall caused the lahars, and they did not cause a dramatic rise in the AFM registry because precipitation accumulation was gradual and did not begin with abrupt, intense rainfall accumulation, as did the February 12, 2005 event.

At Tungurahua rain-generated lahars commonly have a high concentration of sand, pebbles, and cobbles, and relatively few boulders. Granulometric analysis by Williams et al. (2008) on three samples of the February 12, event showed that the sand-size fraction was 53–66 wt.% followed by pebbles and cobbles with 29–42 wt.%. One of the few examples of large clasts being transported during the February 12, event was a mound of meter-scale boulders left around a bridge pier in the Vazcún stream.

6.4.1 Discussion

Primary lahars are often life-threatening events since they can travel very fast, occupy entire channels, sometimes even entire valleys, and destroy important transport routes that are needed for evacuations. As has been shown by study of Cotopaxi lahars, discharge can surpass 40,000 m^3/s, and lahars can arrive at populated areas within an hour of onset.

In areas where potential lahars are a threat, the safety of the local residents depends on them knowing their local geography, being informed of the hazard zones, and knowing the location of their home with respect to these zones. Safety also depends on residents knowing where nearby temporary safe areas are located and how to get to them quickly. Civil protection groups, with the aid of experienced volcanologists and geophysicists, typically provide information to citizens prior to activity that could produce lahars. Volcanic hazard maps need to be made and distributed well in advance, and scientists and local civil defense leaders need to begin working together before a crisis situation develops (Peterson, 1988).

Lahars have the potential to take many lives because of their far-downstream reach. Even with the clear delineation of high-risk zones on volcanic-hazard maps, governmental planning agencies frequently do not restrict urban sprawl on older high-risk lahar surfaces because they perceive

FIGURE 6.16 One of the largest lahar events ever measured at Tungurahua occurred on February 12, 2005. It flowed down the Vazcún *quebrada* at 3–4 m/s. An early warning was given over the Civil Defense Radio System, so that people in the El Salado bathing facility could evacuate. Photographs: (a) and (b) Evidence of the quickly rising slurry. *(Photos courtesy Civil Defense, Baños.)* (c) Graph of the progression of the AFM registry and when the alert was given. The dip in the peak may represent a blockage by a landslide, which briefly impeded flow. The lahar was triggered by a localized thunderstorm. AFM, acoustic-flow-monitor.

that restrictions would be politically unpopular and difficult to enforce. The tragedy at Armero, Colombia, where 23,000 people died from the impacts of lahars generated by Nevado del Ruiz volcano, could have been prevented had residents and public officials been informed about the potential for lahar inundation, its danger, and how to respond to that danger. A similar magnitude lahar event had devastated Armero 140 years earlier in 1845 (Voight, 1990; Voight et al., 2013).

Towns, housing subdivisions, and an array of important infrastructure and lifelines now unfortunately occupy many areas around Cotopaxi inundated by lahars during the 1877 eruption. Urban sprawl in lahar-hazard zones has increased despite distribution of three generations of volcanic-hazard maps to the mayors and planning offices of Latacunga and San Rafael, and despite numerous presentations explaining lahar hazards by volcanologists in each municipality.

Given their involvement in the analysis of volcanic hazards and monitoring data and given the implications these may have across a broad geographic and social spectrum, volcanologists are the individuals best suited for translating scientific information to public authorities involved with risk management. Volcanologists have to be prepared to work actively alongside risk managers and may have to repeat training when authorities and the "trained" risk managers are replaced. Risk managers should appreciate that practicing volcanologists know the eruptive patterns and lahar history of their local volcano and that they have a knowledge base and institutional memory about volcanic hazards. Ideally volcanologists and risk managers can form a close working relationship based on trust and confidence.

6.4.2 Conclusions

Steep slopes, voluminous loose volcanic debris, and abundant water sources on volcanoes combine to make lahars common during or after eruptions. Large snow-capped stratocones with extreme relief, like Cotopaxi, are the perfect candidates to generate voluminous far-traveled primary lahars. When primary lahars follow a long repose period (i.e., hundreds of years), population centers and infrastructure are frequently built upon old lahar deposits, which present good building sites. Unfortunately, these sites are also likely to be inundated by future lahars.

Both Cotopaxi and Tungurahua volcanoes are well-monitored with a broad suite of modern instruments. Cotopaxi has not erupted for 137 years. Consequently, the eruptive sequence and mitigation interval between the first signs of unrest and the onset of pyroclastic flows and lahars is poorly known. However, study of the volcano's stratigraphy tells us that a likely future eruption scenario may be a combination of seismic unrest and initial Strombolian–Vulcanian eruptive activity that evolves to subplinain- or plinian-style eruptions—a pattern much like those of historical VEI 3–4 eruptions at Cotopaxi (Hall and Mothes, 2008; Pistolesi et al., 2011).

Monitoring of a newly awakened glacier-clad volcano involves many uncertainties, including timing of the generation of extensive pyroclastic flows, the extent of lahar formation, and the downstream evolution and potential impacts of these flows on nearby populations. How will the population be alerted and will the alert be effective? Of particular concern in Ecuador is the safety of the approximately 100,000 people who live within or near lahar pathways in the Los Chillos and Latacunga valleys. What happens in the case of a large lahar that occurs during an eruption preceded by few precursors? How will people in hazard zones respond to warnings and how will they get out of harm's way?

Hazard prevention work at Cotopaxi has continued at varying levels for three decades with IGEPN scientists working alongside local and national civil defense groups. Nonetheless, a number of factors contribute to a potentially unfavorable social scenario and perhaps high loss of life and property should Cotopaxi reactivate quickly and produce lahars similar in size or even smaller than those of the June 1877 event. These factors include: insufficient local understanding of Cotopaxi's potential lahar hazards, insufficient mitigation of the potential impact of widespread lahar inundation, and a common local attitude to the effect that "the volcano will not reactivate in my lifetime."

In contrast, Tungurahua has been erupting for 15 years and has generated secondary lahars from the beginning. Although mostly of small size, these flows cut through populated areas and roads and have destroyed numerous bridges. But because they are a common occurrence during rainy periods, people acknowledge them and adjust their lives accordingly. Given their frequent occurrence, relative small sizes, and restriction to narrow valleys, the local threat posed by Tungurahua lahars is far less than the threat posed by voluminous, far-traveled Cotopaxi lahars, whose eventual recurrence will likely have a grave and long-term (1–10 years) negative effect on the national economy of Ecuador.

ACKNOWLEDGMENTS

The authors thank the Office of Foreign Disaster Assistance of U.S.A.I.D and the U.S. Geological Survey's (USGS) Volcano Disaster Assistance Program (VDAP) for funding to study huge lahars from Cotopaxi; VDAP for valued collaboration with the IGEPN during unrest at Cotopaxi and Tungurahua volcanoes; John Pallister and Jon Major of USGS for their constructive reviews of this manuscript. The SENPLADES program in Ecuador provides funding and contributes to running the IGEPN's geophysical monitoring networks. Help of IG staff and keen volunteers in database upkeep and assistance in 24/7 monitoring at IGEPN and OVT is greatly appreciated. The authors acknowledge Steve Schilling of USGS for instructing IG personnel in the use of LAHARZ, lahar modeling program and Jorge Ordoñez for help in preparing figures. The VUELCO project, which is revisiting some of the Cotopaxi geophysical data from the 2001–2002 seismic crisis, has been a great help. The authors also especially thank Minard L. Hall for his support in many ways.

REFERENCES

Auker, M.R., Sparks, R.S.J., Siebert, L., Crosweller, H.S., Ewert, J., 2013. A statistical analysis of the global historical volcanic fatalities record. J. Appl. Volcanol. 2, 2. http://dx.doi.org/10.1186/2191-5040-2-2.

Hadley, K.C., LaHusen, R.G., 1995. Technical Manual for the Experimental Acoustic Flow Monitor. U.S. Geological Survey. Open-File Report 95−114 25 pp.

Hall, M.L., Mothes, P., 2008. The rhyolitic-andesitic eruptive history of Cotopaxi, volcano, Ecuador. Bull. Volcanol. 70 (6), 1−28. http://dx.doi.org/10.1007/s00445-007-0161-2.

Hall, M.L., Samaniego, P., Le Pennec, J.-L., Johnson, J.B., 2008. Ecuadorian Andes volcanism: a review of late pliocene to present activity. J. Volcanol. Geotherm. Res. 176, 1−6.

Hall, M.L., Steele, A.L., Mothes, P., Ruiz, M.C., 2013. Pyroclastic density currents (PDC) of the 16−17 August 2006 eruptions of Tungurahua volcano, Ecuador: geophysical registry and characteristics. J. Volcanol. Geotherm. Res. 265, 78−93.

IGEPN internal report, 2005. Weekly Report from the Tungurahua Volcano Observatory. (07−14 February, 2005). Report No. 260. Instituto Geofisico, Quito Ecuador. See web page: www.igepn.edu.ec (Informes-_Volcánicos).

Le Pennec, J.-L., Jaya, D., Samaniego, P., Ramón, P., Moreno, S., Egred, J., van der Plicht, J., 2008. The AD 1300−1700 eruptive periods at Tungurahua volcano, Ecuador, revealed by historical narratives, stratigraphy and radio-carbon dating. J. Volcanol. Geotherm. Res. 176, 70−81.

Meyer, H., 1907 (publ. In Spanish. Anal. Univ. Central., Quito, vols. LX LXI, LXII). In: den Hochanden von Ecuador.

Mothes, P., Hall, M., Janda, R., 1998. The enormous Chillos valley lahar: an ash-flow generated debris flow from Cotopaxi volcano, Ecuador. Bull. Volcanol. 59, 233−244.

Mothes, P., Hall, M., Andrade, D., Samaniego, P., Pierson, Th C., Ruiz, G.A., Yepes, H., 2004. Character, stratigraphy and magnitude of historical lahars of Cotopaxi volcano, Ecuador. Acta Vulcanol. 16 (1−2), 85−108.

Mothes, P., 2010. Variations in Lahar Matrices at Cotopaxi Volcano and Their Hazard Implications. Abstract NH13C-05 presented at 2010 Fall Meeting, American Geophysical Union Fall meeting, San Francisco, Calif. 13−17 Dec.

Ordoñez, J., Samaniego, P., Mothes, P., Schilling, S., 2013. Las Potenciales Zonas de Inundación por Lahares en el Volcán Cotopaxi (The Potential Lahar Inundation Zones of Cotopaxi Volcano) (IGEPN internal report, Quito, Ecuador).

Peterson, D.W., 1988. Volcanic hazards and public response. J. Geophys. Res. 93, 4161−4170.

Pierson, Th C., Janda, R., Thouret, J.C., Borrero, C., 1990. Perturbation and melting of snow and ice by the 13 November 1985 eruption of Nevado del Ruiz, Colombia, and consequent mobilization, flow and deposition of lahars. J. Volcanol. Geotherm. Res. 41, 17−66.

Pistolesi, M., Rosi, M., Cioni, R., Cashman, K.V., Rossotti, A., Aguilera, E., 2011. Physical volcanology of the post-twelfth-century activity at Cotopaxi volcano, Ecuador: behavior of an andesitic central volcano. Geol. Soc. Am. Bull. 123, 1193−1215. http://dx.doi.org/10.1130/B30301.1.

Ruiz, M.C., Lees, J.M., Johnson, J.B., 2006. Source constraints on Tungurahua explosion events. Bull. Volcanol. 68, 480−490.

Samaniego, P., Ordoñez, J., Schilling, S., Mothes, P., 2011. Los posibles escenarios eruptivos del Cotopaxi y la delimitación de zonas inundadas por lahares: Aplicación de LAHARZ en el drenaje Sur. (Possible eruptive scenarios of Cotopaxi and delimitation of lahar inundation zones: Application of LAHARZ en the southern drainage). Unpublished report for the InterAmerican Development Bank, Quito. 34 p.

Smith, G.A., Fritz, W.J., 1989. Volcanic influences on terrestrial sedimentation. Geology 17, 375–376.

Thouret, J.-C., 1990. Effects of the November 13, 1985 eruption on the snow pack and ice cap of Nevado del Ruiz volcano, Colombia. J. Volcanol. Geotherm. Res. 41, 17–66.

Vallance, J.W., 2000. Lahars. In: Sigurdsson, H., Houghton, B.F., McNutt, S.R., Rymer, H., Stix, J. (Eds.), Encyclopedia of Volcanoes. Academic Press, London, pp. 601–616.

Vallance, J.W., Scott, K.M., 1997. The osceola mudflow from Mount Rainier: sedimentology and hazard implications of a large clay-rich debris flow. Geol. Soc. Am. Bull. 109 (2), 143–163.

Vallance, J.W., 2005. Volcanic debris flows. In: Jakob, M., Hungr, O. (Eds.), Debris Flow Hazards and Related Phenomena. Praxis-Springer, Berlin and Heidelberg, pp. 247–274.

Voight, B., 1990. The 1985 Nevado del Ruiz volcano catastrophe: anatomy and retrospection. J. Volcanol. Geotherm. Res. 42 (1), 151–188.

Voight, B., Calvache, M.L., Hall, M.L., Monsalve, M.L., 2013. Nevado del Ruiz volcano, Colombia 1985. In: Bobrowsky, P.T. (Ed.), Encyclopedia of Natural Hazards. Springer, Berlin Heidelberg London, pp. 732–738.

Williams, R., Stinton, A.J., Sheridan, M.F., 2008. Evaluation of the Titan2D two-phase flow model using an actual event: case study of the 2005 Vazcún valley Lahar. J.Volcanol. Geotherm. Res. 177, 760–766.

Wolf, T., 1878. Memoria Sobre el Cotopaxi and su Última Erupción Acaecida el 16 de Junio de 1877 (Memoir of Cotopaxi and its Final Eruption of 26 June, 1877). Imprenta del Comercio, Guayaquil, Ecuador.

Chapter 7

In situ Volcano Monitoring: Present and Future

Gilberto Saccorotti [1], Masato Iguchi [2] and Alessandro Aiuppa [3,4]

[1] *Istituto Nazionale di Geofisica e Vulcanologia e Sezione di Pisa, via U. Della Faggiola 32, Pisa, Italy,* [2] *Kyoto University, Disaster Prevention Research Institute. Sakurajima Volcano Research Center, Sakurajima-Yokoyama, Kagoshima, Japan,* [3] *Università di Palermo, DISTEM. Via Archirafi Palermo, Italy,* [4] *Istituto Nazionale di Geofisica e Vulcanologia – Sezione di Palermo. Via La Malfa, Palermo, Italy*

ABSTRACT

During the last couple of decades, volcanology has evolved significantly, allowing for an improved understanding of volcanic processes preceding, accompanying and following eruptive events. Key elements to these achievements are the huge amounts of high quality data being collected by networks of increasingly sensitive instruments deployed at active volcanoes. The diffusion of continuous, precise measurements of: (1) wide-band ground displacement; (2) flux and chemistry of volatile emissions; and (3) the spatio-temporal variations of potential fields (e.g., gravity) now permit imaging the mechanism that controls mass transfer underneath volcanoes to an unprecedented level of detail. Joined to the establishment of a solid theoretical framework, these progresses are contributing to the evolution of the short- to medium-term forecast of volcanic eruptions from a phenomenological approach to a forecasting based on understanding of the physics of the causative processes.

7.1 INTRODUCTION

Volcanic eruptions are preceded by the migration of multiphase mixtures through a subsurface fracture network, which can be detected at the surface in terms of changes in geophysical (ground deformation, seismicity, gravity) and geochemical observables. A primary goal in monitoring active volcanoes is therefore to capture signs of such movements, and to interpret the former in terms of precursors to an eruption. However, changes in geophysical and/or geochemical tracers are not always related to the dynamics of magma bodies alone, rather reflecting the complex interplay between tectonic and gravity forces acting at both regional and local scales,

Volcanic Hazards, Risks, and Disasters. http://dx.doi.org/10.1016/B978-0-12-396453-3.00007-1

and the dynamics of hydrothermal and magmatic systems. Understanding the relationships between these processes is therefore one of the major goals to be attained toward a quantitative assessment of volcano dynamics and eruption precursors. Experiences worldwide now show that this objective is best achieved by integrating a combination of approaches, rather than relying on any single method or precursory indicator. Following this premise, this chapter intends to present a balanced overview of the current status of multidisciplinary, in situ volcano observations, in turn indicating the most promising directions for improving the significance of monitoring data in terms of a forecasting tool.

7.2 GROUND DEFORMATION

The observation of ground deformation at volcanoes is one of the most relevant methods for in situ volcano monitoring. During the precursory stages of volcanic eruptions, inflation of the edifice is observed associated with seismicity increase (Nishi et al., 1999; Ukawa et al., 2000; Yamashina and Shimizu, 1999). Ground deformation allows us to estimate the volume of intruded magma. Negative ground deformation (deflation) is also observed to be associated with volcanic eruptions and, occasionally, volcanic tremor (TR) (Iguchi, 2013; Iguchi et al., 2013). Volumes of ejecta by volcanic eruptions are comparable to negative volume changes as inferred from the ground deflation. As a monitoring method, well-installed deformation sensors are more sensitive than detection of volcanic earthquake resulting from rock failure. Increase in seismicity is generally not detected prior to middle or minor size eruptions, whereas recent ground deformation measurement techniques can detect inflation and deflation cycles even for such minor-scale eruptions, independently from the recording of volcanic earthquakes. In addition, the onset of ground deformation often precedes the onset of seismicity of volcanic earthquakes (Iverson et al., 2006).

7.2.1 Precise Leveling

Precise leveling is a traditional method to detect relative vertical displacement between two points by using a leveling instrument and two staffs. Reading precision in the vertical is 0.01 mm and vertical displacement measured by leveling is on the order of 1 mm. Precision leveling has a long history in volcano monitoring. Beside tide gauge measurement, vertical displacement at the Campi Flegrei caldera (Italy) has been monitored by precise leveling since 1905 (Del Gaudio et al., 2010). The leveling survey is useful to grasp a long-term activity.

An eruption starting on January 12, 1914 at the Sakurajima volcano was the largest in the twentieth century in Japan, and 1.34 km^3 of lava and 0.52 km^3 of pumice and ash were ejected from the craters. The first leveling

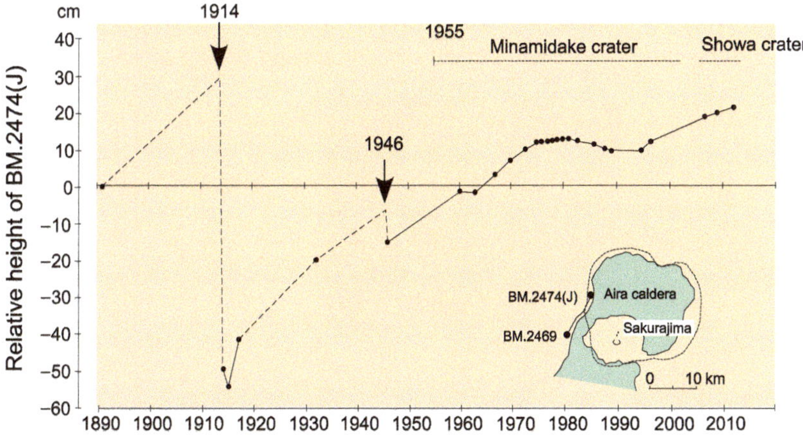

FIGURE 7.1 The temporal change of elevation of benchmark 2474(J) at the west rim of the Aira caldera, referred to benchmark 2469 for the period 1891–2012. Dashed lines indicate the estimation of uplifting and subsidence, considering tide gauge data. After 1914, three eruptive episodes were recorded; the effusion of lava in 1946, vulcanian eruptivity at the Minamidake crater during the period from 1955 to 2002, and minor vulcanian eruptions at the Showa crater since 2006.

survey in south Kyushu was conducted in 1891. After the 1914 eruption, a second leveling survey was conducted and a significant ground subsidence was detected all over south Kyushu. By combining tide gauge data with precise leveling, subsidence of benchmark BM2474 at the western rim of the Aira caldera, north of Sakurajima, was estimated to be on the order of 0.8 m. The center of the ground subsidence is not located beneath Sakurajima but rather near the center of the Aira caldera. A model of elastic deformation induced by a spherical pressure source embedded in a homogeneous half space was applied to the spatial distribution of the vertical displacement. The deflation pressure source was estimated to be located at a depth of 10 km near the center of the Aira caldera. This indicates that the magma reservoir, as the origin of the magma product ejected by the 1914 eruption, was not located beneath Sakurajima, but rather beneath the Aira caldera. After the 1914 eruption, uplift of the Aira caldera has continued except the 1946 eruption and frequent repetition of vulcanian eruptions at the Minamidake crater during the period from 1974 to 1992 (Figure 7.1). This indicates that magma has been stored in the magma reservoir beneath the Aira caldera at a rate of 10^7 m^3/year.

7.2.2 Electronic Distance Measurement

Electronic distance measurement (EDM) is a technique to measure slope distances along flanks of volcanoes, similarly to a range-finder. The slope distance is obtained by measuring phase differences between modulated

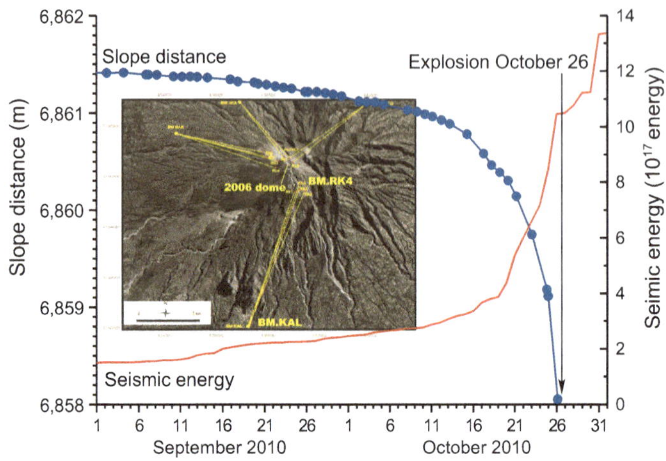

FIGURE 7.2 The temporal change of slope distance from the flank (BM. KAL) to the summit (BM. RK4) prior to the eruption on October 26, 2010, at Merapi volcano, Indonesia. The seismic energy of volcanic earthquakes is shown by the red curve. Ground deformation and seismicity accelerated starting on mid-October. The deformation rate attained 1 m per day immediately before the eruption. *Center for Volcanology and Geological Hazard Mitigation, Geological Agency, ESDM, Republic of Indonesia.*

light beams radiated from the measuring device and that reflected by a mirror installed at a target. For precise measurement, the recorded values are corrected for atmospheric condition (temperature, pressure, and humidity). The effectiveness of EDM deformation data was demonstrated during the preparatory phase of the catastrophic eruption on May 18, 1980 at Mt. St. Helens (USA). The observed distance change exceeded 3 m prior to the eruption.

Figure 7.2 shows slope distance change prior to the eruption on October 26, 2010 at Merapi volcano, Indonesia. Minor but continuous shortening of the slope distance began in May 2009 and the ground deformation rate gradually increased from mid-October, 2010. The increase of ground deformation was associated with increase in the rate of volcanic earthquakes. The deformation rate attained 1 m per day immediately before the eruption. The accelerating pattern is different from the slope distance change prior to the 2006 eruption. The 2006 eruption was effusive; however the 2010 eruption was more violent. Ballistic bombs fell within a distance of 1.5 km and pyroclastic flow reached a distance of 8 km from the summit, accompanying pyroclastic surge.

7.2.3 Global Navigation Satellite System

Ground deformation of volcanoes can be detected by receiving radio waves from Global Navigation Satellite System (GNSS). The benefit of GNSS is 3D positioning (latitude, longitude and elevation), longer-distance measurement,

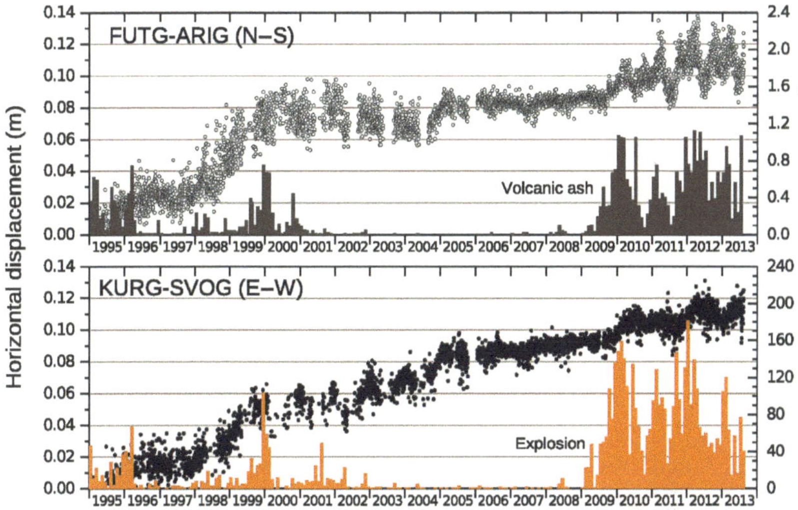

FIGURE 7.3 Temporal change in horizontal distance of baselines along the north–south and east–west directions measured by GPS during the period 1995–2013. Note that the fluctuations of data points get smaller after replacement of GPS receivers and antennas in June 2005. Temporal changes of eruptive activity are shown by monthly weight of volcanic ash (top) and monthly number of vulcanian eruptions (bottom). The eruptive activity at the Showa crater began in June 2006 and increased in 2009. The monthly weight and numbers before 2006 are based on the eruptivity at the Minamidake crater.

higher sampling measurement, and stable continuation of long-term observation. The GPS (Global Positioning System) launched by USA is the first GNSS widely used for volcano monitoring and it is now a standard observation method of ground deformation of volcano. The combination with GLONASS operated by Russia and the Galileo system prepared by EU and the European Space Agency is expected to be applied for supplementing GPS and higher precision of positioning.

Figure 7.3 shows the temporal change in the horizontal lengths of the baselines across Sakurajima volcano, measured by GPS during the period from 1995 to 2013. The eruptive activity at the Minamidake crater reached a peak of 104 vulcanian eruptions in December 1999. The extension of baselines along the north–south and east–west directions began 2 years before the peak of activity. Eruptive activity at the Showa crater began in June 2006; however no significant coeruptive ground deformation was detected, whereas a minimal change continued until September 2009. Significant ground deformation began in October 2009 when the number of explosions and the ejected weight of volcanic ash increased. The horizontal distances of the baselines along the north–south direction showed an extension during the period from October 2009 to April 2010. When more than 130 explosions occurred during the period from December 2009 to

March 2010, the extension of baseline along the east–west direction became clear. The east–west baseline showed an extension of 1 cm and the baseline in the north–south direction was extended by 2 cm during the period from October 2009 to April 2010. The east–west baseline showed a disturbance of up to 1 cm after April 2010, whereas the north–south baseline repeated a cycle of contractions and extensions. This turned into a contraction in May 2010 and showed a clear extension in November 2010. The extension turned into a contraction again in April 2011, followed by a restart of an extension in August 2011. The change in baseline ranged from 1 to 2 cm for both the extension and contraction episodes. When the extension of the north–south baseline was observed during the periods from November 2010 to April 2011 and from August to October 2011, the monthly number of explosions and the ejection of volcanic ash showed a tendency to increase or stayed at a high level. In contrast, contractions along the north–south baseline corresponded to a decrease in or low-level activity during the periods from April to July 2010 and from April to July 2011.

7.2.4 Tilt and Strain Observation

Tilt and strain observation can detect minor ground deformation associated with volcanic activity. Resolutions of tiltmeters and strain meters, which are spatial derivatives of vertical displacement and horizontal displacement, respectively, are an order of nanoradian and nanostrain. To detect such minor changes, tiltmeters and strain meters are installed in borehole or underground tunnel. Borehole tiltmeters are used world-wide for volcano monitoring and volumetric strain meters are used at some volcanoes, such as Hekla (Iceland), Monserrat, and Izu-Oshima (Japan) volcanoes. Tilt and strain observations in underground tunnels at Sakurajima volcano show an example of detection of minor precursory inflation immediately before vulcanian eruptions. The tiltmeter is a water-tube type with a length of 28 m, directed toward the crater and tangential to the crater. The strain meter has three sensors to measure linear strain in radial, tangential, and oblique directions. An advantage of three-component strain measurement is the estimation of depth of a pressure source. During the eruptive activity of the summit crater of Minamidake, radial tilt shows upward tilt of the crater side and both radial and tangential strains increase before the occurrence of an explosion (Figure 7.4). The depth of a pressure source is estimated to be located at a depth of 4 km. On the other hand, radial and tangential strains have inverted polarities prior to vulcanian eruptions at the Showa crater, east of Minamidake; contraction radial strain and extension tangential strain. These strain and minor tilt changes indicate a shallow source (c. 1 km). The strain changes range from 3 to 100 nanostrain; nonetheless, precursory inflation strain changes are detected for 90% of the vulcanian eruptions.

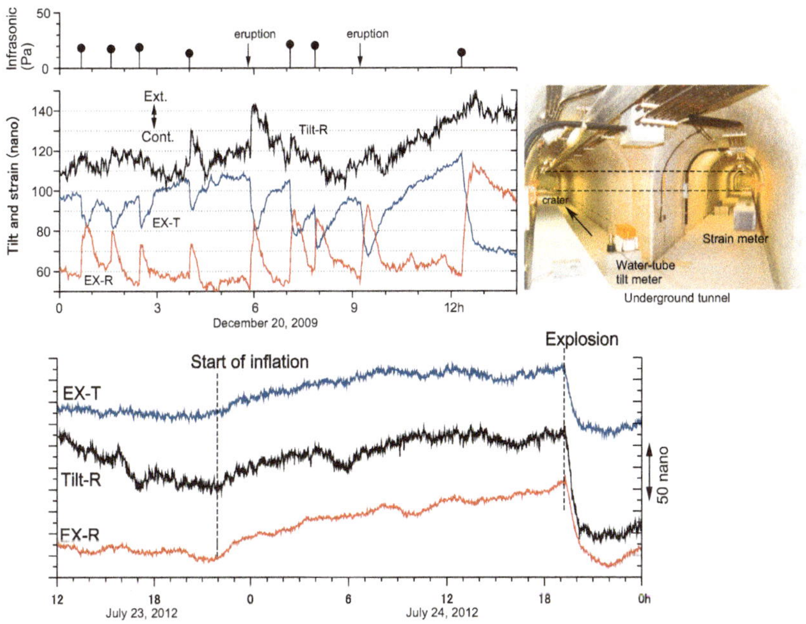

FIGURE 7.4 Tilt and strain changes prior to vulcanian eruptions at the Showa crater (top) and the Minamidake crater (bottom). Dots at the top indicate occurrence of vulcanian eruptions. Radial tilt is indicated by "Tilt-R"; "EX-R"; and "EX-T" indicate strain changes along the radial and tangential directions to the crater, respectively. Note that radial and tangential strains have different polarities associated with eruptions at the Showa crater (top), whereas polarities are the same for eruptions at the Minamidake crater (bottom).

7.3 GRAVITY OBSERVATIONS

Phenomena preceding volcanic eruptions can be simplified in terms of magmatic fluids being transferred from depth toward the surface. Since magma uprising is accompanied by volatiles ex-solution, large density contrast among magma–gas mixtures and the surrounding country rock are established. This may lead to significant variations in the local gravity field, which can, in turn, be detected through gravity measurements.

7.3.1 Discrete 4D (Time-lapse) Gravity Measurements

Since the 1950s, microgravity surveys were shown to be an effective tool for measuring the subsurface mass and/or density changes that occur at different temporal scales during periods of volcanic unrest. Till the late 1990s, the usual approach to microgravity measurements has been that of 4D (three dimensions of space and one of time) surveys (e.g., Locke et al., 2003). In this approach, a single (or a few) instrument(s) is (are) used to measure the local gravity field at a number of benchmarks spanning the area of interest.

Individual measurements are taken relative to a reference point, outside the active area. The whole survey is then repeated at successive times, thus allowing the retrieval of time–gravity changes, relative to the reference point. Gravity data require corrections for the effects of solid earth and ocean-loading tides, atmospheric changes, and instrumental effect ("drift" and calibration shifts; Carbone and Rymer, 1999). Additionally, elevation changes (uplift/subsidence) of the ground must be taken into account, thus requiring the accomplishment of accompanying deformation measurements, such as precise leveling and GPS. Once these effects are removed, the residual gravity changes can be interpreted in terms of subsurface mass and/or density variations occurring on wide time (e.g., months to years) and space (tens of kilometers) scales within volcanic systems (e.g., Rymer and Brown 1989; Battaglia et al., 2006; Carbone et al., 2009).

7.3.2 Continuous Microgravity

An inevitable drawback of the discrete 4D approach is the fact that the time resolution is constrained by the time span (days to years) between two consecutive surveys. However, processes accompanying volcanic unrest are very broad-band, occurring over timescales spanning the 10^0–10^8 s range. To investigate these changes over their full bandwidth, it is therefore necessary to monitor the systems with a high temporal rate and for extended periods of time. In contrast to the discrete 4D spatial microgravity survey, continuous gravity measurements are conducted at a single location for long time intervals. The temporal resolution of these continuous studies is excellent (up to 10 samples/s; e.g., Carbone et al., 2010). However, due to the high cost of instrumentation, continuous measurements are rarely taken simultaneously at more than one to two locations in the same area, a fact that greatly limits the spatial resolution. Furthermore, it is generally difficult to keep spring gravimeters working continuously for extended periods of time under harsh conditions. Nonetheless, the results obtained thus far using only one or two continuously recording instruments, possibly together with other sensors, illustrates the wealth of information on the rates and timing of subsurface movements of mass that can be derived from even a crude attempt at continuous gravity data analysis (e.g., Williams-Jones et al., 2008).

7.3.3 Case Studies

7.3.3.1 Etna Volcano, Italy

Of particular interest is the continuous gravity sequence collected at Etna Volcano (Italy) during a 2.5-month-long time interval, bracketing the onset of the 2002–2003 eruption (Carbone et al., 2007). The start of the eruption on the northeastern flank was marked by a rapid gravity decrease of 400 μGal in less than 1 h, followed immediately by a gravity increase at a rate of 100 μGal/h to near-background levels. This anomaly was interpreted as the consequence of the opening of a fracture system along the northeastern rift and subsequent magma intrusion from the central conduit toward the lower elevation eruptive vents. During the eruption, three smaller-amplitude, negative anomalies (10–30 μGal) were observed to occur simultaneously with increases in the amplitude of volcanic TR. Detailed analysis of these data showed a clear inverse correlation between gravity decrease and TR increase, suggesting a joint source for the two geophysical observables. These anomalies occurred during periods of transition from intense lava fountaining to mild Strombolian activity. Carbone et al. (2006) thus hypothesized that the joint tremor-gravity source could be associated with accumulation of a low-density gas cloud within the shallow volcanic plumbing system (see Figure 7.5 in Seismology Section).

7.3.3.2 Nisyros Volcano, Greece

At Nisyros Volcano (Greece), short-term gravity variations of 20–30 μGal occurred over time scales of 40–60 min, in conjunction with seismic bursts and rapid ground deformation (Gottsmann et al., 2007). Independent data indicate that the local dynamics is governed by shallow aqueous fluid movement within the hydrothermal system. Gottsmann et al. (2007) suggested that

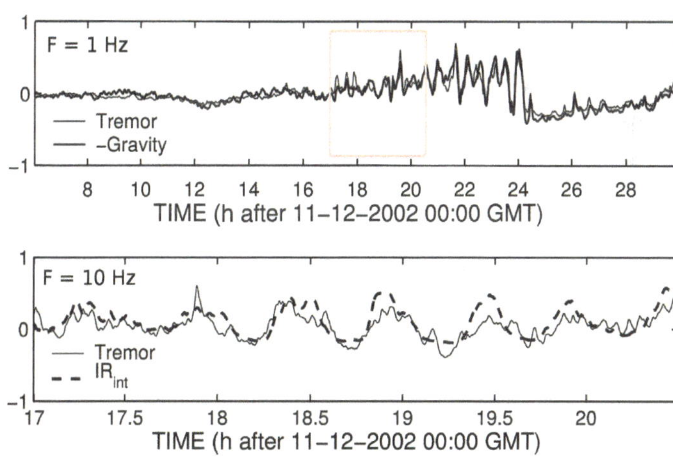

FIGURE 7.5 (Top) Time evolution of tremor effective amplitude at frequency 1 Hz, compared with (sign-reversed) data from a continuous recording spring gravimeter. Data are from the 2002–2003 Etna eruption, and are associated with the transition from lava fountain to pulsating Strombolian activity. After normalization for the respective maximum amplitudes, the two time series overlap almost perfectly. (Bottom) Tremor amplitude at frequency 10 Hz compared with light intensity from an infrared camera pointed to the eruptive vent for the subperiod marked by the gray rectangle in the top panel. Light changes are associated with pulsating Strombolian explosions, whose intensity is well tracked by the high-frequency tremor signal. Gravity and thermal data were provided by Daniele Carbone and Mike Burton, respectively.

the data could be best explained by assuming thermo-hydro-mechanical variations caused by the release and upward migration of hydrothermal fluids during anomalous degassing of a deeper magmatic reservoir. The study by Gottsmann et al. (2007) also highlights a main issue concerning the interpretation of gravity changes during periods of unrest, i.e., the possibility of discriminating between shallow hydrothermal and deep magmatic contributions. To this purpose, additional constraints can be gained from the joint inversion of ground deformation and gravity signals, as demonstrated for the Phlegrean Fields Caldera, Italy, case study (e.g., Battaglia et al., 2006).

7.3.3.3 Kilauea Volcano, Hawaii (USA)

Carbone and Poland (2012) reported gravity oscillations with a period of ~150 s, observed at two continuous stations, on the summit of Kilauea Volcano, Hawai'i. Comparison with simultaneous seismic signals allowed Carbone and Poland (2012) to exclude inertial acceleration as the cause of the gravity fluctuations. Rather, simplified source modeling suggests that the oscillations are caused by density inversions in a magma reservoir located ~1 km beneath the east margin of Halema'uma'u Crater in Kilauea Caldera—a location of known magma storage. This result may thus represent rare evidence that convection in a shallow magma chamber is able to produce measurable geophysical signals, as theoretically postulated by Vassalli et al. (2009). In a subsequent work, Carbone et al. (2013) observed a strong gravity decrease (about 200 μGal) accompanying a marked change in lava level within Kilauea's summit eruptive vent. Using an appropriate model of the eruptive vent geometry, and from comparison with optical recordings of the lava free-surface, Carbone et al. (2013) were thus able to invert the observed gravity changes for lava density, obtaining values as low as $950 \pm 300\ kg/m^3$. Representing the very first *in situ* geophysical determination of a fundamental property of magmatic fluids, these data open important perspectives toward the use of continuous gravity observations for constraining dynamical processes in active magmatic systems.

7.4 IN SITU MONITORING OF VOLCANIC GASES

7.4.1 Premise, Scope of This Section

The last 15 years have seen major advances in our ability to make high-rate (up to 1 Hz), nearly real-time, volcanic gas observations with compact, fully automated, instruments. The routine gas measurements, now made with permanent monitoring networks, have enabled the capture of, in nearly real-time, a variety of precursory chemical changes occurring prior to eruptions. This section aims at briefly summarizing the current state-of-the-art instrumental volcanic gas monitoring; the reader is referred to excellent reviews of Symonds et al. (1994) and Giggenbach (1996) for the interpretation of results of *direct sampling* (conventional laboratory analysis of gas samples directly collected from fumarole vents). We put our accent on those instrumental monitoring techniques that are contributing more to high-rate volcanic gas observations, ultimately opening the way for integration between geophysical and geochemical data sets. Although the term *in situ monitoring* should strictly speaking only apply to *direct sampling/sensing* of volcanic fluids, we also give emphasis in this section to *remote-sensing observations* of volcanic fluid chemistry/flux, which are contributing so effectively to our understanding of volcanic processes. Three main fields of chemical monitoring have progressed more in recent years, as discussed below.

7.4.2 SO_2 Flux

Despite its being only the third component (in order of abundance) in volcanic gases, sulfur has long been of tremendous importance in volcano monitoring (Oppenheimer, 2010; Oppenheimer et al., 2011); especially since when the first volcano applications of the correlation spectrometer in the late 1970s demonstrated the relatively ease with which the volcanic SO_2 flux could be detected (Stoiber et al., 1983), and that its time-variations are markedly related to volcanic activity style and vigor (Malinconico, 1979; McGee and Sutton, 1994; Allard et al., 1994). More recently, with the advent of ultraviolet (UV)-sensitive charge-coupled device detectors (McGonigle et al., 2002; Galle et al., 2003), the first networks of fully autonomous scanning spectrometer systems (Edmonds et al., 2003; Burton et al., 2009; Salerno et al., 2009; Galle et al., 2010) far improved the temporal resolution and the continuity of volcanic SO_2 flux observations. Novel UV imaging techniques (UV cameras; Mori and Burton, 2006) are now enabling remote SO_2 sensing with a temporal resolution of up to ~1 Hz, and, hence, the study of rapid SO_2 flux variations associated with either passive degassing (Tamburello et al., 2013), or Strombolian (Mori and Burton, 2009; Dalton et al., 2010; Tamburello et al., 2012) and Vulcanian (Yamamoto et al., 2008; Holland et al., 2011; Nadeau et al., 2011) explosions.

After more than 40 years of research, it is now reasonably established that unrests of intermediate to silicic volcanoes are frequently associated with detectable precursory SO_2 flux variations. During their long-lived dormant periods, silicic volcanoes are pervaded by low-temperature (H_2S-dominated) hydrothermal fluids; as a new unrest cycle commences, the influx of hotter, more oxidized (and, therefore, SO_2-richer) magmatic fluids is likely to leave a clear, measurable track in the SO_2 flux record. The giant 1991 eruption of Pinatubo, for instance, was preceded by an order of magnitude SO_2 flux increase over ~2 weeks, which was interpreted at that time as an evidence of magma intrusion at shallow depth (Daag et al., 1996). On a far smaller eruption scale, the transition from quiescence to explosive (Strombolian/ Vulcanian) activity at Tungurahua, Ecuador (Hidalgo et al., 2013) has been shown to be anticipated (typically by days) by SO_2 flux increases. There have also been recent reports (Olmos et al., 2007; Conde et al., 2013) for that small phreatic eruptions, such as at Santa Ana (El Salvador) in 2005 and Turrialba (Costa Rica) in 2010, can be preceded by long-lasting (months) large (factor ~10–30 increases) SO_2 degassing unrests (Figure 7.6). During dome-building eruptions, SO_2 degassing trends have been particularly significant, although problematic to interpret: while posteffusive phases (e.g., after cessation of dome-building activity) have typically shown decaying SO_2 fluxes, such as after the eruptions of Mount St. Helens in 1980 (Casadevall et al., 1983), Augustine in 1986–1987 (Symonds et al., 1990), Redoubt in 1989–1990 (Casadevall et al., 1994) and 2009 (Werner et al., 2013) and Galeras in 1986–1995 (Zapata et al., 1997), more erratic, large SO_2 flux variations have been observed during long-lived dome-building phases. At Soufrière Hills volcano in Montserrat, for example, sustained (~500 t/day on average; Christopher et al., 2010) but fluctuating SO_2 emissions have been maintained from 1995 to 2010, as extrusive phases alternated with periods of quiescence. Research has shown that a complex interplay between: (1) variable rates of mafic magma supply at depth; (2) permeability drops in upper conduit/dome, due to cooling, "sealing" and collapse of networks of fracture (leading to abrupt decreases in SO_2 flux; Fischer et al., 1994; Edmonds et al., 2003); or (3) permeability increases after large explosions or rock falls dome-collapse events (producing the opposite effect), were the primary controls on rates of SO_2 degassing.

The SO_2 flux has also extensively been measured at mafic volcanoes, where particularly complete data sets have for instance been obtained on Etna (Caltabiano et al., 2004) and Kilauea (Sutton et al., 2001), among others. These studies have proved that SO_2 is a particularly powerful tool to estimate magma degassing budgets, e.g., to evaluate rates of magma supply and degassing (Allard et al., 2006) at pressures <100–150 MPa (pressure below which S is typically degassed from mafic magma; Métrich and Wallace, 2008; Métrich and Mandeville, 2010). It has experimentally been proved that changes in the SO_2 flux regime broadly correlate with changes in the rates of

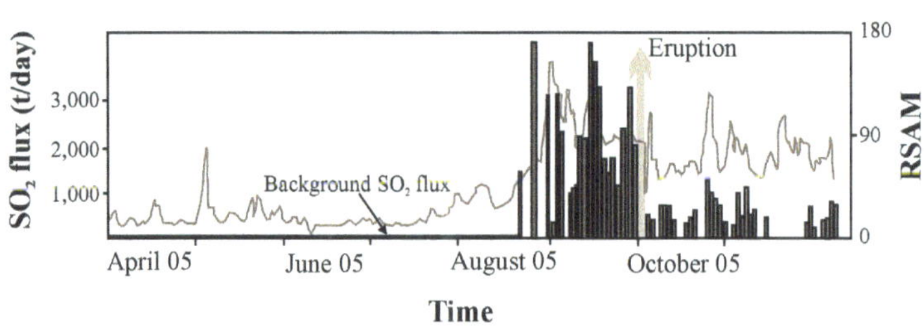

FIGURE 7.6 Degassing unrest of Santa Ana volcano (El Salvador) in 2005 as captured by improved ultraviolet scanning spectrometer networks (NOVAC, Galle et al., 2010). The SO_2 flux was seen to increase by factor 10–30 in the months prior to a phreatic explosion on October 1, 2005. Gray line indicates the Real-time Seismic-Amplitude Measurement (RSAM). *Redrawn from Olmos et al. (2007).*

magma extrusion (e.g., eruption intensity), with exceptionally high SO_2 emissions observed *during* paroxysmal eruptions (Caltabiano et al., 2004; Kazahaya et al., 2004). Precursory SO_2 variations prior to eruptions have been more difficult to identify, however, probably because, in contrast to silicic volcanoes (where SO_2 emissions can be low, or even zero, during dormant periods), open-vent basaltic volcanoes are strong, persistent SO_2 emitters during noneruptive phases, too (as due to continuous degassing-driven magma overturning of low-viscosity magma in conduits; Shinohara, 2008). These continuous background emissions may potentially mask any signal related to fast-decompressing new magma prior to eruption. New horizons to exploring pre-eruptive processes at basaltic volcanoes have fortunately been opened by the measurement of volcanic CO_2 (see below).

7.4.3 Chemistry of Volcanic Plumes: Fourier-transform infrared and MultiGAS at Play

An era of renovated interest in volcanic gas research started in the mid-90s, when the first successful reports on the volcano application of open-path Fourier-transform infrared (FTIR) spectrometers (Mori et al., 1993; Mori and Notsu, 1997; Francis et al., 1995, 1998; Love et al., 1998; Oppenheimer et al., 1998) motivated an explosion of new studies on volcanic plumes released by open-vent (mafic) volcanoes. A major breakthrough arose when incandescent rocks/magma were used as source of IR radiation to obtain fast (1 Hz), real-time observation of $H_2O-CO_2-SO_2-HCl$ gas compositions during lava-fountaining episodes (Allard et al., 2005), gas piston and lava spattering events (Edmonds and Gerlach, 2007), Strombolian explosions (Burton et al., 2007; Oppenheimer et al., 2006), or lava lake activity (Oppenheimer et al., 2004; Oppenheimer and Kyle, 2008; Sawyer et al., 2008a,b). Used in tandem with results of volatile saturation models (e.g., Moretti et al., 2003) and the determinations of pre-eruptive volatile contents in magmas (Blundy and Cashman, 2008; Metrich and Wallace, 2008), FTIR observations contributed to new important discoveries, such as identification of source region, and regimes of segregation and loss, of gas feeding basaltic explosions (Allard et al., 2005; Burton et al., 2007; Edmonds and Gerlach, 2007) and experimental verification of pulsatory magma supply and degassing behavior of active lava lakes (Oppenheimer et al., 2009). FTIR observations have also increasingly been used for volcano monitoring porpoises, such as in Japan (Notsu and Mori, 2010), at Masaya in Nicaragua (Horrocks et al., 2003), and Etna in Italy (Burton et al., 2003). In most cases, however, measurements have remained too sparse and discontinuous to unravel clear compositional patters prior to eruption. Duffell et al. (2003) reported of a temporary change (factor >2 increase) of the otherwise very stable SO_2/HCl ratio of the Masaya plume prior to a minor phreatic explosion in 2001, which they interpreted as evidence of permeability drop and pressure build-up in the conduit prior to the event. An increase of the gas plume SiF_4/SO_2 ratio was reported before eruption at Popocatepétl (Love et al., 1998), though this has remained a single observations to date, and the processes causing this compositional change are not entirely understood (Stremme et al., 2011). In the time being, although high promises are offered by recent first attempts to run fully autonomous FTIR prototype systems, such as at Stromboli (La Spina et al., 2013), far simpler, and cheaper technical solutions, such as the use of the Multi-component Gas Analyser System (Multi-GAS; Shinohara, 2005; Aiuppa et al., 2005), have already demonstrated the value in continuous volcanic gas plume observations.

The Multi-GAS technique stands on the simple idea of assembling together different infra-red (for CO_2, H_2O) and electrochemical (SO_2, H_2S, H_2, HCl) gas sensors into a single, compact sensor kit unit. The concept is therefore not new in itself, since it is perceptibly reminiscent of the sensor assemblages long used for airborne profiling of gas plume compositions (McGee, 1992; Gerlach et al., 1998). The major step ahead brought about by the Multi-GAS method has been the possibility to assemble small, compact prototypes to expedite surveys on volcano summits (e.g., Shinohara and Witter, 2005; Aiuppa et al., 2006, 2008, 2012; Shinohara et al., 2008); and, even more importantly, the demonstration that fully automated permanent Multi-GAS instruments can be field-deployed, and run unattended for months/years to acquire, continuously, and in nearly real-time, the composition of volcanic gas plumes (Aiuppa et al., 2007, 2009, 2010a,b; Shinohara, 2013). The main target of permanent observations has been measurement of the volcanic gas plume CO_2/SO_2 ratio; which, as numerical simulations had insinuated (Allard et al., 2005; Aiuppa et al., 2007, 2009; Burton et al., 2007) was considered a potential indicator of pre-eruptive degassing of deeply rising (or deeply-stored) magma (in view of the large solubility contrast between the two volatiles in melts; Moretti et al., 2003). These expectations have in fact been confirmed when, thanks to the improved temporal resolution of Multi-GAS networks, precursory increases of the volcanic gas CO_2/SO_2 ratio were first detected prior to basaltic eruptions at Etna (Aiuppa et al., 2007, 2010a) and Stromboli (Aiuppa et al., 2009, 2010b, 2011). Even more importantly, the combination of simultaneously measured CO_2/SO_2 ratios (by Multi-GAS) and SO_2 fluxes (by UV spectroscopy) opened the way to obtaining unprecedented time-series of the magmatic CO_2 flux, and allowed first documentation of the large, precursory CO_2 flux variations (Figure 7.7) occurring before basaltic paroxysmal events (Aiuppa et al., 2010b, 2011). These precursory changes in CO_2 flux degassing regime can be common to several basaltic volcanoes, and may be triggered by increasing supply rate of mafic magmas from the mantle source (e.g., Kilauea; Poland et al., 2012), replenishment of shallow (0–3 km a.s.l.) magma storage zones with CO_2-rich bubbles, originated from degassing of deep-rising magma (e.g., Etna; Aiuppa et al., 2010a; Patanè et al., 2013), or precursory leakage of a deeply stored CO_2-rich foam (e.g., Stromboli; Aiuppa et al., 2010b, 2011). Intruding

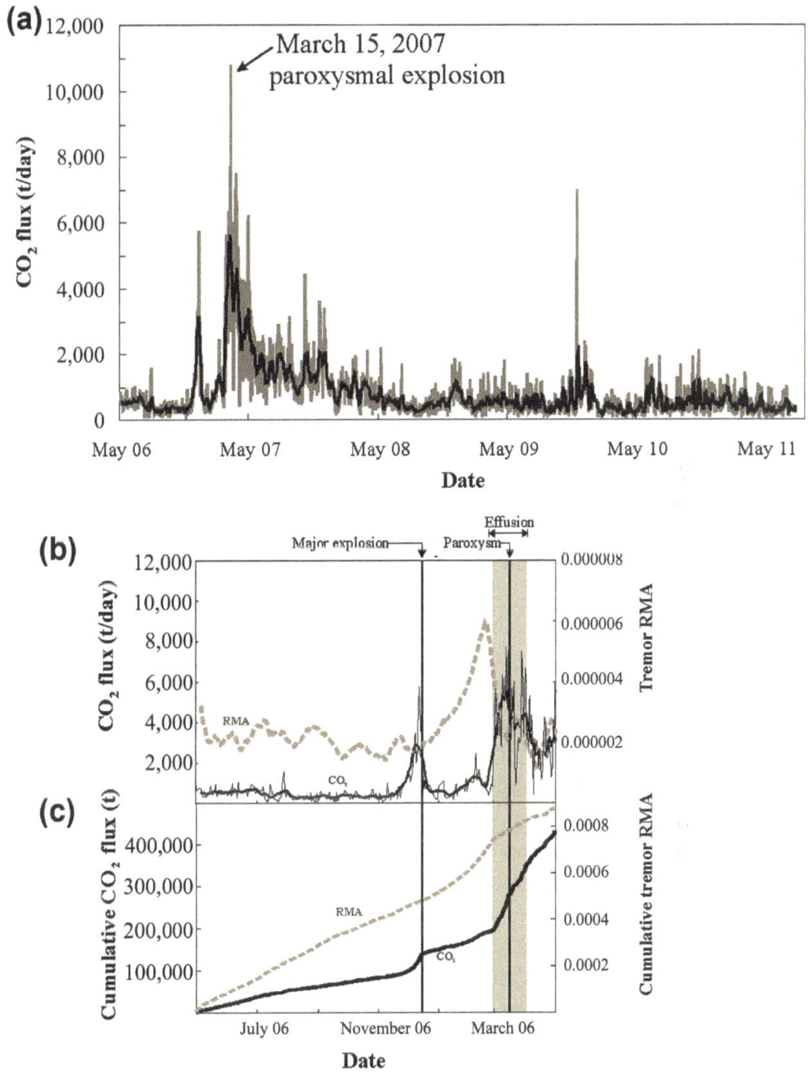

FIGURE 7.7 (a) Five years of continuous daily record of the CO_2 output from Stromboli volcano *(redrawn from Aiuppa et al., 2010b, 2011)*, demonstrating the extraordinary large CO_2 emissions prior to the March 15, 2007 paroxysmal explosion (the most explosive event on the volcano since 2003). A detail of the May 2006–May 2007 period is shown in (b) and (c) (a cumulative version of (b)). This shows that the precursory CO_2 increase started concurrently with onset of the 2007 lava effusion event, and with cessation of Strombolian activity, as demonstrated by the abrupt drop of seismic tremor mean amplitude (RMA). *Data courtesy of Maurizio Ripepe, Università di Firenze.*

mafic magma at depth has also been recently implicated as dominating the volatile budget of intermediate (andesitic) arc volcanoes (Edmonds et al., 2010, 2013), and this mafic magma eventually contributes to triggering as well as sustaining the eruptions. The pulses in CO_2-rich gas recently detected prior to the 2009 Redoubt eruption in Alaska (Werner et al., 2012, 2013) are supporting these interpretations.

7.4.4 Soil CO_2 Flux

It has been known since pioneering work by Baubron et al. (1990) and Allard et al. (1991) that, in addition to crater plume emissions, high amounts of CO_2 are also passively released—as diffuse soil emissions—by the flanks of active volcanoes. With the advent of compact, portable, relatively low-cost CO_2 analyzers, periodic soil CO_2 surveys have increasingly complemented gas studies at, for instance, Campi Flegrei (Chiodini et al., 2010), Etna (Giammanco et al., 1995), Yellowstone (Werner et al., 2000), Long Valley (Bergfeld et al., 2006), Mammoth Mountain (Farrar et al., 1995), Teide (Hernández et al., 1998; Pérez et al., 2013), Miyakejima (Hernández et al., 2001a), Vesuvio (Granieri et al., 2010), Vulcano Island (Chiodini et al., 1996; Diliberto et al., 2002), and many others (see Chiodini et al., 2005, 2008; Pérez et al., 2012, 2013; Burton et al., 2013; for a list of references). These studies have allowed mapping the spatial heterogeneity of soil CO_2 flux regime in volcanic edifices, demonstrating a main structural control on degassing (Finizola et al., 2002). Periodic soil surveys have also allowed the first characterization of long-term trends in diffuse CO_2 degassing (e.g., Chiodini et al., 2010, 2012), and have prompted design, setting up, and field deployment of automatic instruments for continuous soil CO_2 observations, based on either the "accumulation chamber" (Chiodini et al., 1998; Granieri et al., 2003) or "dynamic concentration" (Gurrieri and Valenza, 1988; Gurrieri et al., 2008) technique. Even with this improved temporal resolution, however, clear relations with volcanic activity state have often remained difficult to identify, because: (1) the CO_2 flux in volcanic soil is normally sustained by not only volcanic-hydrothermal sources, but also from external (nonvolcanic) contributions, including biogenic activity and crustal carbon. To filter out the effect of nonvolcanic CO_2 flux, the use of either statistical techniques (Chiodini et al., 1998) or results of carbon isotopic composition measurement (Chiodini et al., 2008; Parks et al., 2013) has been attempted; (2) soil CO_2 flux variations are strongly influenced from, and in many instances modulated by, external factors, such as the barometric pressure, air and soil temperature and humidity, wind speed, and rainfall. Although filtering technique can sometimes be used to smooth/remove the effect of external factors (e.g., Granieri et al., 2003), their contribution can sometimes overwhelm any volcanic signal present.

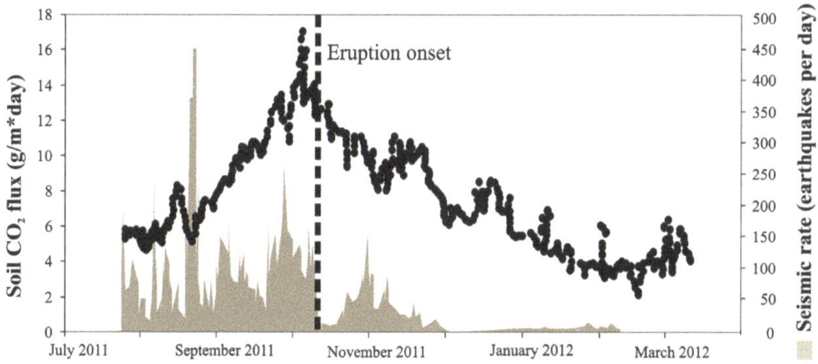

FIGURE 7.8 Precursory diffuse soil CO_2 flux increases prior to the 2011–2012 El Hierro submarine eruption, in the Canary Islands (Spain). *Re-drawn from Pérez et al. (2012).*

Despite the limitations above a number of demonstrations from both surveys and continuous measurements with automated instruments, the observations of soil CO_2 flux can valuably contribute to monitoring volcanic activity. Hernández et al. (2001b) were among the first to report a precursory diffuse soil CO_2 flux increase before eruptive activity (the March 2000 Usu eruption in Japan). Later work has shown that continuous soil CO_2 flux measurements bring promise as eruption precursors at other volcanoes such as on Stromboli (Italy), where strong increases in CO_2 flux (peaking at 80,000 g/m^2/day) have been detected before the onset of the 2002–2003 effusive eruption (Carapezza et al., 2004) and a paroxysmal explosion in 2007 (Inguaggiato et al., 2011). More recently, precursory pulses of diffuse magmatic gas (CO_2, H_2S, and He) release have been detected by continuous soil gas networks (Pérez et al., 2012) and surveys (Padrón et al., 2013) prior to the 2011–2012 El Hierro submarine eruption, in the Canary Islands (Figure 7.8).

7.5 SEISMOLOGICAL OBSERVATIONS

Changes in the amplitude and/or rate of occurrence of seismic events have been observed to precede and accompany eruptions at most volcanoes worldwide, making the observation of ground motion one of the prominent tools for monitoring volcanic activity. Over the last 20 years, tremendous progress in sensor and data transmission technologies, the increase of computational power, and the establishment of a solid modeling framework improved insights into the role played by magmatic and hydrothermal fluids in generating seismic waves for a variety of volcanic environments. For a comprehensive overview of the theoretical framework beneath modern volcano seismology, the reader can refer to several recent review papers and books (e.g., Chouet, 1996a,b, 2003; Chouet and Matoza, 2013; Kawakatsu and

Yamamoto, 2007; Kumagai, 2009; Neuberg, 2011; Zobin, 2012, and references therein). This section, instead, aims at presenting a general picture of operative observational seismology in volcanic environments, with a particular emphasis on the open issues and future trends.

7.5.1 Seismo-Volcanic Sources

The transport of magma and hydrothermal fluids and their interaction with solid rock produce elastic waves that are recorded at the surface over a wide range of frequencies. Typical volcanic source processes include: (1) brittle failure reflecting the response of the rock to stress changes induced by magma movement; and (2) pressure oscillations accompanying the dynamics of liquids and/or gases in conduits and cracks. Associated with these sources are different types of signals, whose classification—though not being yet unique (Chouet, 1996a,b; McNutt, 2005)—is now entered into the lexicon of observatory practice.

7.5.1.1 Volcano-Tectonic Earthquakes

Sources of type (1) produce volcano-tectonic (VT) earthquakes, which share waveform and spectral characteristics with pure tectonic earthquakes. These events are manifestations of shear failure within the volcano in response to the regional stress field, to magmatic forcing, or a combination of both. VT earthquakes map inelastic damage within the volcanic edifice; although several relevant eruptions have been preceded by intense VT swarms (see Chouet and Matoza, 2013), a definitive law has not yet been extracted from VT seismicity to identify changes in fluid/magma states in terms of their pressure, volume, location, and gas/liquid content (Traversa and Grasso, 2010). Hence increased VT activity alone is not a sufficient condition for making informed decisions about any impending eruption. A notable example of this is the long-lasting earthquake swarm which, during 1983–1984, accompanied the unrest of Phlegrean Fields Caldera, Italy, without being followed by any eruption (Barberi et al., 1984).

7.5.1.2 Very-Long-Period, Long-Period, and TR Signals

Seismic signals originating in the dynamics of magmatic and hydrothermal fluids (source type (2)) typically include long- and very-long-period (LP, VLP) events and TR. LP events share with the TR type similar spectral signature (dominant peaks spanning the 0.5–5 Hz range), but differ in the temporal behavior of the signals (transient vs sustained, respectively; e.g., McNutt, 2005; Nishimura and Iguchi, 2011). TR signals are a challenging subject of study, as they may result from the contribution of different source mechanisms acting at different locations. Nonetheless, the growing availability of multidisciplinary measurements indicates the potential of TR observations for tracking both degassing processes of deep origin (e.g., Carbone et al., 2006; Zuccarello et al., 2013), and the intensity of shallow explosive activity (see Figure 7.5).

For both LP and TR signals a common source process is generally invoked, which consists of acoustic resonance in fluid-filled cavities or cracks. Resonance processes in the conduit system are favored by the marked impedance contrasts that establish at the fluid/rock interface once gas exsolution induces a dramatic lowering of both acoustic velocity and density of the magmatic/hydrothermal fluids. These considerations imply that simple observables such as the signal's dominant frequency and quality factor may be used for gaining inferences on the nature and properties of the fluids involved into the resonance process (e.g., Kumagai and Chouet, 1999). More complex is to imagine the mechanisms triggering such resonances, for which a variety of processes have been invoked: these include magmatic—hydrothermal interactions; passive or explosive magma degassing; and the brittle failure of magmatic melts (e.g., Neuberg et al., 2006).

Very-long-period (VLP) signals, with dominant periods spanning the 2- to 100-s interval, are typically attributed to inertial forces associated with perturbations in the flow of magma and gases through conduits. VLP wavelengths span the 5—100 km range; therefore, path effects are generally negligible, and waveform modeling can be attempted even in the absence of a detailed knowledge of the velocity structure of the volcano's interior (see below). Thanks to the advent of low-cost, broad-band seismological instruments, the analysis and modeling of VLP waveforms are now performed at a number of volcanoes worldwide and, in a few cases, these procedures are part of routine observatory practice (Auger et al., 2006).

7.5.2 Location

With their emergent onsets and lack of clear body-wave phase arrivals, LP/TR and VLP signals cannot be located using standard methods based on the inversion of arrival times estimated at stations from a distributed network. Alternative methods exploit either the multichannel coherence at a number of stations spaced by a fraction of the dominant wavelength (arrays), or the decay in signal's amplitude with increasing distance from the source. Though not being included yet in most observatories practice, these nonconventional approaches are able to provide rather precise estimates of source location and its evolution through time.

7.5.2.1 Seismic Arrays

For station spacing significantly shorter than the wavelength of interest, the signal recorded by the different receivers can be assumed to be the same except for a phase delay due to propagation (*common-waveform model*).

Therefore, the differential travel-times at the array stations may be retrieved independently from the visual detection of individual phases, thus allowing solution of the location (or direction-finding) problem even for sustained and/or emergent wave-packets. The core of any array-based location procedure consists in seeking that source location for which an array-averaged measure of signal power or multichannel coherence takes a maximum. At Stromboli Volcano, a real-time implementation of this procedure allows the continuous tracking of the location and depth of the explosive source (Auger et al., 2006).

If the source-array distance is large enough and the underground beneath the array is homogeneous, then the plane-wave assumption holds; in that case, the differential times are linearly related to the horizontal slowness, which can be determined either by maximizing the beam's power in a grid-search procedure, or by fitting a plane wave to the interchannel differential times. Constrained estimates on source location can only be retrieved from back-projecting of seismic rays determined by different arrays through an assumed velocity structure (e.g., Almendros et al., 2001; see Figure 7.9).

7.5.2.2 Amplitude Inversion

A further location method consists in seeking those source position and strength for which the difference between observed and predicted amplitudes is minimized. Theoretical amplitudes at individual stations are calculated using standard attenuation models; the precision of this method is therefore strongly affected by: (1) the (characteristically poor) knowledge about the distribution of velocity and quality factors within the volcanic edifice; (2) uncertainties regarding the dominant wave-types at individual stations, for which different geometrical spreading models are required; and (3) the influence of near-field effects affecting the amplitude-distance curve.

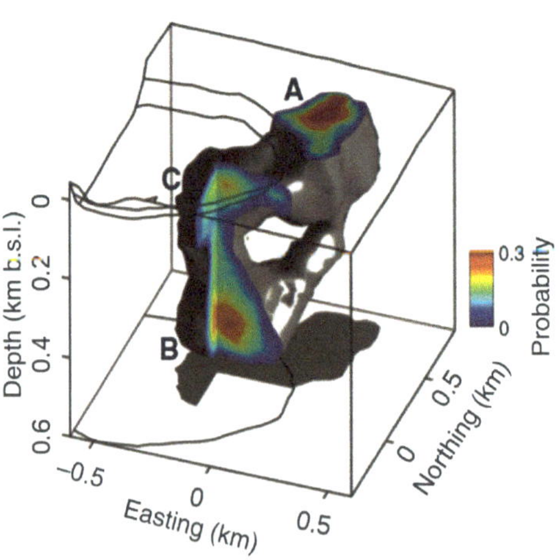

FIGURE 7.9 The location and extent of the source region of long-period seismicity and tremor beneath the Halemaumau pit crater (Kilauea Volcano, Hawaii) obtained from joint inversion of slowness measurements from three small-aperture seismic arrays. (A, B, C) Three subregions of maximum clustering of hypocenters; subregion A coincides with the location of tremor signals. Colors indicate the probability of source location. *Modified after Almendros et al. (2001).*

Notwithstanding these difficulties, successful applications of the amplitude-inversion technique were reported for locating volcanic TR and/or LP events at a variety of volcanoes worldwide (e.g., Bromo volcano, Indonesia; Gottschämmer and Surono, 2000; Mt. Etna, Italy; Carbone et al., 2008; Kilauea, Hawaii; Battaglia et al., 2003).

7.5.3 Source Mechanism

Seismic sources at volcanoes can be described in terms of a moment tensor (MT) and single forces, which are all linearly related to the ground displacement at an arbitrary location via the Green's functions (impulse responses) of the propagation medium in between the source and the observer. A variety of computational methods now exist for calculating the Green's tensor for realistic, 3-D heterogeneous Earth's models including topography. Once the Green's functions are known, the recorded ground displacement can thus be inverted for the time history of the force system acting at the source, allowing retrieval of the source geometry, mechanism and volume change. The main issue related to full-waveform inversions concerns the physical significance and reliability of the solution, as the accuracy of the prediction operators (i.e., the Green's functions) can be strongly influenced by our poor knowledge of the propagation medium (e.g., Bean et al., 2008). In this respect, efforts should be addressed toward improving both the knowledge on the internal structure of volcanoes and station's density in the near field of the potential sources. Notwithstanding these limitations, full-waveform MT inversion permits deriving reliable estimates of source volume changes, and hence on the mass budget of the volcano during either quiescent or eruptive stages. In a recent work, Kazahaya et al. (2011) combined high-rate SO_2 flux measurements with seismological data from Mt. Asama Volcano (Japan), obtaining a linear relationship between VLP pulse moment and SO_2 emission. Similar analysis were performed by Zuccarello et al. (2013), who compared VLP volumetric changes with surface SO_2 emissions from high-rate FTIR acquisitions at Etna Volcano (Italy), for a 4-month long time interval (Figure 7.10). These latter results indicate that the linear scaling between VLP volumetric changes and SO_2 gas flux establishes only for a particular location of the VLP source, suggesting that conduit geometry might play a significant role in the VLP-generating process, as also suggested by analogue laboratory models (James et al., 2006).

7.5.4 Changes in the Elastic Properties of the Medium

Pressurization or depressurization of a magmatic system affects the stress state of the crust around a volcanic edifice, which in turn may affect: (1) the alignment of fluid-filled microcracks and pore space; and (2) the bulk elastic parameters of the propagation medium, and associated variations in seismic

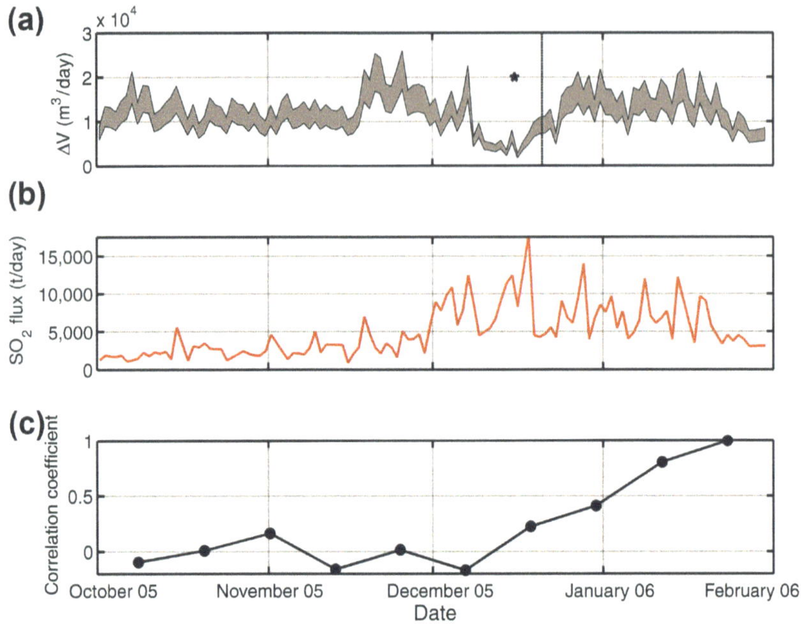

FIGURE 7.10 The time series of seismological and geochemical data at Etna Volcano, Italy, for the October 1, 2005–January 31, 2006 time interval. (a) Cumulative daily volume changes of VLP source; asterisk and dashed line, respectively, indicate the timing of a strong explosion from the summit craters, and the migration of the VLP source some 200 m upward. (b) Daily fluxes of SO_2 as estimated from FTIR measurements. (c) The zero-lag correlation coefficient between the two time series in (a) and (b) calculated over nonoverlapping, 2-week long time windows. Note the establishment of a marked correlation after the VLP source migration. SO_2 flux data and VLP volume changes were contributed by Mike Burton and Luciano Zuccarello, respectively. VLP, very-long-period; FTIR, Fourier-transform infrared.

velocities. Specific lines of investigation have thus been aimed at exploring these phenomena, in light of their relevance as precursor to renewing eruptive activity.

7.5.4.1 Seismic Anisotropy

Changes in the density and orientation of dry and/or wet microcracks result in seismic anisotropy, for which shear waves with subvertical incidence are split into two nearly perpendicular components propagating with different velocities. Observations of the fast polarization direction and the delay time between the fast and slow S wavelets provide estimates of the direction and relative strength of the main horizontal stress axis for a region surrounding the ray path. Changes in the shear-wave splitting polarization and delay times have been observed to track crustal stress variations prior to the large 2001 Etna eruption (Bianco et al., 2006), and accompanying the pre- to posteruptive stages at Ruapehu Volcano, New Zealand (Gerst and Savage, 2004).

7.5.4.2 Interferometric Detection of Velocity Changes

Seismic interferometry consists essentially in the comparison of two signals generated by the same source at different times, and recorded by the same seismic station. If a perturbation in the elastic properties of the medium occurred during the time span in between the two events, then the corresponding seismograms should exhibit tiny temporal shifts associated with the relative velocity changes in between the two events. Seismic interferometry is applied to multiply-scattered waves composing the later portions (coda) of either earthquake multiplets, or noise correlation function for a pair of stations. Temporal shifts are retrieved by calculating moving-window cross-correlation between a reference and successive recordings. For the 2002–2003 Mt Etna eruption, Zaccarelli et al. (2009) analyzed earthquake doublets using both shear-wave splitting and coda-wave interferometry. Zaccarelli et al. (2009) observed syneruptive, anticorrelated changes in delay times and seismic velocity, which were interpreted in terms of a depressurization of the system, heralding the termination of the eruption, which occurred three days later.

7.5.4.3 4-D Travel-Time Tomography

Local-earthquake travel time tomography is the most common method for deriving the spatial distribution of velocity heterogeneity within a volcano. When large catalogs are available, separate inversions of data from different epochs can be carried out independently, and the differences in the derived structures can be assumed to represent temporal variations in the seismic velocity of the medium. By repeating three-dimensional tomographic inversions before and during the 2002–2003 Mt. Etna eruption, Patanè et al. (2006) observed the appearance of well-defined V_P/V_S anomalies, which were interpreted in terms of the fast (few months) emplacement of volatile-rich basaltic magma within the upper portion of the volcano's plumbing system. Recently, however, the validity of such approach has been questioned due to the influence of uneven ray coverage on the results of the independent inversions (Julian and Foulger, 2010).

7.6 INFRASONIC

Measurement of acoustic waves in the 0.05–100 s period range (atmospheric infrasound) is a recent technique that is widely employed for the detection and quantification of volcanic processes occurring at the interface between the Solid Earth and the atmosphere. Infrasonic signals are associated with a variety of processes, including steady-state degassing in shallow basaltic magmas, Strombolian and Vulcanian explosions, magmatic–hydrothermal interactions, rockfalls and dome collapses, and pyroclastic flows. Seismic and acoustic signals provide complementary information, and their mutual correlation can greatly help in constraining near-surface degassing mechanisms. Techniques for the analysis of infrasonic data are similar to those adopted for seismological observations; further advantages of acoustic signals are: (1) the source processes are commonly amenable to direct observation; and (2) path effects are generally easier to assess, as they are essentially related to the physical conditions of the atmosphere, wind flow and ground topography asperities. For additional information, readers are referred to one of the recent reviews on the argument (e.g., Johnson and Ripepe, 2011, and references therein).

7.7 CONCLUSIONS

The tremendous technological progresses of the last 20 years led to significant advances in volcano monitoring techniques, which now allow imaging the geometry of the plumbing system to an unprecedented level of detail, and to infer possible physical processes driving the mass transfer before and during eruptive episodes. In particular, the most significant advances regarded: (1) the possibility of conducting high-rate, nearly real-time, volcanic gas observations with compact, fully automated, instruments; (2) the deployment of dense digital networks of broad-band seismic sensors; and (3) the acquisition of high-frequency, high-precision ground deformation and gravitational field data. The recent history of these improved monitoring systems evidenced the potential of in situ observations not only for tracking the ongoing processes, but also for the determination of material properties, whose temporal changes may reflect variations in the stress state of the volcanic edifice. Further technological improvements are still needed, and they mostly regard the need of increasing both bandwidth and precision of the observations. Particularly promising are the emerging optical technologies applied to the determination of volatile emissions, and to the deployment of advanced systems for measuring the gravity and displacement fields (e.g., de Angelis et al., 2009; Belfi et al., 2012).

Key points for turning these progresses into improved forecasting abilities are recognized in: (1) further development and consolidation of the on-line processing of observational data, now made possible by the continuous increase in real-time observations and computational power; (2) the joint analysis of data from different disciplines, in order to gain deeper insights into the causative processes; and (3) the consolidation of a comprehensive modeling framework, for the forward-generation of those synthetic signals which are necessary to the interpretation of the observed ones.

Notwithstanding the complexity and intrinsic nonlinearity of volcanic systems, attainment of the above goals will hopefully increase our preparedness toward the growing demand for confident hazard and risk evaluation around active volcanoes worldwide.

ACKNOWLEDGMENTS

Daniele Carbone is sincerely acknowledged for his careful revision of the gravity section. FTIR and infrared data discussed in the seismicity section were contributed by Mike Burton.

REFERENCES

Aiuppa, A., Moretti, R., Federico, C., Giudice, G., Gurrieri, S., Liuzzo, M., Papale, P., Shinohara, H., Valenza, M., 2007. Forecasting Etna eruptions by real-time observation of volcanic gas composition. Geology 12, 1115–1118. http://dx.doi.org/10.1130/G24149A.

Aiuppa, A., Burton, M., Allard, P., Caltabiano, T., Giudice, G., Gurrieri, S., Liuzzo, M., Salerno, G., 2011. First observational evidence for the CO_2-driven origin of Stromboli's major explosions. Solid Earth 2 (2), 135–142.

Aiuppa, A., Burton, M., Caltabiano, T., Giudice, G., Guerrieri, S., Liuzzo, M., Murè, F., Salerno, G., 2010b. Unusually large magmatic CO_2 gas emissions prior to a basaltic paroxysm. Geophys. Res. Lett. 37 (17) art. no. L17303.

Aiuppa, A., Federico, C., Giudice, G., Gurrieri, S., Liuzzo, M., Shinohara, H., Favara, R., Valenza, M., 2006. Rates of carbon dioxide plume degassing from Mount Etna volcano. J. Geophys. Res. 111, B09207. http://dx.doi.org/10.1029/2006JB004307.

Aiuppa, A., Cannata, A., Cannavò, F., Di Grazia, G., Ferrari, F., Giudice, G., Gurrieri, S., Puglisi, G., 2010a. Patterns in the recent 2007–2008 activity of Mount Etna volcano investigated by integrated geophysical and geochemical observations. Geochem. Geophys. Geosyst. 11 (9) art. no. Q09008.

Aiuppa, A., Federico, C., Giudice, G., Giuffrida, G., Guida, R., Gurrieri, S., Liuzzo, M., Moretti, R., Papale, P., 2009. The 2007 eruption of Stromboli volcano: Insights from real-time measurement of the volcanic gas plume CO_2/SO_2 ratio. J. Volcanol. Geotherm. Res. 182, 221–230. http://dx.doi.org/10.1016/j.jvolgeores.2008.09.013.

Aiuppa, A., Federico, C., Giudice, G., Gurrieri, S., 2005. Chemical mapping of a fumarolic field: La Fossa crater, vulcano island (Aeolian islands, Italy). Geophys. Res. Lett. 32 (13), 1–4.

Aiuppa, A., Giudice, G., Gurrieri, S., Liuzzo, M., Burton, M., Caltabiano, T., McGonigle, A.J.S., Salerno, G., Shinohara, H., Valenza, M., 2008. Total volatile flux from Mount Etna. Geophys. Res. Lett. 35, L24302. http://dx.doi.org/10.1029/2008GL035871.

Aiuppa, A., Giudice, G., Liuzzo, M., Tamburello, G., Allard, P., Calabrese, S., Chaplygin, I., McGonigle, A.J.S., Taran, Y., 2012. First volatile inventory for Gorely volcano, Kamchatka. Geophys. Res. Lett. 39 (6) art. no. L06307.

Allard, P., Carbonnelle, J., Dajlevic, D., Le Bronec, J., Morel, P., Robe, M.C., Maurenas, J.M., Zettwoog, P., 1991. Eruptive and diffuse emissions of CO_2 from Mount Etna. Nature 351 (6325), 387–391.

Allard, P., Burton, M.R., Mure, F., 2005. Spectroscopic evidence for a lava fountain driven by previously accumulated magmatic gas. Nature 433 (7024), 407–410. http://dx.doi.org/10.1038/nature03246.

Allard, P., Carbonnelle, J., Métrich, N., Loyer, H., Zettwoog, P., 1994. Sulphur output and magma degassing budget of Stromboli volcano. Nature 368, 326–329.

Allard, P., Behncke, B., D'Amico, S., Neri, M., Gambino, S., 2006. Mount Etna 1993–2005: anatomy of an evolving eruptive cycle. Earth Sci. Rev. 78, 85–114.

Almendros, J., Chouet, B., Dawson, P., 2001. Spatial extent of a hydrothermal system at Kilauea Volcano, Hawaii, determined from array analyses of shallow long-period seismicity 2. Results. J. Geophys. Res. 106, 13,581–13,597. http://dx.doi.org/10.1029/2001JB000309.

de Angelis, M., Bertoldi, A., Cacciapuoti, L., Giorgini, A., Lamporesi, G., Prevedelli, M., Saccorotti, G., Sorrentino, F., Tino, G.M., 2009. Precision gravimetry with atomic sensors. Meas. Sci. Technol. 20. http://dx.doi.org/10.1088/0957-0233/20/2/022001.

Auger, E., D'Auria, L., Martini, M., Chouet, B., Dawson, P., 2006. Real-time monitoring and massive inversion of source parameters of very long period seismic signals: an application to Stromboli Volcano, Italy. Geophys. Res. Lett. 33, 04301. http://dx.doi.org/10.1029/2005GL024703.

Barberi, F., Corrado, G., Innocenti, F., Luongo, G., 1984. Phlegraean Fields 1982–1984: brief chronicle of a volcano emergency in a densely populated area. Bull. Volcanol. 47, 175–185.

Battaglia, J., Got, J.L., Okubo, P., 2003. Location of long-period events below Kilauea Volcano using seismic amplitudes and accurate relative relocation. J. Geophys. Res. 108 (B12), 2553. http://dx.doi.org/10.1029/2003JB002517.

Battaglia, M., Troise, C., Obrizzo, F., Pingue, F., De Natale, G., 2006. Evidence for fluid migration as the source of deformation at Campi Flegrei caldera (Italy). Geophys. Res. Lett. 33. http://dx.doi.org/10.1029/2005GL024904.

Baubron, J.C., Allard, P., Toutain, J.-P., 1990. Diffuse volcanic emissions of carbon dioxide from Vulcano Island, Italy. Nature 344 (6261), 51–53.

Bean, C., Lokmer, I., O'Brien, G., 2008. Influence of near-surface volcanic structure on long-period seismic signals and on moment tensor inversions: simulated examples from Mount Etna. J. Geophys. Res. 113. http://dx.doi.org/10.1029/2007JB005468.

Belfi, J., Beverini, N., Bouhadef, B., Carelli, G., Cuccato, D., Di Virgilio, A., Licciardi, A., Maccioni, E., Ortolan, A., Saccorotti, G., Stefani, F., 2012. Laser gyroscopes for very high sensitive applications. In: IEEE Proc. European Frequency and Time Forum (EFTF), 2012, pp. 448–452.

Bergfeld, D., Evans, W.C., Howle, J.F., Farrar, C.D., 2006. Carbon dioxide emissions from vegetation-kill zones around the resurgent dome of Long Valley caldera, eastern California, USA. J. Volcanol. Geotherm. Res. 152 (1–2), 140–156.

Bianco, F., Scarfı, L., Del Pezzo, E., Patanè, D., 2006. Shear wave splitting changes associated with the 2001 volcanic eruption on Mt Etna. Geophys. J. Int. 167, 959–967.

Blundy, J., Cashman, K., 2008. Petrologic reconstruction of magmatic system variables and processes. Rev. Mineral. Geochem. 69, 179–239.

Burton, M., Allard, P., Murè, F., La Spina, A., 2007. Depth of slug-driven Strombolian explosive activity. Science 317, 227–230.

Burton, M.R., Caltabiano, T., Murè, F., Randazzo, D., 2009. SO_2 flux from Stromboli during the 2007 eruption: results from the FLAME network and traverse measurements. J. Volcanol. Geotherm. Res. 182, 214–220.

Burton, M., Allard, P., Murè, F., Oppenheimer, C., 2003. FTIR remote sensing of fractional magma degassing at Mount Etna, Sicily. Geol. Soc. Spec. Publ. 213, 281–293.

Burton, M.R., Sawyer, G.M., Granieri, D., 2013. Deep carbon emissions from volcanoes. Rev. Mineral. Geochem. 75, 323–354.

Caltabiano, T., Burton, M., Giammanco, S., Allard, P., Bruno, N., Murè, F., Romano, R., 2004. Volcanic Gas emissions from the Summit craters and flanks of Mt. Etna, 1987–2000. In: Bonaccorso, A., Calvari, S., Coltelli, M., Del Negro, C., Falsaperla, S. (Eds.), Mt. Etna: Volcano Laboratory, AGU Geophysical Monograph Series, vol. 143.

Carapezza, M.L., Inguaggiato, S., Brusca, L., Longo, M., 2004. Geochemical precursors of the activity of on open-conduit volcano: the Stromboli 2002–2003 eruptive events. Geophys. Res. Lett. 31. L07620.

Carbone, D., Rymer, H., 1999. Calibration shifts in a LaCoste & Romberg gravimeter: comparison with a Scintrex CG-3M. Geophys. Prospect. 47 (1), 73–83.

Carbone, D., Zuccarello, L., Saccorotti, G., Greco, F., 2006. Analysis of simultaneous gravity and tremor anomalies observed during the 2002–2003 Etna eruption. Earth Planet. Sci. Lett. 245, 616–629.

Carbone, D., Budetta, G., Greco, F., Zuccarello, L., 2007. A data sequence acquired at Mt. Etna during the 2002–2003 eruption highlights the potential of continuous gravity observations as a tool to monitor and study active volcanoes. J. Geodyn. 43, 320–329.

Carbone, D., Zuccarello, L., Saccorotti, G., 2008. Geophysical indications of magma uprising at Mt Etna during the December 2005 to January 2006 non-eruptive period. Geophys. Res. Lett. 35, L06305. http://dx.doi.org/10.1029/2008GL033212.

Carbone, D., D'Amico, S., Musumeci, C., Greco, F., 2009. Comparison between the 1994–2006 seismic and gravity data from Mt. Etna: new insight into the long-term behavior of a complex volcano. Earth Planet. Sci. Lett. 279 (3), 282–292. http://dx.doi.org/10.1016/j.epsl.2009.01.007.

Carbone, D., Zuccarello, L., Saccorotti, G., Rymer, H., Rapisarda, S., 2010. The effect of inertial accelerations on the higher-frequency components of the signal from spring gravimeters. Geophys. J. Int. 182 (2), 772–780.

Carbone, D., Poland, M.P., 2012. Gravity fluctuations induced by magma convection at Kīlauea Volcano, Hawai'i. Geology 40, 803–806. http://dx.doi.org/10.1130/G33060.1.

Carbone, D., Poland, M.P., Patrick, M.R., Orr, T.R., 2013. Continuous gravity measurements reveal a low-density lava lake at Kīlauea Volcano, Hawai'i. Earth Planet. Sci. Lett. 376, 178–185.

Casadevall, T., Rose, W., Gerlach, T., Greenland, L.P., Ewert, J., Wunderman, R., Symonds, R., 1983. Gas emissions and the eruptions of Mount St. Helens through 1982. Science 221 (4618), 1383–1385.

Casadevall, T.J., Doukas, M.P., Neal, C.A., McGimsey, R.G., Gardner, C.A., 1994. Emission rates of sulfur dioxide and carbon dioxide from Redoubt Volcano during the 1989–1990 eruptions. J Volcanol. Geotherm. Res. 62, 519–530.

Chiodini, G., Caliro, S., Cardellini, C., Avino, R., Granieri, D., Schmidt, A., 2008. Carbon isotopic composition of soil CO_2 efflux, a powerful method to discriminate different sources feeding soil CO_2 degassing in volcanic-hydrothermal areas. Earth Planet. Sci. Lett. 274 (3–4), 372–379.

Chiodini, G., Cioni, R., Guidi, M., Raco, B., Marini, L., 1998. Soil CO_2 flux measurements in volcanic and geothermal areas. Appl. Geochem. 13 (5), 543–552.

Chiodini, G., Frondini, F., Raco, B., 1996. Diffuse emission of CO_2 from the Fossa crater, Vulcano Island (Italy). Bull. Volcanol. 58 (1), 41–50.

Chiodini, G., Granieri, D., Avino, R., Caliro, S., Costa, A., Werner, C., 2005. Carbon dioxide diffuse degassing and estimation of heat release from volcanic and hydrothermal systems. J. Geophys. Res. B Solid Earth 110 (8), 1–17 art. no. B08204.

Chiodini, G., Caliro, S., Cardellini, C., Granieri, D., Avino, R., Baldini, A., Donnini, M., Minopoli, C., 2010. Long-term variations of the Campi Flegrei, Italy, volcanic system as revealed by the monitoring of hydrothermal activity. J. Geophys. Res. 115, B03205. http://dx.doi.org/10.1029/2008JB006258.

Chiodini, G., Caliro, S., De Martino, P., Avino, R., Ghepardi, F., 2012. Early signals of new volcanic unrest at Campi Flegrei caldera? Insights from geochemical data and physical simulations. Geology 40, 943–946.

Chouet, B., 1996a. Long-period volcano seismicity: its source and use in eruption forecasting. Nature 380, 309–316.

Hernández, P.A., Pérez, N.M., Salazar, J.M., Nakai, S., Notsu, K., Wakita, H., 1998. Diffuse emission of carbon dioxide, methane, and helium-3 from Teide volcano, Tenerife, Canary Islands. Geophys. Res. Lett. 25 (17), 3311–3314.

Hernández, P.A., Salazar, J.M., Shimoike, Y., Mori, T., Notsu, K., Pérez, N., 2001a. Diffuse emission of CO_2 from Miyakejima volcano, Japan. Chem. Geol. 177 (1–2), 175–185. http://dx.doi.org/10.1016/S0009-2541(00)00390-9. Cited 31 times.

Hidalgo, S., Battaglia, J., Steele, A., Arellano, S., Bernard, B., Ruiz, M., Galle, B., 2013. Open and closed system eruptive dynamics at Tungurahua volcano constrained by SO_2 emission rate and seismo-acoustic intensity measurements. In: Proceeding IAVCEI 2013 Scientific Assembly, July 20–24, Kagoshima, Japan, Abstract 4A2_2I-011.

Holland, P.A.S., Watson, M.I., Phillips, J.C., Caricchi, L., Dalton, M.P., 2011. Degassing processes during lava dome growth: insights from Santiaguito lava dome, Guatemala. J. Volcanol. Geotherm. Res. 202, 153–166.

Horrocks, L., Oppenheimer, C., Burton, M., Du¡ell, H., 2003. Compositional variation in tropospheric volcanic gas plumes: evidence from ground-based remote sensing. In: Oppenheimer, C., Pyle, D.M., Barclay, J. (Eds.), Volcanic Degassing, Geol. Soc. London Spec. Publ, vol. 213, pp. 349–369.

Iguchi, M., 2013. Magma movement from the deep to shallow Sakurajima volcano as revealed by geophysical observations. Bull. Volcanol. Soc. Japan 58 (1), 1–18.

Iguchi, M., Tameguri, T., Ohta, Y., Ueki, S., Nakao, S., 2013. Characteristics of volcanic activity at Sakurajima volcano's Showa crater during the period 2006 to 2011. Bull. Volcanol. Soc. Japan 58 (1), 115–135.

Inguaggiato, S., Vita, F., Rouwet, D., Bobrowski, N., Morici, S., Sollami, A., 2011. Geochemical evidence of the renewal of volcanic activity inferred from CO_2 soil and SO_2 plume fluxes: the 2007 Stromboli eruption (Italy). Bull. Volcanol. 73 (4), 443–456.

Iverson, R.M., Dzurisin, D., Grardner, C.A., Gerlach, T.M., LaHusen, R.G., Lisowski, M., Major, J.J., Malone, S.D., Messerich, J.A., Moran, S.C., Pallister, J.S., Qamar, A.I., Schilling, S.P., Vallance, J.W., 2006. Dynamics of seismogenic volcanic extrusions at Mount St Helens in 2004–2005. Nature 444, 439–443.

James, M.R., Lane, S.J., Chouet, B., 2006. Gas slug ascent through changes in conduit diameter: laboratory insights into a volcano-seismic source process in low-viscosity magmas. J. Geophys. Res. 111, B05201. http://dx.doi.org/10.1029/2005JB003718.

Johnson, J.B., Ripepe, M., 2011. Volcano infrasound: a review. J. Volcanol. Geotherm. Res. 206, 61–69.

Julian, B.R., Foulger, G.R., 2010. Time-dependent seismic tomography. Geophys. J. Int. 182, 1327–1338. http://dx.doi.org/10.1111/j.1365-246X.2010.04668.x.

Kawakatsu, H., Yamamoto, M., 2007. Volcano seismology. In: Kanamori, H., Schubert, G. (Eds.), Treatise on Geophysics. Earthquake Seismology, vol. 4. Elsevier, New York, pp. 389–420.

Kazahaya, K., Shinohara, H., Uto, K., Odai, M., Nakahori, Y., Mori, H., Iino, H., Miyashita, M., Hirabayashi, J., 2004. Gigantic SO_2 emission from Miyakejima volcano, Japan, caused by caldera collapse. Geology 32, 425–428. http://dx.doi.org/10.1130/G20399.1.

Kazahaya, R., Mori, T., Takeo, M., Ohminato, T., Urabe, T., Maeda, Y., 2011. Relation between single very-long-period pulses and volcanic gas emissions at Mt. Asama, Japan. Geophys. Res. Lett. 38, L11307. http://dx.doi.org/10.1029/2011GL047555.

Kumagai, H., 2009. Source quantification of volcano seismic signals. In: Meyers, R.A. (Ed.), Encyclopedia of Complexity and Systems Science. Springer-Verlag, New York, pp. 9899–9932.

Kumagai, H., Chouet, B.A., 1999. The complex frequencies of long-period seismic events as probes of fluid composition beneath volcanoes. Geophys. J. Int. 138, F7–F12.

La Spina, A., Burton, M.R., Harig, R., Mure, F., Rusch, P., Jordan, M., Caltabiano, T., 2013. New insights into volcanic processes at Stromboli from Cerberus, a remote-controlled open-path FTIR scanner system. J. Volcanol. Geotherm. Res. 249, 66–76.

Locke, C.A., Rymer, H., Cassidy, J., 2003. Magma transfer processes at persistently active volcanoes: insights from gravity observations. J. Volcanol. Geotherm. Res. 127, 73–86.

Love, S.P., Goff, F., Counce, D., Siebe, C., Delgado, H., 1998. Passive infrared spectroscopy of the eruption plume at Popocatepetl volcano, Mexico. Nature 396, 563–566.

Malinconico, L.L., 1979. Fluctuations in SO_2 emission during recent eruptions of Etna. Nature 278, 43–45.

McGee, K.A., Sutton, A., 1994. Eruptive activity at Mount St. Helens, Washington, USA, 1984–1988 – a gas geochemistry perspective. Bull. Volcanol. 56, 435–446. http://dx.doi.org/10.1007/s004450050053.

McGee, K.A., 1992. The structure, dynamics, and chemical composition of noneruptive plumes from Mount St. Helens, 1980–1988. J. Volcanol. Geotherm. Res. 51 (3), 269–282.

McGonigle, A.J.S., Oppenheimer, C., Galle, B., Mather, T., Pyle, D., 2002. Walking traverse and scanning DOAS measurements of volcanic gas emission rates. Geophys. Res. Lett. 29 (20), 1985. http://dx.doi.org/101029/2002GL015827.

McNutt, S.R., 2005. Volcanic seismology. Ann. Rev. Earth Planet. Sci. 32, 461–491.

Metrich, N., Wallace, P.J., 2008. Volatile abundances in basaltic magmas and their degassing paths tracked by melt inclusions. In: Putirka, Keith D., Tepley III, Frank J. (Eds.), Minerals, Inclusions and Volcanic Processes, Rev Mineral, vol. 69, pp. 363–402.

Métrich, N., Mandeville, C.W., 2010. Sulfur in magmas. Elements 6, 81–86.

Moretti, R., Papale, P., Ottonello, G., 2003. A model for the saturation of C–O–H–S fluids in silicate melts. Geol. Soc. Spec. Publ. 213, 81–101.

Mori, T., Notsu, K., 1997. Remote CO, COS, CO_2, SO_2 and HCl detection and temperature estimation of volcanic gas. Geophys. Res. Lett. 24, 2047–2050.

Mori, T., Burton, M., 2006. The SO_2 camera: a simple, fast and cheap method for ground-based imaging of SO_2 in volcanic plumes. Geophys. Res. Lett. 33, L24804. http://dx.doi.org/10.1029/2006GL027916.

Mori, T., Burton, M., 2009. Quantification of the gas mass emitted during single explosions on Stromboli with the SO_2 imaging camera. J. Volcanol. Geotherm. Res. 188, 395–400. http://dx.doi.org/10.1016/j.jvolgeores.2009.10.005.

Mori, T., Notsu, K., Tohjima, Y., Wakita, H., 1993. Remote detection of HCl and SO_2 in volcanic gas from Unzen volcano, Japan. Geophys. Res. Lett. 20, 1355–1358.

Nadeau, P.A., José, L.P., Waite, G.P., 2011. Linking volcanic tremor, degassing, and eruption dynamics via SO_2 imaging. Geophys. Res. Lett. 38, L01304. http://dx.doi.org/10.1029/2010GL045820.

Neuberg, J.W., 2011. Earthquakes, volcanogenic. In: Gupta, H.K., et al. (Eds.), Encyclopedia of Solid Earth Geophysics. Springer, pp. 261–269.

Neuberg, J.W., Tuffen, H., Collier, L., Green, D., Powell, T., Dingwell, D., 2006. The trigger mechanism of low-frequency earthquakes on Montserrat. J. Volcanol. Geotherm. Res. 153, 37–50.

Nishi, K., Ono, H., Mori, H., 1999. Global positioning system measurements of ground deformation caused by magma intrusion and lava discharge: the 1990–1995 eruption at Unzendake volcano, Kyushu, Japan. J. Volcanol. Geotherm. Res. 89, 23–34.

Nishimura, T., Iguchi, M., 2011. Volcanic Earthquakes and Tremor in Japan. Kyoto University Press.

Notsu, K., Mori, T., 2010. Chemical monitoring of volcanic gas using remote FT-IR spectroscopy at several active volcanoes in Japan. Appl. Geochem. 25 (4), 505–512.

Olmos, R., Barrancos, J., Ivera, C.R., Barahona, F., López, D.L., Henriquez, B., Hernández, A., Galle, B.O., 2007. Anomalous emissions of SO_2 during the recent eruption of Santa Ana volcano, El Salvador, Central America. Pure Appl. Geophys. 164 (12), 2489–2506.

Oppenheimer, C., Kyle, P.R., 2008. Probing the magma plumbing of Erebus volcano, Antarctica, by openpath FTIR spectroscopy of gas emissions. J. Volcanol. Geotherm. Res. 177 (3), 743–754. http://dx.doi.org/10.1016/j. jvolgeores.2007.08.022.

Oppenheimer, C., Scaillet, B., Martin, R.S., 2011. Sulfur degassing from volcanoes: source conditions, surveillance, plume chemistry and earth system impacts. In: Sulfur in Magmas and Melts: Its Importance for Natural and Technical Processes. Rev Mineral Geochem, vol. 73, pp. 363–422. http://dx.doi.org/10.2138/rmg.2011.73.13.

Oppenheimer, C., Lomakina, A.S., Kyle, P.R., Kingsbury, N.G., Boichu, M., 2009. Pulsatory magma supply to a phonolite lava lake. Earth Planet. Sci. Lett. 284, 392–398.

Oppenheimer, C., 2010. Ultraviolet sensing of volcanic sulfur emissions. Elements 6, 87–92.

Oppenheimer, C., Bani, P., Calkins, J.A., Burton, M.R., Sawyer, G.M., 2006. Rapid FTIR sensing of volcanic gases released by Strombolian explosions at Yasur volcano, Vanuatu. Appl. Phys. B Lasers Opt. 85 (2–3), 453–460.

Oppenheimer, C., Francis, P., Burton, M., Maciejewski, A.J.H., Boardman, L., 1998. Remote measurement of volcanic gases by Fourier transform infrared spectroscopy. Appl. Phys. B Lasers Opt. 67 (4), 505–515.

Oppenheimer, C., McGonigle, A.J.S., Allard, P., Wooster, M.J., Tsanev, V.I., 2004. Sulfur, heat and magma budget of Erta 'Ale lava lake, Ethiopia. Geology 32, 509–512.

Padrón, E., Pérez, N.M., Hernández, P.A., Sumino, H., Melián, G.V., Barrancos, J., Nolasco, D., Nagao, K., 2013. Diffusive helium emissions as a precursory sign of volcanic unrest. Geology 41 (5), 539–542.

Parks, M.M., Caliro, S., Chiodini, G., Pyle, D.M., Mather, T.A., Berlo, K., Edmonds, M., Raptakis, C., 2013. Distinguishing contributions to diffuse CO_2 emissions in volcanic areas from magmatic degassing and thermal decarbonation using soil gas ^{222}Rn-δ ^{13}C systematics: application to Santorini volcano, Greece. Earth Planet. Sci. Lett. 377–378, 180–190.

Patanè, D., Aiuppa, A., Aloisi, M., Behncke, B., Cannata, A., Coltelli, M., Di Grazia, G., Salerno, G., 2013. Insights into magma and fluid transfer at Mount Etna by a multiparametric approach: a model of the events leading to the 2011 eruptive cycle. J. Geophys. Res. B Solid Earth 118 (7), 3519–3539.

Patanè, D., Barberi, G., Cocina, O., De Gori, P., Chiarabba, C., 2006. Time-resolved seismic tomography detects magma intrusions at Mount Etna. Science 313, 821–823. http://dx.doi.org/10.1126/science.1127724.

Pérez, N.M., Padilla, G.D., Padrón, E., Hernandez, P.A., Melin, G.V., Barrancos, J., Dionis, S., Nolasco, D., Rodríguez, F., Calvo, D., Hernandez, I., 2012. Precursory diffuse CO_2 and H_2S emission signatures of the 2011–2012 El Hierro submarine eruption, Canary Islands. Geophys. Res. Lett. 39. http://dx.doi.org/10.1029/2004GL019614.

Pérez, N.M., Hernández, P.A., Padrón, E., Melián, G., Nolasco, D., Barrancos, J., Padilla, G., Calvo, D., Rodríguez, F., Dionis, S., Chiodini, G., 2013. An increasing trend of diffuse CO_2 emission from Teide volcano (Tenerife, Canary Islands): geochemical evidence of magma degassing episodes. J. Geol. Soc. 170 (4), 585–592.

Poland, M.P., Miklius, A., Sutton, J.A., Thornber, C.R., 2012. A mantle-driven surge in magma supply to Kīlauea Volcano during 2003–2007. Nat. Geosci. 5 (4), 295–300.

Rymer, H., Brown, G.C., 1989. Gravity changes as a precursor to volcanic eruptions at Poás volcano, Costa Rica. Nature 342, 902–905.

Salerno, G.G., Burton, M., Oppenheimer, C., Caltabiano, T., Randazzo, D., Bruno, N., Longo, V., 2009. Three-years of SO_2 flux measurements of Mt. Etna using an automated UV scanner array: comparison with conventional traverses and uncertainties in flux retrieval. J. Volcanol. Geotherm. Res. 183, 76–83.

Sawyer, G.M., Carn, S.A., Tsanev, V.I., Oppenheimer, C., Burton, M.R., 2008a. Investigation into magma degassing at Nyiragongo volcano, Democratic Republic of the Congo. Geochem. Geophys. Geosyst. 9, Q02017. http://dx.doi.org/10.1029/2007GC001829.

Sawyer, G.M., Oppenheimer, C., Tsanev, V.I., Yirgu, G., 2008b. Magmatic degassing at Erta 'Ale volcano, Ethiopia. J. Volcanol. Geotherm. Res. 178, 837–846. http://dx.doi.org/10.1016/j.jvolgeores.2008.09.017.

Shinohara, H., 2005. A new technique to estimate volcanic gas composition: plume measurements with a portable multi-sensor system. J. Volcanol. Geotherm. Res. 143 (4), 319–333.

Shinohara, H., 2008. Excess degassing from volcanoes and its role on eruptive and intrusive activity. Rev. Geophys. 46 (4) art. no. RG4005.

Shinohara, H., 2013. Composition of volcanic gases emitted during repeating Vulcanian eruption stage of Shinmoedake, Kirishima volcano, Japan. Earth Planets Space 65 (6), 667–675.

Shinohara, H., Aiuppa, A., Giudice, G., Gurrieri, S., Liuzzo, M., 2008. Variation of H_2O/CO_2 and CO_2/SO_2 Ratios of Volcanic Gases Discharged by Continuous Degassing of Mount Etna Volcano, Italy, vol. 113 (B9), 9203. http://dx.doi.org/10.1029/2007JB005185.

Shinohara, H., Witter, J.B., 2005. Volcanic gases emitted during mild Strombolian activity of Villarrica volcano, Chile. Geophys. Res. Lett. 32 (20), 1–5 art. no. L20308.

Stoiber, R.E., Malinconico Jr., L.L., Williams, S.N., 1983. Use of the correlation spectrometer at volcanoes. In: Tazieff, H., Sabroux, J.-C. (Eds.), Forecasting Volcanic Events. Elsevier, Amsterdam, pp. 425–444.

Stremme, W., Ortega, I., Siebe, C., Grutter, M., 2011. Gas composition of Popocatépetl Volcano between 2007 and 2008: FTIR spectroscopic measurements of an explosive event and during quiescent degassing. Earth Planet. Sci. Lett. 301, 502–510.

Sutton, A.J., Elias, T., Gerlach, T.M., Stokes, J.B., 2001. Implications for eruptive processes as indicated by sulfur dioxide emissions from Kilauea Volcano, Hawai'i, 1979–1997. J. Volcanol. Geotherm. Res. 108 (1–4), 283–302.

Symonds, R.B., Rose, W.I., M Gerlach, T., H Briggs, P., Harmon, R.S., 1990. Evaluation of gases, condensates, and SO_2 emissions from Augustine volcano, Alaska: the degassing of a Cl-rich volcanic system. Bull. Volcanol. 52 (5), 355–374.

Symonds, R.B., Rose, W.I., Bluth, G., Gerlach, T.M., 1994. Volcanic gas studies: methods, results, and applications. In: Carroll, M.R., Holloway, J.R. (Eds.), Volatiles in Magmas: Mineralogical Society of America Reviews in Mineralogy, vol. 30, pp. 1–66.

Tamburello, G., Aiuppa, A., McGonigle, A.J.S., et al., 2013. Periodic volcanic degassing behavior: the Mount Etna example. Geophys. Res. Lett. 40, 4818–4844. http://dx.doi.org/10.1002/grl.50924.

Tamburello, G., Aiuppa, A., Kantzas, E.P., McGonigle, A.J.S., Ripepe, M., 2012. Passive vs. active degassing modes at an open-vent volcano (Stromboli, Italy). Earth Planet. Sci. Lett. 359–360, 106–116.

Traversa, P., Grasso, J.R., 2010. How is volcano seismicity different from tectonic seismicity? Bull. Seismol. Soc. Am. 100, 1755–1769. http://dx.doi.org/10.1785/0120090214.

Ukawa, M., Fujita, E., Yamamoto, E., Okada, Y., Kikuchi, M., 2000. The 2000 Miyakejima eruption: crustal deformation and earthquakes observed by the NIED Miyakejima observation network. Earth Planets Space 52 (8) xix–xxvi.

Vassalli, M., Longo, A., Montagna, C.P., O'Brien, G.S., Bean, C.J., Bisconti, L., Papale, P., Saccorotti, G., 2009. An integrated method to model volcanic processes and associated geophysical signals. In: Bean, C.J., et al. (Eds.), The VOLUME Project, VOLcanoes: Understanding Subsurface Mass Movement: Dublin, Ireland, the Volume Consortium, pp. 162–174.

Werner, C., Brantley, S.L., Boomer, K., 2000. CO_2 emissions related to the Yellowstone volcanic system 2. Statistical sampling, total degassing, and transport mechanisms. J. Geophys. Res. B Solid Earth 105 (5), 10,831–10,846.

Werner, C., Evans, W.C., Kelly, P.J., McGimsey, R., Pfeffer, M., Doukas, M., Neal, C., 2012. Deep magmatic degassing versus scrubbing: elevated CO_2 emissions and C/S in the lead-up to the 2009 eruption of Redoubt Volcano, Alaska. Geochem. Geophys. Geosyst. 13 (3) art. no. Q03015.

Werner, C., Kelly, P.J., Doukas, M., Lopez, T., Pfeffer, M., McGimsey, R., Neal, C., 2013. Degassing of CO_2, SO_2, and H_2S associated with the 2009 eruption of Redoubt Volcano, Alaska. J. Volcanol. Geotherm. Res. 259, 270–284.

Williams-Jones, G., Rymer, H., Mauri, G., Gottsmann, J., Poland, M., Carbone, D., 2008. Toward continuous 4D microgravity monitoring of volcanoes. Geophysics 73, WA19–WA28. http://dx.doi.org/10.1190/1.2981185.

Yamamoto, H., Watson, I.M., Phillips, J.C., Bluth, G.J., 2008. Rise dynamics and relative ash distribution in vulcanian eruption plumes at Santiaguito Volcano, Guatemala, revealed using an ultraviolet imaging camera. Geophys. Res. Lett. 35 (8) art. no. L08314.

Yamashina, K., Shimizu, H., 1999. Crustal deformation in the mid-May 1991 crisis preceding the extrusion of a dacite lava dome at Unzen volcano, Japan. J. Volcanol. Geotherm. Res. 89, 43–55.

Zaccarelli, L., Pandolfi, D., Bianco, F., Saccorotti, G., Bean, C.J., Del Pezzo, E., 2009. Temporal changes in seismic wave propagation towards the end of the 2002 Mt Etna eruption. Geophys. J. Int. 178, 1779–1788. http://dx.doi.org/10.1111/j.1365-246X.2009.04219.x.

Zapata, G., Calvache, M.L., Cortés, G.P., Fischer, T.P., Garzon, V., Gómez, M.D., Narváez, M.L., Ordoñez, V.M., Ortega, E.A., Stix, J., Torres, C.R., Williams, S.N., 1997. SO_2 fluxes from Galeras volcano, Colombia 1989–1995: progressive degassing and conduit obstruction of a Decade Volcano. J. Volcanol. Geotherm. Res. 77, 195–208.

Zobin, V., 2012. Introduction to Volcanic Seismology, second ed. Elsevier.

Zuccarello, L., Burton, M.R., Saccorotti, G., Bean, C.J., Patanè, D., 2013. The coupling between very long period seismic events, volcanic tremor, and degassing rates at Mount Etna volcano. J. Geophys. Res. Solid Earth 118, 4910–4921. http://dx.doi.org/10.1002/jgrb.50363.1.

Chapter 8

Using Multiple Data Sets to Populate Probabilistic Volcanic Event Trees

Chris G. Newhall[1] and John S. Pallister[2]

[1] *Earth Observatory of Singapore, Nanyang Technical University, Singapore,* [2] *Volcano Disaster Assistance Program, US Geological Survey, Vancouver, WA USA*

ABSTRACT

Bayesian event trees are now constructed before or during many volcanic crises, to reflect multiple possible scenarios and the relative and absolute likelihoods of each scenario. Group discussion followed by formal expert elicitation is an excellent way to estimate probabilities at each branch of a tree, but commonly neither sufficient time exists nor a large group of experts is available for a new formal elicitation. This chapter describes a simpler, easily understandable method that can be used in advance of more formal elicitations. This "method of multiple data sets" intentionally uses a number of different data sets at each node. At some nodes, quite different types of data are used, e.g., seismic, gas, and dates of historical eruptions. At other nodes, data sets differ only in the time periods or subsets of volcanoes that are considered. The goal is to define a robust range of plausible probabilities at each branch. The best estimate is probably within this range, and the spread between high and the low values is a rough indication of the uncertainty. We offer a generic Excel-based template that can be modified by any user as needed, case histories of several applications in past crises, and ideas for future improvements and uses of event trees.

8.1 INTRODUCTION

The key parameters one needs to forecast outcomes of volcanic unrest are hidden kilometers beneath the Earth's surface, and volcanic systems are so complex that invariably stochastic elements will occur in the evolution of any unrest. Fortunately, sufficient regularity occurs in behavior that some, perhaps many, eruptions can be forecast with enough certainty for populations to be evacuated and kept safe. Volcanologists charged with forecasting eruptions must try to understand each volcanic system well enough

Volcanic Hazards, Risks, and Disasters. http://dx.doi.org/10.1016/B978-0-12-396453-3.00008-3

that unrest can be interpreted in terms of pre-eruptive process, but must simultaneously recognize and convey uncertainties in their assessment. We have found that the use of event trees helps to focus discussion, integrate data from multiple sources, reach consensus among scientists about both pre-eruptive process and uncertainties, and in some cases, to explain all of this to officials. Figure 8.1 shows a generic volcanic event tree from Newhall and Hoblitt (2002) that can be modified as needed for each specific volcano. This chapter reviews how we and our colleagues have used such trees during a number of volcanic crises worldwide, for rapid hazard assessments in situations in which more formal expert elicitations could not be conducted. We describe how *Multiple Data Sets* can be used to estimate probabilities at each node and branch. We also present case histories of probability estimation during crises, how the estimates were used by public officials, and some suggestions for future improvements.

8.2 PROBABILISTIC VERSUS DETERMINISTIC FORECASTS

In the world of earthquakes and volcanic eruptions, scientists historically drew a dichotomy between probabilistic and deterministic forecasts. Early probabilistic forecasts relied exclusively on the statistics of recurrence. Long-range probabilistic forecasts relied on the compilation of geologic and recorded histories of earthquakes and eruptions; short-range probabilistic forecasts relied on the correlation of patterns of unrest and subsequent earthquakes or eruptions, without the explanation of physical processes. In contrast, purely deterministic forecasts avoided statistics of recurrence and invoked measurements and models of processes leading to earthquakes and eruptions. In principle, events could be forecast if failure thresholds and the physical phenomena that precede them could be measured or inferred. Probabalists dismissed determinists as hopelessly optimistic, and determinists dismissed probabilists as overly pessimistic and lacking in curiosity.

We and others reject this false dichotomy (Marzocchi et al., 2007, 2012; Woo, 2010; Jordan et al., 2011; Selva and Sandri, 2013). We do not believe that forecasts must be EITHER probabilistic OR deterministic. Rather we believe that the best forecasts must be probabilistic AND deterministic. They must be hybrids. It is scientifically prudent to express forecasts in probabilistic terms, which include all of the possible outcomes and their relative and absolute likelihoods (noted, among others, by IAVCEI Subcommittee for Crisis Protocols, 1999). But it would also be irresponsible for forecasters to ignore what can be learned about physical processes and thresholds for failure. A Bayesian framework reconciles the two forecast approaches. We start with initial (a priori) probabilities based on historical statistics and then update them into a posteriori probabilities that are based on interpretation of monitoring data and of physical and chemical processes that are thought to be controlling the system.

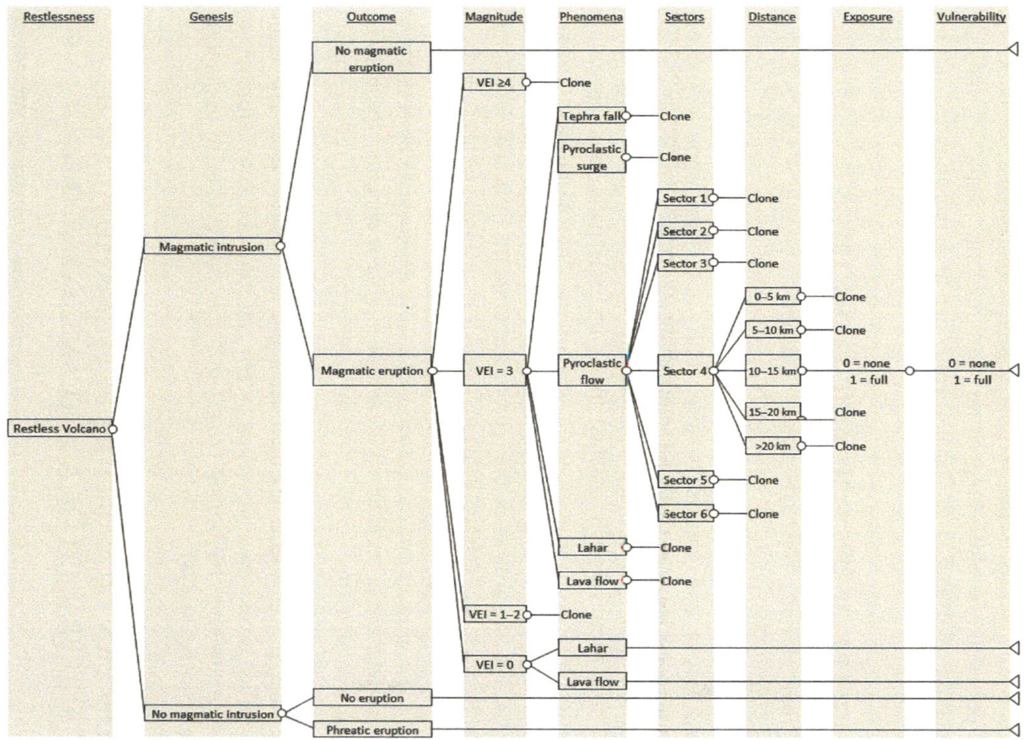

FIGURE 8.1 Generic event tree, as outlined in Newhall and Hoblitt (2002), which extends from onset of unrest (at left) to potential risk of death (at right). Relative and cumulative probabilities are estimated at each branch and propagated through the tree. Each estimate must be supported by data, usually compiled in a separate portion of the event tree document.

An added benefit exists to expressing long- and short-range forecasts in probabilistic terms—one can then estimate hazard and risk quantitatively, compare the geologic risk to more familiar risks and to thresholds of risk acceptability (e.g., Table 5 of Newhall and Hoblitt, 2002; Figure 2.5 of Aspinall, 2006), and make cost-benefit analyses of various risk mitigation options, e.g., evacuations. Based on a review of the psychological literature and an online survey of scientists and emergency managers, Doyle et al. (2014) provided helpful suggestions on how to improve end-user understanding of probabilistic assessments, and to avoid admitted pitfalls in their use.

8.3 CONCEPT OF THE VOLCANIC EVENT TREE

Newhall and Hoblitt (2002) described a generic volcanic event tree that progresses from the onset of volcanic unrest through subsequent, increasingly more specific phenomena such as eruption, eruptive phenomena, distances and sectors, and impacts. The probability of the final, right-most event in this tree is the *product* of the probability at node 1 (e.g., onset of unrest) and the *conditional* probabilities on each branch leading to the right-most event. If that same result is reached along multiple paths, e.g., involving pyroclastic flows from eruptions of various volcanic explosivity indices (VEI), the final products from each path are summed. A pyroclastic flow is a pyroclastic flow, regardless of the VEI of the eruption from which it came. Multiply through the strictly volcanic events to quantify *hazard*, or from hazard onward through exposure, vulnerability, and value to quantify *risk*. The nodes and branches of this generic tree were designed to be relevant at most volcanoes, but at each new volcano or crisis for which a tree is to be developed, scientists may want to modify the generic tree to reflect special behavior or knowledge of that volcano. Most of the case examples that we describe later in this paper are modified from the generic tree. The basic concept of an event tree underlies nearly all probabilistic estimates of evolving hazard and risk during crises; different approaches to populating trees are considered next.

8.4 HOW CAN PROBABILITIES BE ESTIMATED AT EACH NODE AND BRANCH OF A VOLCANIC EVENT TREE?

In general, multiple data sets can be used to estimate probability. Volcanologists may differ, though, on the degree of confidence or weight that they would give to each data set. These differences arise due to: bias toward one's own experiences and specialization or ignorance of others'; different conceptual models of how volcanoes work; and differences in weight for data from the specific volcano in question versus data from a larger set of analogous volcanoes. These differences are natural and mostly healthy.

However, the scientific team must strive to reach and present a simplified consensus, to avoid confusing officials and the public with conflicting assessments. Several ways exist to reach consensus; we note two possible approaches in practice today:

1. The team discusses all pertinent data and reaches a verbal consensus about appropriate data set(s) and probability to be used at each branch of an event tree. The USGS' Volcano Disaster Assistance Program (VDAP) team has used this approach in a number of crises in which neither time nor a large group of experts was available for more formal estimation of probabilities. An added attraction of this approach is that the data set used to estimate probability at each node is easily seen, and the basic concept of multiplication through the branches of a tree to reach final probabilities is easily understood by public officials. The reliability of this approach depends heavily on the choice of the data sets to use at each node, and on the experience and personal dynamics of the team.
2. A formal, expert elicitation gathers probability estimates from all team members. In most cases, estimates are weighted for reliability of either parameters or individual experts. The product is a calculated set of probabilities and uncertainties, which the team members agree to present as a consensus. In volcanology, two popular methods of expert elicitation are the Cooke method, in which the credibility of each expert is anonymously tested and weighted (Cooke, 1991; Aspinall et al., 2003; Aspinall, 2006), and the BET-EF (Bayesian Event Tree - Eruption Forecasting) method in which the diagnostic value of each monitoring parameter is weighted (Marzocchi et al., 2008). The Marzocchi et al. approach debates and assigns diagnostic weights to monitoring parameters at each volcano long before a crisis, to then be applied objectively during a crisis. The Cooke/Aspinall method could weight experts long before a crisis, independently from specifics of the particular volcano, but in practice, the weighting has typically been done during a volcano-specific crisis or exercise.

The increasingly prominent issue of scientists' legal liability in forecasts of natural events favors use of probabilistic rather than purely deterministic forecasts. Probabilities are just that—probabilities, not precise predictions. Event trees acknowledge various possible outcomes *including* null events. Therefore, although scientists who estimate probabilities might be faulted for not considering a particular data set, or a particular outcome or scenario, the explicit use of probabilities conveys uncertainty and multiple possible outcomes to officials and the public. Expert elicitations in which individual votes are anonymous may offer some extra legal protection to individuals.

In our past applications of probabilistic event trees during crises, we have in almost all cases considered multiple data sets and then estimated a single probability at each node that was somewhere between our high and low estimates, but not rigorously related to either. Uncertainty was loosely described in terms of plus or minus in a potentially large, but undefined range. At Mount St. Helens in the early 1980s (see below), we informally checked the sensitivity of probabilities to various assumptions and data sets, and told officials that when we used the most and the least conservative data sets early in a crisis, probability values varied by up to two orders of magnitude higher or lower than our "best guess," and that the range decreased to ± 1 order of magnitude as an eruption drew closer. One colleague at the time said estimates from this method are "guesses that are not intuitively objectionable, bracketed by higher or lower estimates that evoke visceral rejection." Today's legal environment will require more rigor. If authorities want scientists to choose just one "preferred" data set at each node, the scientific team might try to do so but must present it as just one estimate within a range of possibilities. Uncertainty must be spelled out more explicitly than in the past, ideally including distinctions between that which is knowable but just not yet known (epistemic uncertainty), and that which can never be fully known because of the natural complexity and variability of volcanic systems (aleatoric uncertainty).

Within the VDAP team, we now plan to use a variant of the first approach for the future: retaining focus on transparency and data sets but carrying ALL plausible data sets forward and using the range of results as a measure of uncertainty. The final result will be presented as a range of probability at the ending node, with or without a preferred value. We will approximate uncertainty along every pathway by multiplying all of the highest plausible values obtained from multiple data sets, then all of the lowest plausible values, and defining uncertainty as the spread between the two. Estimated uncertainty will vary from one pathway to the next, but we will probably use the wider of these spreads, especially if they are for relatively likely events. The authors described this variant in the 2013 IAVCEI Scientific Assembly, with application to crises when neither time nor a large group of experts is available (Newhall and Pallister, 2013). Although we called it the DIY ("do it yourself") method in that abstract, a better term would be the "Multiple Data Sets" method. We stress that this is a plan, not yet formally tested. The closest we have come to formal application was at Mount St. Helens in 1982–1984, as described above, and in a more limited way during analyses of Sinabung, Indonesia in 2014, in which high and low values were assigned to nodes for which we could not reach consensus.

Where time and expertise are available, we recommend that both expert elicitation and multiple data sets be used. The method of multiple data sets (without formal statistical analysis of uncertainties) offers advantages of easy, transparent tracking of the basis for probability estimates, but can be criticized for the lack of statistical rigor. The group elicitation methods have statistical rigor and track individual inputs—but they also manipulate input (by weighting and averaging) in ways that may be difficult for some users to follow. Table 8.1 compares the three above-mentioned methods. We

TABLE 8.1 Comparison of Three Approaches to Probabilistic Volcanic Hazard and Risk Assessment

	Cooke/Aspinall	BET-EF	Multiple Data Sets
Suitable for: Long-term estimate	Y	Y	Y
Short-term estimate	Y	Y	Y
Well-studied volcano	Y	Y	Y
Sudden crisis at previously unstudied volcano	Y	Not intended	Y
Weights diagnostic value of parameters	Indirect	Direct	Indirect
Weights experts	Y	N	N
Requires group of experts	Y, simultaneously	Y, but not necessary simultaneously	Preferred, but not necessary
Requires expert facilitator	Y	Y	N
Tracks all input	Y	Y	Y
Basis for each probability is easy to see and reconstruct	N	Relation of activity to predefined thresholds and bins is clear; basis for those thresholds bins, and weights is in prior publication	Y
Formal statistical uncertainty	Y	Y	N
Gives single preferred value, with uncertainty	Y	Y	Y, but not necessary
Legal protection if uncertainty clearly stated	Strong	Strong	Strong

recommend intercomparisons of results from these approaches, similar to interlaboratory comparisons.

8.5 A HANDY EXCEL-BASED TOOL FOR BUILDING YOUR OWN TREE

Our earliest uses of trees at volcanoes were on paper, or using clunky and quickly obsolete software for creating probability trees. One of us (JP) developed a

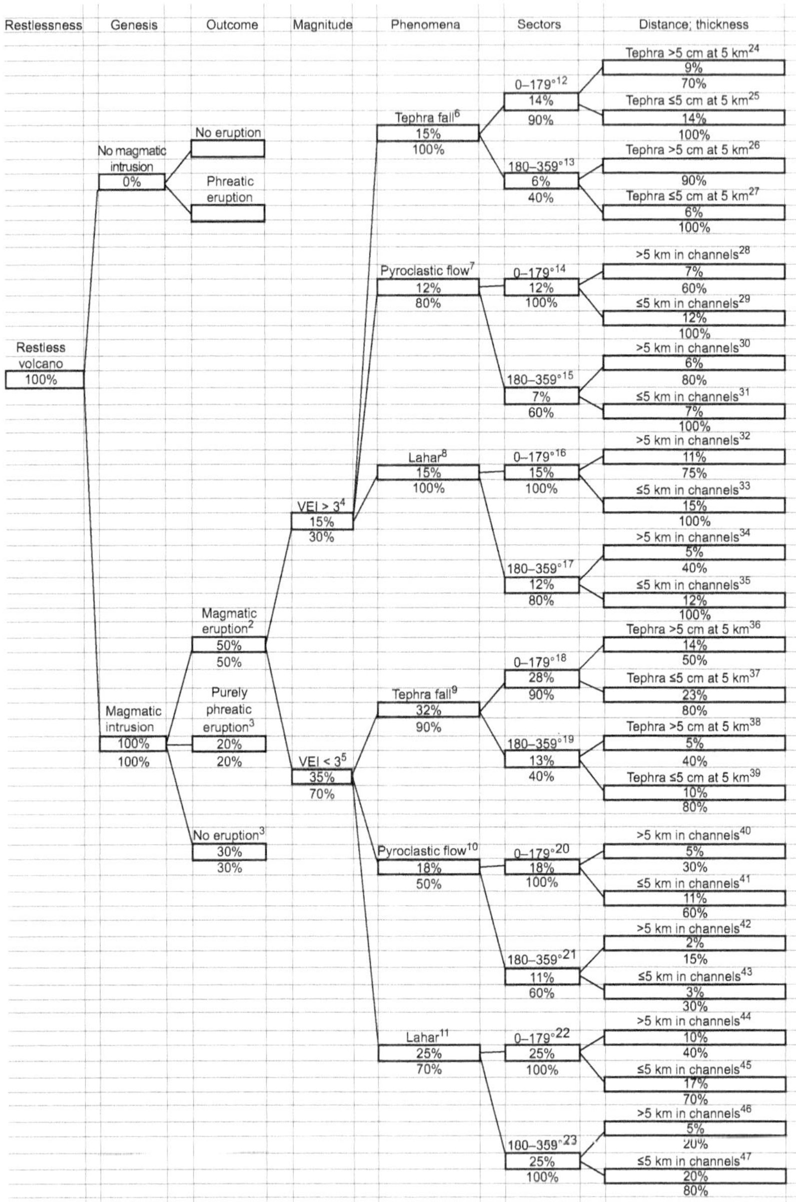

FIGURE 8.2 Generic event tree, in Excel format, that can be modified as desired for any volcano and populated interactively. Probabilities within bold boxes are cumulative. Superscript numbers above text labels correspond or point to supporting data and discussion in a linked Excel sheet.

simple method using simple drawing tools within an Excel spreadsheet (Figure 8.2). The starting node may include a single cell with an initial probability estimate, e.g., the probability of a volcano at rest becoming restless (although during crisis response, this node is often omitted or assigned a probability of 1.0). From that, as many branches are drawn as there are logical or reasonable possibilities, e.g., two branches for magmatic or nonmagmatic origin of that unrest. At that second node, the same number of boxes (e.g., two) are drawn and those cells are filled with a formula that multiplies the prior (cumulative) probability and the conditional probability of each branch (magmatic or nonmagmatic), given the prior event (unrest). For easy visualization, we put the conditional probability in the cell immediately beneath each box that contains the calculated cumulative probability. Retaining the conditional probability is important as that is what must be changed as time progresses. The assignment of probabilities continues with more branches at the next node, each with the conditional probabilities immediately beneath the cumulative products. After the first few nodes, the number of branches may become cumbersome or impossible to keep on a single screen, but one can simply build continuations in cells to the right of, or below, the first ones.

You can start with our generic template (available at http://vhub.org and as Electronic Supplement 1) and modify your tree to suit your particular volcano and questions. If you modify the tree, take care to maintain correct cell references for the cell-based calculations. Once a spreadsheet is set up, it is easy to change estimates as new data require. Excel will automatically recalculate the final probabilities after any change in the Chart. The generic template presented in Electronic Supplement 1 uses a single probability value per branch. The chosen value can be the high, low, or preferred value. To approximate uncertainty in any probability estimate, calculate the spread between the high and low estimates, as described above.

8.6 IMPORTANCE OF DOCUMENTING THE BASIS FOR ALL PROBABILITY ESTIMATES

In the rush to develop a tree during a crisis, it is easy to forget to document the data and reasoning that you have used. Yellow sticky notes and scraps of paper will get lost during a crisis. We have used two slightly different methods. In one, we create a companion text file and, node by node, use narrative text to explain the data sets and reasoning that were used. The resulting document can be many pages long, but because it follows the exact same sequence as branches of the tree, it is easy to follow. Our other approach has been to footnote each conditional probability on the Excel spreadsheet and, in a linked spreadsheet of the same file, give the same narrative of the data and reasoning behind each individual number. A text file may be easier to prepare and edit but can leave some ambiguity about how numbers at specific branches were estimated. The footnotes and companion Excel chart are a bit more cumbersome to prepare but have the advantage of being in the same file with explicit links between each footnote on the graphical tree and the corresponding basis for that estimate. One may also include data tables as linked Excel spreadsheets, such as tables showing numbers of volcanoes that did or did not eruption after periods of unrest (see the Mount St. Helens and Garbuna examples, included as Electronic Supplements 2 and 3 of this paper).

Sometimes, years after a crisis, we have wanted to revisit a particular tree or to recall the basis for a particular probability estimate. Either the companion text file or the footnoted, linked narrative in the Excel file make it easy to recover that information—much easier, we contend, than trying to reconstruct how a particular number was estimated in an expert elicitation.

Though we hope to never see such a day, there may be litigation in the future that requires point-by-point reconstruction of how every probability estimate was made. Either approach—ours or more formal expert elicitations—must retain documentation of every step in the probability estimation process.

8.7 REMOTE PARTICIPATION IN DEVELOPMENT OF PROBABILITY TREES

Although it is generally best for probabilities to be estimated by scientists who are onsite at a crisis and who participate in all of the discussion of data and interpretations, it is sometimes helpful to bring other colleagues into the discussion. The Excel spreadsheet and accompanying documentation of data and reasoning can be a way to do this. We have, on a number of occasions, iterated these files back and forth between colleagues at home or in other countries. We have also developed trees within the VDAP team to assist in reaching an internal consensus, and then as a basis for discussion with colleagues in foreign observatories. In such long-distance discussions it is critical to keep track of versions and to have every participant in the discussion document every addition and change. Online and secure cloud-based document sharing software may be useful for such applications.

8.8 APPLICATIONS OF THE MULTIPLE DATA SETS METHOD, BY VDAP AND OTHERS

Mount St. Helens, USA, 1982 and 1984: Mount St. Helens had a massive sector collapse and lateral blast in May 1980 that killed 57 and flattened over 600 km^2 of forest. We started to use probabilities at Mount St. Helens in 1981, in response to feedback from public officials and corporate users that USGS warnings about ongoing, smaller eruptions were hard to relate to other risks they faced. At the time, USGS warnings used terms like high and low hazard, or higher and lower hazard. Both social science research and our own experience showed that such qualitative adjectives were subject to wide differences of understanding. What geologists consider high hazard might be very different than that considered "high" by a logger or by a public official. We needed a way to calibrate our perceptions—to "get on the same page"—and the best solution seemed to be to quantify the risk of death from the volcano so that users could compare that to more familiar risks.

The first attempt at quantification considered intermediate- to long-range risk, overcoming months and years (Newhall, 1982), using mainly the geologic and historical record of Mount St. Helens and some analogous volcanoes. Such long-term probabilistic analyses of volcanic hazards are still of great value and similar methods have been widely used (e.g., Nathenson et al., 2012). In the Mount St. Helens case, probabilities of further eruptions (in 1981–1982, after the large 1980 sector collapse and blast) were based mainly on eruption chronologies of Mount St. Helens and its analogues. The estimation of probable flow runouts and tephra thicknesses versus distance was based on geologic and historical records of those parameters. Zones of roughly equal, order-of-magnitude risk were delineated, and users could see that risk decreased sharply with distance from the volcano, and more rapidly along ridges than down valleys. A table of familiar risks was presented and loggers were able to compare their risk from the volcano (if they were to return and salvage downed timber) to their risk from logging itself. When the volcanic risk from dominantly dome–growth activity turned out to be similar to or less than their normal risk, they agreed to salvage the timber.

The second task was to estimate changes in hazard, and thereby risk with varying states of volcanic unrest, i.e., to make short-range estimates of risk (Newhall, 1984). Five simplified states of the volcano were recognized (from quiet through mild and severe unrest to dome growth and explosive eruption) and risks were calculated for each. No attempt was made to estimate the increase in probability of eruption from specific patterns of seismicity or other unrest but, rather, the increase in risk given experience in these five states of the volcano. Probability distributions for flow runouts and tephra fall remained the same as in the 1982 report, but levels of occupation (exposure) were assumed to decrease sharply after warnings were issued. The main conclusions of this second report were that risk varied by orders of magnitude depending on the short-term state of the volcano. The land managers (principally, the US Forest Service and the State of Washington) decided that they could allow access near the volcano yet maintain a reasonable degree of safety by simply expanding the "no-go" zone during brief periods of elevated unrest. The US Forest Service built its Johnston Ridge Visitor Center and invited visitors during most summer and autumn days, with only brief closures. Figure 8.2 of the 1984 report shows zones with order-of-magnitude differences in risk, and the difference in levels of risk in each zone during different states of the volcano and different assumptions regarding occupancy. In retrospect, this figure and the discussion behind it were difficult for many users to understand.

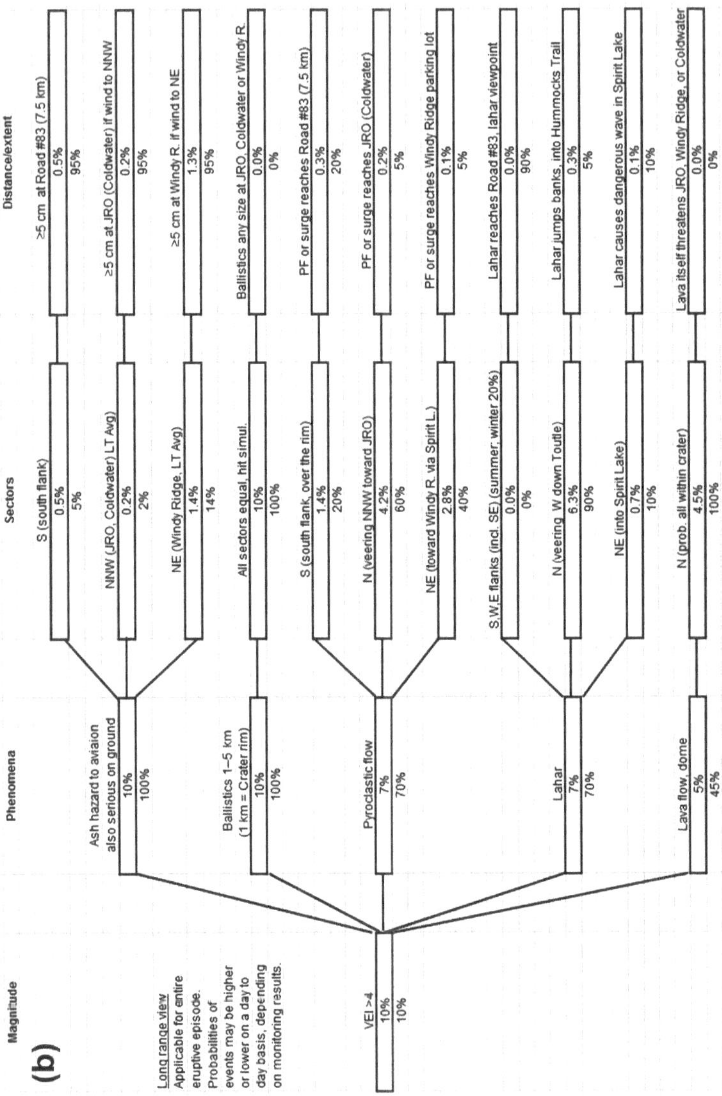

FIGURE 8.3 cont'd

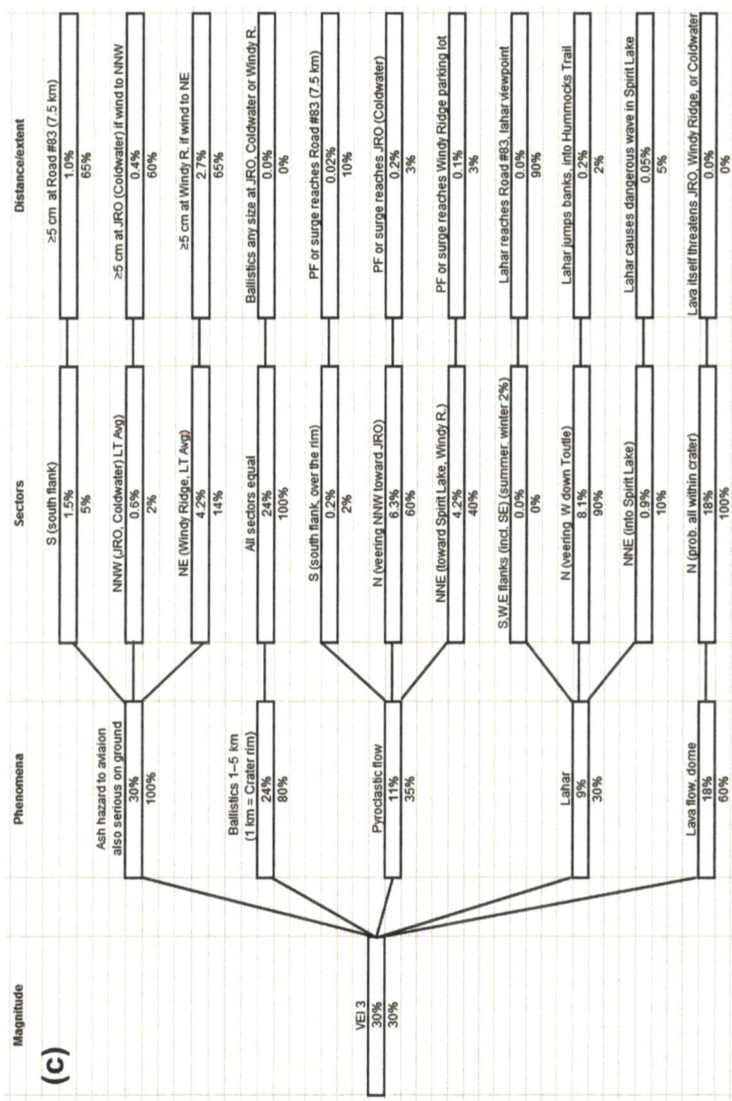

FIGURE 8.3 cont'd

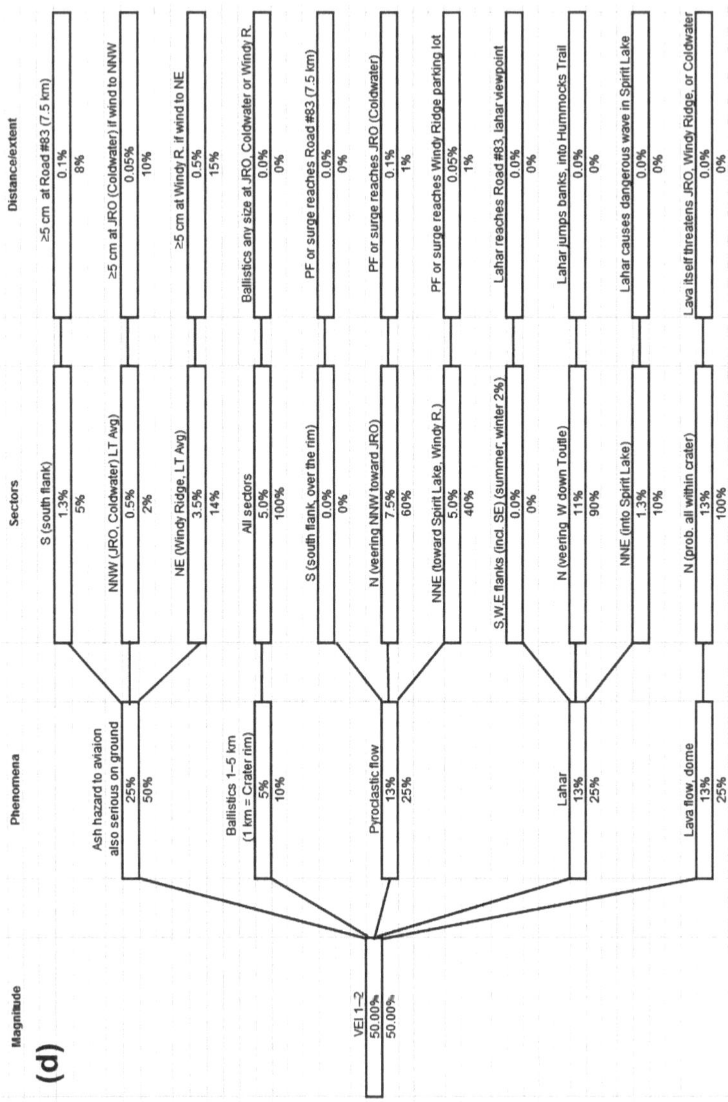

FIGURE 8.3 cont'd

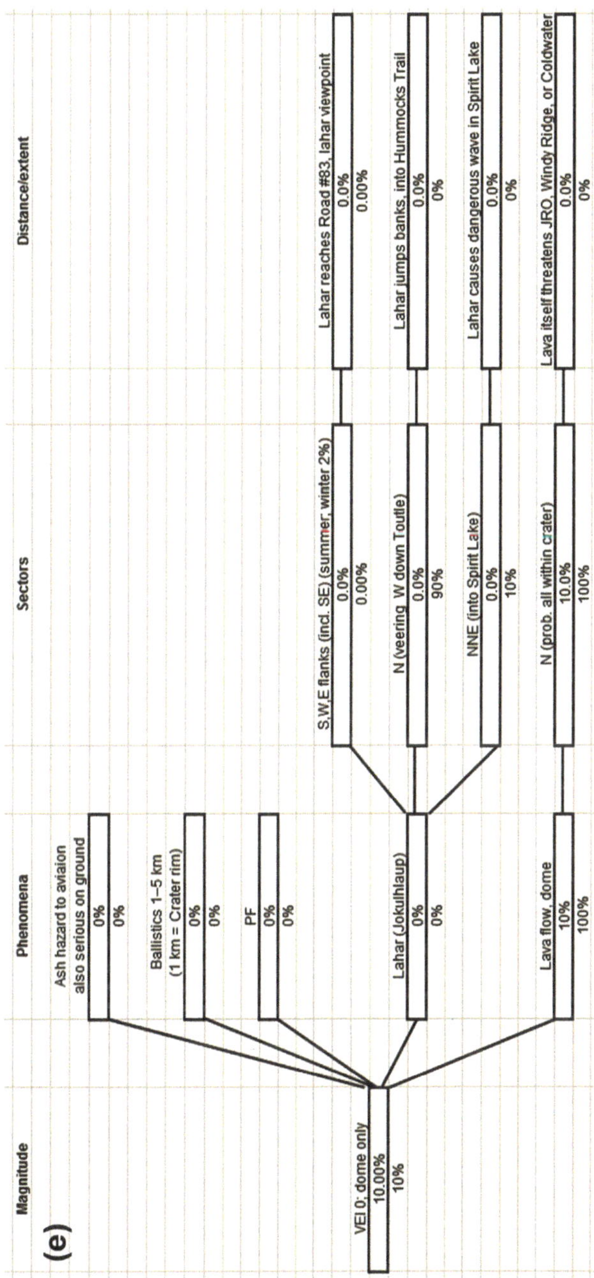

FIGURE 8.3 cont'd

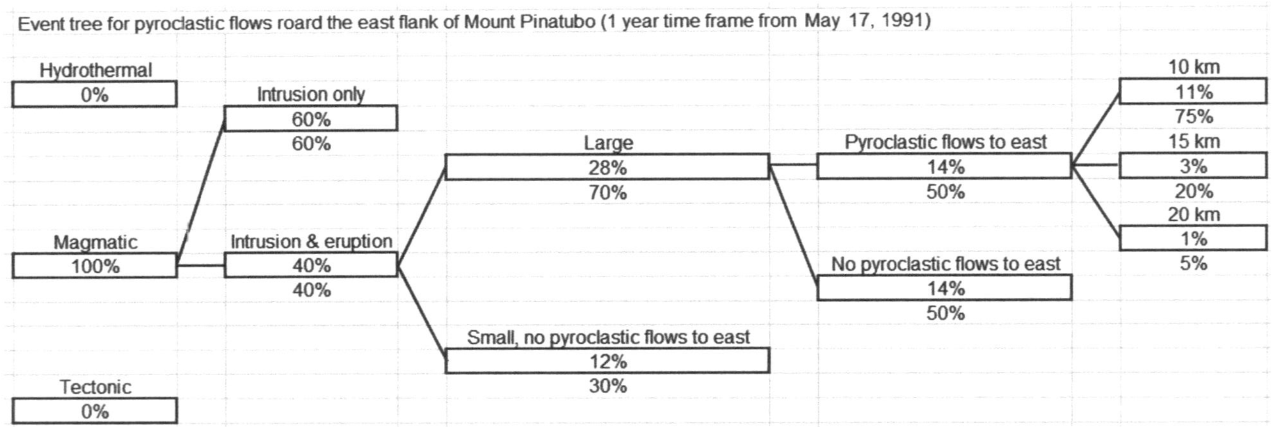

FIGURE 8.4 Simplified event tree developed at Pinatubo Volcano, May 17, 1991, to quantify the probability that pyroclastic flows would reach Clark Air Base (nearest point, 15 km east of the summit). Distances in the rightmost node were maximum distances, i.e., up to 10 km, from 10–15 km, and from 15–20 km.

roughly 3 percent chance of a pyroclastic flow reaching onto Clark Air Base within the coming months. That percentage might sound low to a layman, but it was seen as alarmingly high by military officials. We needed every tool we could muster to convince skeptics, and we combined this probability for a pyroclastic flow with graphic images of pyroclastic flows in the IAVCEI video Understanding Volcanic Hazards (by the late Maurice Krafft). A 1 percent chance of "becoming toast" is a stronger message than simply a 1 percent chance of an unfamiliar phenomenon called pyroclastic flow.

Probability estimates at Pinatubo were based partly on very preliminary information about its eruptive history; indeed, we were learning day by day how often eruptions had occurred in preceding millennia and how far pyroclastic flows of those eruptions had reached. No attempt was made to estimate probabilities of roof collapse and death from tephra fall, though if that had been included we might have anticipated the extra weight of ash that is wetted by heavy rain. Neither were probabilities of lahars estimated, as they are dominantly a posteruption phenomenon and we were preoccupied with pre-eruption questions. The largest change in probability estimates of May 17 and June 10 occurred when magma reached the surface. On May 17, we estimated the probability of eruption (of any type) to be 40 versus 60 percent that magma would not erupt. By June 10, magma was already erupting, so the probability was 100 percent. Our May 17 estimates were useful; those of June 10 were practically moot. Starting in early June, unrest escalated sharply and we used a different tool, eruption alert levels, to raise warnings that an eruption was likely to begin within two weeks, then 24 h. The highest alert level was sounded when the eruption became explosive.

At the time, we did not think of linking the probability trees to the alert level scheme or to the hazard maps. Given more time and data they might have been linked, but the buildup to eruption was so fast and the PHIVOLCS-USGS team was so stretched that we contemplate this only in retrospect!

8.8.2 Garbuna, Papua New Guinea, 2005–2006

Garbuna is the central of three closely spaced volcanoes, Krummel, Garbuna, and Welcker, which rise 455–1,063 m above sea level at the south end of the Willaumez Peninsula, West New Britain Province (WNB), Papua New Guinea. Reconnaissance field work and geochronology (McKee et al., 2005) indicate that pyroclastic flow-producing explosive eruptions from the complex took place approximately 1,800 and 4,600 years ago. The provincial capital Kimbe lies about 17 km SE from the main vent on Garbuna and approximately 20,000 people work and live within 20 km.

Seismic swarms and several relatively small phreatic eruptions suggestive of magmatic intrusion took place in late October and mid-November, 2005—the first known activity at Garbuna in the past 1,800 years. These small eruptions caused widespread concern among the population owing to the memory of the extensive impacts that a small eruption of nearby Pago Volcano had on the population and economy in 2002. In late December 2005, following about a month of quiet, Garbuna experienced another earthquake swarm and incandescence was reported at the October–November vents.

The Rabaul Volcano Observatory (RVO), VDAP, and colleagues from Geoscience Australia (GA) responded to the activity at Garbuna during October–November 2006 and again in January–February 2006. The RVO-VDAP-GA team installed a real-time seismic network, started SO_2 monitoring, conducted reconnaissance geologic investigations, and evaluated potential hazards. An important objective of the response was to assist the local emergency managers to prepare to cope with the range of possible eruptive behavior. To do so, the team created a simplified event tree (Figure 8.5) with probability estimates for future eruptive activity and potential areas affected. The event tree was not used directly in public communications, but instead it was used to achieve a consensus among the small on-site response team of four or five volcanologists and with three remotely located experts in volcano seismology, volcanology, and the geology of Garbuna. It was then used in discussions with members of the Provincial Disaster Committee, particularly Major Paul Kaliop, the committee chair. The committee, in consultation with scientists and staff from RVO and local land managers, then created an eruption response plan and conducted an education program with the at-risk population.

Much of the population in the region is concentrated in the town of Kimbe, and at extensive palm oil plantations near the volcano. Consequently, a major focus of the education program was focused on the plantation workers and managers, and the response plan utilized vehicles and other resources provided by the plantation managers. As activity had diminished by February 2006 and an effective response plan was in place, key questions facing the community and its plantation-based economy were: "what is the probability of renewed activity within the coming months to years, and what areas are most susceptible to eruption impacts?" Consequently, a relatively long time frame of 3 years was chosen for an event tree (Figure 8.5). A digital copy of the tree and its accompanying notes, which explain the reasons for selection of probabilities at each node of the tree are included as Electronic Supplement 3 to this chapter.

Because of the limited geologic and monitoring data that were available for Garbuna, we relied heavily on analogous data from other volcanoes (i.e., the "global basis" noted in the tree) to estimate the probability of additional magmatic intrusions and the subsequent probability of eruptions within the following 3 years ($\sim$75 percent probability of additional intrusions; then $\sim$70 percent ($\sim$50 percent cumulative) for a subsequent eruption within $\sim$2 months of the intrusive episode). Lacking a fully populated global database on precursors to intrusive and eruptive episodes, such as WOVOdat will become (e.g., Venezky and Newhall, 2007; Ratdomopurbo et al., 2012), we instead had to rely on the combined experience of the VDAP team. From this experience we were able to list 26 cases of multiple intrusions in a 2- to 4-year

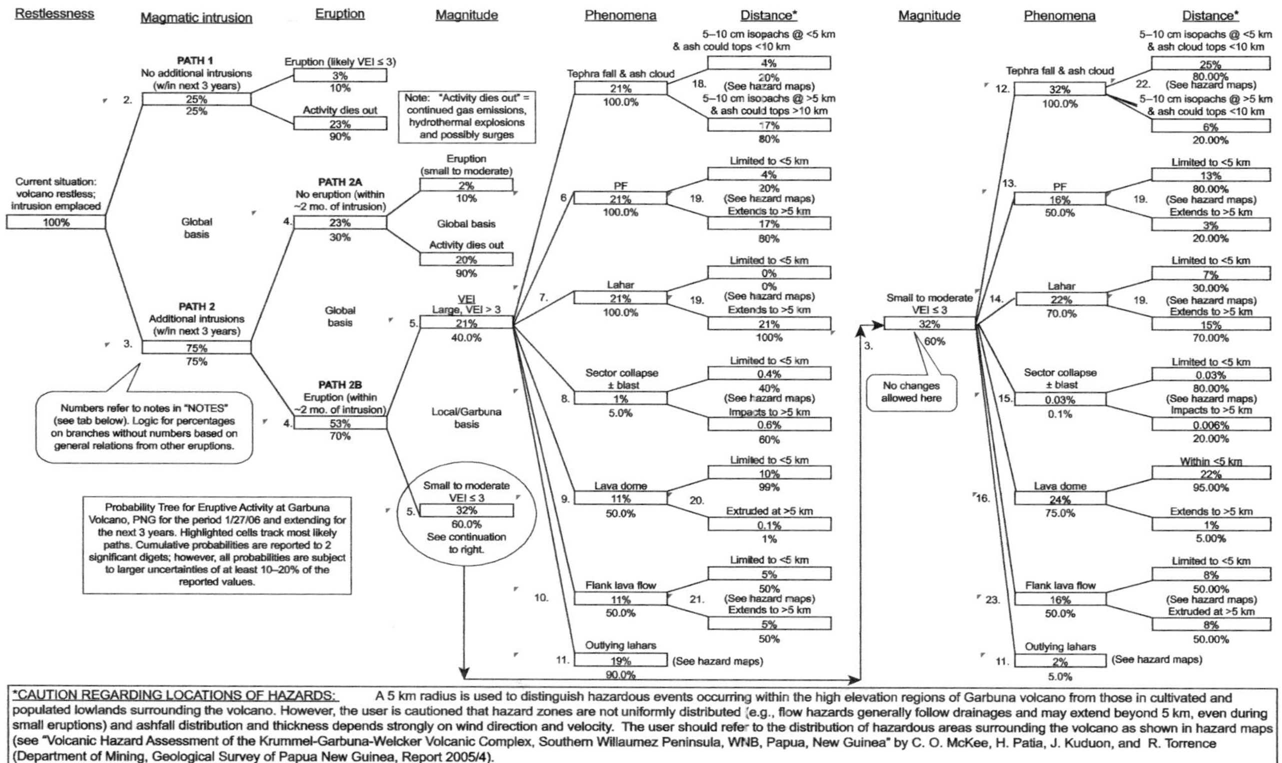

FIGURE 8.5 Event tree for Garbuna volcano, Papua New Guinea for a period of 3 years following February 13, 2006. A digital version of this tree (in MS Excel) is in Electronic Supplement 3. It includes linked spreadsheets (tabs) that explain the basis for probabilities assigned at each node of the tree.

time period with or without eruptions, and nine cases of singular intrusive episodes; yielding the figure of ~75 percent probability of additional intrusions. The subsequent probability of an additional intrusion leading to an eruption was also based on VDAP experience, which suggested that multiple episodes of intrusive seismicity of the intensity seen at Garbuna have led to eruptions in about 40–60 percent of cases at other andesitic stratovolcanoes. We therefore chose a higher percentage of 70 percent (50 percent cumulative) because the volcano had already experienced two intrusive episodes accompanied by small phreatic eruptions in the past 4 months, had an active fumarole field at the summit, and had shown recent evidence of magma close to the surface (glow from the summit). We then proceeded to evaluate the subsequent probabilities of eruptions with explosivities of >VEI 3 and ≤VEI 3 (40 and 60 percent, Figure 8.5), based on the relative abundance of pyroclastic deposits in the geologic record at Garbuna and the extended time period since prior eruptions—adequate time to replenish the magmatic reservoir system and to build up a large gas load. We then estimated the probabilities of the various hazardous events and their extents (tephra fall and ash cloud, pyroclastic flows and surges (PF), lahars, sector collapse ± blast, lava dome, lava flows on flanks, and outlying lahars). Such events were based on factors such as plume heights and deposit volumes for various explosivities and on expected runouts for flowage deposits given the topography of the Garbuna area. The geographic extents of the hazards were presented in terms of the RVO hazard map for Garbuna in discussions with emergency managers and the public.

Over the 3 years following finalization of the event tree (February 13, 2006) additional seismic swarms and small eruptions occurred in March 2008 and March 2009. The latter resulted in ash fall in Kimbe town. These events suggest that additional intrusive activity took place, although the ash eruptions were small and likely only phreatic, much like those of 2005–2006. Our event tree showed a 37 percent cumulative probability of VEI ≤3 eruptions. Subsequent events of 2008 and 2009 were indeed ≤3, in fact very small (VEI ~1 or 2) events without any VEI 3 event or pyroclastic flow. In retrospect, our estimate of the probability of VEI 3 and greater events may have been biased upward because cases of magmatic intrusions that do not lead to eruptions (so called "failed eruptions") are invariably underreported in the global record (Moran et al., 2011). This fact and the need to improve forecasting at volcanoes that have had few if any historic eruptions, such as at Garbuna, stand as a compelling reason for WOVOdat, a comprehensive data set on episodes of global volcanic unrest and subsequent outcomes.

Merapi, Indonesia, April–June 2006: Merapi volcano (central Java, Indonesia) is one of the most active and hazardous volcanoes in the world. It is known for frequent small to moderate eruptions, pyroclastic flows produced by lava-dome collapse, and the large population settled on and around the flanks of the volcano. Its usual behavior for the last decades abruptly changed in late October and early November 2010, when the volcano produced its largest and most explosive eruptions in more than a century, displacing 400,000 people, and claiming nearly 400 lives. Despite the challenges involved in forecasting any "hundred-year eruption," the magnitude of precursory signals was proportional to the large size and intensity of the eruption (Surono et al., 2012). A probabilistic event-tree analysis during the preceding 2006 eruption reminded everyone that Merapi can produce unusually large eruptions, and forecasts made in 2010 by the Indonesian Center of Volcanology and Geological Hazard Mitigation (CVGHM) helped save 10,000–20,000 lives.

The eruption of Merapi that began on April 26, 2006 attracted much local and some international attention. The eruption began like many other recent episodes of dome growth at Merapi, with small-scale collapses of the summit lava dome to the southwest into areas that were accustomed to such activity. However, mid-way through the eruption a septum in the summit region was breached and collapses began to send pyroclastic flows toward populated areas of the south flank. Then, a strong tectonic earthquake shook Yogyakarta and Merapi on May 27, killing ~5,000 people and apparently increasing the frequency of dome collapses. Concern increased about what more might follow. A first event tree was prepared for internal discussion between CVGHM and VDAP on May 7, and updated on May 29 and June 22. The nodes of these trees differ from the generic format shown in Figure 8.2. Early nodes considered the rate of lava extrusion and evidence for pressurization as factors in estimating the likelihood of eruptions of various VEI. Highly explosive eruptions like those in 1872 and 1930 AD were deemed more likely if effusion rate was high and magma (gas) pressure was increasing. Concern about the recurrence of such larger eruptions was emphasized in a collection of papers on Merapi in 2000 (e.g., Voight et al., 2000; Newhall et al., 2000; Andreastuti et al., 2000), and re-emphasized in the event trees of 2006. The event trees also utilized new scenarios and hazard maps developed by the Merapi volcano observatory for dome collapse into various geographic sectors depending on initial eruption site within the summit region and depending on eruption magnitude. Lava extrusion in 2006 was faster than in other recent eruptions (Ratdomopurbo et al., 2013). Our assessment that rapid extrusion rates increase the probability of explosive eruption was based on experience on Montserrat and at other volcanoes where rapid extrusion rates (more correctly, rapid ascent rates) were associated with increased conduit pressure and explosivity. In the end, moderately large dome-collapse pyroclastic flows (including one with two fatalities) occurred, but the 2006 eruption did not turn explosive. In retrospect, after the unusually explosive 2010 eruption, we wonder whether rapid extrusion in 2006 was an early indicator of what was to follow in 2010.

To our knowledge, the 2006 probability trees were not used in discussions with Indonesian civil defense or local officials. However, they helped to focus discussion on how fresh, fast-rising gas-rich magma can change the normal activity associated with degassing of magma in the conduit. The use of extrusion rate and evidence for conduit pressurization in the 2006 tree

from which to choose, and a submenu of possible Boolean searches that could be run.

- Prepare volcano-specific trees in advance of crisis at all high-risk volcanoes. The process of preparing the trees would guide those most familiar with each volcano to develop one or more conceptual models of how that volcano works, and related ways to interpret new monitoring data. Trees prepared in advance will then be ready for use when the volcano reawakens. For example, trees have been prepared in advance of crises for the volcanoes from Arenal to Vesuvius in the list above. In some of these cases, the relative influence of supporting data is weighted and these parameter weights are listed in the event tree documentation.
- Improve links from trees to WOVOdat, a database of global volcanic unrest (www.wovodat.org).

  Links to the Smithsonian's Volcanoes of the World database of historical eruptions (www.volcano.si.edu) can also be improved, as can links to other databases with information on flow runouts, tephra distributions, lava dome eruptions, etc. For example, data and generalizations from WOVOdat can be used at node 2 of the generic tree (Figures 8.1 and 8.2) to estimate the probabilities of magmatic versus nonmagmatic origin of unrest, given the nature of freshly observed unrest. Statistical information from WOVOdat can similarly be used to estimate the probabilities of eruptions of various VEI after specific patterns of unrest. As WOVOdat becomes more complete, Boolean searches will generate new subsets of pertinent data. We have long lost track of the number of tally sheets on paper that we have used over the years; it would be much more efficient and also better for tracking and documentation purposes to be able to do most of these searches online. Indeed, one of the major rationales for WOVOdat is to make monitoring data from many volcanoes available on short notice during volcanic crises. Where internet connections and access to online databases are good, it may be better to leave the searching until actually required. However, where internet connections are still unreliable, some of these data sets can also be provided for stand-alone use and/or some common searches might be run in advance.
- Increase use and rigor of the method of Multiple Data Sets. The first step is simply to use the method more consistently than we have in the past. Use multiple data sets at each node and estimate uncertainties by looking at the spread between high and low estimates. If a single data set is favored over others, document the rationale for this preference and include the other data sets in supporting documentation.
- Make rigorous, retrospective intercomparisons of results from various methods to estimate probabilities. Consider a variety of pre- and syn-crisis episodes at various types of volcanoes. One part of this intercomparison should be a comparison of uncertainties. Because uncertainties in the Cooke-Aspinall, Marzocchi et al., and the present (Multiple Data Sets)

methods are estimated in completely different ways, a higher or lower uncertainty does not tell us that one method is better than another. Try to judge which epistemic and aleatoric uncertainties are captured and which are not.

- As WOVOdat grows, use it to estimate not just probabilities of occurrence, as described above, but also to evaluate uncertainties and the diagnostic value of various precursory indicators. Retrospective study of precursors and outcomes, irrespective of whether probabilities were estimated, and/or eruptions were forecast, is key to judging the diagnostic value of precursory indicators.
- Evaluate how well the actual outcomes of volcanic crises reflect prior estimates of probabilities. Success can be claimed if the probabilistically estimated "most likely scenarios" occur at a rate that is significantly higher than that expected had alternatives at each node been weighted equally.
- Consider linking alert levels to probabilities at nodes 1–4 of the generic event tree. An ongoing study of how well alert levels capture the state and forecast the behavior of volcanoes (Winson et al., 2012) is finding that alert levels of the past have embarrassingly low success rates and need to be improved.

REFERENCES

Acocella, V., Puglisi, G., 2013. How to cope with volcano flank dynamics? A conceptual model behind possible scenarios for Mt. Etna. J. Volcanol. Geotherm. Res. 251, 137–148.

Andreastuti, S.D., Alloway, B.V., Smith, I.E.M., 2000. A detailed tephrostratigraphic framework at Merapi Volcano, Central Java, Indonesia: implications for eruption predictions and hazard assessment. J. Volcanol. Geotherm. Res. 100 (1–4), 51–67.

Aspinall, W.P., Loughlin, S.C., Michael, F.V., Miller, A.D., Norton, G.E., Rowley, K.C., Sparks, R.S.J., Young, S.R., 2002. The Montserrat Volcano Observatory: its evolution, organization, role and activities. In: Druitt, T.H., Kokelaar, B.P. (Eds.), The Eruption of Soufrière Hills Volcano, Montserrat, from 1995 to 1999, vol. 21. Geological Society of London, Memoir, pp. 71–91.

Aspinall, W.P., Woo, G., Voight, B., Baxter, P.J., 2003. Evidence-based volcanology: application to volcanic crises. J. Volcanol. Geotherm. Res. 128, 273–285.

Aspinall, W.P., 2006. Structured elicitation of expert judgement for probabilistic hazard and risk assessment in volcanic eruptions. In: Mader, H.M., Coles, S.G., Connor, C.B., Connor, L.J. (Eds.), Statistics in Volcanology. IAVCEI Publications. ISBN:978-1-86239-208-3, pp. 15–30.

Baxter, P.J., Aspinall, W.P., Neri, A., Zuccaro, G., Spence, R.J.S., Cioni, R., Woo, G., 2008. Emergency planning and mitigation at Vesuvius; a new evidence-based approach. J. Volcanol. Geotherm. Res. 178, 454–473. http://dx.doi.org/10.1016/j.jvolgeores.2008.08.015.

Brancato, A., Gresta, S., Alparone, S., Andronico, D., Bonforte, A., Caltabiano, T., Cocina, O., Corsaro, R.A., Cristofolini, R., Di Grazia, G., Distefano, G., Ferlito, C., Gambino, S., Giammanco, S., Greco, F., Napoli, R., Sandri, L., Selva, J., Tusa, G., Viccaro, M., 2011. Application of BET_EF to Mount Etna: a retrospective analysis (years 2001–2005). Ann. Geophys. 54, 642–661. http://dx.doi.org/10.4401/ag-5346.

Brancato, A., Gresta, S., Sandri, L., Selva, J., Marzocchi, W., Alparone, S., Andronico, D., Caltabiano, T., Cocina, O., Corsaro, R.A., Cristofolini, R., Di Grazia, G., Distefano, G.,

Ferlito, C., Gambino, S., Giammanco, S., Greco, F., Napoli, R., Tusa, G., Viccaro, M., 2012. Quantifying probabilities of eruption at a well-monitored active volcano: an application to Mt. Etna (Sicily, Italy). Boll. Geofis. Teor. Appl. 53, 55–74.

Constantinescu, R., Thouret, J.C., Sandri, L., Irimus, I.A., Stefanescu, R., 2012. BET_VH probabilistic assessment of pyroclastic flows hazard at El Misti volcano, South Peru, based on geological record and numerical simulations with TITAN2D (abstract). Geophys. Res. Abstr. 14. EGU2012-8997-1, EGU General Assembly 2012.

Cooke, R.M., 1991. Experts in Uncertainty: Opinion and Subjective Probability in Science. Oxford University Press, New York, 321 p.

Doyle, E.E.H., McClure, J., Johnston, D.M., Paton, D., 2014. Communicating likelihoods and probabilities in forecasts of volcanic eruptions. J. Volcanol. Geotherm. Res. 272, 1–15.

Fedde, I.D., 2009. Application of Probabilistic Tools and Expert Elicitation for Hazard Assessment at Volcan de Colima, Mexico (unpublished MS thesis). Michigan Technological University, 100 pp.

Freundt, A., Kutterolf, S., Schmincke, H.-U., Hansteen, T., Wehrmann, H., Perez, W., Strauch, W., Navarro, M., 2006. Volcanic hazards in Nicaragua; past, present, and future. Geol. Soc. Am. Spec. Pap. 412, 141–165. http://dx.doi.org/10.1130/2006.2412(08).

Garcia-Aristizabal, A., Selva, J., Fujita, E., 2013. Integration of stochastic models for long-term eruption forecasting into a Bayesian event tree scheme; a basis method to estimate the probability of volcanic unrest. Bull. Volcanol. 75, 1–13. http://dx.doi.org/10.1007/s00445-013-0689-2.

Hoblitt, R.P., Crandell, D.R., Mullineaux, D.R., 1980. Mount St. Helens eruptive behavior during the past 1,500 yr. Geology 8 (11), 555–559.

IAVCEI Subcommittee for Crisis Protocols, 1999. Professional conduct of scientists during volcanic crises. Bull. Volcanol. 60, 323–334.

Jordan, T.H., Chen, Y.-T., Gasparini, P., Madariaga, R., Main, I., Marzocchi, W., Papadopoulos, G., Sobolev, G., Yamaoka, K., Zschau, J., 2011. Operational earthquake forecasting: state of knowledge and guidelines for implementation. Ann. Geophys. 54, 315–391. http://dx.doi.org/10.4401/ag-5350.

Komorowski, J.-C., Legendre, Y., Caron, B., Boudon, G., 2008. Reconstruction and analysis of sub-plinian tephra dispersal during the 1530 A.D. Soufriere (Guadeloupe) eruption; implications for scenario definition and hazards assessment. J. Volcanol. Geotherm. Res. 178, 491–515. http://dx.doi.org/10.1016/j.jvolgeores.2007.11.022.

Kushendratno, Pallister, J.S., Kristianto, Ruskanda Bina, F., McCausland, W., Carn, S., Haerani, N., Griswold, J., Keeler, R., 2012. Recent explosive eruptions and volcano hazards at Soputan volcano—a basalt stratovolcano in north Sulawesi, Indonesia. Bull. Volcanol. 74, 1581–1609. http://dx.doi.org/10.1007/s00445-012-0620-2.

Lindsay, J., Marzocchi, W., Jolly, G., Constantinescu, R., Selva, J., Sandri, L., 2010. Towards real-time eruption forecasting in the Auckland Volcanic Field: application of BET_EF during the New Zealand national disaster exercise 'Ruaumoko'. Bull. Volcanol. 72, 185–204. http://dx.doi.org/10.1007/s00445-009-0311-9.

Marti, J., Aspinall, W.P., Sobradelo, R., Felpeto, A., Geyer, A., Ortiz, R., Baxter, P.J., Cole, P.D., Pacheco, J.M., Blanco, M.J., Lopez, C., 2008. A long-term volcanic hazard event tree for Teide-Pico Viejo stratovolcanoes (Tenerife, Canary Islands). J. Volcanol. Geotherm. Res. 178, 543–552. http://dx.doi.org/10.1016/j.jvolgeores.2008.09.023.

Marzocchi, W., Neri, A., Newhall, C.G., Papale, P., 2007. Probabilistic volcanic hazard and risk assessment. Eos Trans. Amer. Geophys. Union 88 (32), 318. http://www.agu.org/eos_elec/2007/32-318.html.

Marzocchi, W., Newhall, C.G., Woo, G., 2012. The scientific management of volcanic crises. J. Volcanol. Geotherm. Res. 247–248, 181–189.

Marzocchi, W., Sandri, L., Gasparini, P., Newhall, C., Boschi, E., 2004. Quantifying probabilities of volcanic events: the example of volcanic hazard at Mount Vesuvius. J. Geophys. Res. 109, B11201. http://dx.doi.org/10.1029/2004JB003155.

Marzocchi, W., Sandri, L., Selva, J., 2008. BET_EF: a probabilistic tool for long- and short-term eruption forecasting. Bull. Volcanol. 70, 623–632.

McKee, C.O., Patia, H., Kuduon, J., Torrence, R., 2005. Volcanic Hazard Assessment of the Krummel-Garbuna-Welcker Volcanic Complex, Southern Willaumez Peninsula, WNB, Papua New Guinea. Department of Mining, Geological Survey of Papua New Guinea. Report 2005/4, 39 p.

Meloy, A.F., 2006. Arenal-type pyroclastic flows; a probabilistic event tree risk analysis Arenal Volcano, Costa Rica; magma genesis and volcanological processes. J. Volcanol. Geotherm. Res. 157, 121–134. http://dx.doi.org/10.1016/j.jvolgeores.2006.03.048.

Moran, S.C., Newhall, C., Roman, D.C., 2011. Failed magmatic eruptions: late-stage cessation of magma ascent. Bull. Volcanol. 73, 115–122.

Mullineaux, D.R., 1986. Summary of pre-1980 tephra-fall deposits erupted from Mount St. Helens, Washington State, USA. Bull. Volcanol. 48 (1), 17–26.

Nathenson, M., Clynne, M.A., Muffler, L.J.P., 2012. Eruption Probabilities for the Lassen Volcanic Center and Regional Volcanism, Northern California. USGS Sci. Invest. Rpt. 2012-5176-B, 23 p.

Neri, A., Aspinall, W.P., Cioni, R., Bertagnini, A., Baxter, P.J., Zuccaro, G., Andronico, D., Barsotti, S., Cole, P.D., Esposti Ongaro, T., Hincks, T.K., Macedonio, G., Papale, P., Rosi, M., Santacroce, R., Woo, G., 2008. Developing an event tree for probabilistic hazard and risk assessment at Vesuvius. J. Volcanol. Geotherm. Res. 178, 397–415. http://dx.doi.org/10.1016/j.jvolgeores.2008.05.014.

Newhall, C.G., 1982. A Method for Estimating Intermediate and Long Term Risks from Volcanic Activity, with an Example from Mount St. Helens, Washington. U.S. Geological Survey Open file Report 82–396, 59 p.

Newhall, C.G., 1984. Semiquantitative Assessment of Changing Volcanic Risk at Mount St. Helens, Washington. U.S. Geological Survey Open File Report 84–272, 29 p.

Newhall, C., Bronto, S., Alloway, B., Banks, N.G., Bahar, I., del Marmol, M.A., Hadisantono, R.D., Holcomb, R.T., McGeehin, J., Miksic, J.N., Rubin, M., Sayudi, S.D., Sukhyar, R., Andreastuti, S., Tilling, R.I., Torley, R., Trimble, D., Wirakusumah, A.D., 2000. 10,000 years of explosive eruptions of Merapi Volcano, Central Java: archaeological and modern implications. J. Volcanol. Geotherm. Res. 100, 9–50.

Newhall, C.G., Hoblitt, R., 2002. Constructing event trees for volcanic crises. Bull. Volcanol. 64, 3–20.

Newhall, C.G., Pallister, J.S., 2013. What to Do When There Is Neither Time Nor a Group of Experts for Expert Elicitation? Abstract 3P2_4C-O21. IAVCEI Scientific Assembly 2013, Kagoshima, Japan.

Orsi, G., di Vito, M.A., Selva, J., Marzocchi, W., 2009. Long-term forecast of eruption style and size at Campi Flegrei caldera (Italy). Earth Planet. Sci. Lett. 287 (1–2), 265–276.

Pallister, J.S., Hoblitt, R.P., Crandell, D.R., Mullineaux, D.R., 1992. Mount St. Helens a decade after the 1980 eruptions: magmatic models, chemical cycles, and a revised hazards assessment. Bull. Volcanol. 54, 126–146.

Pallister, J.S., Schneider, D.J., Griswold, J.P., Keeler, R.H., Burton, W.C., Noyles, C., Newhall, C.G., Ratdomopurbo, A., 2013. Merapi 2010 eruption—chronology and extrusion rates monitored with satellite radar and used in eruption forecasting. J. Volcanol. Geotherm. Res. 261, 144–152.

Punongbayan, R.S., Bautista, M.L.P., Harlow, D.H., Newhall, C.G., Hoblitt, R.P., 1996. Pre-eruption hazard assessments and warnings. In: Newhall, C.G., Punongbayan, R.S. (Eds.), Fire and

Mud: Eruptions and Lahars of Mount Pinatubo, Philippines. Philippine Institute of Volcanology and Seismology, and Seattle, University of Washington Press, Quezon City, pp. 67–85.

Queiroz, M.G., Pacheco, J.M., Gaspar, J.L., Aspinall, W.P., Guest, J.E., Ferreira, T., 2008. The last 5000 years of activity at Sete Cidades Volcano (Sao Miguel Island, Azores); implications for hazard assessment. J. Volcanol. Geotherm. Res. 178, 562–573. http://dx.doi.org/10.1016/j.jvolgeores.2008.03.001.

Ratdomopurbo, A., Widiwijayanti, C., Win, N.-T.-Z., Chen, L.-D., Newhall, C., 2012. WOVOdat Progress 2012: installable DB template for volcano monitoring database. Geophys. Res. Abstr. 14. EUG2012-6742-1, 2012.

Ratdomopurbo, A., Beauducel, F., Subandriyo, J., Nandaka, I.G.M.A., Newhall, C.G., Suharna, Sayudi, D.S., Suparwaka, H., Sunarta, 2013. Overview of the 2006 eruption of Mt. Merapi, J. Volc. Gotherm. Res. 261, 87–97.

Sandri, L., Guidoboni, E., Marzocchi, W., Selva, J., 2009. Bayesian event tree for eruption forecasting (BET_EF) at Vesuvius, Italy: a retrospective forward application to the 1631 eruption. Bull. Volcanol. 71 (7), 729–745.

Sandri, L., Jolly, G., Lindsay, J., Howe, T., Marzocchi, W., 2012. Combining long- and short-term PVHA with cost-benefit analysis to support decision making in a volcanic crisis from the Auckland Volcanic Field, New Zealand. Bull. Volcanol 74 (3), 705–723. http://dx.doi.org/10.1007/s00445-011-0556-y.

Sandri, L., Thouret, J.-C., Constantinescu, R., Biass, S., Tonini, R., 2014. Long-term multi-hazard assessment for El Misti volcano (Peru). Bull. Volcanol. 76. http://dx.doi.org/10.1007/s00445-013-0771-9.

Selva, J., Marzocchi, W., Papale, P., Sandri, S., 2012. Operational eruption forecasting at high-risk volcanoes; the case of Campi Flegrei, Naples. J. Appl. Volcanol. 1. http://dx.doi.org/10.1186/2191-5040-1-5.

Selva, J., Sandri, L., 2013. Probabilistic seismic hazard assessment: combining Cornell-like approaches and data at sites through Bayesian inference. Bull. Seismol. Soc. Am. 103, 1709–1722. http://dx.doi.org/10.1785/0120120091.

Surono, Jousset, P., Pallister, J., Boichu, M., Buongiorno, M.F., Budisantoso, A., Costa, F., Andreastuti, S., Prata, F., Schneider, D., Clarisse, L., Humaida, H., Sumarti, S., Bignami, C., Griswold, J., Carn, S., Oppenheimer, C., 2012. The 2010 explosive eruption of Java's Merapi volcano — a '100-year' event. Journal of Volcanology and Geothermal Research 241–242, 121–135.

Vallance, J.W., Schneider, D.J., Schilling, S.P., 2008. Growth of the 2004–2006 lava-dome complex at Mount St. Helens, Washington. In: Sherrod, D.R., Scott, W.E., Stauffer, P.H. (Eds.), A Volcano Rekindled: The Renewed Eruption of Mount St. Helens, 2004–2006. U.S. Geological Survey Professional Paper 1750, pp. 169–208.

Venezky, D.Y., Newhall, C.G., 2007. WOVOdat Design Document: The Schema, Table Descriptions, and Create Table Statements for the Database of Worldwide Volcanic Unrest (WOVOdat Version 1.0). U.S. Geological Survey Open File report 2007–1117.

Voight, B., Constantine, E.K., Siswowidjoyo, S., Torley, R., 2000. Historical eruptions of Merapi Volcano, Central Java, Indonesia, 1768–1998. J. Volcanol. Geotherm. Res. 100, 69–138.

Winson, A., Newhall, C., Costa, F., Woo, G., November 2012. How Successful Are Volcanic Alert Levels at Anticipating Eruptions (abstr.). Cities on Volcanoes 7, Colima, Mexico.

Woo, G., 2010. Operational earthquake forecasting and risk management. Seismol. Res. Lett. 81, 778–782. http://dx.doi.org/10.1785/gssrl.81.5.778.

Yamaguchi, D.K., 1985. Tree-ring evidence for a two-year interval between recent prehistoric explosive eruptions of Mount St. Helens. Geology 13 (8), 554–557.

Chapter 9

Operational Short-term Volcanic Hazard Analysis: Methods and Perspectives

Jacopo Selva [1], Warner Marzocchi [2], Laura Sandri [1] and Antonio Costa [1]
[1] *Istituto Nazionale di Geofisica e Vulcanologia, Sezione di Bologna, via D. Creti 12, 40128 Bologna, Italy,* [2] *Istituto Nazionale di Geofisica e Vulcanologia, Sezione Roma 1, via di Vigna Murata 605, 00143 Roma, Italy*

ABSTRACT

Probabilistic Volcanic Hazard Assessment (PVHA) is the most complete scientific contribution for planning rational strategies aimed at mitigating the risk posed by volcanic activity at different time scales. The definition of the space–time window for PVHA is related to the kind of risk mitigation actions that are under consideration. During volcanic unrest episodes or eruptions, short-term PVHA is essential for crisis management, since tephra may heavily affect building stability, public health, transportations and evacuation routes (airports, trains, road traffic), and lifelines (electric power supply). In this study, we propose a new methodology for the short-term PVHA based on the Bayesian Event Tree model, in which measures from the monitoring system are used to routinely update the probabilities of ash fall loading in the surroundings of the volcano. As an example, we apply this procedure to the 1982–1984 volcanic unrest at Campi Flegrei, Italy.

9.1 INTRODUCTION

Volcanic hazard and risk analysis represent the most important scientific contributions for planning rational strategies aimed at reducing the impact of volcanic eruptions at different time scales (for a recent review see Marzocchi et al. (2012a)). Notwithstanding its importance, volcanic hazard is often loosely defined, ranging from purely qualitative to more or less quantitative assessments. Ideally, the volcanic hazard is defined as the probability to have one specific kind of volcanic event (a fixed amount of ash, the arrival of a pyroclastic flow, the eruption onset, and so on) in a given space and time interval. However, until now the most popular way to represent the volcanic

Volcanic Hazards, Risks, and Disasters. http://dx.doi.org/10.1016/B978-0-12-396453-3.00009-5

hazard is by means of maps reporting areas with different colors representing the different hazard levels (for example, low, medium, and high). Such maps are used in different ways, i.e., for land-use planning and for evacuation plans.

Marzocchi et al. (2012a) argued that the use of semiqualitative maps may be very helpful in many situations where the risk at stake is not particularly high, but they are clearly inappropriate to manage situations where the volcanic risk is high, as for volcanoes close to high-urbanized areas, nuclear repositories, or nuclear power plants, or, more generally, critical infrastructures. In this case, any kind of mitigation action turns out to be often significantly expensive. A procedure to rationally allocate the available funds for risk reduction plans demands a careful and quantitative comparison of different factors, such as the level of threats, the feasibility of mitigation actions, and their direct and indirect costs and benefits. The adoption of a systematic and rational decision-support procedure based on quantitative assessment has the advantage of providing a transparent audit trail, which reduces the degree of subjective opinion in volcanological communication to civil authorities. Should a risk mitigation action turn out to be a false alarm, this audit trail would provide a direct means for tracking the decision process in any subsequent formal enquiry. Eventually, a transparent decision-making protocol based on quantitative procedures allows the general public to understand that alarms are sounded because their lives are highly valued, even if the alarms prove later to be false.

A complete decision-making protocol requires several competences that are not in the usual background of a scientist, because it should also entail economic, engineering, social, and political expertise. Paraphrasing Jordan et al. (2011) who paved the way for operational earthquake forecasting in seismology, the role of scientists in the decision-making process is to provide authoritative information about the time evolution of the hazard to help communities prepare for potentially destructive events. Taking into account the ubiquitous presence of reducible (epistemic) and irreducible (aleatory) uncertainties, a quantitative volcanic hazard analysis must be expressed in terms of probabilities. Besides being fundamental for establishing rational decision-making protocols for managing volcanic risk, we argue that the definition of volcanic hazards in terms of probability allows us to compare risks of different nature; this is of paramount importance for decision makers that have to allocate funds for risk-reduction strategies across a wide range of possible threats (Marzocchi et al., 2012b).

The definition of the space–time window for hazard assessment is related to the kind of risk mitigation actions that are under consideration. For instance, a long time interval (years to decades) is usually pertinent for land-use planning. Shorter intervals (days to weeks) are important for short-term risk mitigation actions like the evacuation of a volcanic area. Here we focus our attention on the short-term volcanic hazard analysis and its operational implementation. As a tutorial example we report the retrospective operational hazard assessment of short-term tephra fall in the surroundings of Campi Flegrei at different stages of the 1982–1984 unrest. This information is crucial for crisis management because tephra may strongly affect building stability, public health, transportation, and evacuation routes (airports, trains, road traffic), and lifelines (electric power supply). Theoretically, this assessment should be based on sound modeling procedures stemming from frequently updated meteorological forecast and information about the evolution of the crisis. In addition, the relevance of epistemic uncertainties arising from the forecast of the future eruption dynamics and wind conditions, and from the tephra dispersal model, should be estimated.

We rely on the model BET_stVH (Bayesian Event Tree for short-term Volcanic Hazard; Selva et al., submitted for publication; Selva et al., 2013). BET_stVH is based on the Event Tree in Figure 9.1, being nodes 1–5 relative to the eruption forecasting (e.g., Marzocchi et al., 2008), and nodes 6–7–8 to the impact assessment (e.g., Marzocchi et al., 2010). BET_stVH is able to include information about an ongoing unrest, considering real-time monitoring measures to provide a short-term assessment of the probability of unrest, magmatic unrest, and eruption occurrence, as well as of vent positions, and updated weather forecasts to compute short-term probabilistic hazard due to tephra dispersion. Then, it combines short- and long-term assessments, weighting them with their reliability (Marzocchi et al., 2008). With respect to other BET models in the literature (e.g., Marzocchi et al., 2008, 2010), BET_stVH includes specific improvements to enable quantitative short-term hazard analysis, such as:

- a time function to describe how the probability of eruption evolves within the time window defined for the forecast, together with an explicit management of the temporal evolution of the reliability of short-term forecasts;
- the production of hazard curves, enabling a complete description of Probabilistic Volcanic Hazard Assessment (PVHA) and its uncertainties.

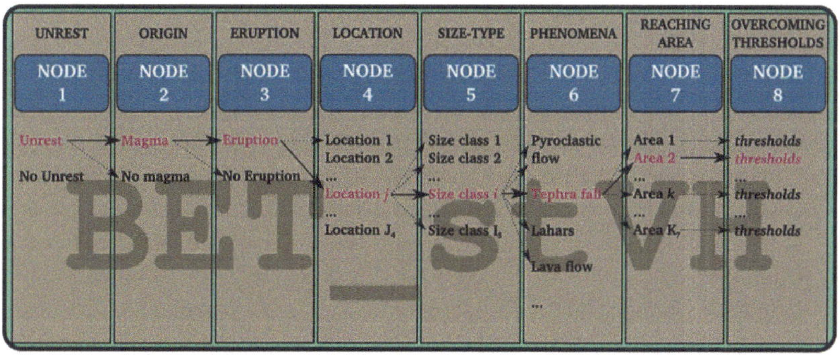

FIGURE 9.1 Event Tree scheme used in this study for short-term operational PVHA (BET_stVH model, Selva et al., submitted for publication).

The goal of this application is to show what, and how, pieces of scientific knowledge could be operationally transferred to decision makers if, during the 1982–1984 Campi Flegrei crisis, scientists knew what we know today about this volcano.

9.2 THE BRADYSEISMIC CRISES AT CAMPI FLEGREI IN 1982–1984

The Campi Flegrei caldera is a complex and active structure that has been showing signs of unrest since the 1950s, with two major unrest episodes (1969–1972 and 1982–1984) which generated a net vertical ground uplift of about 3.5 m nearby the town of Pozzuoli (e.g., Orsi et al., 1999). The 1982–1984 unrest episode is the most significant recorded at Campi Flegrei in recent times, and it has been widely studied in literature. A general review of this episode may be found, among others, in Barberi et al. (1984), de Natale et al. (1991), and Orsi et al. (1999).

Due to the high risk posed by Campi Flegrei, in the last decade two sequential scientific projects were funded by the Italian Department of Civil Protection (INGV-DPC 2005, 2007). Within this framework, a series of 5-year long elicitation sessions were carried out in order to identify what monitoring parameters, their weight, and their anomaly thresholds were indicative of unrest (node 1), of an active magma role in unrest (node 2) and of an eruption occurrence (node 3) at Campi Flegrei. A pool of about 30–40 experts was formally elicited each time, and each session was preceded by seminars, analysis of previous elicitation results, and debate, followed by public discussion. A complete picture of the results of this elicitation process is given in Selva et al. (2012b), whereas in Figure 9.2 we display enlargements of the resulting probability distributions for such crises.

In this chapter, we analyze three different periods of the 1982–1984 crisis, from a tutorial perspective. In particular, we focus on three different stages in which the amount of evidence brought by the monitoring data is different:

- July 8–14, 1982: this was the beginning of the unrest (Figure 9.2(b)), as detected by BET_EF_CF code set during the 5-year long elicitation experiment (Selva et al., 2012b); the unrest was not complete (the maximum degree of unrest was about 20%), nonetheless it was detected because of a slightly anomalous ground uplift rate. However, neither significant anomalous seismicity was detected at this point, nor are fractures
- March 18–24, 1983: this period belongs to a phase of clear unrest (Figure 9.2(c)) due to anomalous ground uplift, uplift rate, and VT events; further, there were some VT epicenters available that we used to define a spatial probability map for vent opening for this week, in order to delineate the areas most prone to be the vent location in case of eruption.

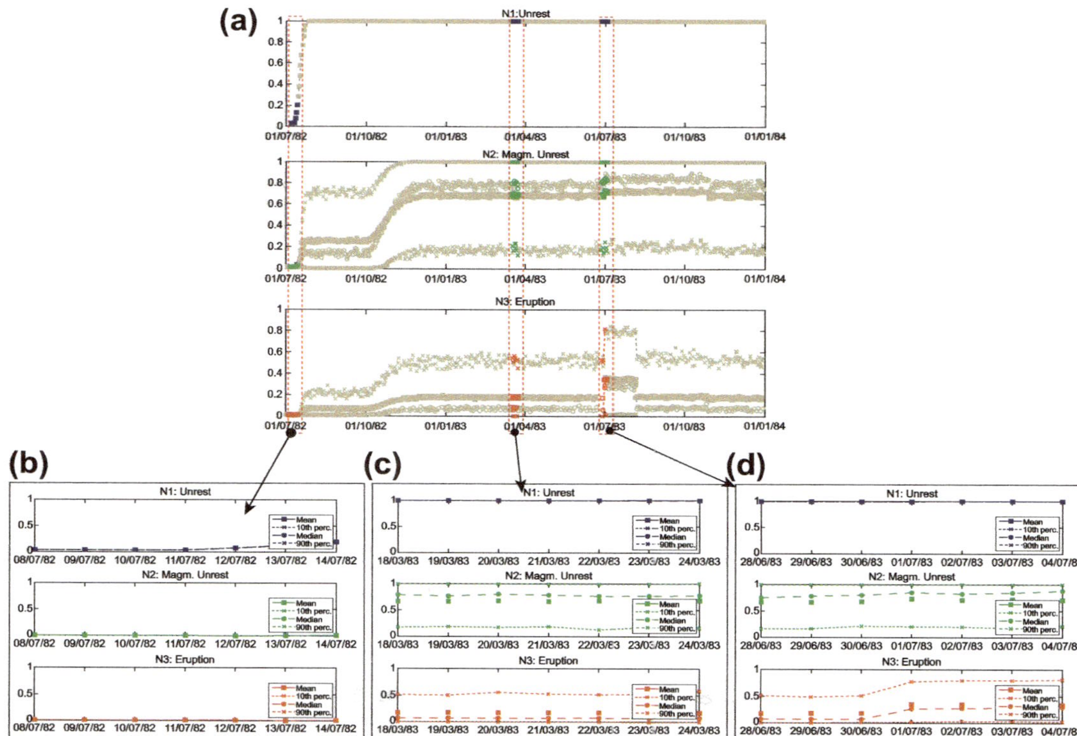

FIGURE 9.2 (a) Time series (in gray symbols) of the monthly probability of unrest (top), magmatic unrest (middle), and magmatic eruption (bottom). Crosses show 10th and 90th percentiles, circles show the mean, and squares show the median. In red dashed lines we highlight the three periods we analyzed for short-term PVHA: July 8–14, 1982 (b), March 18–24, 1983 (c) and June 28 to July 4, 1983 (d).

- June 28–July 4, 1983: this is the beginning of the period during which the probability of eruption was highest (Figure 9.2(d)), according to BET_EF_CF applied to the whole 1981–2009 period (Selva et al., 2012b). It was characterized by anomalous ground uplift and uplift rate, VT seismicity, and by the opening of a fracture in the Solfatara area (Orsi et al., 1999); we made use of the localization of the fracture and of some VT epicenters to define the spatial probability map for vent opening for this week, as above.

Operational activities require that forecasts are regularly updated every time we get new observations. Here, for the sake of conciseness for each of these three periods, we show the tephra fall probabilistic short-term (24 h) volcanic hazard analysis for one specific day (the fourth day of each period, respectively, July 11, 1982, March 22, 1983, July 2, 1983). It is noteworthy that the final assessment accounts for all uncertainties and the time evolution of the probability of unrest, magmatic unrest, and eruption occurrence (nodes 1–2–3); location of the vent (node 4); size of the possible eruption (node 5); production of tephra (node 6); transport, dispersal, and deposition of tephra according to the eruptive conditions and wind field (nodes 7–8).

In this example we neglected some sources of uncertainty such as internal variability of eruptive conditions within a given eruptive size (the so-called intrasize variability, Sandri et al., in preparation), or the variability among different models to describe tephra transport and dispersal (a source of epistemic uncertainty) that we did not explore because we made use of only one model (i.e., HAZMAP, Macedonio et al., 2005; Pfeiffer et al., 2005). However, we did consider a subjective measure of reliability of the model as we associated with its results an equivalent number of data (or equivalent sample size, see Marzocchi et al. (2008)).

9.3 SHORT-TERM BET_VH SETTING FOR CAMPI FLEGREI

What follows are the basic settings of BET_stVH for this application (Table 9.1).

9.3.1 Nodes 1–2–3

At these nodes, we used the settings established during the elicitation 5-year long process. A complete summary of settings for nodes 1, 2, and 3 is reported in Selva et. al. (2012b; Table 4).

An important issue that emerged is the inertia of anomalies. Briefly, it was important to define, after an anomaly was observed, for how long the information it carries must be taken into account. Inertial time depends on the parameter (e.g., counts of events such as earthquakes, or uplift rates) and on the node (node 1 has longer inertial times, node 3 much shorter due to the

TABLE 9.1 BET_VH Settings at All the Nodes

	Prior		Likelihood	Source
	Θ (Mean Probabilities)	Λ (Equivalent Number of Data)	Past Data	
Node 1	0.5	1	$N_1 = 306$ $y_1 = 7.4$	Selva et al. (2012b)
Node 2	0.5	1	$N_2 = 0$ $y_2 = 0$	Selva et al. (2012b)
Node 3	0.33	1	$N_3 = 3.7$ $y_3 = 0$	Selva et al. (2012b)
Node 4	Structural model	2	27 eruptions	Selva et al. (2012a)
Node 5	Effusive: 0.05 Small explosive: 0.79 Medium explosive: 0.13 Large explosive: 0.03	1	$y_{Eff} = 3$ $y_{S\text{-}Expl} = 15$ $y_{M\text{-}Expl} = 6$ $y_{L\text{-}Expl} = 1$	Orsi et al. (2009)
Node 6	Effusive: 0 Small explosive: 1 Medium explosive: 1 Large explosive: 1	–	–	Selva et al. (2010)
Nodes 7–8	From HAZMAP simulations	10	–	Selva et al. (2010) and further modifications as in Sandri et al. (in preparation)

expected shorter lead time of eruption "precursors," Marzocchi et al. (2004)). In the present application, we rely on the inertia definitions provided by experts during the INGV-DPC projects (Selva et al. (2012b); their Table 4). In summary:

- for parameters based on counting of events (e.g., seismic), the number of observed events is divided by the number of days from the observation,
- for the Boolean parameters and the temperature of Pisciarelli fumarole, the anomaly persists for the whole inertia time window, and
- for the uplift rates, the rate is computed daily on a 2-week baseline, and the maximum taken over the inertia time window.

Another important aspect of BET_stVH is that it must deal with the time frames required for operative PVHA. One of the goals of PVHA is to provide the probability of exceedance of several levels of the hazard (i.e., loading due to tephra fall) due to eruption occurring within the 24 h following the observation of new monitoring data. In this case we are interested in forecasts on a daily basis, that is, to assess the probability of eruption *within the next 24 h*. However, the settings reported in Table 9.1 refer to a time window $(t_0, t_0 + \tau)$, where t_0 is the time of observation and τ is the time window selected, that is set to 1 month in this case. Therefore, BET_stVH needs to model the evolution of the monthly probability on a shorter time window, as it cannot be assumed as uniform within τ. Indeed, a forecast based on measures at t_0 is stronger close to the time at which the information is collected, and weaker away from t_0. For this reason, BET_stVH models the probability of eruption, conditioned to a magmatic unrest, in a given time interval dt (contained in the time range $(t_0, t_0 + \tau)$ as

$$\theta_3(t, t + dt) = F(t, t + dt) \cdot \theta_3(t_0, t_0 + \tau) \tag{9.1}$$

so that the total probability of eruption over $(t_0, t_0 + dt)$ becomes

$$\theta_E(t, t + dt) = \theta_1 \cdot \theta_2 \cdot \theta_3(t, t + dt) \tag{9.2}$$

that is, the absolute probability of eruption in $(t, t + dt)$ is the product of the probability of the first three nodes, θ_1, θ_2, θ_3, and $F(t, t + dt) = F(t + dt) - F(t)$, where $F(t)$ is a cumulative density function (cdf) that represents what proportion of $\theta_3(t_0, t_0 + \tau)$ is concentrated in $(t, t + dt)$. Its derivative $f(t)$ represents the probability density function (pdf) describing the evolution in time of θ_3. As the strength of the forecast is expected to drop rather quickly after t_0, the probability of eruption is expected to be high in the early days and then to drop. In other words, we expect that $f(t)$ may be well represented with an exponential form, so that $f(t) = k \exp(-t/T)$, where T is the characteristic time (we set it to 1 week in our case) and k is a normalization constant to impose that $F(t_0, t_0 + \tau) = 1$, i.e.,

$$k = \{T[1 - \exp(-\tau/T)]\}^{-1} \tag{9.3}$$

Figure 9.3 shows the daily probability of eruption for the 1982–1984 crisis, with enlargements of the three periods considered, that we obtain by applying Eqn (9.1) to every day of the whole 1982–1984 interval with a dt of 24 h from the time t corresponding to the observation of the daily monitoring data.

9.3.2 Node 4

For Campi Flegrei, we base our evaluation of the spatial probability of vent opening on two pieces of evidence (Marzocchi et al. (2008) and in the Electronic Supplementary Material): one is the long-term map based on geomorphological and structural features of the caldera and on past eruptive activity

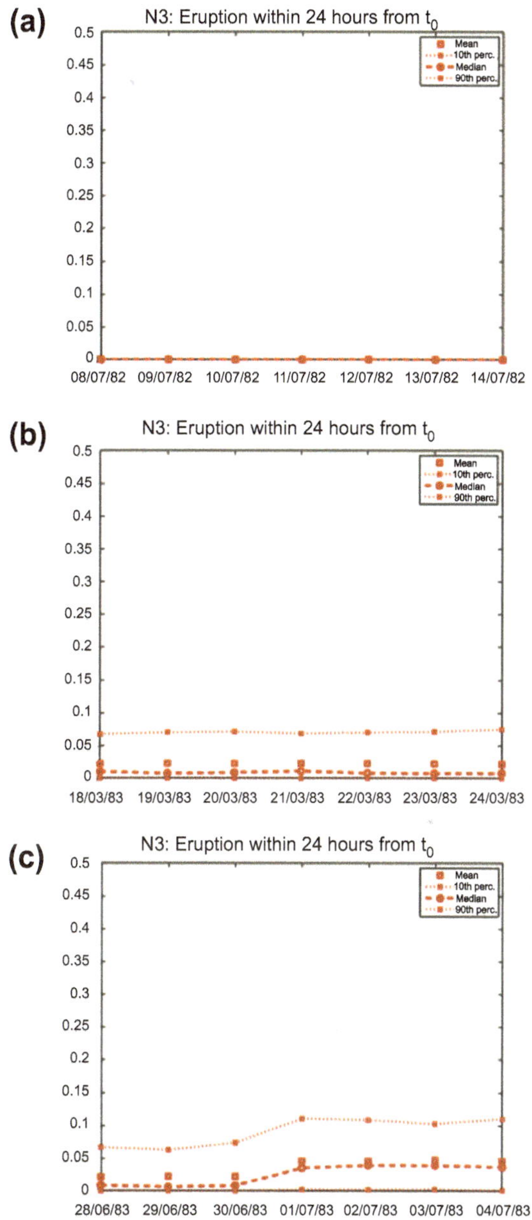

FIGURE 9.3 Same as panels b, c, and d of Figure 9.2, but showing the probability of eruption on the following 24 h from the day indicated on the x-axis. (a) July 8–14, 1982; (b) March 18–24, 1983; (c) June 28 to July 4, 1983.

(Selva et al., 2012a); the other is based on short-term evidence brought by monitoring data. The total number of potential vent position is 700.

9.3.3 Node 5

For the size probability distribution, we consider four eruptive classes (Effusive, Small-, Medium-, and Large-explosive; Orsi et al., 2009). Based on past activity of Campi Flegrei, we modeled each size through a "reference scenario" that is assumed as a representative of that size class (Selva et al., 2010). This is done in order to guarantee the near-real-time feasibility of the assessments, because this choice reduces by far the number of simulations required.

9.3.4 Node 6

We focus on the hazard from tephra. At this node, we simply model the probability of tephra production, ranging from no significant tephra production for the effusive events to tephra from explosive eruptions. For simplicity, we also assume that tephra production is not affected where vent locations are submerged.

9.3.5 Nodes 7−8

Our goal is to model all possible kinds of fallout deposit produced by sustained columns at Campi Flegrei, i.e., to cover all the possible combinations of vent position and size class:

- As a first approximation, we neglect orography effects. This is reasonable because topographic differences are much smaller than eruption columns.
- We model dispersion and sedimentation of volcanic particles released from a sustained eruption column of short duration (few days) in the medium range (i.e., at distances <100 km) with HAZMAP (Macedonio et al, 2005; Pfeiffer et al., 2005), a semianalytical, fast running model that assumes a constant and horizontally uniform wind field and neglects the vertical diffusion coefficient and vertical wind component.
- We model the tephra dispersion for all the three size classes producing tephra (Small-, Medium-, and Large-Explosive), but using only one single vent location. Then, we translate the virtual vent to each of all the possible 700 vent locations. We use the Agnano Monte Spina, the Astroni 6 and the Averno 2 eruptions as representative of respectively Small-, Medium-, and Large-Explosive eruptive sizes, following Orsi et al. (2004), Costa et al. (2009), and Orsi et al. (2009). The input parameters of these three reference eruptions can be found in Selva et al. (2010), their Table 3).
- In order to compute the short-term tephra fall hazard for a given day, we take the reanalysis of the wind field (from National Oceanic and

Atmospheric Administration (NOAA) reanalysis, Costa et al., 2009) of that day and assume that it coincides with the forecast of the wind field that was provided on the day before for that day.

- To evaluate the total tephra loading, we assume a dry condition.

Results of the simulated fallout deposits provide the exceedance probability for different points (on a grid of 500 m spacing, 60 × 60 km, coordinates of the lower left point LON 394,880, LAT 464,480 in UTM WGS84, zone 33 N), experimenting different tephra loads (in kPa), given an Eruptive Scenario (ES, meaning an eruption occurring in one specific vent position and with one specific size; Selva et al., 2010). Then, the exceedance probabilities of different tephra-load thresholds from all the possible ESs are computed to obtain a complete PVHA, combining each ES according to its specific probability of occurrence.

Results from previous studies on PVHA are represented as probability maps, in which, for every point on a grid, the probability of overcoming a selected intensity measure (in the case of tephra fall, the tephra load in kPa) in a given time window is displayed, either conditional to the occurrence of an eruption of a specific size from a specific vent (e.g., Macedonio et al., 2008) or considering every possible size and vent location (see for example Selva et al., 2010; Sandri et al., 2012, in press). These maps show a best evaluation value (e.g., the mean) or provide in terms of selected percentiles, to give an idea of the dispersion due to the epistemic uncertainty. Such maps synthesize the hazard evaluation, but much more information is available beyond these maps.

Hazard curves (Sandri et al., in preparation; Costa et al., 2013; Tonini et al., 2013; Tonini et al., submitted for publication) show, for a given point of a grid, the exceedance probability (on the y axis) as a function of the threshold intensity measure (on the x axis) of the considered phenomena. The hazard curves represent a significant improvement since:

- They preserve more detailed hazard information, providing the probability of overcoming different threshold values for each spatial point or area.
- They are the most common output from hazard investigations for quantitative risk assessment, as they can be successfully coupled to fragility analysis.
- They permit production of statistical hazard maps, where the intensity corresponding to a predefined value of the probability of exceedance is plotted (e.g., 10% in 50 years in the case of long term hazard, in analogy with the standard seismic hazard).

Here we compute hazard curves for all the points of the considered target grid. We report the whole curve only for one strategic point (the Osservatorio Vesuviano) and, as BET_stVH is a Bayesian tool, we also portray the uncertainty associated with these curves. The results are then displayed in probability and hazard maps for the whole area, derived from the hazard curves at each grid point.

9.4 OPERATIONAL SHORT-TERM PVHA: THE ROLE OF REAL-TIME MONITORING DATA IN BET_VH

Long-term probability of eruption (nodes 1, 2, and 3), vent position (node 4), and tephra loading (nodes 7 and 8) can be regularly updated based on the updated monitoring data for the three periods considered, and on the forecast of the wind field, retrieved day by day from NOAA. The BET_stVH settings at nodes 5 (eruption's size) and 6 (tephra production) are largely independent of evidence from monitoring. At each node, short-term (when available) and long-term assessments are finally merged, weighting them by their subjective credibility (Marzocchi et al., 2008).

For long-term assessments, we rely on previous studies (Selva et al., 2010, 2012a,b; Orsi et al., 2009). For short-term assessments, at nodes 1, 2, and 3, we use the monitoring measures identified through elicitation (Selva et al., 2012a). Tables 9.2, 9.3, and 9.4 show daily monitoring data for the three nodes (respectively) and the three periods, as well as the same data as treated for the inertia issue. For every day we show the real input to BET_EF code that we use to keep into account the "memory" of the anomaly. Figure 9.2 shows resulting probability distributions for nodes 1–3 (absolute monthly probabilities), while the corresponding daily absolute probabilities (Eqn (9.2)) are in Figure 9.3.

For node 4, we rely on the location of some of the VT epicenters (Bianco, 2013; personal communication), and on the localization of fractures at Solfatara (Orsi et al., 1999). We discard the deformation measures as we do not have a spatially complete information, only leveling along a few profiles and nothing from the submerged part of caldera. Obviously, the short-term information varies in time: the probability maps of vent opening for the fourth day of each period analyzed (respectively, July 11, 1982; March 22, 1983; July 2, 1983) in this study are shown in Figure 9.4.

For nodes 7 and 8, we retrieve the forecast on the wind field, as mentioned above, day by day from NOAA, and we analyze the results of the selected tephra model dispersion (HAZMAP) for such wind fields.

These inputs represent the contributions of monitoring evidence to the short-term PVHA, which combines both short- and long-term information (see Marzocchi et al. (2008), ESM).

9.5 OPERATIONAL SHORT-TERM PVHA: RESULTS

The most complete way to provide a probabilistic hazard assessment is by plotting a hazard curves for every grid point, obtained by running the BET_stVH model with the monitoring observations and including the results from HAZMAP simulations based on the forecasted wind field. As an example, Figure 9.5 shows the hazard curve for the grid point near Osservatorio Vesuviano in Naples, an obvious key infrastructure in case of a volcanic

TABLE 9.2 Daily Monitoring Data at Node 1 for the Three Periods Considered in this Application. The Top Part Shows the Measured Data, Whereas the Bottom Part Shows the Same Data but Smoothed With the Inertia as Defined in Section 3.1. Zero Values Do Not Necessarily Mean Zero Measures, but They Might Mean (As in the Case of the Temperature of Pisciarelli Fumarole) that We Are Below Any Anomaly Threshold

Year	Month	Day	#VT	#LP/ VLP/ ULP	Uplift cm	Uplift Rate cm/ Month	Fract.	Degas.	Acid Gases	T Pisciarelli c
					Node 1 Top					
1982	7	8	0	0	0.18	0.62	0	0	0	0.00
1982	7	9	0	0	0.23	0.66	0	0	0	0.00
1982	7	10	0	0	0.28	0.70	0	0	0	0.00
1982	7	11	0	0	0.33	0.74	0	0	0	0.00
1982	7	12	0	0	0.38	0.78	0	0	0	0.00
1982	7	13	0	0	0.42	0.82	0	0	0	0.00
1982	7	14	0	0	0.47	0.86	0	0	0	0.00
1983	3	18	8	0	15.64	5.40	0	0	0	0.00
1983	3	19	8	0	15.67	5.47	0	0	0	0.00
1983	3	20	8	0	15.71	5.54	0	0	0	0.00
1983	3	21	8	0	15.75	5.61	0	0	0	0.00
1983	3	22	8	0	15.78	5.68	0	0	0	0.00
1983	3	23	8	0	15.82	5.75	0	0	0	0.00
1983	3	24	8	0	15.85	5.81	0	0	0	0.00
1983	6	28	19	0	13.01	4.39	0	0	0	0.00
1983	6	29	19	0	12.95	4.43	0	0	0	0.00
1983	6	30	19	0	12.93	4.46	0	0	0	0.00
1983	7	1	35	0	12.98	4.57	1	0	0	0.00
1983	7	2	35	0	13.03	4.63	1	0	0	0.00
1983	7	3	35	0	13.08	4.70	1	0	0	0.00
1983	7	4	35	0	13.13	4.77	1	0	0	0.00

Continued

TABLE 9.2 Daily Monitoring Data at Node 1 for the Three Periods Considered in this Application. The Top Part Shows the Measured Data, Whereas the Bottom Part Shows the Same Data but Smoothed With the Inertia as Defined in Section 3.1. Zero Values Do Not Necessarily Mean Zero Measures, but They Might Mean (As in the Case of the Temperature of Pisciarelli Fumarole) that We Are Below Any Anomaly Threshold—cont'd

Year	Month	Day	#VT	#LP/ VLP/ ULP	Uplift cm	Uplift Rate cm/ Month	Fract.	Degas.	Acid Gases	T Pisciarelli c
					Node 1 Bottom					
1982	7	8	0.00	0	0.18	0.62	0	0	0	0.00
1982	7	9	0.00	0	0.23	0.66	0	0	0	0.00
1982	7	10	0.00	0	0.28	0.70	0	0	0	0.00
1982	7	11	0.00	0	0.33	0.74	0	0	0	0.00
1982	7	12	0.00	0	0.38	0.78	0	0	0	0.00
1982	7	13	0.00	0	0.42	0.82	0	0	0	0.00
1982	7	14	0.00	0	0.47	0.86	0	0	0	0.00
1983	3	18	36.35	0	15.64	5.52	0	0	0	0.00
1983	3	19	36.59	0	15.67	5.52	0	0	0	0.00
1983	3	20	36.82	0	15.71	5.54	0	0	0	0.00
1983	3	21	37.04	0	15.75	5.61	0	0	0	0.00
1983	3	22	37.25	0	15.78	5.68	0	0	0	0.00
1983	3	23	37.46	0	15.82	5.75	0	0	0	0.00
1983	3	24	37.65	0	15.85	5.81	0	0	0	0.00
1983	6	28	97.63	0	13.01	6.29	0	0	0	0.00
1983	6	29	97.92	0	12.95	6.25	0	0	0	0.00
1983	6	30	98.21	0	12.93	6.21	0	0	0	0.00
1983	7	1	114.54	0	12.98	6.18	1	0	0	0.00
1983	7	2	122.84	0	13.03	6.14	1	0	0	0.00
1983	7	3	128.45	0	13.08	6.10	1	0	0	0.00
1983	7	4	132.73	0	13.13	6.06	1	0	0	0.00

TABLE 9.3 As Table 9.2, but for Node 2

Year	Month	Day	Deep VT	Deep LP	#VLP/ ULP	Tremor	Deep Tremor	Uplift cm	Fract.	Macr. Variat.	Acid Gases	Var. in Gases
						Node 2 Top						
1982	7	8	0	0	0	0	0	−3.59	0	0	0	0
1982	7	9	0	0	0	0	0	−3.54	0	0	0	0
1982	7	10	0	0	0	0	0	−3.48	0	0	0	0
1982	7	11	0	0	0	0	0	−3.43	0	0	0	0
1982	7	12	0	0	0	0	0	−3.38	0	0	0	0
1982	7	13	0	0	0	0	0	−3.33	0	0	0	0
1982	7	14	0	0	0	0	0	−3.28	0	0	0	0
1983	3	18	0	0	0	0	0	32.47	0	0	0	0
1983	3	19	0	0	0	0	0	32.68	0	0	0	0
1983	3	20	1	0	0	0	0	32.89	0	0	0	0
1983	3	21	0	0	0	0	0	33.10	0	0	0	0
1983	3	22	0	0	0	0	0	33.31	0	0	0	0
1983	3	23	0	0	0	0	0	33.52	0	0	0	0
1983	3	24	0	0	0	0	0	33.73	0	0	0	0

Continued

TABLE 9.3 As Table 9.2, but for Node 2—cont'd

Year	Month	Day	Deep VT	Deep LP	#VLP/ ULP	Tremor	Deep Tremor	Uplift cm	Fract.	Macr. Variat.	Acid Gases	Var. in Gases
Node 2 Top												
1983	6	28	0	0	0	0	0	48.00	0	0	0	0
1983	6	29	0	0	0	0	0	48.15	0	0	0	0
1983	6	30	0	0	0	0	0	48.30	0	0	0	0
1983	7	1	0	0	0	0	0	48.52	1	0	0	0
1983	7	2	0	0	0	0	0	48.74	1	0	0	0
1983	7	3	0	0	0	0	0	48.96	1	0	0	0
1983	7	4	1	0	0	0	0	49.18	1	0	0	0
Node 2 Bottom												
1982	7	8	0.00	0	0	0	0	−3.59	0	0	0	0
1982	7	9	0.00	0	0	0	0	−3.54	0	0	0	0
1982	7	10	0.00	0	0	0	0	−3.48	0	0	0	0
1982	7	11	0.00	0	0	0	0	−3.43	0	0	0	0
1982	7	12	0.00	0	0	0	0	−3.38	0	0	0	0

1982	7	13	0.00	0	0	0	0	−3.33	0	0	0	0
1982	7	14	0.00	0	0	0	0	−3.28	0	0	0	0
1983	3	18	0.03	0	0	0	0	32.47	0	0	0	0
1983	3	19	0.03	0	0	0	0	32.68	0	0	0	0
1983	3	20	1.03	0	0	0	0	32.89	0	0	0	0
1983	3	21	0.52	0	0	0	0	33.10	0	0	0	0
1983	3	22	0.36	0	0	0	0	33.31	0	0	0	0
1983	3	23	0.27	0	0	0	0	33.52	0	0	0	0
1983	3	24	0.22	0	0	0	0	33.73	0	0	0	0
1983	6	28	0.50	0	0	0	0	48.00	0	0	0	0
1983	6	29	0.45	0	0	0	0	48.15	0	0	0	0
1983	6	30	0.41	0	0	0	0	48.30	0	0	0	0
1983	7	1	0.37	0	0	0	0	48.52	1	0	0	0
1983	7	2	0.35	0	0	0	0	48.74	1	0	0	0
1983	7	3	0.32	0	0	0	0	48.96	1	0	0	0
1983	7	4	1.30	0	0	0	0	49.18	1	0	0	0

TABLE 9.4 As Table 9.2, but for Node 3

Year	Month	Day	ACC EQS	ACC RSAM	Vert Disp Hypoc km	Tremor	Macr. Var	Migr Max Uplift	Fract.	Acid Gases	PHR. Activity
					Node 3 Top						
1982	7	8	0	0	0.00	0	0	0	0	0	0
1982	7	9	0	0	0.00	0	0	0	0	0	0
1982	7	10	0	0	0.00	0	0	0	0	0	0
1982	7	11	0	0	0.00	0	0	0	0	0	0
1982	7	12	0	0	0.00	0	0	0	0	0	0
1982	7	13	0	0	0.00	0	0	0	0	0	0
1982	7	14	0	0	0.00	0	0	0	0	0	0
1983	3	18	0	0	0.00	0	0	0	0	0	0
1983	3	19	0	0	0.00	0	0	0	0	0	0
1983	3	20	0	0	0.00	0	0	0	0	0	0
1983	3	21	0	0	0.00	0	0	0	0	0	0
1983	3	22	0	0	0.00	0	0	0	0	0	0
1983	3	23	0	0	0.00	0	0	0	0	0	0
1983	3	24	0	0	0.00	0	0	0	0	0	0

1983	6	28	0	0	0.00	0	0	0	0	0	0
1983	6	29	0	0	0.00	0	0	0	0	0	0
1983	6	30	0	0	0.00	0	0	0	0	0	0
1983	7	1	0	0	0.00	0	0	0	1	0	0
1983	7	2	0	0	0.00	0	0	0	1	0	0
1983	7	3	0	0	0.00	0	0	0	1	0	0
1983	7	4	0	0	0.00	0	0	0	1	0	0
Node 3 Bottom											
1982	7	8	0	0	0.00	0	0	0	0	0	0
1982	7	9	0	0	0.00	0	0	0	0	0	0
1982	7	10	0	0	0.00	0	0	0	0	0	0
1982	7	11	0	0	0.00	0	0	0	0	0	0
1982	7	12	0	0	0.00	0	0	0	0	0	0
1982	7	13	0	0	0.00	0	0	0	0	0	0
1982	7	14	0	0	0.00	0	0	0	0	0	0
1983	3	18	0	0	0.00	0	0	0	0	0	0
1983	3	19	0	0	0.00	0	0	0	0	0	0
1983	3	20	0	0	0.00	0	0	0	0	0	0

Continued

TABLE 9.4 As Table 9.2, but for Node 3—cont'd

Year	Month	Day	ACC EQS	ACC RSAM	Vert Disp Hypoc km	Tremor	Macr. Var	Migr Max Uplift	Fract.	Acid Gases	PHR. Activity
					Node 3 Bottom						
1983	3	21	0	0	0.00	0	0	0	0	0	0
1983	3	22	0	0	0.00	0	0	0	0	0	0
1983	3	23	0	0	0.00	0	0	0	0	0	0
1983	3	24	0	0	0.00	0	0	0	0	0	0
1983	6	28	0	0	0.00	0	0	0	0	0	0
1983	6	29	0	0	0.00	0	0	0	0	0	0
1983	6	30	0	0	0.00	0	0	0	0	0	0
1983	7	1	0	0	0.00	0	0	0	1	0	0
1983	7	2	0	0	0.00	0	0	0	1	0	0
1983	7	3	0	0	0.00	0	0	0	1	0	0
1983	7	4	0	0	0.00	0	0	0	1	0	0

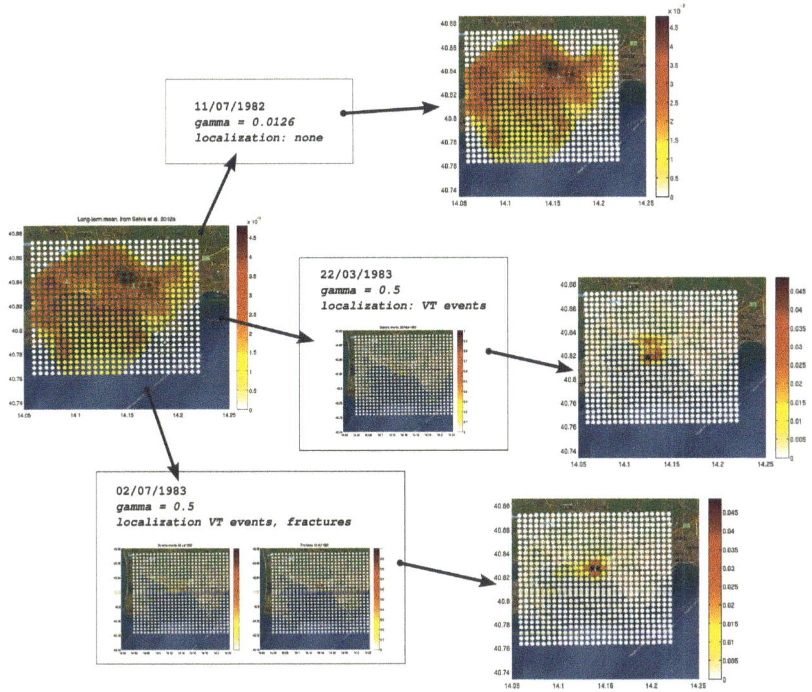

FIGURE 9.4 Spatial probability of vent opening. The long-term map (from Selva et al. (2012a)) is given on the leftmost panel. From that, we update the map on the basis of the monitoring data for each of the analyzed periods. For simplicity, we only show the forecast on the middle day of every periods, i.e., July 11, 1982 (top) in which no anomalous seismicity or fractures were reported; March 22, 1983 (middle) in which some localized VT events are available; July 2, 1983, in which some localized VT events and one fracture at Solfatara are available.

emergency at Campi Flegrei. The curves show the exceedance probability, i.e., the probability of experiencing a tephra load larger than a series of increasing values (in kPa) at the Vesuvius observatory due to eruptions occurring in the 24 h after the observation of the monitoring data and forecast of the wind field (Figure 9.5(a,b,c), respectively, show the results made on July 11, 1982, March 22, 1983, and July 2, 1983 for the following 24 h). The 10th, 50th, and 90th percentiles show uncertainties around the best estimate curve.

By cutting the curves of all the grid points at a selected tephra load threshold, x, we can reconstruct the probability maps by plotting the corresponding probability values that we read on the y-axis (Figure 9.6). Each probability map shows the best evaluation probability (mean) for every grid point to experience a tephra load larger than or equal to 3 kPa due to eruptions occurring in the 24 h after the observation of the monitoring data and forecast of the wind field (Figure 9.6(a,b,c) similar to Figure 9.5(a,b,c) respectively).

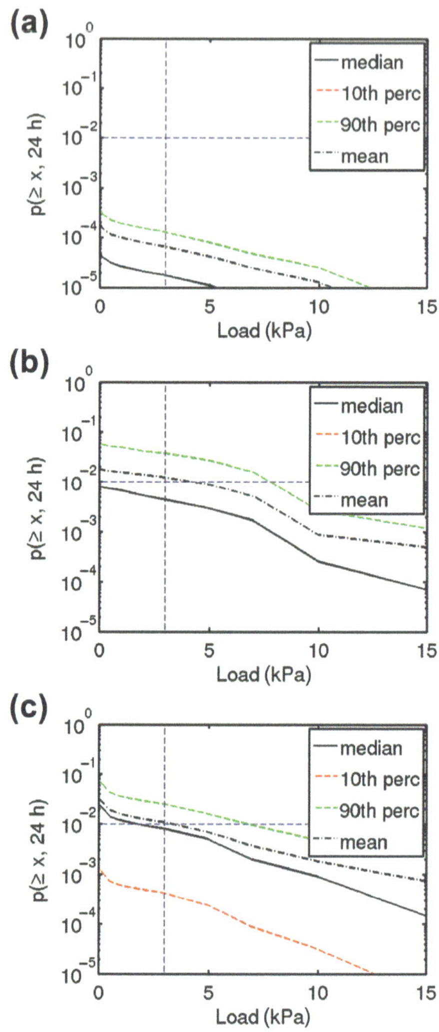

FIGURE 9.5 Hazard curves for the grid point closest to Osservatorio Vesuviano in Naples, for July 11, 1982 (a), March 22, 1983 (b), and July 2, 1983 (c), displaying the exceedance probability (on the y-axis) for a set of increasing tephra load thresholds (in kPa, on the x-axis). Note that, to facilitate the analysis of temporal evolutions, we adopted the same scale for y axes in the three panels. Epistemic variability is represented through 10th and 90th percentiles, median and mean of the hazard curve distribution. The horizontal and vertical blue dashed lines show, respectively, how we cut the hazard curve to produce the probability and hazard maps, by plotting the corresponding mean probability (Figure 9.6) and tephra load (Figure 9.7).

Conversely, by cutting the hazard curves of all the grid points at a selected exceedance probability value y, we can construct hazard maps (Figure 9.7) by plotting the corresponding tephra-load threshold that we read on the x-axis. These maps represent another way to provide a hazard evaluation in analogy to the standard for seismic hazard, that is to provide a hazard map. Each hazard map shows at every grid point the tephra load having a best evaluation probability (mean) larger than 1% in the 24 h after the observation of the monitoring data and forecast of the wind field (Figure 9.7(a,b,c) similar to Figure 9.5(a,b,c) respectively). Note that these maps include information about

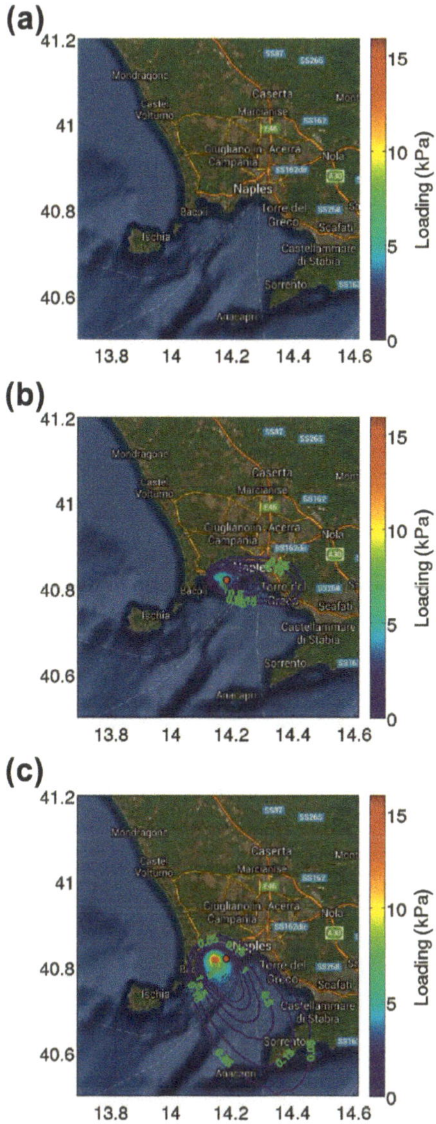

FIGURE 9.6 Hazard maps for July 11, 1982 (a), March 22, 1983 (b), and July 2, 1983 (c), displaying the tephra load with a probability (mean value of the posterior distribution) equal to 1%, from an eruption of any size and vent position in the 24 h. The red dot in the map represents the position of the Osservatorio Vesuviano, for which hazard curves are reported in Figure 9.5.

eruption probability and thus, when this probability is <1%, the whole map region is identically 0 (Figure 9.5(a)).

Figures 9.5, 9.6, and 9.7 show the absolute probability. Similarly, we can compute and show the probability map, hazard map, and hazard curves for the conditional probability, that is the hazard estimated conditional to the occurrence of an eruption of a given size and/or from a given vent position. Such maps can be in principle compared to existing "hazard" maps (e.g., Macedonio

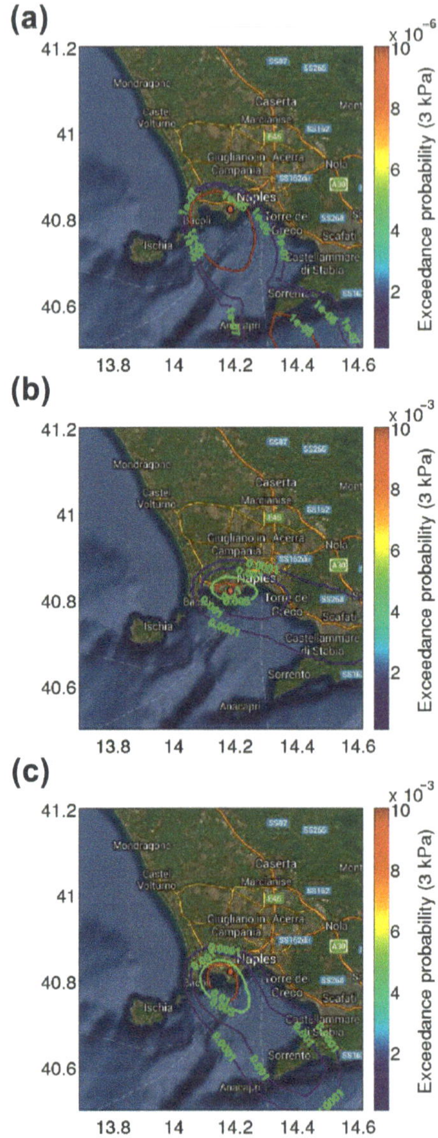

FIGURE 9.7 Probability maps for July 11, 1982 (a), March 22, 1983 (b), and July 2, 1983 (c), displaying the probability (mean value of the posterior distribution) of exceeding a tephra load of 3 kPa due to eruptions occurring in the 24 h of any size and vent position. The red dot in the map represents the position of the Osservatorio Vesuviano, for which hazard curves are reported in Figure 9.5.

et al., 2008) that are based on the assumption of the occurrence of an eruption with specific eruptive conditions (i.e., a specific scenario). However, in order to use the results of operational PVHA as scientific advice for decision makers, absolute probabilities are more appropriate, as they are the scientific input to probabilistic risk and multirisk assessment (Marzocchi et al., 2012b), or to cost−benefit analysis (Marzocchi and Woo, 2009; Sandri et al., 2012).

9.6 FINAL REMARKS

In this chapter, we have described an example of how scientists can provide information to assist decision makers in managing the volcanic risk in real-time. Scientific information is provided in terms of probabilities and their uncertainties. In our view, this attitude represents the ideal situation for establishing a rational risk management protocol.

In this chapter, we adopt the novel BET_stVH model (Selva et al., submitted for publication). This approach has some aspects worth noting:

1. Models, data, and expert opinion are all included in a model that has been set up before the crisis. In this way, the model is open to any scrutiny by the scientific community before its operational use. Moreover, scientists have much more time to establish rules and to find a consensus. This also reduces the stress of scientists during an emergency when cognition can be biased by intense pressure.
2. The model can work in near real time, updating the output as long as we collect new data from the monitoring system. This makes the model particularly useful for operational purposes during a volcanic unrest where the time evolution of the threat cannot be easily managed by repeated meetings of experts.
3. The model accounts for the most relevant uncertainties that can be shown at every step of the analysis and its application. This is particularly important because it allows scientists to show the degree of confidence on the estimation provided to decision makers.

To our knowledge, this approach represents one of the most advanced, short-term, hazard-analysis techniques in the volcanological field. More important, we hope it represents a useful milestone for future similar activities. But, ash fall represents only one of the possible threats posed by a volcano, so we encourage the development of similar models for pyroclastic flows, lahars, deformation of the ground, etc., in different time windows.

Finally, for a full operational implementation, we emphasize the need and importance of negotiating the kind and format of the hazard model output with the interested stakeholders. Not all our scientific knowledge is useful during a volcanic unrest, and only scientific information in a proper format can be effectively and rationally used to reduce the risk.

REFERENCES

Barberi, F., Corrado, G., Innocenti, F., Luongo, G., 1984. Phlegraean fields 1982–1984: brief chronicle of a volcano emergency in a densely populated area. Bull. Volcanol. 47, 175–185.

Costa, A., Dell'Erba, F., Di Vito, M.A., Isaia, R., Macedonio, G., Orsi, G., Pfeiffer, T., 2009. Tephra fallout hazard assessment at the Campi Flegrei caldera (Italy). Bull. Volcanol. 71, 259–273. http://dx.doi.org/10.1007/s00445-008-0220-3.

Costa, A., Folch, A., Macedonio, G., Sandri, L., Selva, J., July 2013. New Methods of Probabilistic Hazard Assessment of Tephra Dispersal: Application to the Neapolitan Area. IAVCEI General Assembly, Kagoshima, Japan.

projects funded within the 'Convenzione Istituto Nazionale di Geofisica e Vulcanologia e Dipartimento Protezione Civile',http://progettosv.rm.ingv.it/Progetti/Vulcanologici/vulcanologici_con_frame.htm http://progettosv.rm.ingv.it/Progetti/Vulcanologici/vulcanologici_con_frame.htm.

projects funded within the 'Convenzione tra Istituto Nazionale di Geofisica e Vulcanologia e Dipartimento Protezione Civile', http://istituto.ingv.it/l-ingv/progetti/progetti-finanziati-dal-dipartimento-di-protezione-civile-1/progetti-dpc-convenzione-2007-2009/progetti_v/progetti-vulcanologici

Jordan, T.H., Chen, Y.T., Gasparini, P., Madariaga, R., Main, I., Marzocchi, W., Papadopoulos, G., Sobolev, G., Yamaoka, K., Zschau, J., 2011. Operational earthquake forecasting: state of knowledge and guidelines for implementation. Ann. Geophys. 54 (4) http://dx.doi.org/10.4401/ag-5350.

Macedonio, G., Costa, A., Longo, A., 2005. A computer model for volcanic ash fallout and assessment of subsequent hazard. Comput Geosci 31, 837–845.

Macedonio, G., Costa, A., Folch, A., 2008. Ash fallout at Vesuvius: numerical simulations and implication for hazard assessment. J. Volcanol. Geotherm. Res. 178, 366–377.

Marzocchi, W., Woo, G., 2009. Principles of volcanic risk metrics: theory and the case study of Mount Vesuvius and Campi Flegrei, Italy. J. Geophys. Res. 114, B03213. http://dx.doi.org/10.1029/2008JB005908.

Marzocchi, W., Sandri, L., Gasparini, P., Newhall, C., Boschi, E., 2004. Quantifying probabilities of volcanic events: the example of volcanic hazard at Mount Vesuvius. J. Geophys. Res. 109, B11201. http://dx.doi.org/10.1029/2004JB003155.

Marzocchi, W., Sandri, L., Selva, J., 2008. BET_EF: a probabilistic tool for long- and short-term eruption forecasting. Bull. Volcanol. 70, 623–632. http://dx.doi.org/10.1007/s00445-007-0157-y.

Marzocchi, W., Sandri, L., Selva, J., 2010. BET_VH: a probabilistic tool for long-time volcanic hazard assessment. Bull. Volcanol. 72, 705–716. http://dx.doi.org/10.1007/s00445-010-0357-8.

Marzocchi, W., Newhall, C.G., Woo, G., 2012a. The scientific management of volcanic crises. J. Volcanol. Geotherm. Res. 247-248, 181–189.

Marzocchi, W., Garcia-Aristizabal, A., Gasparini, P., Mastellone, M.L., Di Ruocco, A., 2012b. Basic principles of multi-risk assessment: a case study in Italy. Nat. Hazards 62, 551–573.

de Natale, G., Pingue, F., Allard, P., Zollo, A., 1991. Geophysical and geochemical modelling of the 1982–1984 unrest phenomena at Campi Flegrei caldera (southern Italy). J. Volcanol. Geotherm. Res. 48 (1 – 2), 199–222.

Orsi, G., Civetta, L., Del Gaudio, C., de Vita, S., Di Vito, M.A., Isaia, R., Petrazzuoli, S.M., Ricciardi, G.P., Ricco, C., 1999. Short-term ground deformations and seismicity in the resurgent Campi Flegrei caldera (Italy): an example of active block-resurgence in a densely populated area. J. Volcanol. Geotherm. Res. 91, 415–451.

Orsi, G., Di Vito, M.A., Isaia, R., 2004. Volcanic hazard assessment at the restless Campi Flegrei caldera. Bull. Volcanol. 66, 514–530. http://dx.doi.org/10.1007/s00445-003-0336-4.

Orsi, G., di Vito, M.A., Selva, J., Marzocchi, W., 2009. Long-term forecast of eruption style and size at Campi Flegrei caldera (Italy). Earth Planet. Sci. Lett. 287, 265–276.

Pfeiffer, T., Costa, A., Macedonio, G., 2005. A model for the numerical simulation of tephra fall deposits. J. Volcanol. Geotherm. Res. 140, 273–294.

Sandri, L., Jolly, G., Lindsay, J., Howe, T., Marzocchi, W., 2012. Combining long- and short-term probabilistic volcanic hazard assessment with cost-benefit analysis to support decision making in a volcanic crisis from the Auckland volcanic field, New Zealand. Bull. Volcanol. 74, 705–723. http://dx.doi.org/10.1007/s00445-011-0556-y.

Sandri, L., Thouret, J.C., Constantinescu, R., Biass, S., Tonini, R. Long-term multi-hazard assessment for El Misti volcano (Peru). Bull. Volcanol., in press.

Sandri, L., Selva, J., Costa, A. Long-term Tephra Fallout Hazard in Naples: The Importance of Uncertainty, in preparation.

Selva, J., Costa, A., Marzocchi, W., Sandri, L., 2010. BET_VH: exploring the influence of natural uncertainties on long-term hazard from tephra fallout at Campi Flegrei (Italy). Bull. Volcanol. 72 (6), 705–716.

Selva, J., Orsi, G., Di Vito, M., Marzocchi, W., Sandri, L., 2012a. Probability hazard map for future vent opening at the Campi Flegrei caldera, Italy. Bull. Volcanol. 74, 497–510.

Selva, J., Marzocchi, W., Papale, P., Sandri, L., 2012b. Operational eruption forecasting at high-risk volcanoes: the case of Campi Flegrei, Naples. J. Appl. Volcanol. 1 (5). http://dx.doi.org/10.1186/2191-5040-1-5.

Selva, J., Costa, A., Sandri, L., Macedonio, G., Marzocchi, W. Probabilistic short-term volcanic hazard in phases of unrest: a case study for tephra fallout, submitted for publication.

Selva, J., Costa, A., Sandri, L., Folch, A., Macedonio, G., July 2013. Short-term Probabilistic Hazard Assessment of Tephra Dispersal: Application to the Neapolitan Area. IAVCEI General Assembly, Kagoshima, Japan.

Tonini, R., Sandri, L., Costa, A., Selva, J. Brief Communication: The effect of submerged vents on probabilistic hazard assessment for tephra fallout, submitted for publication.

Tonini, R., Sandri, L., Selva, J., Marzocchi, W., July 2013. An Operational Approach to Generate and Visualize Bayesian Probabilistic Volcanic Hazard Curves and Maps by Means of BET_VH Tool. IAVCEI General Assembly, Kagoshima, Japan.

Chapter 10

Human and Structural Vulnerability to Volcanic Processes

Giulio Zuccaro [1], Daniela De Gregorio [1] and Peter Baxter [2]
[1] *Dipartimento di Strutture per l'Ingegneria e l'Architettura, Università di Napoli Federico II, Napoli, Italy,* [2] *Department of Public Health and Primary Care, University of Cambridge, Institute of Public Health, Cambridge, UK*

ABSTRACT
The main value of studying buildings in the aftermath of eruptions is that their damaged or destroyed remains preserve the record of the fleeting destructive forces involved. Forensic investigation of any tragic human victims can add importantly to the evidence. Using these insights from some recent major eruptions, computer modeling has advanced enough to begin to successfully simulate the eruptive dynamics in some examples and provide confidence in developing more powerful methods to illuminate future emergency planning and mitigation measures for buildings and people. Quantifying the vulnerability of building stock in areas of future volcanic hazard and to provide this necessary input to scenario development are an integral part of this multidisciplinary scientific task.

10.1 INTRODUCTION

Volcanic vulnerability is the probability that elements at risk (people, buildings, settlements) suffer injury, damage, or other changes in the status quo following impacts from volcanic hazards. In contrast to single catastrophic natural events (such as tectonic earthquakes, landslides, etc.), during a volcanic eruption, several phenomena may be generated (lava flows, earthquakes, ashfall, pyroclastic flows, ballistics, debris flows, tsunami, and lahars) in different spatial areas and at different times. The sequence of these separate hazards in an eruption may modify the resistance of the exposed element at each stage and, in consequence, the vulnerability evaluation may require sequential analyses focusing on cumulative damage or changes (Zuccaro and De Gregorio, 2013). For brevity, in this chapter, we confine the analysis to the

Volcanic Hazards, Risks, and Disasters. http://dx.doi.org/10.1016/B978-0-12-396453-3.00010-1

main volcanic hazards in urban areas, namely, earthquake, ashfall, and pyroclastic flows.

Anomalous seismic activity generally accompanies the unrest phase of a volcano preceding its eruption. It is considered an eruption precursor, together with ground uplift and geochemical changes in fumaroles indicative of an increasing output of magmatic gas. Compared with tectonic earthquakes, earthquakes associated with a volcanic process have lower intensity and shallower depth. Therefore, their impacts are usually confined to a localized area. However, since the seismic waves are less damped, a volcanic earthquake can cause more damage than a tectonic one of equal magnitude. Volcanic earthquakes, such as tectonic ones, can produce damage to buildings and consequently to people who are inside the building or close to it.

Commonly in explosive eruptions, the initial eruptive phase is characterized by the formation of a sustained eruptive column of gas–solid pyroclastic products (tephra) dispersed in a plume. Tephra of various sizes falls to ground from the plume in its dispersal by the winds at different altitudes, typically in an elliptical shape that is elongated in the direction of wind in the stratosphere. Grain size and thickness of ashfall deposits decrease with the distance from the eruptive vent and it is generally uniform over small areas. Up to tens of kilometers from the vent, ashfall can accumulate to sufficient depth on roofs to cause building damage, with accompanying deaths and injuries (Spence et al., 2005a).

Pyroclastic density currents (PDCs) are the most dangerous phenomena produced by an explosive eruption. At major eruptions of Vesuvius, in AD 79 and 1631, for example, they were generated by the gravitational collapse of the eruptive column at the end of the ashfall phase and this is likely to occur again in a future large eruption. PDCs are clouds containing a mixture of hot gases and dispersed solid particles of various sizes capable of flowing down the volcanic slopes at high speeds. Casualties and physical building damage from PDCs are caused by the temperature, pressure, and particle load of the flow (Spence et al., 2007).

10.2 HUMAN VULNERABILITY AND BUILDINGS

The recent growing interest in human vulnerability in eruptions actually had its origins in the eruption of Mount St Helens, USA, in 1980, with the forensic study of victims of the great May 18 pyroclastic surge (Eisele et al., 1981). The local coroner understood straightaway the scientific importance of uncovering the mystery of how the 57 people died by arranging for postmortem examinations on the small number of bodies which the search and rescue teams were able to find in the devastated area. The pathologists attempted to match up their findings and the locations of the bodies with the field work that was begun almost immediately as volcanologists, who were just as curious about the phenomenon, mapped the surge impacts. This medical initiative was not purely of scientific interest, for many loggers continued to work in the area after the eruption, as did other groups such as foresters and volcanologists, and their safety was of prime importance. For example, was there personal protective equipment these workers could wear or carry to help them survive another unexpected eruption?

Mount St Helens is situated in a forested wilderness, but what if the eruption had occurred close to a city and the millions of flattened and broken trees had been people (Newhall, 2000)? And what would be the effects on the city's buildings? Accurate forecasting of the size and occurrence of an eruption leading to timely evacuation of populations at risk is the primary goal of much current research and volcano crisis management, but it is unlikely that it will be achieved for highly dangerous eruptions at volcanoes like Mount St Helens in the near future. Thus a comprehensive assessment of the risk to lives and to the urban environment from volcanic impacts is also an important aspect of disaster preparedness, as well as for devising mitigation measures. This task has to begin by quantifying whenever possible the vulnerability of life and property for the foreseeable eruption scenarios.

10.2.1 Developments in Human Vulnerability Studies

10.2.1.1 Computer Simulation Modeling

Nowhere in the world has this work been more anticipated than in Italy at two of the world's most dangerous volcanoes, Vesuvius and Campi Flegrei, whose future eruptions threaten the conurbation of at least 3 million people around Naples. Civil defense planning is benefiting from advances in computer simulation of future eruptions of Vesuvius, based on modeling of the magma ascent along the volcano conduit and the subsequent dispersion of the eruptive mixture in the atmosphere, including the development and collapse of the eruptive column and propagation of the associated PDCs. Unlike most previous modeling attempts, the outputs from the model devised by Dobran et al. (1994) included key parameters for human and building risk assessment, namely temperature, dynamic pressure, and particle concentration, which can all vary during a single PDC run-out and according to the size of the eruption (Baxter et al., 1998). Initially presented as a two-dimensional model, supercomputer technology later made visualization of topography in three dimensions possible; for example, in an impressive simulation of the May 18, 1980, surge and its behavior over the topography in the sector of forest that was impacted at Mount St Helens (Esposti Ongaro et al., 2012). The modeling assumptions fitted well into the view that the surge had indeed moved as a gravity current for most of its run-out (Belousov et al., 2007).

10.2.1.2 Eruption of Soufrière Hills Volcano, Montserrat

If Mount St Helens in 1980 opened a new chapter in human vulnerability, a rare opportunity to study the impacts on structures of the temperature and dynamic pressure in dilute PDCs (surges) arose during the eruption of Soufrière Hills volcano (SHV), Montserrat, in 1997, when three surges struck three evacuated villages over the course of a few months (Baxter et al., 2005). This event also provided vital insights in to how the vulnerability of buildings could influence and govern the eruption impacts on their human occupants. Houses built on the island since Hurricane Hugo in 1980 had been constructed to a high standard to resist hurricanes and so a range of destruction was found in accordance with the raised dynamic pressures of the surges, but they were also damaged by fires which had ignited the furniture left behind in most if not all of the buildings and gutted their wooden roofs. The surge on Boxing Day (December 26) was the most violent due to the dome exploding, and the building damage ranged from slight outside thermal damage in the lateral periphery of the impacted area to complete obliteration of structures in the central axis of the surge where dynamic pressure was highest. A dynamic pressure scale of building damage was produced (Baxter et al., 2005) and provided an important stimulus for the computer simulation work at Vesuvius and certain other European volcanoes in the EXPLORIS project (Baxter et al., 2008). The dynamic pressures estimated by field surveys of building damage due to the blast were also reproducible in a numerical simulation model of the Boxing Day surge (Esposti Ongaro et al., 2008).

10.2.1.3 Eruption of Merapi Volcano, Java, 2010

Merapi volcano in October/November 2010 was the violent scene of the latest examples of PDC destructiveness in inhabited areas in an eruption, which comprised eight stages of major activity over only 11 days (Jenkins et al., 2013; Pallister et al., 2013). There were two major surges: one that caused extensive blast damage to buildings, but at the same time had limited thermal impacts compared to the surges in the SHV eruptions, and a detached surge that had weak dynamic pressure but caused high mortality from its thermal effects alone. Unusually rapid lava dome growth and inflation was accompanied by dome explosions and collapses together with the two laterally directed high-energy PDCs occurring on October 26 and November 5. The latter damaged or destroyed more than 1,300 evacuated buildings within a 22 km^3 sector on the southern flank. The buildings were built to a much lower standard than on Montserrat and readily broke up at lower dynamic pressures, but it was still possible to estimate from the damage levels and the findings on the transport of missiles in the PDC the gradient of dynamic pressure all the way to the end of the run-out at 8 km (Jenkins et al., 2013). In contrast to the Montserrat events, thermal effects were not much in evidence. Photographs taken by rescuers and journalists were placed on social media and these provided an important record of the different eruption phases; for example, the surge impacts on buildings on October 26 had been wiped away by the PDCs that followed and would have been unavailable to scientists arriving after the end of the eruption (Jenkins et al., 2013). These images provided indicators of dynamic pressures and confirmed the absence of significant thermal damage in this surge event, too.

Another important feature of the climactic phase on November 5 was the dense PDC (pyroclastic flow) that traveled for 15.5 km along the Gendol River valley and led to most of the 200 deaths in the eruption (Jenkins et al., 2013). During its long course, this flow spilled over into some small villages located along the banks where people did not have enough notice to evacuate in time and were killed in their houses, many of which also caught fire. At the village of Bronggang, a dilute PDC detached from the flow in the main valley at a turn of the river near the front of a SABO dam, and flowed down the bank, coming to halt inside the village. The impacted houses were not badly damaged as the velocity and dynamic pressures were low, but melted plastics were found inside and outside of the houses and vegetation was destroyed by the heat. Out of 59 villagers in the surge area in- and outside their houses at the time, only four survived exposure to an estimated temperature of 250–300 °C, even though the duration of exposure was quite brief (1–2 min). The hot ash readily penetrated houses made vulnerable by the local custom of building houses with ventilation gaps in the walls and a gap around the house between the roof and the outside walls. Had these gaps been filled in, the people inside would undoubtedly have survived when the houses were enveloped by the low momentum surge which left the buildings undamaged (Jenkins et al., 2013).

10.2.2 Human Vulnerability to PDCs

Today, people are regularly reminded by the ruins and the plaster casts of Pompeii and the skeletons of Herculaneum of the vulnerability of human beings and their buildings in the face of a volcano's fury. This view was reinforced by the destruction of the city of St Pierre, Martinique, in 1902, with the loss of almost the entire population of 28,000 people in a fireball triggered by what they then called a nuée ardente (Baxter, 1990). Later, a French geologist, Alfred Lacroix, began a landmark investigation to piece together from the ruins of St Pierre what had happened to the buildings and their occupants in the eruption, and so began the first scientific study of the impacts of a pyroclastic surge, or dilute PDC—the modern terms for nuées ardentes. Lacroix's work (Lacroix, 1904) languished on library shelves until the eruption of Mount St Helens mentioned above which drove a new interest in volcanic processes and their impacts.

Lacroix was the first scientist to grasp the role of dynamic pressure and temperature in a PDC and he was the first to photograph one in December 1902, as it rolled down the flank of Mont Pelée (Lacroix, 1904). The human body is not readily injured by the low to moderate dynamic pressures found in weak surges, and can resist higher levels unless the victims are knocked over or violently thrown into obstacles, or the house they are inside is blown down and they are killed and injured by being struck by the falling roof and walls, similar to when houses collapse in earthquakes. Heat is another matter, however, and hot ash at or above the temperature of boiling water coming into direct contact with the skin will cause scalding and full-thickness burns in a fraction of a second. Even thin cotton clothing can be protective for a time, but above 300 °C the danger is that clothes will ignite spontaneously with almost invariably fatal consequences. Mortality is closely related to the percentage of the skin surface that is burned, so that over 40 percent of total body surface area the victim will need hospital treatment to help them survive. A high risk of immediate death or raised mortality in the hours afterward arises when fine, hot ash gets inhaled into the lungs. As far as vulnerability is concerned, victims with extensive skin burns have also been found to be more likely to incur inhalation thermal injuries.

Will people inside houses which resist the PDC impact survive the experience? In dilute PDCs of brief duration, the chances are indeed good, but it is not recommended as a basis for mitigation measures by building bunkers or sealing up houses in areas under volcanic threat. Nevertheless, the expedient of protecting the openings with shutters or boards of wood when evacuating may make all the difference between returning to an intact house or a burned-out shell. Human tolerance to hot air while at rest in a still room begins to be exceeded over 50 °C when hyperthermia will eventually limit exposure time. Above 121 °C, skin pain will limit exposure to less than 10 min. Theoretical tolerance curves can be used to predict survival times in occupants sitting inside well-sealed buildings and protected from infiltrating ash while being enveloped outside by a continuous flowing hot PDC (Spence et al., 2004a).

Outside in the open is a different story, particularly if the surge is fast moving when the convective heat against the body rapidly increases with the velocity and augments the radiant heat flux from the surge cloud penetrating and heating the skin. A typical victim of a PDC has "fourth-degree" burns that extend below the skin into the musculature and the victim lies dead in a fixed posture known as the pugilistic attitude, with the burned body covered in pieces of tattered clothing (Baxter, 1990).

At high temperatures or in fast-moving flows, as seen in victims at Mount St Helens, in which the dynamic pressure is raised, the ash is forced into the mouths and lungs, making breathing impossible and death comes about rapidly and even instantaneously from suffocation combined with the internal effects of the sudden intense heat on the body affecting the heart and brain (Eisele et al., 1981).

Building openings are highly vulnerable to low dynamic pressures, particularly glazed or lightly shuttered windows. Studies of the 1997 surges on Montserrat showed unequivocally how windows were blown in by the dynamic pressure followed by the hot ash which, at around 400 °C, spontaneously ignited the furnishings inside. This finding explained how the fireball that killed nearly the whole city population was generated in St Pierre at the catastrophic eruption on May 8, 1902. At 8.02 AM, when the surge struck St Pierre, the unprepared population would have had the shutters of most buildings wide open, allowing the ash at over 400 °C access to furnishings and other flammables, which ignited all at once over the city and rapidly coalesced into a huge fireball. At high dynamic pressures, other parts of the building will give way to the lateral force, and powerful surges can raze buildings to their foundations.

The scene of destruction has only one other possible analog: nuclear blasts (Baxter et al., 2005). Unlike these, though, there is no shock front in the moderate-sized surges that have occurred in recent times, despite some being generated by dome explosions with blast waves that extend only for short distances, but the dynamic pressure wave does resemble the powerful, fast moving wind that follows the shock wave (peak overpressure) in nuclear explosions, in having a slower rate of rise (and longer duration) than in conventional explosions. Conventional explosions cannot typically duplicate the directionality and sweeping displacement of houses into the ocean that was seen in the Boxing Day surge on Montserrat, for example. The fires ignited by the radiant heat of a nuclear explosion have a more extensive damage range from ground zero compared to its blast and dynamic pressure, and carry a high chance of developing into a mass fire (conflagration or firestorm), which may spread outward beyond the blast-damaged area. The hazard of surges triggering mass fires, which then coalesce and spread beyond the surge run-outs, also needs to be considered in planning for cities, as invoked by the fireball at St Pierre.

10.2.3 Human Vulnerability to Tephra Fall

Lying buried in the deposits of the AD 79 eruption of Vesuvius are the ruins of houses that collapsed under the fallout of several meters of coarse-grained pumice during the early hours of the eruption before the phase of column collapse began that generated the major PDCs (Cioni et al., 2000). On May 7, 1902, one day before the more notorious eruption of Mont Pelée on Martinique, La Soufriere volcano on St Vincent erupted with radially directed PDCs of similar destructiveness (Hovey, 1902), but they were confined to a more sparsely populated area and with less loss of life (1,350 deaths). In contrast, though, the tephra fallout was extensive around the island with ground deposits 60–70 cm deep in Georgetown, 5 miles from the crater (Russell, 1902). The material comprised scoria and abundant stones (dense lithics) with 12–15 cm in diameter, which also caused a substantial number of deaths and injuries, as well as impacting on roofs. Many of the stones were hot and still glowing when they struck and were capable of causing fires. The less-dense lapilli and scoria fell as far as Kingstown, 12 miles from the crater. The stones fell on Georgetown, damaging buildings and breaking numerous windows, which paradoxically faced away from the volcano. This was explained by the trade winds that had first blown the erupted tephra toward the west, but as the plume rose in the atmosphere, the higher-altitude winds were blowing west to east toward Georgetown and on falling down toward the ground they were being blown from east to west again toward the windows facing away from the volcano (Russell, 1902). The inhabitants had first thought they would escape the worst of the eruption, being on the east of the volcano, but they had not taken into account the direction of the winds at altitude. A similar kind of tephra impact had been observed in Soufrière's last eruption in 1812 and once again later at Vesuvius (Perret, 1924), to be described next.

In 1906, about 300 people died in the eruption of Vesuvius as a result of tephra accumulation causing roofs to collapse in the neighboring towns of Ottaviano and San Giuseppe. Lava fountaining occurring overnight on April 7 directed a storm of lapilli and clasts toward Ottaviano, forcing people to escape the town during the bombardment. By daybreak when the eruption stopped, the fallout was 80 cm deep, roofs had collapsed on almost every house and the town was in ruins (e.g., Chester and Duncan, 2007). In the last eruption of Vesuvius in 1944, the fallout was much smaller and directed by winds to a southerly sector, where there were 21 deaths from roof collapses in adobe houses (Chester and Duncan, 2007). These eruptions at Vesuvius are called Violent Strombolian and they do not generate PDCs (Cioni et al., 2008). Disaster planning at Vesuvius includes modeling and mapping the vulnerability of the buildings at risk from such eruptions in the future (Baxter et al., 2008).

The need to extend this modeling to larger eruptions and their ashfalls was dramatically illustrated at the Plinian eruption of Mt Pinatubo, Philippines, on June 15, 1991. About 60,000 inhabitants inside a radius of 30 km of the volcano escaped certain death in PDCs by timely evacuation and warnings provided by the authorities and the scientists monitoring the volcano, but 300 fatalities and 279 serious injuries occurred due to roofs collapsing in the loading of wet ash beyond this hazard zone (Spence et al., 1997), which had not been foreseen (or that Typhoon Yunya and heavy rain would pass through at the same time as the eruption was happening). A survey of 51 buildings conducted 2 weeks after the eruption in Castillejos, one of the worst affected cities with an ash deposit of 20 cm, found that one-third had suffered partial or complete roof collapse, and the ash loading on the roof supporting structure or the roof sheets (or both) had exceeded typical design loads for most tropical roofs. The authors recognized the wider lessons, though, that the full range of eruption impacts and the extent of the vulnerability of the local building types in a volcanic area need to be incorporated into composite hazard and risk maps, and that there was a need to develop simple and rapid building- and settlement-survey techniques to identify buildings at risk and to classify them according to their vulnerability to each of the eruption hazards (Spence et al., 1997).

10.2.4 Incorporating Human and Building Vulnerability into Disaster Planning

An impact model is a tool for the estimation of property damage and human casualties resulting from defined volcanic hazard scenarios. In crisis planning, these are not intended to be interpreted strictly as forecasts or predictions of what will happen in the next eruption, but are a scoping exercise in the development of a range of scenarios to examine the vulnerability of assets in the impact area of study and to produce mitigation solutions, as well as investigating the resilience of the community, if key infrastructure is destroyed, or key escape routes are blocked by collapsing debris, for example. The vulnerability of infrastructure is integral to studies of building impacts, but is covered in another chapter (Wilson et al., 2014 in this volume).

The first step is to draw up an event tree that contains in its step-by-step branches all the potential eruptive events for the study volcano, beginning with the state of unrest (Newhall and Hoblitt, 2002). This is an easy recommendation to make, but it can take volcanologists months of meetings to complete, if you have a complicated volcano like Vesuvius. But once produced everyone, including nonscientists, can view and follow a future eruption crisis depending on what path the volcano may take, which cannot be foretold in advance. Another name for an event tree is a logic tree and it does indeed introduce a rational structure that has been arrived at by consensus using a wide range of data from many sources. The main value of this process here is that it enables emergency planners to use the event tree to devise a range of eruption scenarios in time and space according to how the eruption evolves. A tricky and more controversial next step, which is not essential to scenario development, is adding probabilities to the different branches at the nodes where they subdivide into different paths of eruptive activity, and formal methods of elicitation to derive these by rational consensus have been used at Vesuvius (Neri et al., 2008). The use of the probabilities in volcanic hazard assessment and volcanic crises is described in detail elsewhere (Marzocchi et al., 2004; Marzocchi et al., 2010; Marzocchi et al., 2012).

One of the eruptive events on the event tree may be selected as the reference eruption for planning and risk assessment purposes (Baxter et al., 2008). The computer model, which simulates the volcanic eruption and its defined hazard scenarios with the appropriate hazard output parameters, incorporates the exposure data for buildings and population, as well as the vulnerability data for the three principal hazards affecting building damage and human casualties for PDCs, ashfall, and earthquake ground shaking.

From the event tree, a volcanic hazard scenario is selected and then defined in each zone by a single physical parameter for each of the three principal hazards affecting building damage (Spence et al., 2005a). For tephra fall, the parameter chosen is vertical gravitational load (Pa, measured in N/m^2) acting on roofs applied as an average for the whole of the impact zone, which is not often available in models and so tephra depth is estimated, with assumptions made about the density of the fallen tephra, which may or may not be wet. For PDC pressure, measured in Pa, an average value for the whole of the impact zone dynamic pressure at the level of ground-floor windows (N/m^2) is taken, and assumptions are made for the vertical profile of the PDC and its impact on upper-floor windows, or the building as a whole. For earthquakes, the effects on buildings are reasonably well defined by the same macroseismic intensity scales used by seismologists.

Finally, the numbers of casualties are derived from the effects at every intensity level for each of the three hazards by incorporating vulnerability relations (tables or charts) specific to a building class and are applicable to the whole area. A joint probability approach may be adopted to sum impacts from the separate hazards, treating them as separate or sequential events.

The uncertainties in the modeling are taken into account by allowing for each parameter (hazard, vulnerability, and building classification) to be input as a probability distribution rather than a single value, so that a range of impacts can be derived from multiple runs (Spence et al., 2005b).

The early attempts to differentiate survival in building occupants in a dilute PDC impact area was based on the chances that the dynamic pressure varied enough in the course of its run-out from the eruptive vent that in more fortunate locations the failure of openings (windows and doors) would not be universal but would be confined to one or two rooms only, providing occupants of other rooms in the house, especially those rooms with windows facing away from the volcano, a good possibility of survival if the duration of the current was brief (Spence et al., 2004). Some houses would remain entirely intact, depending on their vulnerability to dynamic pressure. It was realistic to assume from our field observations in Montserrat that robust modern European buildings might remain intact in some dilute PDCs and actually protect the occupants until the PDC stopped erupting. But if the PDC was of long duration (over many minutes), the hot PDC would actually gradually infiltrate the structure of an average building and present a secondary threat to life from hyperthermia and skin burns due to the heat hazard (Spence et al., 2004; Spence et al., 2005a,b). Intensely hot PDCs with temperatures in excess of 350–400 °C would also trigger rapid ignition of furnishings in the room or rooms invaded by PDC (ash) deposits and all the occupants would be at risk from fire spreading through the house. In all of these assumptions, we have no human evidence from eruptions on which to base the vulnerability functions and so they were indirectly drawn from studies of houses in settlements on Montserrat that were first evacuated because of the danger and were subsequently struck by PDCs, as described above. Tropical housing on Montserrat, though of robust construction, would not be highly resistant to infiltration in long-duration PDCs, and Javanese building types studied after the more recent eruption of Merapi in 2010 were neither robust nor able to resist the rapid ingress of a highly lethal PDC moving slowly at weak dynamic pressure and at a temperature too low to ignite furniture. It was also reasonable to assume that anyone caught in the open would be killed outright, but in the 1994 eruption of Merapi, there were survivors among those who were outdoors, including some who had suffered major skin burns.

Again, in the absence of more epidemiological data on deaths and injuries from actual eruptions, assumptions based on casualties caused by building collapses in earthquakes were used as analogs for modeling human impacts due to roof collapses caused by tephra loading (Spence et al., 2005a,b). In summary, for the conurbation around Vesuvius:

- Single-story buildings: one-third of occupants of the whole building are trapped, of whom half are killed and half are seriously injured.
- Two-story buildings: 10 percent of occupants of the whole building are trapped, of whom half are killed and half are seriously injured.
- Three or more stories: 20 percent of those on the top floor are trapped, of whom half are killed and half are seriously injured.

For multistory buildings, deaths and injuries are therefore assumed to be confined to the top floor only. These estimates were made for the housing around Vesuvius, but they could be a starting point for modeling in other urban areas (Spence et al., 2005b). However, for rural areas, other estimates have been used, e.g., in a building survey in the Azores (Pomonis et al., 1999).

It is clearly important for future mitigation research that the evidence base for human and building vulnerability is added to with every major eruption. It is too soon to advocate generalizable values for human survival or percentages of deaths and injuries in inhabited areas impacted by PDCs or tephra for the slavish adoption by future workers, even though values may be found in the references cited. Instead, researchers need to make their own valuations (including sensitivity analyses) according to the type of volcano and eruption under consideration, and the characteristics of the local housing. Another aspect which is not included is human avoidance behavior, for example, when a house is under heavy ashfall, people are forced to stay inside and may become aware that the roof is under strain as the ash piles up and then take evasive action, like go under a table or another protected part of the house. Rapid rescue for those trapped under a collapsed roof could also be crucial for survival. The work so far has been based on static scenarios, which is not to say that they are without value when planning mitigation measures. Modeling performed by one scientist is the equivalent of another scientist's experimental laboratory, and even simplifying by sticking to a few key determinants can often usefully supplement complex numerical simulations that include many more functions. Thus, there are other features of a PDC in addition to its temperature that may contribute to its lethality in humans, but this is by far the main one; and by not including the presence of missiles, the modeling of a violent PDC using dynamic pressure alone may partly underestimate its impact on buildings.

10.3 BUILDING VULNERABILITY IN MAIN VOLCANIC PROCESSES

10.3.1 Preliminary Remark

Building vulnerability is the degree to which a system (entire building), subsystems (walls, frames, roofs, etc.), or system components (beams, columns, infill panels, windows, doors, etc.) are likely to experience damage due to exposure to hazards.

The sensitivity of a building to damage from a volcanic phenomenon is a function of the elements that constitute it. Resistance and technological aspects of structural elements (such as walls, frames, floors, roofs, etc.) and nonstructural elements (infill panels, windows, openings, etc.) of which the building is constructed strongly influence the vulnerability of the building itself.

In this chapter, the "vulnerability factors," which influence a building's volcanic vulnerability, are analyzed with the aim of developing a methodology capable of deriving a "global" volcanic vulnerability index (as a value between zero and one) for a geographic area and for a given hazard severity level.

The methodology adopts multicriteria approaches (Hwang and Yoon, 1981; Saaty, 1980), in which a vulnerability judgment is produced on the basis of indicators of performance (called "criteria"), and derived from buildings impact analyses on the effects of single volcanic phenomenon (earthquake, ashfall and pyroclastic flows). We use here the vulnerability curves developed by the authors in the framework of volcanic risk assessment focused on the example of Mount Vesuvius (Naples, Italy) (Zuccaro and De Gregorio, 2013).

Typical "vulnerability curves" are shown in Figure 10.1. They express the probability that a given "building vulnerability class" (with similar behavior with respect to the individual volcanic phenomenon) exceeds a certain level of damage, given a level of hazard magnitude.

The examination of the effects of the past eruptions has shown that the eruptive phenomena (especially pyroclastic flows) produce significant damage to nonstructural elements such as openings (doors and windows) and infill panels of framed buildings (Baxter et al., 2005). For this reason, the damage scale adopted (Table 10.1) includes the effects to nonstructural elements (infill panels, I and openings, O), in addition to those structural ones (S).

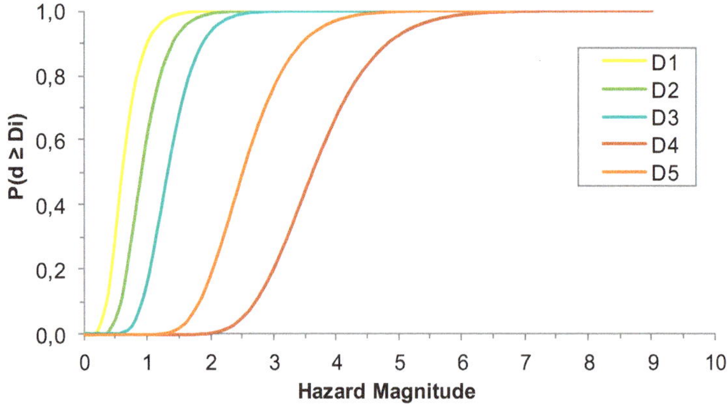

FIGURE 10.1 Typical shapes of vulnerability curves refer to a certain class of vulnerability and to a specific hazard.

TABLE 10.1 Damage Scale

Damage State		Damage Description	
D0	No damage	Absence of damage	
D1	Slight damage	S:	Negligible damage to structural elements
		I:	Negligible damage to infill panels
		O:	Breakthrough of large or weak openings
D2	Moderate damage	S:	Moderate damage to structural elements
		I:	Moderate damage to the infill panels in reinforced concrete buildings
		O:	Breakthrough of windows mildly resistant
D3	Heavy damage	S:	Severe damage to structural elements
		I:	Severe damage to infill panels in reinforced concrete buildings. In few cases, total collapse of infill panels
		O:	Breakthrough of strong windows
D4	Partial collapse	S:	Partial of structural elements
		I:	Breakthrough of strong infill
D5	Collapse	S:	Total collapse of structural elements

S, structural elements; I, infill walls; O, openings—doors and windows.

10.3.2 Vulnerability Classes for Earthquake

The behavior of buildings under seismic action is influenced by different factors (vertical structures, materials, regularity in plan and elevation, floors stiffness, etc.). A simple way to group the buildings with similar behavior under the action of earthquakes is by the combination of vertical structures (classified according to strength) and horizontal structures (classified according to stiffness).

In Table 10.2, a possible vulnerability classification is shown (Zuccaro et al., 2008). They have been identified for the Vesuvian area (Italy), but they can be easily adapted to building types in other geographic areas.

For each vulnerability class, seismic vulnerability curves (as in Figure 10.1) can be calculated from the following: analyses of damage observed after past earthquakes (empirical methods), computational analyses (analytical or mechanical methods), combination of postearthquake damage statistics with analytical simulations using mathematical models of the building typologies under consideration (hybrid methods).

TABLE 10.2 Vulnerability Classes with Reference to Seismic Events (s)

Earthquake					
	Horizontal Structures				
	Poor Stiffness	Poor Technology	Medium Stiffness	Medium–High Stiffness	High Stiffness
Vertical Structures	Metal Sheet, Vaults and/or Wooden Floor (without Ties)	(For Example, "SAP" Floor[a])	Vaults and/or Wooden Floor (without Ties)	Iron Beam Floor	Reinforced Concrete and Steel Floors
Weak masonry Rubble masonry neglected	As	As	As	As	As
Medium quality Rubble masonry maintained	As	As	Bs	Bs	Bs
Good masonry Squared masonry	As	As	Bs	Bs	Cs
Framed structures (reinforced concrete or steel)	–	Bs	–	–	Ds

[a] *SAP floor (self-supporting floor) is a typical Italian horizontal structure made of clay/cement mix with smooth bars at intrados. This technology is considered very dangerous because the cement casting superior slab does not cover the reinforcement bars inserted in the hollow tile.*

For the Vesuvian area, the seismic vulnerability curves have been assessed through an empirical approach founded on numerous in situ damage distribution surveys (about 170,000 buildings) related to past seismic events (Zuccaro, 2004; Zuccaro et al., 2008).

10.3.3 Vulnerability Classes for Ashfall

The action of ashfall on structures can be considered as a distributed gravitational load that depends on density and dampness of the deposit (Macedonio and Costa, 2012; Macedonio et al., 2008; Spence et al., 2005). The action on the roof is comparable to that of snow loading (De Gregorio et al., 2010). The weight of ash, by increasing the vertical load on buildings, could cause collapse of structural elements belonging to vertical or horizontal structures. However, the observation of past eruptions (Blong, 2003) has shown that the most vulnerable elements of a building exposed to ashfall hazard are the roofs. Therefore, under the action of ashfall, as for snow, buildings can be classified on the basis of strength and slope of roofs.

In Table 10.3, possible vulnerability classes are shown (Spence et al., 2005). They have been identified for the Vesuvian area (Italy), but they can be easily adapted to building types of other geographic areas.

TABLE 10.3 Vulnerability Classes with Reference to Ashfall (r)

Ashfall	
Type	**Description**
Ar	Weak pitched wooden roof
Br	Flat standard wooden roof
	Reinforced concrete flat roof—SAP type
	Weak steel flat roof
C1r	Old flat reinforced concrete roof
	Weak pitched steel roof
C2r	Recent flat reinforced concrete roof
	Recent flat steel roof
Dr	Recent pitched reinforced concrete roof
	Recent pitched steel roof

For each vulnerability class in the Vesuvian area, ashfall vulnerability curves (type Figure 10.1) have been calculated through a hybrid method characterized by the following steps.

1. A robust data set was collected by survey in the study area (about 19,000 roofs). It was elaborated statistically with the aim to assess the statistical distribution of roof typologies (main structures, materials, slopes, dimensions, etc.) in the Vesuvian area.
2. A representative sample of roof typologies was generated on the basis of their main characteristics (main structures, materials, slopes, dimensions, etc.). It was developed using a Monte Carlo simulation, in accordance with the statistical information obtained by the data set of step 1.
3. The collapse load of each roof generated by the Monte Carlo simulation was determined and compared with experimental tests developed on different typologies of roofs located in the Vesuvian area (Spence et al., 2005).
4. The vulnerability curves were obtained as lognormal cumulative distributions of collapse load calculated in step 3 (Spence et al., 2005; Zuccaro et al., 2008).

10.3.4 Pyroclastic Density Currents

The experience from the 1997 Montserrat eruption (UK Caribbean Islands) has indicated that a robust building can survive under moderate PDC dynamic pressure (1−5 kPa) if its integrity survives; although, if one or more openings (windows and doors) fail, allowing hot gas and ash to gain access inside the structure, the whole building may be destroyed by fire (Baxter et al., 2005). In this case, the contents of the construction and any timber structure are likely to catch fire; at the same time the principal structural walls and roofing will suffer a combination of internal and external pressures, which will cause partial or total failure (Spence et al., 2004a).

In general, the first elements to collapse are the glass windows and the shutters. However, they can be easily protected by more resistant panels. Nevertheless, the lateral resistance of a building to a PDC strongly depends on the design criteria applied to resist ordinary load conditions: of course, an aseismic building has larger strength and stiffness capabilities than a non-aseismic building.

The structural behavior of buildings under the action of PDCs is not comparable to that induced by earthquakes, because the horizontal action pressure is not inertial and cyclic. So, the structural response is less influenced by the ductility for dissipating energy. Furthermore, the PDC pressure can be assumed to be a surface force (earthquake is a mass force), so the effect of the mass is stabilizing, whereas the effect of the slimness is an important vulnerability factor in PDCs.

TABLE 10.4 Vulnerability Classes with Reference to Pyroclastic Flows (p)

Pyroclastic Flows			
Structural Elements		Nonstructural Elements	
Type	Description	Type	Description
Ap	Weak masonry buildings of 3–4 stories with deformable floor	Ap*	Window glasses of ordinary buildings
	Weak or strong masonry buildings with more than 4 stories	Bp*	Aluminum window in bad condition
Bp	Medium masonry buildings of 1–2 stories with deformable floor	Cp*	Aluminum window in good condition
	Strong masonry buildings of 3 or more stories with rigid floor	Dp*	Old wooden door
Cp	Strong masonry buildings of 1–2 stories with rigid floor	Ep*	Yellow tuff masonry wall
Dp	Nonaseismic reinforced concrete buildings of more than 6 stories (high)	Fp*	Old wooden window
Ep	Nonaseismic reinforced concrete buildings of 4–6 stories (medium)	Gp*	Terra-cotta tile infill panel without window
Fp	Nonaseismic reinforced concrete buildings of 1–3 stories (low)	Hp*	Terra-cotta tile infill panel with window

For these reasons, with reference to the Vesuvian area, two different vulnerability scales (Table 10.4) have been defined by Spence et al. (2004a,b) and Zuccaro et al. (2008).

The first scale, depending on structural elements, is constituted by six classes (Ap, Bp, ..., Fp; p = PDC), in function of the vertical and horizontal structures and number of stories. The second scale, depending on the nonstructural elements, is constituted by eight classes (Ap*, Bp*, ..., Hp*; p = PDC), which identify typologies of openings (windows and doors) and infill panels. For each structural class, Ap–Fp, vulnerability functions (type Figure 10.1) have been defined (Zuccaro et al., 2008). As for the ashfall case, they have been determined through a hybrid method based on typological analysis of about 90,000 buildings surveyed in the Vesuvian area (Spence et al., 2004a,b; Zuccaro et al., 2008).

For nonstructural classes, Ap*–Fp*, the vulnerability curves are not yet available, but the collapse load R has been determined; thanks to experimental tests carried out in situ on typical windows, doors, and infill panels of the Vesuvian area (Spence et al., 2004b) and damage studies in the Montserrat eruption (Baxter et al., 2005).

10.3.5 Vulnerability Index

10.3.5.1 Criteria and Weights

In the previous sections, it was noted that the building vulnerability is strongly linked to the vulnerability typological classes. The methodology illustrated here proposes a "global" volcanic vulnerability index (as a value between zero and one) to be used to define building vulnerability "attitude" for a whole geographic area and for given hazard severity levels. The aim is to supply a quick method to compare typologies of buildings in various geographic areas differentiating the expected response to several hazards.

The vulnerability judgment is based on the analysis of n criteria C_j chosen as an element able to influence the volcanic vulnerability of buildings. The criteria represent the measure of the influence of the typological components in the evaluation of vulnerability of the buildings under the action of each volcanic hazard considered.

The criteria adopted are the 23 vulnerability classes (As, Bs, …, Ds, Ar, Br, …, Dr, Ap, Bp, …, Fp, Ap*, Bp*, …, Hp*) shown in Tables 10.2, 10.3 and 10.4. For each criterion, the weights w_j have been calculated like the measure of the priorities assigned to the criterion on the vulnerability assessment for a given hazard severity level.

The weights calculation has been carried out on the basis of binary comparisons $a$ij between criteria C_i e C_j organized in the so-called *weights matrix* $\mathbf{A}_{n \times n}$. In particular, fixed beforehand the level of the three hazards H considered (macroseismic intensity $\overline{I}$ (EMS-98), vertical ashfall load $\overline{q}$, horizontal pyroclastic pressure $\overline{p}$), the generic element $a$ij is calculated as in the following ratio.

$$a_{ij} = \frac{P_{C_j}(D \geq D5 | H = (I \text{ or } q \text{ or } p))}{P_{C_i}(D \geq D5 | H = (I \text{ or } q \text{ or } p))} \tag{10.1}$$

Where: P_{C_i} and P_{C_j} are the probabilities that the damage D exceeds the damage $D5$ for a given level of hazards, respectively for the criteria C_i and C_j.

For example, with reference to volcanic eruptions of different scale, the probabilities P_{C_i} are shown in Table 10.5.

The eruptions assumed are two and they relate to possible explosive events in areas near the vent. The former eruption, called medium (M), is characterized by the sequence of phenomena with the following medium-hazard levels: $\overline{I} = \mathrm{VIII(EMS-98)}$, $\overline{p} = 6\ \mathrm{kPa}$, and $\overline{q} = 15\ \mathrm{kPa}$. The latter eruption, called small (S), is characterized by the sequence of phenomena with the following medium-hazard levels: $\overline{I} = \mathrm{V(EMS-98)}$, $\overline{p} = 2\ \mathrm{kPa}$, and $\overline{q} = 3\ \mathrm{kPa}$.

The probabilities (P_{C_i} and P_{C_j}) shown in Table 10.5 have been calculated through the vulnerability curves described in paragraphs 10.3.2–10.3.4.

TABLE 10.5 Probability [0–1] for Each Criterion that the Damage *D* Exceeds the Damage *D5* for Two Given Eruptions, Medium M and Small S, Characterized by Assigned Level of Seismic Intensity *I*, Vertical Ash Fall Load *q*, and Horizontal Pyroclastic Pressure *p*

Criteria		Medium Eruption (M)			Small Eruption (S)		
		$I=8$ (EMS-98)	$p=6$ kPa	$q=15$ kPa	$I=5$ (EMS-98)	$p=2$ kPa	$q=3$ kPa
Seism	As	0.103	–	–	0.006	–	–
	Bs	0.018	–	–	0.001	–	–
	Cs	0.003	–	–	0.000	–	–
	Ds	0.001	–	–	0.000	–	–
Pyroclastic flows	Ap	–	0.879	–	–	0.000	–
	Bp	–	0.080	–	–	0.000	–
	Cp	–	0.016	–	–	0.000	–
	Dp	–	1.000	–	–	0.001	–
	Ep	–	1.000	–	–	0.001	–
	Fp	–	0.570	–	–	0.001	–
	Ap*	–	1.000	–	–	1.000	–

Continued

TABLE 10.5 Probability [0–1] for Each Criterion that the Damage *D* Exceeds the Damage *D*5 for Two Given Eruptions, Medium M and Small S, Characterized by Assigned Level of Seismic Intensity *I*, Vertical Ash Fall Load *q*, and Horizontal Pyroclastic Pressure *p*—cont'd

		Medium Eruption (M)			Small Eruption (S)		
Criteria		$I=8$ (EMS-98)	$p=6$ kPa	$q=15$ kPa	$I=5$ (EMS-98)	$p=2$ kPa	$q=3$ kPa
	Bp*	–	1.000	–	–	1.000	–
	Cp*	–	1.000	–	–	0.000	–
	Dp*	–	1.000	–	–	0.000	–
	Ep*	–	1.000	–	–	0.000	–
	Fp*	–	1.000	–	–	0.000	–
	Gp*	–	1.000	–	–	0.000	–
	Hp*	–	0.000	–	–	0.000	–
Ashfall	Ar	–	–	1.000	–	–	0.920
	Br	–	–	1.000	–	–	0.500
	C1r	–	–	1.000	–	–	0.050
	C2r	–	–	1.000	–	–	0.000
	Dr	–	–	0.850	–	–	0.000

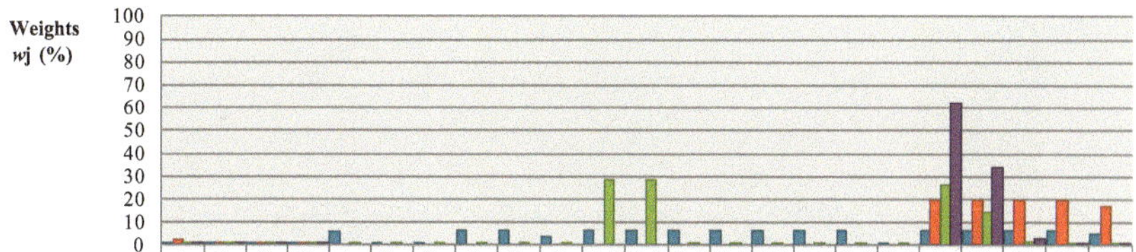

	As	Bs	Cs	Ds	Ap	Bp	Cp	Dp	Ep	Fp	Ap*	Bp*	Cp*	Dp*	Ep*	Fp*	Gp*	Hp*	Ar	Br	C1r	C2r	Dr
M (EQ + PF + AF)	0.7	0.1	0.0	0.0	5.7	0.5	0.1	6.4	6.4	3.7	6.4	6.4	6.4	6.4	6.4	6.4	6.4	0.0	6.4	6.4	6.4	6.4	5.5
M (EQ + AF)	2.1	0.4	0.1	0.0	0.0	0.0	0.0	0.0	0.0	0.0	0.0	0.0	0.0	0.0	0.0	0.0	0.0	0.0	20.1	20.1	20.1	20.1	17.1
S (EQ + PF + AF)	0.2	0.0	0.0	0.0	0.0	0.0	0.0	0.0	0.0	0.0	28.7	28.7	0.0	0.0	0.0	0.0	0.0	0.0	26.4	14.4	1.4	0.0	0.0
S (EQ + AF)	0.4	0.1	0.0	0.0	0.0	0.0	0.0	0.0	0.0	0.0	0.0	0.0	0.0	0.0	0.0	0.0	0.0	0.0	62.3	33.9	3.4	0.0	0.0

Criteria *Cj* (vulnerability classes)

FIGURE 10.2 Weights for each criterion calculated for two given eruptions, medium M and small S, and for two different sequences of volcanic phenomena: earthquake, pyroclastic flows, and ashfall (EQ + PF + AF); and earthquake and ashfall (EQ + AF).

The weights w_j assigned to each criterion C_j are calculated by dividing the geometric mean Mg_i of each criterion for the sum of the geometric means of the criteria:

$$w_j = \frac{Mg_i}{\sum_{i=1}^{n} Mg_i} = \frac{(a_{1j} \cdot \ldots \cdot a_{nj})^{\frac{1}{n}}}{\sum_{i=1}^{n} Mg_i} \qquad (10.2)$$

The Figure 10.2 summarizes the weights calculated for two volcanic eruptions assumed (M and S) and for two different sequences of volcanic phenomena: earthquake, pyroclastic flows, and ashfall (EQ + PF + AF); and earthquake and ashfall (EQ + AF).

Analyzing the values of weights of Figure 10.2, it is observed that the criteria with the most influence on the vulnerability assessment are the vulnerability classes related to pyroclastic flows and to ashfall, while the seismic ones are negligible. This is due to the fact that, during volcanic eruptions, seismic events are of low magnitude and have much less impact than large ashfalls and PDCs.

10.3.5.2 Alternatives and Decision Matrix

Once the weights have been defined, it is necessary to identify the geographic zones for comparison, called "alternatives *A*" in multicriteria approaches.

The volcanic areas examined here are the following four (Figure 10.3). They have been investigated by the authors through visual survey activity in situ.

1. The destroyed city of Herculaneum in Italy is today an archaeological site preserved in the ashes of the eruption of Vesuvius (79 AD). Buildings in the old Herculaneum were made of various types of local stone (tuff and sandstone), covered with plaster, and lavishly decorated when finances are allowed. A typical technology used to erect residences for middle- and

FIGURE 10.3 Pictures of volcanic areas investigated.

lower-class citizens is the "opus craticium." It consists of walls constituted by brick columns and wooden frame filled by "opus incertum" stones.

Typical roofs are made of tile. They are roofs with slope from 25 to 35 percent, covered with two layers of tiles (bottom and covering), sometimes laid in mortar. The tiles layers generally lie on a wooden frame, and sometimes on one or more cross or barrel vaults.

In the old Herculaneum, houses were built very close together, and right up against the street. The streets were recessed, leaving raised sidewalks. There were stepping-stones available for crossing the street, so as to avoid the filth that might be found there. The floors were made of concrete, and windows were usually closed by wooden shutters (though some glass panes have been found).

2. Herculaneum today is a city of about 55,000 people located near the old city. The data collected during the survey activity indicate that the main

vertical structures consist of sack masonry or tuff masonry, or buildings made of monodirectional or bidirectional frames of reinforced concrete, with infill panels made of tuff blocks or hollow bricks. The main horizontal structures are floors with steel beams and vaults or hollow flat blocks without tie beams, whereas the remaining horizontal structures consist of concrete block or RC slabs and vaults without tie beams. It was observed that the roofs are predominantly horizontal and they are made out of steel beams with hollow clay or concrete blocks. Vaults were also recorded in some cases. The openings and their protections are made out of wood or aluminum and are in good condition. In a representative sample of buildings, the openings cover between 25 and 50 percent of the building facades.

3. Arequipa city is an important commercial center of Peru with a population of about 900,000 people. It is located in the shadow of the El Misti volcano. The data collected during the survey activity indicate that the main vertical structures consist of masonry of lapil-concrete, terra-cotta or ignimbrite blocks, or reinforced concrete buildings generally realized without seismic design. The main horizontal structures are concrete and masonry flooring systems and vaults without tie beams. It was observed that the roofs are predominantly horizontal and they are made of metal sheets or concrete and masonry flooring systems. Vaults were also recorded in some cases. The openings are made of glass mounted on a weak iron frame and they are mostly deprived of protection (persian, damper, etc.).
4. Santorini (Greece) is an island in the southern Aegean Sea with about 15,000 inhabitants. It is the largest island of a small, circular archipelago, which bears the same name and is the remnant of a volcanic caldera. The data collected during the survey activity indicate that the main vertical structures consist of masonry made of red or black volcanic stones, masonry mixed to RC, and reinforced concrete buildings generally realized without seismic design. The main horizontal structures are floors with concrete and masonry flooring or RC slabs and vaults of masonry or RC. It was observed that the roofs are horizontal (with concrete and masonry system or RC slabs) or vaults of masonry or RC. The openings and their protections are made of wood or aluminum and are in good condition. In a representative number of buildings, the openings cover between 25 and 50 percent of the building facades.

Once one location A (Old Herculaneum, Herculaneum 2013, Arequipa, Santorini) has been selected, the measures d_{ij} of the criterion C_j with respect to the alternative i are collected in the decision matrix **Dm×n** ($n =$ number of criteria; $m =$ number of towns), whose generic terms (d_{ij}) are the percentage distribution of criterion C_j (vulnerability class) in the alternative i (town).

Subsequently, the matrix terms should be normalized and combined with the criteria weights, obtaining the new matrix **Vm×n**, whose generic terms are achieved as it follows.

$$v_{ij} = \frac{w_j \cdot d_{ij}}{\sqrt{\sum_{i=1}^{n} d_{ij}^2}} A \tag{10.3}$$

10.3.5.3 Calculation of the Orders Among the Alternatives: The Vulnerability Index

According to the Technique for Order Preference by Similarity to Ideal Solution multicriterion method (Hwang and Yoon, 1981), aiming at obtaining the final ranking through the vulnerability indexes, the ordering of each alternative A is required.

The real alternatives are characterized by the rows of the decision matrix normalized and weighted (**V**), each one is represented by the following row vector.

$$A_i = \{v_{i1}, v_{i2}, \ldots, v_{in}\} \tag{10.4}$$

Therefore it is possible to determine two virtual solutions: the positive ideal (A^+) and negative ideal (A^-) alternatives, represented by the following vectors.

$$A^+ = \{\max v_{ij}, j = 1, 2, \ldots, n\} = \{v_1^+, v_2^+, \ldots, v_n^+\} \tag{10.5}$$

$$A^- = \{\min v_{ij}, j = 1, 2, \ldots, n\} = \{v_1^-, v_2^-, \ldots, v_n^-\} \tag{10.6}$$

The alternatives A can be considered as points in an n-dimensions space (n being the criteria), where the j-axis is the measure of the performance normalized and weighted (v_{ij}) of the alternatives with respect to the j-criterion. In this view, the Euclidean distances between the real alternative A_i and the ideal ones can be calculated as follows.

$$S_i^+ = |A_i - A^+| = \sqrt{\sum_{j=1}^{n} \left(v_{ij} - v_j^+\right)^2} \tag{10.7}$$

$$S_i^- = |A_i - A^-| = \sqrt{\sum_{j=1}^{n} \left(v_{ij} - v_j^-\right)^2} \tag{10.8}$$

In conclusion, the relative distance of each alternative from ideal ones is the reference parameter for their classification, such as the vulnerability index. In particular, on the basis of the Euclidean distances calculated, the vulnerability index is expressed through the following relationship:

$$I_V(A_i) = \frac{S_i^-}{\left(S_i^+ + S_i^-\right)} \tag{10.9}$$

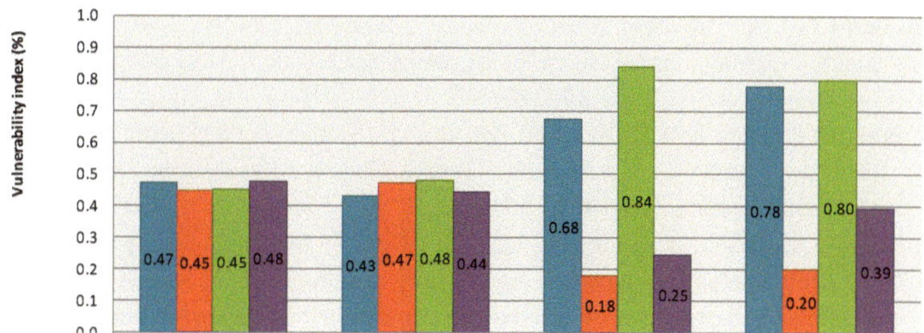

	M (EQ + PF + AF)	M (EQ + AF)	S (EQ + PF + AF)	S (EQ + AF)
Old herculaneum (Italy)	0.47	0.43	0.68	0.78
Herculaneum 2013 (Italy)	0.45	0.47	0.18	0.20
Arequipa (Peru)	0.45	0.48	0.84	0.80
Santorini Island (Greece)	0.48	0.44	0.25	0.39

Volcanic phenomena

FIGURE 10.4 Vulnerability indexes assessed for four alternatives (Old Herculaneum, Herculaneum 2013, Arequipa, Santorini), for two given eruptions, medium M and small S, and for two different sequences of volcanic phenomena (EQ + PF + AF and EQ + AF).

The vulnerability index ranges between zero and one. In fact if the alternative A_i is coincident with the negative ideal solution A^-, S_i^- is equal to zero and therefore the vulnerability index is equal to zero. To the contrary, if the alternative A_i is coincident with the positive ideal solution A^+, S_i^+ is equal to zero and therefore the vulnerability indexes are equal to one. For the four towns examined, the vulnerability indexes determined through the methodology are indicated in Figure 10.4.

Analyzing the values of volcanic vulnerability indexes of Figure 10.4, it is observed that for most violent eruptions (M), the different towns have almost equal indexes (about 0.5). This is due to the negligible influence of typological characteristics of buildings for high values of the magnitude of the hazards, whereas in the case of minor eruptions, the weight of the typological characteristics becomes important; in fact, the vulnerability indices have different values for each geographic area. In particular, roofs and openings with poor resistance make the Old Herculaneum and Arequipa more vulnerable than Herculaneum 2013 and Santorini. It is important to note that this methodology may also be applied by taking individual buildings as alternatives to compare. In this case, the measures d_{ij} of the criterion C_j with respect to the alternative i can be 1 or 0 when for the ith building the criterion C_j is present or absent, respectively.

REFERENCES

Baxter, P.J., 1990. Medical effects of volcanic eruptions; I-main causes of death and injury. Bull. Volcanol. 52 (7), 532–544.

Baxter, P.J., Neri, A., Todesco, M., 1998. Physical modelling and human survival in pyroclastic flows. Nat. Hazards 17, 165–176.

Baxter, P.J., Cole, P.D., Spence, R., Zuccaro, G., Boyd, R., Neri, A., 2005. The impacts of pyroclastic density currents on buildings during the eruption of the Soufrière Hills volcano, Montserrat. Bull. Volcanol. 67, 292–313.

Baxter, P.J., Aspinall, W.P., Neri, A., Zuccaro, G., Spence, R.J.S., Cioni, R., Woo, G., 2008. Emergency planning and mitigation at Vesuvius: a new evidence-based approach. J. Volcanol. Geotherm. Res. 178, 454–473.

Belousov, A., Voight, B., Belousova, M., 2007. Directed blasts and blast-generated pyroclastic density currents: a comparison of the Bezymianny 1956, Mount St Helens 1980, and Soufrière Hills volcano, Montserrat 1997 eruptions and deposits. Bull. Volcanol. 69 (7), 701–740.

Blong, R., 2003. Building damage in Rabaul, Papua New Guinea, 1994. Bull. Volcanol. 65, 43–54. http://dx.doi.org/10.1007/s00445-002-0238-x.

Chester, D.K., Duncan, A.M., 2007. Lieutenant- Colonel Delmée-Radcliffe's report on the 1906 eruption of Vesuvius. J. Volcanol. Geotherm. Res. 166, 204–216.

Cioni, R., Gurioli, L., Sbrana, A., Vougioukalakis, G., 2000. Precursory phenomena and destructive events related to the Late Bronze Age Minoan (Thera, Greece) and AD 79 (Vesuvius, Italy) Plinian eruptions; inferences from the stratigraphy in the archaeological areas. In: McGuire, W.G., Griffiths, D.R., Hancock, P.L. (Eds.), The Archaeology of Geological Catastrophes, vol. 171. Geological Society, London, pp. 123–141. Special Publications.

Cioni, R., Bertagnini, A., Santacroce, R., Andronico, D., 2008. Explosive a activity and eruption scenarios at Somma–Vesuvius (Italy): a review. J. Volcanol. Geotherm. Res. 178, 331–346.

De Gregorio, D., Faggiano, B., Formisano, A., Mazzolani, F.M., 2010. Air fall deposits due to explosive eruptions: action model and robustness assessment of the Vesuvian roofs. In: Mazzolani, F.M. (Ed.), Urban Habitat Construction under Catastrophic Events. Vol. COST C26 Final Conference. CRC Press, pp. 507–512. ISBN:9780415606851.

Dobran, F., Neri, A., Todesco, M., 1994. Assessing the pyroclastic flow hazard at Vesuvius. Nature 367, 551–554.

Eisele, J.W., O'Halloran, R.L., Reay, D.T., Lindhock, G.R., Lewman, L.V., Brady, W.J., 1981. Deaths during the May 18, 1980, eruption of Mount St Helens. N. Engl. J. Med. 305, 931–936.

Esposti Ongaro, T., Clarke, A.B., Neri, A., Voight, B., Widiwijayanti, C., 2008. Fluid dynamics of the 1997 Boxing Day volcanic blast on Montserrat, West Indies. J. Geophys. Res. 113 (B3), B03221.

Esposti Ongaro, T., Clarke, A.B., Voight, B., Neri, A., Widiwijayanti, C., 2012. Multiphase flow dynamics of pyroclastic density currents during the May 18, 1980 lateral blast of Mount St Helens. J. Geophys. Res. 117 (B6), B06208.

Hovey, E.O., 1902. The eruptions of La Soufrière, St Vincent in May 1902. Nat. Geogr. Mag. 13 (12), 444–459.

Hwang, C.L., Yoon, K., 1981. Multiple Attribute Decision Making Methods and Applications. Springer Verlag, Berlin, Heidelberg, New York.

Jenkins, S., Komorowski, J.-C., Baxter, P.J., Spence, R., Picquout, A., Lavigne, F., Surono, S., 2013. The Merapi 2010 eruption: an interdisciplinary impact assessment methodology for studying pyroclastic density current dynamics. J. Volcanol. Geotherm. Res. 261, 316–329.

Lacroix, A., 1904. La Montagne Pelée et ses eruptions. Masson, Paris.

Macedonio, G., Costa, A., 2012. Brief communication. Rain effect on the load of tephra deposits. Nat. Hazards Earth Syst. Sci. 12, 1229–1233. http://dx.doi.org/10.5194/nhess-12-1229-2012.

Macedonio, G., Costa, A., Folch, A., 2008. Ash fallout scenarios at Vesuvius: numerical simulations and implications for hazard assessment. J. Volcanol. Geotherm. Res. 178, 366–377.

Marzocchi, W., Sandri, L., Gasparino, L., Newhall, C., Boschi, E., 2004. Quantifying probabilities of volcanic events: the example of volcanic hazard at Mount Vesuvius. J. Geophys. Res. 109, B11201. http://dx.doi.org/10.1029/2004JB003155.

Marzocchi, W., Sandri, L., Selva, J., 2010. BET_VH: a probabilistic tool for long-term volcanic hazard assessment. Bull. Volcanol. 72, 705–716.

Marzocchi, W., Newhall, C., Woo, G., 2012. The scientific management of volcanic crises. J. Volcanol. Geotherm. Res. 247-248, 181–189.

Neri, A., Aspinall, W.P., Cioni, R., Betagnini, A., Baxter, P.J., Zuccaro, G., Andronico, D., Barsotti, S., Cole, P.D., Esposti Ongaro, T., Hincks, T.K., Macedonio, G., Papale, P., Rosi, M., Sanatacroce, R., Woo, G., 2008. Developing an event tree probabilistic hazard and risk assessment at Vesuvius. J. Volcanol. Geotherm. Energy 178, 397–415.

Newhall, C.G., 2000. Mount St Helens, master teacher. Science 288 (1181), 1183.

Newhall, C.G., Hoblitt, R.P., 2002. Constructing event trees for volcanic crises. Bull. Volcanol. 64, 3–20.

Pallister, J.S., Schneider, D.J., Griswold, J.P., Keeler, R.H., Burton, W.C., Noyles, C., Newhall, C.G., Ratdomopurbo, A., 2013. Merapi 2010 eruption – chronology and extrusion rates monitored with satellite radar and used in eruption forecasting. J. Volcanol. Geotherm. Res. 261, 144–152.

Perret, F., 1924. The Vesuvius Eruption of 1906. Carnegie Institution of Washington, Washington, D.C.

Pomonis, A., Spence, R., Baxter, P., 1999. Risk assessment of residential buildings for an eruption of Furnas volcano, Sao Miguel, the Azores. J. Volcanol. Geotherm. Res. 92, 107–131.

Russell, I.C., 1902. Volcanic eruptions on Martinique and St Vincent. Nat. Geogr. Mag. 13 (12), 415–435.

Saaty, T.L., 1980. The Analytic Hierarchy Process. McGraw Hill, New York.

Spence, R., Pomonis, A., Baxte, P., Coburn, A., 1997. Building damage caused by the Mount Pinatubo eruption of June 15th, 1991. In: Newhall, C.G., Punongbayan, R.S. (Eds.), Fire and Mud: Eruptions and Lahars of Mt Pinatubo. Washington University Press, Philippines, pp. 1005–1012.

Spence, R.J.S., Baxter, P.J., Zuccaro, G., 2004a. Building vulnerability and human casualty estimation for a pyroclastic flow: a model and its application to Vesuvius. J. Volcanol. Geotherm. Res. 133, 321–343.

Spence, R., Zuccaro, G., Petrazzuoli, S., Baxter, P.J., 2004b. The resistance of buildings to pyroclastic flows: theoretical and experimental studies in relation to Vesuvius. ASCE Nat. Hazards Rev. 5, 48–50. ISSN:1527-6988.

Spence, R., Kelman, I., Baxter, P.J., Zuccaro, G., Petrazzuoli, S., 2005a. Residential building and occupant vulnerability to tephra fall. Nat. Hazards Earth Syst. Sci. 5, 477–494.

Spence, R.J.S., Kelman, I., Calogero, C., Toyos, G., Baxter, P.J., Komorowski, J.-C., 2005b. Modelling expected physical impacts and human casualties from explosive volcanic eruptions. Nat. Hazards Earth Syst. Sci. 5, 1003–1015.

Spence, R., Kelman, I., Brown, A., Toyos, G., Purser, D., Baxter, P.J., 2007. Residential building and occupant vulnerability to pyroclastic density currents in explosive eruptions. Nat. Hazards Earth Syst. Sci. 7, 219–230.

Wilson, T.M., Jenkins, S., Stewart, C., 2014. Impacts from Volcanic Ash Fall. In: Papale, P. (Ed.), Volcanic Hazards, Risks and Disasters. Elsevier, pp. 47–86.

Zuccaro, G. (2004). Final Report SAVE. Project Link to the Web Page of the 'Save' Project in 'Atlantis'. http://gndt.ingv.it/Att_scient/Prodotti_attesi_2004/Dolce_Zuccaro/Mappe/Start2.htm.Roma.

Zuccaro, G., De Gregorio, D., 2013. Time and space dependency in impact damage evaluation of a sub-Plinian eruption at Mount Vesuvius. Nat. Hazards 68, 1399–1423. http://dx.doi.org/10.1007/s11069-013-0571-.

Zuccaro, G., Cacace, F., Spence, R.J.S., Baxter, P.J., 2008. Impact of explosive eruption scenarios at Vesuvius. J. Volcanol. Geotherm. Res. 178, 416–453.

Chapter 11

Cost–Benefit Analysis in Volcanic Risk

Gordon Woo
Risk Management Solutions, London

ABSTRACT

Any volcanic crisis in a densely populated area is intrinsically also a regional economic crisis. Every risk mitigation action involving the evacuation of people at risk involves an economic cost both to the evacuees and to the regional civic authority. The primary tool for economic decision-making is cost–benefit analysis (CBA). The application of CBA in the context of volcanic risk is discussed and compared with prevailing approaches used for dealing with volcanic crises. CBA is shown to be especially relevant for segmenting a population according to evacuation cost. This methodology helps to rationalize a major dilemma during volcanic crises: the reluctance to evacuate farmers and others whose livelihoods are perceived to be at stake if they choose to evacuate.

11.1 ASSESSING CRISIS MANAGEMENT STRATEGIES

The challenge during volcanic crises of making decisions on public safety can be met in various ways. Every crisis has some special characteristics that confound direct comparisons with other volcanic crises. There can be significant differences in the style of eruptive behavior, the level and extent of monitoring, learning experience of previous crises, population education on hazard awareness, and so on. Volcano observatories operate largely independent of any international regulatory oversight, and without much effort at collective assessment or performance critique of operational experience. Without drawing attention to any specific shortcomings, a general global review of the crisis management performance of volcano observatories indicates a tendency toward greater hesitation in raising alert levels and calling for evacuation than might be optimal for mitigating casualties (Winson et al., 2012). This suggests some scope for adoption of a more systematic and objective crisis management approach. The purpose here is to explain, justify, and advocate the cost–benefit analysis (CBA) approach as a practical tool for volcanic crisis risk management.

Volcanic Hazards, Risks, and Disasters. http://dx.doi.org/10.1016/B978-0-12-396453-3.00011-3

Before discussion of CBA, some common strategies are reviewed. A tentative decision-maker, and indeed his or her volcanological advisers, might be inclined to wait until there is a very high probability or near certainty that a dangerous eruption is imminent. Unless specifically trained in dealing with risk, scientists' ingrained response to uncertainty is to gather information to reduce it. By avoiding rushing to judgment amid uncertainty, this strategy minimizes the chance that the decision-maker and the advisory team make a false positive error. Furthermore, since only a small proportion of unrest periods ever leads to a dangerous eruption, this strategy also minimizes the frequency of decision-maker errors in general.

The strategy of deferring action until there is a high probability of an imminent eruption might also serve as a satisfactory strategy for society, in circumstances where the evacuation time is sufficiently short, so that the casualty consequences of any false negative error were limited. This might be the case where the population at risk is small, ample transport to effect a rapid and orderly evacuation is available, and traffic flows well. But for many volcanic regions, this is a much idealized emergency response situation that may well descend into chaos. The population at risk might well be large, and even if it is not, the transport infrastructure might be inadequate for all the inhabitants at risk to be evacuated in a reasonable time. Traffic is liable to become gridlocked if roads are congested with fleeing vehicles, or blocked with volcanic debris.

Faced with obvious difficulty and complexity in the logistics of rapid evacuation under crisis conditions, and aware of the possible casualty consequences, a decision-maker might adopt a patently precautionary approach, and make an early call for evacuation. This might be justified on the grounds of risk aversion of the decision-maker, or that the risk exposure of the population was unduly excessive in comparison with norms associated with other societal risks, e.g., exposure to toxic chemicals.

However, a call for early evacuation may not take sufficient account of the economic disruption and loss of livelihood that an evacuation would entail, especially to local livestock farmers, anxious over what might happen if their animals were untended for any significant length of time. Especially if there are only slight sensory cues, such as felt tremors or visible smoke, as signs of an imminent eruption, a widespread reluctance to evacuate may occur, as well as public resentment and hostility toward volcanologists at any false positive error.

Many evolving volcanic crises occur where the advantages and disadvantages of mitigating actions are quite finely balanced. Weighing the respective costs and benefits of evacuation presents a major burden on decision-makers. Corporate-style executive decisions are often made by them as senior individuals or serving on advisory panels, without necessarily any attempt at formal rigorous quantitative evidence-based justification. Accordingly, executive decisions are prone to all manner of cognitive biases, which a psychologist could catalog after any decision error is committed.

CBA has the positive virtues of transparency and auditability as special commendations for guiding decision-makers. It is widely adopted as an explicit methodology for deciding on cost-effective natural hazard mitigation schemes worldwide (e.g., Red Cross Federation, 2012). Implicitly, in the course of weighing pros and cons, decision-makers may adopt some informal qualitative approach to CBA.

11.2 THE ROOTS OF VALUE-BASED DECISION-MAKING

Formal CBA methods of decision analysis for volcanological crises have only been explored in depth in the twenty-first century (Marzocchi and Woo, 2009). But even though there is an acknowledged lack of historical volcanological experience in the formal use of explicit CBA, nevertheless substantial long-term evidence exists for the optimality of the rational strategy of weighing costs and benefits in hazard warning. This evidence emerges from the process of evolution itself, and the daily imperative for animals to balance the need to avoid predators with the need to forage for food. Compared with some variant of CBA, both the strategy of waiting to be almost certain that a predator is near and the precautionary strategy of hiding from any vague threat would be far more likely to drive a species to extinction.

Every member of a population at risk from a volcano is a human being who has evolved to balance risks and rewards. Indeed, it has been said that life itself consists of a series of decisions in time, which are taken based on the potential costs and benefits associated with possible courses of action (Shadlen et al., 2007). The innate human sense of decision-making in crisis situations is a factor in any call for evacuation.

The threat from volcanoes is just one of the diverse hazards that human beings have to deal with in their pursuit of a long and prosperous life. In order to place into context a decision-making protocol for volcano risk, it is instructive to review the evolution of decision-making for survival. Important insights into the survival decision-making process have recently emerged from the developing interdisciplinary field of *neuroeconomics*, which addresses the neural foundations of value-based decision-making in humans and other nonhuman primates (Rangel et al., 2008). An animal engages in value-based decision-making whenever it makes a choice from several alternatives on the basis of a subjective value that it places on them. Decisions are made, which involve weighing the costs and benefits for each behavioral choice. Keen anthropological interest exists for discerning the underlying neural mechanisms that govern these neurobiological cost–benefit calculations.

It is obvious that adaptive behavioral decisions are essential for survival and reproductive success. Animals choose from several alternatives, which need to be evaluated before the most desirable option is selected (Herberholz and Marquart, 2012). To determine what behavior is most desirable in any given situation, the nervous system must integrate external conditions (e.g., prevailing predator threat) with current internal conditions (e.g., hunger state), thus trading off the costs and benefits of different alternatives before deciding which one to choose. Human reactions have evolved out of the primeval life-preserving decisions required to instinctively distinguish predator from prey. Amid environmental uncertainty over what may be lurking, an optimally fast neurobiological signal processing algorithm exists (Bogacz, 2009), related to the firing of neurons, some of which signal the predator, others signal the prey. The ambiguity over interpreting contrasting signals is resolved by gathering more sensory information until a threshold value R is attained of the ratio of the likelihood of the data given by a lurking predator to the likelihood of the data given by some lurking prey. In mathematical terms: $R = P(sensory\ evidence|predator)/P(sensory\ evidence|prey)$.

This type of sequential probability ratio test has served the human race well during the course of evolution. For natives still eking out a daily hunting and gathering existence in the wilderness, their very survival for many thousands of years is the evidence of successful and effective decision-making in evading predators. However, with rising twenty-first century population densities in volcanic regions, decision support is increasingly needed for those living in the shadow of active volcanoes. Whereas a primitive hunter could hone daily during adolescence the sensory signal processing skills to evade predators and survive in the wilderness, the skills at processing occasional observable volcanic signals, some of which may be just feebly sensed, are possessed only by professional volcanologists, equipped with modern monitoring instruments.

The brain has evolved to generate the best behavior it can, subject to the constraints imposed by the body and the environment (Todorov, 2007). Ad hoc individual strategies for avoiding danger would not have survived the selection pressures of evolution. As with other existential human threats, it would be in keeping with instinctive human nature for an optimal framework for volcanic hazard decision-making to be based on the explicit weighing of costs and benefits.

The evolutionary trade-off between life and livelihood has always been an integral part of human existence, and is still evaluated instinctively today by individuals during times of volcanic crisis. Volcanology has advanced enormously during the past century, but people essentially behave the same as their great-grandparents, have the same emotions, and share the same hopes and fears. Thus when mandatory evacuations are ordered, some individuals will nonetheless insist on ignoring the instruction to leave, reckoning this not to be in their overall best interest. Safeguarding of property and farm animals is typically cited as the reason for staying, despite the volcanic danger. To give just one salient example, in the mandatory evacuation from the Copahue volcano in the Andes Cordillera, at the end of May 2013, one resident reported that she did not want to go to a shelter and leave her chickens. To an outside observer, especially a city-dweller, no size of chicken coop loss would come close to comparison with a single human loss, if the likelihood of an imminent eruption were significant—as it undoubtedly was. Ultimately, reluctant farmers were persuaded to evacuate upon the government guaranteeing care would be provided for animals. Such care tipped the individual cost–benefit balance toward evacuation.

In making their own personal decision as to whether to evacuate or not, every individual is entitled to receive the maximum scientific information upon which to make their own informed choice. Furthermore, the information should be expressed in language they can understand. This information should include not just the best estimate of the likelihood of a dangerous eruption, but also the associated uncertainty. Ultimately, people, such as farmers, with their precious livelihoods at stake, cannot be forced against their will to evacuate, they need to be educated about volcanic hazards, persuaded by scientific evidence, and "nudged" in the direction of personal safety.

Insights into effective ways of accomplishing this hazard communication objective come from the domain of social psychology. Sheena Iyengar (2010), a social psychologist, has examined a wide range of aspects of how people choose what to do. She points out that, whether consciously or subconsciously, we tend to organize our lives to display our identity as accurately as possible. Our choices reflect not just who we are and what we want, but also how those choices will be interpreted by others. We survey our social environment to figure out what others think of particular choices available. Thus the actions that a person chooses to take do not depend solely on what is perceived to be in the individual's best interest but also on the attitude of others to his or her choice.

Before deciding how to react to a hazard warning, an individual will seek to discuss the alternatives with family, friends, neighbors, colleagues, and so on. The extent of peer pressure in reacting to hazard warnings will vary from one group to another. This process of social interaction in individual decision-making engenders a significant degree of self-organization supporting the capability of individuals to take their own decisions in situations of perceived danger. The public dissemination of hazard advisories should recognize this intrinsic societal mechanism for dealing with a communal threat, which encourages a more cohesive public response. Furthermore, the proliferation of online social media increases enormously the degree of social interaction governing individual choice.

11.3 THE APPLICATION OF CBA

The purpose of a decision support tool is to assist the decision-maker in arriving at his or her decisions—most definitely not to make the decisions themselves. In its decision support function, such a tool should be transparent, auditable, and simple enough to be comprehended by those without formal training in risk management, and perhaps lacking specific computational skills. Fitting these requirements is the foundation of decision analysis: CBA, which has its historical roots in nineteenth-century welfare economics, balancing societal benefits against market forces (OECD, 2006). In times of volcanic crisis, decisions on societal risk mitigation have to weigh the safety benefits of evacuation against the socioeconomic costs of an evacuation process that may last for weeks or even months.

It has been argued that although the costs of many actions can be easily evaluated, there are intangible factors that are hardly quantifiable. An extension of CBA is the use of multicriteria analysis (MCA), which considers a broader range of evaluation criteria, including variables that are less tangible and have no market price, e.g., psychological aspects of individual well-being, such as emotional stress and unhappiness. For this extension to be practically viable, it would be necessary for decision-makers and stakeholders to have the opportunity to express their valuation of the relative importance of the various criteria. Although laudable as an exercise in participatory decision-making, the reluctance and discomfort of decision-makers and stakeholders in articulating complex issues of preference render this approach especially prone to excessive subjectivity and further ambiguity. The simpler use of CBA is thus more practical as a bridge between weighing pros and cons approximately and undertaking an elaborate MCA.

The essential argument underlying the recommended use of basic CBA is that this captures the core elements in the evacuation decision. Living in the shadow of an active volcano has manifest advantages, but is inextricably linked with some measure of expected loss due to future volcanic activity. Nobody in harm's way wants to become a casualty; but, then nobody wants life to be disrupted unnecessarily. The loss to a community from its exposure to a natural hazard should be minimized. To take action now to mitigate the potential loss from a natural hazard falls within a common important category of economic decisions: pay a sum now to avert paying a larger sum later, contingent on the occurrence of an uncertain hazard event. The economic character of this class of decisions is exemplified by the basic cost-loss model.

Consider a hazardous situation where a decision-maker has to choose between two actions: either protect or do not protect. The cost of protection is C. In the absence of protection, the decision-maker incurs a loss L, which exceeds C if an adverse hazard state arises. The time interval between the act of protection and the occurrence of the adverse hazard state is assumed to be sufficiently short that financial discounting is negligible. Let the probability of the adverse hazard state arising, within a specified time window, be denoted by p. This can be evaluated within a Bayesian framework (Selva et al., 2015) using probabilistic event trees (Newhall and Pallister, 2015). If the expected expense is to be minimized, then the optimal policy is to protect, if p exceeds C/L, but not to protect if p is less than C/L. The minimal expense is then $\min\{C,pL\}$.

For an evacuation decision, the principal protection cost C is the economic dislocation, which may last for weeks or even months. This may be estimated from GDP per capita for the country concerned. An explicit financial realization of this protection emerged in 2009 from a seismic volcanic crisis in

Saudi Arabia, which is a country of rather modest volcanic hazard—and substantial national oil wealth. On May 19, 2009, up to 30 tremors measuring between 3 and 5.4 on the Richter scale shook the Arabian Peninsula region. Al-Ais, a town 550 km northeast of Jeddah and on the fringe of a volcano field, had been suffering tremors for a month. When they attained the magnitude threshold of five for the first time, the Saudi Government moved 15,000 residents of Al-Ais and surrounding villages to the neighboring cities of Yanbu and Medina, where they stayed in apartments paid for by the state. Families also received between 2,000 and 3,000 riyals ($500–$800) in financial assistance. Three months later, after enduring further seismic alarms, the evacuees were allowed to return.

For a nonevacuation decision, the principal loss incurred is that of human life. No decision-maker wants to lose a single life from a hazard crisis. Accordingly, it is a *minimum* value of a human life that is relevant for crisis applications of CBA. No decision should be made that significantly undervalues human life. This might happen if an evacuation is called too late for people to travel through congested and gridlocked streets and find safe refuge. In particular, for citizens with mobility problems, evacuation may have to be prioritized, as perhaps it might have been for the calamitous Japanese tsunami of March 11, 2011, which disproportionately drowned the less mobile. Indeed, CBA provides a systematic framework for segmenting evacuation by geography and community, and assisting also in delineating hazard exclusion zones (Marzocchi and Woo, 2009).

In CBA applications, human loss is usually measured using the economic concept of Willingness to Pay for Life Saved (WPLS). WPLS represents the amount of money available to reduce a specific lethal risk per person. It is necessarily a bounded value because resources are finite, and risks cannot all be eliminated. In some volcanic high-risk zones, such as the red hazard zone around Vesuvius, residents may be encouraged to relocate through the offer of a resettlement allowance, which may be converted to an equivalent WPLS estimate.

Decision support might consist of bulletins being presented to the decision-maker, making explicit the trade-offs involved in an evacuation call. As with other types of safety analysis, the decision-maker can undertake or commission an independent CBA. Starting from a basic CBA calculation, the decision-maker is free to add other factors into the mix of considerations leading to the ultimate decision. One of these is comparative risk aversion to societal economic disruption as against the avoidance of danger. A delay in an evacuation call may be understood as indicating an enhanced risk aversion to societal disruption, which might eventually prove unwarranted. On the other hand, an early evacuation call would signify a heightened risk aversion to human casualties. In principle, a decision-maker's risk aversion might be formally elicited; in practice, this is difficult to achieve, because of the discomfort people feel at being confronted with explicit questions trading off human and economic loss.

11.4 INTERFACE BETWEEN VOLCANOLOGISTS AND DECISION-MAKERS

Whereas a civic leader is empowered with the authority, and usually endowed with experience and some expertise, to make a crucial societal decision such as a call for an evacuation, volcanologists are not. This distinction in roles governs the interface between volcanologists and decision-makers in times of volcanic crisis. Because volcanologists are not accountable to the public in the way that public officials are, their own personal biases in decision-making should not be permitted to affect the advice they offer. As a professional group motivated to advance their science through acquiring data even from precarious positions of personal danger, volcanologists tend to be prepared, even willing, to expose themselves more to volcanic danger than would be acceptable for the general public.

Sometimes a volcanologist may be asked what he or she would do as the authorized and empowered decision-maker. Such a hypothetical question is of interest, but not that relevant to a real crisis, because it cannot be answered without imposing the volcanologist's own sense of risk aversion to false positives and false negatives. Any decision bears the psychological imprint of the decision-maker. An intrinsic part of a rigorous scientific training is an injunction against jumping to conclusions or disseminating results unless the scientist is pretty confident that they are correct. This is an essential component of sound scientific practice; if insufficient evidence exists to support a conjecture then diligence and patience are needed in gathering further evidence. Accordingly, a scientist will typically favor delay in passing judgment to allow accumulation of the additional data needed to provide extra clarification. The possibility that the hazard level might be rising inexorably in the meantime may not be perceived as much a concern to the scientist as it would be to the more risk-averse citizens in harm's way. An individualized CBA for a scientist amid crisis uncertainty might recognize a very high scientific reputational cost in a false positive, but a much lower personal reputational cost for a false negative, which might be robustly excused by the lack of sufficient scientific evidence of an imminent eruption.

Regrettably but realistically, sound scientific protocol may be an unsound practice in crisis risk management. This dilemma may not be appreciated by many scientists, since few receive any training in risk management or work regularly with professional risk managers. If evacuation times are protracted, then evacuation calls may have to be made well in advance of scientific assurance that a dangerous hazard event will materialize. Scientific procrastination may be far from optimal from a risk management perspective. However, from the perspective of an individual scientist or group of scientists, concerned about peer scientific criticism of a false positive alarm, delay to acquire more scientific information may be perceived as optimal, weighing personal costs and benefits.

Across a range of public policy safety issues, a marked regulatory asymmetry may exist in respect of the personal consequences to the decision-makers and their advisers of false positive and false negative errors. Quite apart from the predicament of a scientist with his reputation to consider, a politician may find himself in a quandary; if he is convinced there will be mass public hostility over a false positive error. This may happen in cultural circumstances where a false negative error may be readily tolerated by "an act of God" excuse, but a false positive error may cause an angry mob to descend upon a politician, metaphorically and even physically. In volcanic crises, as in other areas of public policy, personal self-interest and political survival may outweigh the greater public interest.

One of the practical merits of the CBA framework is the possibility of making more explicit, auditable, and transparent the complex basis underlying decision-making in a volcanic crisis. This transparency in itself would serve to assist those in the decision-making process in deflecting criticism for both false positive and false negative errors. Decisions should be made in the best interests of the population served by the decision-makers.

11.4.1 CBA Segmentation

Although it is always simpler for a decision-maker to treat the population at risk as homogeneous, the best interests of the population may be quite heterogeneous among subgroups, and merit a specific assessment of CBA segmentation. Area by area and community by community, calculations can be done to weigh the economic costs of evacuation against the human benefits of avoiding casualties. Regional geographical segmentation makes obvious sense, because of the spatial inhomogeneity of volcanic risk from pyroclastic, lava, and mud flows. But further segmentation by community may also make good sense.

The relevance of community segmentation is that the loss/cost (*L/C*) ratio of risk mitigation may vary significantly from one community to another. Most people would be willing to forego up to several months of earning a livelihood to save their own lives from some imminent, significant, and potentially lethal threat. Furthermore, this willingness to pay to save a life rises with an individual's sense of risk aversion.

Although the *L/C* ratio is typically large for volcanic crises, there are some people for whom the downside uncertainty of the evacuation cost is so large as to present a major psychological barrier to evacuation. Consider livestock farmers. A prolonged period of evacuation might result not in just a brief disruption of their livelihood, but possibly in a long-term decline in standard of living or even loss of livelihood, if farm animals are injured, die, escape, or are stolen. For such a subpopulation, the eventual cost of evacuation may be a very significant proportion of life earnings, which is one measure of WPSL. Four months loss of earnings as a proportion of income over a 40-year working life is only 1/120. However, the cost of restocking a farm, if all animals were lost, might be a much larger proportion of the total lifetime income of a farmer.

Consider the following example, taken from the volcanic territory of East Java, Indonesia. More than 90 percent of beef cattle production in Indonesia is derived from smallholder cattle operations, commonly with only two to three cattle per household (Privanti et al., 2012). For such a smallholding, each head of cattle represents a significant proportion of the household capital. Adding another head of cattle might require several years of farm labor. So if during an evacuation of several months all the cattle were lost through accident, injury, or rustling, the loss as a proportion of lifetime earnings might be as high as a quarter. Accordingly, the cattle farmer might estimate C/L to be as high as a quarter.

Given this financial perspective, even if a cattle farmer perceived fully the imminence of a volcanic eruption; indeed, even if a volcanic eruption were almost certain, i.e., p is perceived to be close to unity, the cattle farmer might prefer to take the gamble that, conditional on an eruption occurring, the chance of dying as a consequence is not inordinately high. Accordingly, some farmers might be prepared, quite rationally, to accept a conditional risk of fatality as high as 25 percent in order to safeguard their precious livelihoods. For very poor rural communities, destitution and death may not be so far apart. Correspondingly, if the likelihood of a major eruption is not almost certain, but still very high, some farmers might be prepared to accept a conditional risk of fatality, which is even higher than 25 percent.

This type of human response to the serious threat of financial loss has parallels elsewhere. It is witnessed in the bold, apparently foolhardy, defiance of unarmed individuals confronted with an armed assault. Criminological analysis indicates that the U.S. lethality rate from a firearm assault is only about 5 percent, allowing for firearm malfunction, shooting inaccuracy, reluctance to fire, and the firearm not being loaded (Ellifritz, 2012). Even so, most people would not resist; but some may resist and risk being shot.

Taking another example, owners of wooden homes built in forest areas may be prepared to risk staying and protecting their homes from wildfire to the maximum extent. Once evacuated, they may then be forced to rely on the bravery of firefighters to try to save their homes. The tragic deaths of 19 firefighters in Arizona in July 2013 led a pyrogeographer (Kolden, 2013) to argue that nobody should die for a house. Mitigation measures for fire risk, such as using fire-resistant materials and vegetation clearing, should be mandatory, whereas wildfire insurance would alter the cost–benefit balance against putting lives at risk.

If farmers do not have high confidence that the benefits of evacuation outweigh the costs, they may be very reluctant to evacuate unless some guarantees can be made over animal protection and possible compensation for loss. Similarly, residents afraid that their homes would be ransacked and looted if vacated, might rationally remain unless adequately satisfied with security measures taken to preserve law and order during an evacuation.

Decision-makers may have to come to terms with the awkward fact that, without security assurances, for a segment of the population at risk, a gamble against becoming victims of an eruption may well be rational and aligned with their own best economic interests.

11.4.2 Treating Epistemic Uncertainty

The results of CBA calculations may be summarized for the decision-maker in a dynamic regional map showing areas and population groups at a particular risk, such that the lives of those remaining there would be at an excessive level of jeopardy, according to the CBA criterion. CBA maps are a tangible quantitative expression of the interface between volcanologists and decision-makers. But as with all hazard maps, some residual epistemic uncertainty exists over the delineation of the zone boundaries.

At any time during a volcanic crisis, the expected L/C ratio can be evaluated for any segment of the population at risk: $R=pL/C$. Both p and the L/C ratio are subject to epistemic uncertainty, which may be represented by parametric probability distributions. Recognizing these epistemic uncertainties, levels of confidence can be associated with R, which is not a simple crisp number.

For any community, the question can be asked: what is the degree of confidence that the safety benefits of evacuation really do exceed the economic loss costs, i.e., that R exceeds unity? Whatever the best estimate of R may be, the degree of confidence that it really does exceed unity is never total. In practice, concern over degree of confidence may only be a substantive issue if the best estimate of R is just marginally greater than unity. For a population segment for which the expected L/C ratio is close to unity, it would be up to each individual to be his or her own decision-maker, and act according to their own sense of risk aversion, economic circumstances, and available options for staying temporarily away from the volcanic hazard zone.

11.5 CONCLUSION

Human beings have evolved to balance the safety benefits of taking evasive action from lurking predators, against the economic costs of missing out on finding prey, and going hungry. People exposed to volcanic hazard would instinctively weigh the pros and cons in assessing their own choice of actions, based on the information available to them. For each individual, as for civil protection authorities, every volcanic crisis may not be just a safety crisis, but also an economic crisis. Accordingly, a key input into decision-making should be CBA, which is the staple tool for economic decision-making.

As an extension of an instinctive process, the application of formal CBA in the context of volcanic crises provides a quantitative, transparent, and auditable way of informing key decisions that might otherwise be made through the exercise of executive judgment, which is notoriously prone to cognitive bias. One of the merits of CBA is the clear scientific role of volcanologists in estimating the probability of a major eruption, outlining the various destructive scenarios that might happen, and mapping their spatial footprints.

The expertise of local social scientists is also needed to gauge the perceived evacuation costs for especially cost-averse communities, such as some farmers and resort owners, whose future livelihoods may be put in jeopardy during an evacuation. Where these evacuation costs are especially onerous, consideration should be given to providing protection guarantees or establishing some form of financial safety net to rebalance individual costs and benefits.

Volcano scientists in general are already keenly aware of the need to issue timely warnings. But without being explicitly conscious of the inherent cost–benefit dimension to decision-making, there may be no specific distinction between a villager whose trade is not much affected by evacuation and a livestock farmer whose very likelihood and economic future may be at stake.

REFERENCES

Bogacz, R., 2009. Optimal decision-making theories. In: Dreher, J.-C., Tremblay, L. (Eds.), Handbook of Reward and Decision Making. Academic Press, Burlington, MA.

Ellifritz, G., 2012. Resistance to armed crime: what does the research show? Active Response Training. http://www.activeresponsetraining.net/resistance-to-violent-crime-what-does-the-research-show.

Kolden, K.A., 2013. Arizona Fire Deaths Prove Nobody Should Die for a House. Washington Post, July 5. http://articles.washingtonpost.com/2013-07-05/opinions/40386441_1_wildfire-vegetation-fires-wildland-firefighters.

Herberholz, J., Marquart, G.D., 2012. Decision making and behavioral choice during predator avoidance. Front. Neurosci. 6, 125. http://dx.doi.org/10.3389/fnins.20012.00125.

Iyengar, S., 2010. The Art of Choosing. Hachette Book Group, New York.

Marzocchi, W., Woo, G., 2009. Principles of volcanic risk metrics: theory and the case study of Mt. Vesuvius and Campi Flegrei. J. Geophys. Res. 114, B03213.

Newhall, C.G., Pallister, J.S., 2015. Probabilistic volcanic event trees. In: Papale, P. (Ed.), Volcanic Hazards, Risks and Disasters, vol.2 - Volcanoes. Elsevier Inc.

OECD, 2006. Cost–Benefit Analysis and the Environment: Recent Developments. Report ISBN: 92-64-01004-1.

Priyanti, A., Hanifah, V.W., Mahendri, I.G.A.P., Cahyadi, F., Cramb, R.A., 2012. Small-scale beef cattle production in East Java. In: Indonesia. Proc. 56th AARES Annual Conference, Fremantle, Western Australia.

Rangel, A., Camerer, C., Montague, P.R., 2008. A framework for studying the neurobiology of value-based decision making. Nat. Rev. Neurosci. 9, 1–13.

Red Cross Federation, 2012. The long road to resilience: Impact and cost–benefit analysis of community-based disaster risk reduction in Bangladesh. IFRC Report.

Selva, J., Marzocchi, W., Sandri, L., Costa, A., 2015. Operational short-term volcanic hazard analysis: methods and perspectives. In: Papale, P. (Ed.), Volcanic Hazards, Risks and Disasters, vol.2 - Volcanoes. Elsevier Inc.

Shadlen, M.N., et al., 2007. The speed and accuracy of a simple perceptual decision: a mathematical primer. In: Doya, K., et al. (Eds.), Bayesian Brain. MIT press, Cambridge, Mass.

Todorov, E., 2007. Optimal control theory. In: Doya, K., Ishii, S., Pouget, A., Rao, R.P.N. (Eds.), Bayesian Brain. MIT Press, Cambridge, Mass.

Winson, A., Costa, F., Newhall, C., Woo, G., 2012. How Successful Are Volcano Alert Levels at Anticipating Eruptions? Cities on Volcanoes 7 Colima, Mexico.

Chapter 12

Volcanic Risks and Insurance

Anselm Smolka[1,2] and Martin Käser[1]
[1] *Munich Reinsurance Company, Munich, Germany,* [2] *now: GEM Foundation, Pavia, Italy*

ABSTRACT
The volcanic eruption of Eyjafjallajökull in spring 2010 has shown the potential consequences of volcanic eruptions in a globalized world. However, insured losses were insignificant. Nevertheless, volcanic eruptions might pose an underestimated threat to the insurance industry that requires measures to keep the insurance risk under control. In principle, the same risk evaluation processes apply as for other natural perils, where risk is a function of hazard, the vulnerability of the exposed objects, and the exposed values. Probabilistic computer models have been developed for the proper assessment of risks since the late 1980s, but are in their infancy for volcanic risk. Here, the instruments developed in the insurance industry to evaluate volcanic risk are described as they form the basis for controlling and financing future losses. Due to the variety of volcanic products and the large range of possible event sizes, volcanic risk presents a particular challenge for risk management. Two types of volcanic threats have to be considered: (1) city scenarios of local or regional extent and (2) rare extreme events with global consequences that even require a discussion on insurability as such. An urgent need exists to improve our knowledge in order to reduce the high uncertainty in actual risk evaluation to an acceptable level. For the industry, it is particularly important to achieve full transparency regarding covered and noncovered risks and to define the limits of insurability for megadisasters in a systematic manner. The Global Volcano Model makes a number of efforts in this direction.

12.1 INTRODUCTION

In spring 2010, the volcanic eruption of Eyjafjallajökull caused unprecedented air traffic chaos. Although it did not cause major direct losses, it nevertheless demonstrated how far-reaching the consequences of even a minor volcanic eruption can be in our globalized world. Air traffic over many parts of northern and central Europe was repeatedly disrupted in the weeks that followed. Although national economies incurred losses totaling billions of euros, the insurance bill was close to zero. Does this mean that volcanic eruptions are a marginal or even negligible risk for insurance? Each year 50–70 volcanoes

Volcanic Hazards, Risks, and Disasters. http://dx.doi.org/10.1016/B978-0-12-396453-3.00012-5

erupt all over the globe (Global Volcanism Program, 2013). And, with the exception of extremely rare major meteorite impacts, there are no other natural events that can devastate such wide areas with comparable intensity and suddenness as volcanic eruptions. What would have happened if Eyjafjallajökull had remained active for several months or even years, as last occurred in 1821–1823, and is not untypical for Iceland? And what would have been the outcome if 122 million tons of sulfur dioxide had been expelled into the atmosphere, as when the Laki volcano erupted on Iceland in 1783, causing global temperatures to decline for a period of several years?

12.2 INSURED LOSSES FROM VOLCANIC ERUPTIONS

Not more than a handful of volcanic eruptions have occurred in the last 40 years that caused a sizable amount of insured losses, as listed in Table 12.1 together with their total economic loss.

This is not an impressive list when compared to the series of one- and two-digit billion US$ insured losses caused by earthquakes, windstorms, and floods that we have experienced in the short period from 2010 to 2012. Nevertheless, the insurance of volcanic risk merits discussion, and this paper aims to show why.

12.3 VOLCANIC ERUPTION—AN INSURABLE RISK?

Strictly speaking, the insurability of the risk emanating from volcanic eruptions can be questioned from the perspective of insurance theory. One reason is that data and tools for assessing the volcanic risk quantitatively lag far behind those for earthquake risk, for example. In addition, the nature of volcanic risk is extremely complex. In consequence, the calculation of an insurance premium and probable maximum loss is fraught with many problems. In contrast to

TABLE 12.1 Volcanic Eruptions Since 1970 that Caused Significant Insured Losses

Location	Year	Economic Loss (US$ million)	Insured Loss (US$ million)
Mount St Helens (USA)	1980	860	31
Pinatubo (Philippines)	1991	750	70
Tavurvur, Vulcan (Papua New Guinea)	1994	300	66
Montserrat	1995–1997	200	100
Merapi (Indonesia)	2010	380	Minor

earthquakes, it is far more difficult to actually estimate volcanic hazard and to classify it, since, on one hand, eruptions are usually too infrequent for a statistical evaluation, and on the other hand, even classification by *event intensity* appears to be complex. Only recently have probabilistic tools for hazard assessment been developed (Jenkins et al., 2012a,b), whereas going beyond hazard into risk is just starting or is based on simplified ranking approaches (Magill and Blong, 2005a,b). More advanced approaches to probabilistic risk modeling have remained unpublished so far (RMS, 2014). Model results which will become available will be hard to test except for very active volcanoes because of the rarity of damaging events. Nevertheless, although risk assessment lags behind, forecasting methodologies are more advanced for volcanic crises, which paradoxically violates a basic principle of insurability, namely the unforeseeability of an event. Finally, the losses of very rare events, corresponding to global recurrence periods of 10,000 years or more, can reach dimensions that definitely exceed the limits of insurability unless countermeasures are taken.

All these considerations, however, seem to be philosophical: insurance practice has already answered the question of insurability. As a matter of fact, volcanic risk is insured around the world. It is all the more important to observe good risk management practices in order to keep the existing risk under control. The procedures are in place and can be transferred from other perils, especially earthquakes, but they have to be adapted to the following peculiar characteristics of volcanic risk: the manifold manifestations of volcanic phenomena, the extremely wide range of potential eruption sizes and durations, the various insurance-related specifics, and the different event definitions.

In the end, the individual (re)insurer must be aware of the burden it would have to bear in the event of a catastrophe and as an average over the years. It is only then that the limits of insurability can be determined. But drawing a line between what at first sight appears to be an insurable peril and what becomes an uninsurable one, once the event exceeds a certain threshold, is a problem that has not yet been considered systematically, let alone solved, by the insurance industry.

12.4 PRACTICE AND PRINCIPLES

In property insurance, coverage for volcanic risk, with few exceptions, is an automatic add-on to the earthquake coverage, which in itself, is usually not granted on a stand-alone basis, but as an additional endorsement to fire coverage. Exceptions exist to this rule, for instance, insurance pools specifically designed for natural perils only, or countries where volcanic coverage is treated separately, as in the Philippines. Such separate treatment would be desirable in principle, as differentiating in underwriting between varying degrees of exposure is only possible through a separate additional cover. In this way, motivation for prevention measures could also be enhanced by accounting for these when setting the premium. The insurer's liability in the event of volcanic eruptions should basically cover all resulting damage and consequential loss within the framework of the classes of insurance covered, in order to provide complete insurance protection on one hand, and to avoid the difficulties in classifying the cause of loss and consequential hazards of a volcanic eruption on the other. This also facilitates rapid and unbureaucratic loss settlement. Beyond property insurance, other branches potentially affected include agriculture, life, health, motor, marine, and aviation insurance.

12.5 MANAGING THE INSURANCE RISK

What are the measures and procedures to manage the insurance risk from volcanic eruptions? In fact they are no different from, and comprise the same steps as for other natural perils: hazard identification, hazard assessment, risk assessment, and finally risk financing and risk control. In insurance, risk is defined as a function of three components (Smolka, 2007): the hazard, the vulnerability of objects exposed to a hazard, and the value of the exposed objects.

The hazard is defined as the exceedance probability of an event of a specified minimum intensity, e.g., the ground shaking in the case of earthquakes. Volcanic hazard presents a peculiar challenge, as a volcanic eruption can manifest itself in many different ways, which have to be identified beforehand in a complete risk analysis: primary effects such as lava, mud, and pyroclastic flows, as well as glowing clouds and ash eruptions, including the release of toxic gases. And then the consequential effects: fire, tsunami, and even global cooling in the case of very large eruptions. As regards primary effects, it is the ashfall which is most amenable to probabilistic hazard assessment analogous to earthquakes (Jenkins et al., 2012a; Magill and Blong, 2005a). For other phenomena, often their historical or hypothetical extension can be mapped, whereas a probabilistic quantification seems ambitious.

The vulnerability is expressed as the expected average loss, as a percentage of the replacement value, and varies significantly for different types of volcanic activity. An interesting study has been performed for the last Merapi eruption of 2010 (Jenkins et al., 2013). As natural disasters of recent years, and especially the Canterbury earthquake series of 2010 and 2011 in New Zealand have shown, the concept of *replacement value* has to be clearly defined and is far from being trivial. The key question being, if replacement is meant as reinstating to the condition ex ante or to an improved condition ex post.

In recent years, resilience has been introduced as a fourth component in a more general risk context. Resilience extends the concept of physical vulnerability to the overall capability of a region or society to withstand a disaster and to cope with its consequences. Resilience can be described in the form of various socioeconomic indices and their combination (Cardona, 2005; Khazai et al., 2013) and though it is of particular relevance to societal volcanic risk in general, it can also have a positive impact on material (insured) loss reduction.

Risk control is the next step after identifying and evaluating the risks. In a broad sense, risk can be controlled by land use and building regulations and by disaster preparedness measures that enhance resilience. In terms of loss of life, early warning plays a particularly important role for volcanic eruptions, as forecasting methods are much more advanced for volcanism as opposed to earthquakes. As regards the financial dimension of losses and their insurance, the mitigating role of early warning is limited, however, except for movable objects and the possibility of diverting lava flows away from assets at risk. The insurance industry uses various measures to keep the risk within affordable limits. These are briefly introduced below with specific reference to volcanism as appropriate:

1. Calculation and application of a risk-adequate price (i.e., the net insurance rate); as mentioned already, volcanism presents a particular challenge due to its various manifestations, the low event frequencies, and the duration of volcanic crises that has to be reflected in event definitions. Furthermore, it must be remembered that events and their ensuing losses can reach regional and even global dimensions.
2. Accumulation control (i.e., monitoring the insured liabilities within a country or region). The potential supraregional and global dimensions have to be duly taken into account, as well as the multiple insurance classes which may be affected.
3. Self-participation of the insured party in the loss; in the form of deductibles that reduce the number and amount of small insurance claims in the event of large disasters. These are particularly relevant for tephra eruptions in remote areas with little ash cover.
4. Liability limits; these reduce the exposure from very large insurance claims from the top (i.e., losses are compensated for only up to the limit).
5. Clear policy wordings in terms of event definitions and insured amounts; as trivial as this may sound, it is worth mentioning that after the Mount St Helens eruption in 1980, losses were indemnified even in the absence of specific volcanic coverage as losses caused by "explosions." Whereas the intended interpretation referred clearly to technical explosions, the policy wordings lacked such detail.
6. Exclusion of particularly exposed areas or certain hazards (e.g., mapped high-hazard zones near active volcanoes). These are comparable to riverbanks that are frequently flooded and therefore do not fulfill one of the principal preconditions of insurability, that is the lack of foreseeability of an event. And even without such a theoretical argument, premiums might become unaffordable in such highly exposed areas.
7. Improved claims settlement; this requires the formulation of contingency plans for regulating sometimes hundreds of thousands or even millions of claims in a single disaster.
8. Reinsurance and retrocession (the equivalent of reinsurance for the reinsurance sector itself); these serve to limit the exposure from the large

number of claims to be expected in great natural disasters by means of reinsurance contracts. This measure is of particular importance for volcanic risk due to the (almost) unlimited size of events. Again, a possible multicountry accumulation of losses has to be taken into account.

9. Catastrophe reserves; premiums from natural disaster insurance cannot be considered as "earned" premiums after 1 year. Great disasters happen rarely; therefore, the relevant premiums have to be set aside to form financial reserves that are then tapped when the event occurs.
10. Loss prevention; the global trend toward dramatically growing losses shows that it is essential to strengthen efforts in the areas of loss prevention and mitigation, i.e., taking proactive measures to reduce losses from future disasters by means of land-use planning, construction techniques, and contingency plans.

12.6 RATING VOLCANIC ERUPTION RISK

The question of assessing the insurance rate for the volcanic risk deserves a more detailed discussion. Two approaches are used nowadays for assessing insurance rates: (1) the actuarial approach, which is based on mathematical treatment of loss experience, (2) probabilistic risk models, where details on event frequency (hazard), vulnerability (experience-based or derived from an analytical approach), and exposed values (distribution in area and per type of asset) are combined to generate synthetic event sets. Such event sets that may contain hundreds of thousands of simulated events ideally give a complete representation of the occurrence probabilities of events of various sizes and can be presented in graphical form. An example of a so-called loss exceedance curve is shown in Figure 12.1. Such curves allow statements on the average annual loss that is equivalent to the pure technical net insurance rate, as well as on the size of rare loss events equivalent to one single point in the loss distribution.

The actuarial approach has limitations in the field of natural perils because of their rare occurrence. There is no way to make meaningful extrapolations from small losses occurring, say, once in 2 or 5 years to large losses occurring once in hundreds or even thousands of years. For earthquakes, small, frequent events are totally absent in many regions, and even more so for volcanic eruptions. A further fact which hinders a direct transfer of classical rating methods to the volcanic risk is the lack of any standardized scale of event intensity corresponding to ground acceleration or macroseismic intensity that would have a relation to damage caused by volcanoes. Nor can any simple relationship be established between loss, distance, and force of the eruption.

Notwithstanding all these problems, a possible procedure for rating volcanic risk is shown below, along with a practical example. In order to arrive at a rate with a sound scientific base, very detailed documentation material on the nature of activity of the volcano involved, as well as on the topography of the local terrain, is required. Calculations can then be based on the different zones

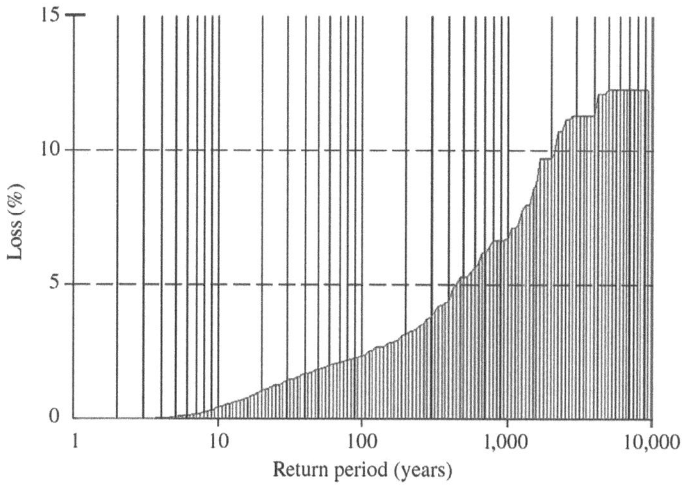

FIGURE 12.1 The loss exceedance curve shows the expected loss as a percentage of the total sum insured versus the average return period for which the particular loss is expected to occur. Such curves are typically the basis for different pricing methods to compute risk-adequate premiums in natural catastrophe insurance.

of exposure, dependent on the factors of the respective peril types and the corresponding intensity of loss. A subdivision into three zones logically suggest itself, namely

- An inner zone, where there is the threat of total destruction;
- A secondary belt, which marks the maximum extent of lava and mudflows; and
- A third zone, exposed only to ashfall.

This is the procedure commonly used to derive volcano hazard maps, making it generally possible to use existing maps to determine insurance rates. We will use Vesuvius as an example. Because of the heterogeneous nature of its activity, i.e., relatively frequent smaller eruptions with lava and mudflows, and small quantities of ash effusion, and also very rare occurrences of violent eruptions with ashfall covering broad areas, Vesuvius is quite a suitable representative example.

Based on the well-documented eruption history since 1631, and assuming a 100 percent loss ratio for areas affected by lava in an inner hazard zone of 3 km radius, an annual net loss expectancy of about 1/380=−2.6‰ can be derived empirically for lava flows alone in that zone. This rate may be roughly doubled for the effects of mudflows and tephra. Analogous considerations lead to a rate of about 0.45‰ for a zone extending from 3 to 8 km in radius, i.e., already an order of magnitude less. For more distant areas up to, say, a distance of 100 km, large eruptions like the 79 AD event have to be taken into account, which may have covered roughly 7 percent of this 100 km circle with up to

10 cm of ash. The corresponding loss ratio is highly uncertain, but assuming 10 percent and an average recurrence period of 3,000 years for such large eruptions results in an expected loss of no more than about 0.0023‰. Even in the inner hazard zones, the contribution of rare large eruptions is insignificant due to their very long recurrence periods.

In summary, and including major eruptions in the premium calculation for the most outward zone, the following net technical rates (without additional underwriting charges or capital loadings) could thus be set for Vesuvius:

- 5–6‰ for the inner zone within a radius of 3 km (zone 1; there are, in fact, few risks located here);
- 0.45‰ for the adjoining belt with a radius of 8 km, larger communities being located at the outer edge of this district (zone 2);
- 0.0023‰ for the zone which is only exposed to ashfall, with the exception of very violent events, at an approximate distance of up to 35 km (zone 3).

It can be noted, in comparison, that an annual loss of approximately 1.5‰ can be estimated for Etna within a radius of 15 km (inner zone). Hazard zone 2 has a radius of 30 km, with a loss expectancy of 0.2–0.3‰, whereas zone 3 is nonexistent, since Etna's ash production is insignificant.

The possibility of active prevention by influencing the path of lava flows artificially has not been included in this sample calculation, and neither have topography or prevailing wind directions. In the case of Vesuvius, it would be more meaningful to establish a premium for the urban district of Naples and neighboring industrial zones, instead of calculating an average rate for the entire region surrounding the volcano. Furthermore, the risk from other volcanic centers would also have to be considered, especially from the Phlegraean Fields. In essence, all these special circumstances should be incorporated into the calculation in each individual case, and lend themselves for statistical treatment in corresponding probabilistic models. As mentioned before, the construction of probabilistic insurance risk models for volcanoes only started in recent years. A principal problem for their development is the lack of a marketplace that would use such models to an extent that would justify their development economically. It is to be hoped that the Global Volcano Model (GVM) initiative (Loughlin, 2013) will provide not only data and data standards, but also a general framework that creates more risk awareness and an ensuing need for such models.

12.7 VOLCANIC ERUPTIONS—AN UNDERESTIMATED RISK?

The previous section talked about rating only. The other aspect, accumulation risk, has not been addressed at all, and it is precisely this aspect that may constitute a problem for the insurance sector. Notwithstanding the previous discussion, it is fair to say that volcanic risk so far has not been in the focus of the insurance realm. One exception may be the Caribbean island of Montserrat, which has become partially uninhabitable in the course of the eruptions of the Soufrière Hills volcano from 1995 to 1997. Other loss-producing events seemingly served more as an assurance that volcanism represents a minor insurance risk. Looking at potential future volcanic disasters, this perception may turn out to be deficient. There are two types of volcanic disasters that deserve closer inspection, not only from an insurance, but also from a general risk viewpoint:

- Local events affecting a large urban area with high ashfall, and possibly lahars or pyroclastic flows with concentrated damage in small zones.
- Global events, whose main loss agents are hemispheric ash distribution and ensuing global cooling.

12.8 LOCAL EVENTS—CITIES AT RISK

Worldwide, around 500 million people live near volcanoes, the majority of them in cities. Figure 12.2 gives a global overview of cities at risk from major ashfall. The inner circle represents the potential reach of mud and lava flows, the second circle an ash height of 1 m, and the outer circle an ash height of 50 cm as reconstructed from the Taupo 186 AD eruption (Wilson and Walker, 1985). Cities are plotted in their corresponding distance and direction from the next volcanic center. The graph is strictly deterministic, and no statements on event probabilities are made (Re, 2007). In the following section, three examples are presented that rank among the top candidates for a large loss-producing event.

The cities are shown in terms of their distance from and direction to the respective volcano located at the center of the diagram. Some cities are exposed to roughly the same threat from several volcanoes and are shown more than once. The two outer zones mark the extent of a 0.5 or 1 m covering of ash in the event of an eruption of the order of magnitude of the Krakatau eruption of 1883. The two isolines (on the right) represent the area actually affected, taking as an example the eruption of the Taupo volcano in New Zealand, which happened in around 130 AD The prevailing winds mean that the ash does not spread symmetrically in a circle. The inner zone indicates the approximate extent of the area that, depending on a volcano's activity, can be affected by phenomena like lava flows, mudflows, and glowing clouds.

- Auckland in New Zealand, for example, lies directly in a volcanic area. However, as a probabilistic risk assessment shows (DEVORA, 2013; Magill and Blong, 2005a,b), it is not the small volcanoes in the urban area that pose the main risk from ashfall but the volcanoes situated some 200 km to the southeast and 260 km to the south of the city, some of them highly explosive—such as Mount Taranaki. The main danger would come from ash deposits following an eruption.

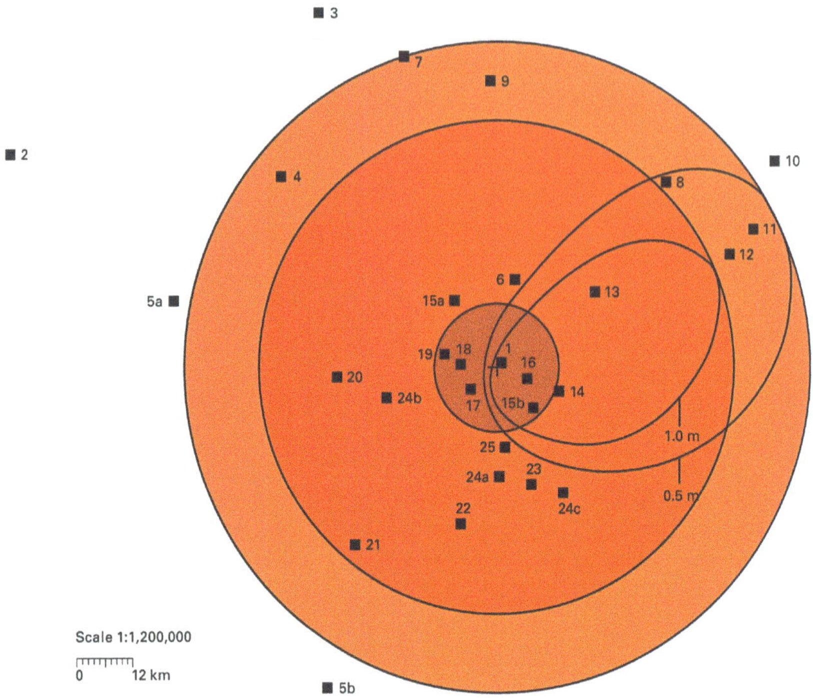

1 Auckland (Auckland volcanic field)
2 Vancouver (Mt. Baker)
3 Seattle (Mt. Rainier)
4 Tacoma (Mt. Rainier)
5 a, b Portland (Mt. Hood, Mt. St. Helens)
6 Mexico City (Xitli)
7 Surabaya (Ardjuno-Welirang)
8 Semarang (Sundoro-Sumbing)
9 Manila (Taal)
10 Tokyo (Fujiyama)
11 Kawasaki (Fujiyama)
12 Yokohama (Fujiyama)
13 Guatemala City (Agua Volcano)
14 San Salvador (San Salvador Volcano)
15 a, b Managua (Masaya-Nindiri, Apoyeque)
16 Quito (Pichincha)
17 Arequipa (El Misti)
18 Kagoshima (Sakurajima)
19 Naples (Vesuvius)
20 Nagasaki (Unzendake)
21 Shizuoka (Fujiyama)
22 Yogyakarta (Merapi)
23 Catania (Etna)
24 a, b, c San José (Barba, Irazú, Poás)
25 Bandung (Tangkuban Prahu)

FIGURE 12.2 Examples of major cities situated near to active volcanoes.

- Urban areas are also at high risk in Japan. The eruption of Mount Fuji in 1707 spread a considerable amount of ash over the Greater Tokyo region over a period of almost 2 weeks. The area in the vicinity of Mount Fuji, where 20 million people spend their holidays each year, would be particularly affected.
- The volcanic threat to the area around Naples in Italy is well known. In the past few centuries, Vesuvius has erupted on average roughly once in 30 years. If there were to be another eruption on the scale of the one that occurred in 79 BC, it is estimated that the property damage would run to tens of billions of US$ (Swiss Re, 1992), but presently the chance of such a large eruption happening seems limited, as opposed to a smaller-sized event such as that in 1631. Whether currently a serious risk exists is

disputed, and the impact of future scenarios is being investigated (DPC, 2013; Zuccaro et al., 2008). However, researchers have been getting more worried since 2001, when they discovered a 400 km^2 magma lake under Vesuvius that extends below the ground to beneath the hills of the Phlegraean Fields in northwest Naples. More than 3 million people live in this region, which means that the Phlegraean Fields probably pose a greater threat than Vesuvius itself. As for Vesuvius, smaller-sized eruptions are much more probable than the last gigantic event 39,000 years ago. They still could cause substantial damage due to the location of the volcanic center right in an urban area.

In most cases, by far the highest proportion of insured damage in urban events would come from property insurance, which might include business interruption for industrial and commercial premises. It is expected that tools being developed under the framework of the GVM initiative (Loughlin, 2013) will significantly improve the status of quantitative loss assessment for city scenarios.

12.9 GLOBAL EVENTS

An event such as the Laki eruption in 1783 would without doubt give rise to consequences extending far beyond what was observed in the year 2010 from Eyjafjallajökull. The historical consequences of the Laki event, for example, have been analyzed to show the vulnerability of our global society (Self, 2006). The other well-known case is the 1815 eruption of Mount Tambora in Indonesia (Self, 2006; Oppenheimer, 2003). Laki and Tambora represent two completely different types of volcanism—a fissure eruption of protracted duration up to several months in the case of Laki, a short-term series of cataclysmic explosions in the case of Tambora. Their common feature is the high emission of SO_2 into the atmosphere. Less well investigated is the Huaynaputina eruption of 1600 in Peru, and in the case of the fourth eruption with a strong SO_2 signal during the second millennium in 1257 or 1258, even the location of the causative volcano is not yet sure, although again Indonesia is a strong candidate (Witze, 2012).

The year following the Tambora explosion went down in history as "the year without a summer." From what we know today and on the basis of data from the past 1,000 years, an event of global impact must be expected on average at least once in every 250–300 years—and the two eruptions mentioned above, Laki and Tambora, were only 32 years apart. The consequences of an eruption on a scale similar to that of the volcano Eyjafjallajökull can still be effectively controlled through suitable technical and organizational measures. Also in the case of major eruptions, the sectors affected and possible interactions must be identified in order to establish a basis for loss prevention programs. Although the effects of the Laki and Tambora eruptions have been relatively well studied, applying this knowledge to our modern globalized world started only recently. The scientific analysis is based on three-dimensional modeling of the ash clouds, as well as on modeling the global spread of the stratospheric aerosol cloud responsible for the effects on global climate (Timmreck et al., 2010).

With global events, property damage is not necessarily the issue, as the volcanic centers of highly explosive volcanoes or fissure eruptions tend to be sufficiently distant from large cities (the Phlegraean Fields near Naples being an exception to the rule). The culprit is to be found in other sectors and, consequently, in different classes of insurance. Four damage foci can be identified beforehand:

- Aviation: Particularly in an Iceland scenario, the customary route over the North Atlantic would be more or less blocked for many months. The obvious response of rerouting flights would not work if aircraft were grounded as in spring 2010. This would have a massive impact on both the tourist industry and the manufacturing industry, according to the degree of dependence on deliveries by air freight. Presently, however, insured loss would only come from the direct loss of aircraft, and the so-called contingent business interruption, where indemnification is not conditional on physical loss at an affected site.
- Shipping: It was severely affected by "dry fog" following the Laki eruption. Today GPS could alleviate the problem to some extent. However, the possibility of signal transmission via satellite being impaired has not yet been studied. Therefore the potential consequences—insured or uninsured—are a matter of speculation.
- Agriculture: The key question is to what extent staple foods, such as rice, soya, and cereals, could withstand a lasting drop in temperature of 2–3 °C over more than one growing season with an accompanying effect on several major farming regions. The amount of direct insured losses from crop insurance depends on the structure of insurance schemes in the affected markets. At present, by far the biggest market is the United States, and insured losses there are capped for private insurers. However, a food shortage, such as that documented after Laki and Tambora, could trigger considerable social upheaval, the financial consequences of which are challenging to estimate, to say the least.
- Health risks: The dry fog containing a high percentage of sulfates that spread over the whole of Europe in 1783–1784 caused considerable damage to health. Again, financial and insured losses are difficult to capture.

12.10 CONCLUSION

Although volcanic eruptions can be considered as an insurable risk, very rare great events are a prime example of extreme natural disasters that pose a threat to the insurance industry. Some (geologically) recent examples are the Toba eruption on Sumatra 74,000 years ago, the eruption in the Phlegraean Fields 39,000 years ago, and the Oruanui eruption in New Zealand 26,500 years ago

(Mason et al., 2004; Self, 2006). The inventory of great volcanic eruptions has grown (Crosweller et al., 2012; Deligne et al., 2010; Mason et al., 2004), but still it can hardly be considered as complete. The GVM project (Loughlin, 2013) has been set up in order to achieve a quantum leap in reducing the risk from volcanic eruptions of all sizes on a global scale by coordinating existing global and regional initiatives and addressing many components of risk assessment and management, including the forecasting of eruptions. But more geological and geophysical fieldwork is needed in order to improve the existing inventories of great eruptions.

The insurance industry's instruments for risk control for more-frequent disasters such as earthquakes, windstorms, or floods are also suited to handle extreme events, but with somewhat differing weightings regarding the various possible choices. This implies the need for exact knowledge of what is covered under existing insurance contracts. However trivial this may sound, it is very important because unclear and misleading terms are not unusual in insurance policy wordings, as the misinterpretation of the term "explosion" after the Mount St Helens eruption in 1980 has shown. As to calculating technical rates, extreme events are so rare that calculating an insurance rate ends up with values close to zero. They are, however, the ultimate expression of low-probability/high-consequence events, and it is therefore absolutely imperative to limit the liability arising from them. A blueprint for such limitation of liability may be found in industrial earthquake insurance in Japan. For specific earthquake zones of high exposure, a limit of indemnity of 30 percent, or in the Tokyo zone, even 15 percent of the loss has been introduced, and in many policies for industrial conglomerates, an even lower indemnity limit of a few percent of the total sum insured comes into effect. Such drastic limitations can prevent the capacity of insurance markets in highly exposed areas from becoming inadequate for the coverage of all risks and can also prevent the limits of insurability from being overstepped. An additional challenge posed by volcanic risk, however, is the cross-border and sometimes even global dimension of very rare large eruptions. In addition, the impact of extreme disasters on the capital markets needs to be considered and planned for, because it reduces the value of the assets and the capital strength of insurance companies. The combined effect of high material losses and reduced asset values was seen after the terrorist attack of September 11, 2001 in New York; although, there was no causal connection between the two phenomena in this case. A systematic discussion on how the insurance sector should handle the risk of great volcanic disasters has not yet taken place, and is indeed overdue.

REFERENCES

Cardona, O.D., 2005. A system of indicators for disaster risk management in the Americas. In: Proc. 250th Anniversary of the 1755 Lisbon Earthquake. LNEC, pp. 80–89.

Crosweller, H.S., Arora, B., Brown, S.K., Cottrell, E., Deligne, N.I., Ortiz Guerrero, N., Hobbs, L., Kiyosugi, K., Loughlin, S.C., Lowndes, J., Nayembil, M., Siebert, L., Sparks, R.S.J.,

Takarada, S., Venzke, E., 2012. Global database on large magnitude explosive volcanic eruptions (LaMEVE). J. Appl. Volcanol. 1 (4), 1–13.

Deligne, N.I., Coles, S.G., Sparks, R.S.J., 2010. Recurrence rates of large explosive volcanic eruptions. J. Geophys. Res. 115, B06203.

DEVORA, 2013. Determining Volcanic Risk in Auckland. Retrieved September 18, 2013, from. http://www.iese.co.nz/our-research/volcanology/devora.

DPC, 2013. Dipartimento Della Protezione Civile. Retrieved September 18, 2013, from. http://www.protezionecivile.gov.it/.

Global Volcanism Program, 2013. Frequently Asked Questions. Retrieved September 18, 2013, from. http://www.volcano.si.edu/faq.cfm.

Jenkins, S.F., Magill, C.R., McAneney, K.J., Blong, R.J., 2012a. Regional ash fall hazard I: a probabilistic assessment methodology. Bull. Volcanol. 74 (7), 1699–1712.

Jenkins, S.F., McAneney, K.J., Magill, C.R., Blong, R.J., 2012b. Regional ash fall hazard II: Asia-Pacific modelling results and implications. Bull. Volcanol. 74 (7), 1713–1727.

Jenkins, S.F., Komorowski, J.-C., Baxter, P., Spence, R., Picquout, A., Lavigne, F., Surono, 2013. The Merapi 2010 eruption: an interdisciplinary impact assessment methodology for studying pyroclastic density current dynamics. J. Volcanol. Geotherm. Res. 261, 316–329.

Khazai, B., Merz, M., Schulz, C., Borst, D., 2013. An integrated indicator framework for spatial assessment of industrial and social vulnerability to indirect disaster losses. Nat. Hazards 67 (2), 145–167.

Loughlin, S., 2013. Global Volcano Model: Progress towards an International Coordinated Network for Volcanic Hazard and Risk. EGU General Assembly 2013. Geophysical Research Abstracts, 15, EGU 2013–9345.

Magill, C.R., Blong, R.J., 2005a. Volcanic risk ranking for Auckland, New Zealand. I. Methodology and hazard investigation. Bull. Volcanol. 67 (4), 331–339.

Magill, C.R., Blong, R.J., 2005b. Volcanic risk ranking for Auckland, New Zealand. II. Hazard consequences and risk calculation. Bull. Volcanol. 67 (4), 340–349.

Mason, B.G., Pyle, D.M., Oppenheimer, C., 2004. The size and frequency of the largest explosive eruptions on Earth. Bull. Volcanol. 66, 735–748.

Oppenheimer, C., 2003. Climatic, environmental and human consequences of the largest known historic eruption: Tambora volcano (Indonesia) 1815. Prog. Phys. Geog. 27, 230–259.

Swiss Re, 1992. Earthquakes and Volcanic Eruptions: A Handbook on Risk Assessment. Swiss Reinsurance Company, Zurich, Switzerland.

Munich Re, 2007. Schadenspiegel: Special Feature Issue, Risk Factor of Earth. Munich, Germany.

RMS, 2014. Oral Communication G.Woo.

Self, S., 2006. The effects and consequences of very large explosive volcanic eruptions. Phil. Trans. R. Soc. A 364, 2073–2097.

Smolka, 2007. Natural disasters and the challenge of extreme events: risk management from an insurance perspective. Phil. Trans. R. Soc. A 364, 2147–2165.

Timmreck, C., Graf, H.F., Lorenz, S.J., Niemeier, U., Zanchettin, D., Matei, D., Jungclaus, J.H., Crowley, T.J., 2010. Aerosol size confines climate response to volcanic super-eruptions. Geophys. Res. Lett. 37, L24705.

Wilson, C.J.N., Walker, G.P., 1985. The taupo eruption, New Zealand I. General aspects. Phil. Trans. R. Soc. Lond. A 314 (1529), 199–228.

Witze, A., 2012. 13th century volcano mystery may be solved. Sci. News 182 (1), 12.

Zuccaro, G., Cacace, F., Spence, R.J.S., Baxter, P.J., 2008. Impact of explosive eruption scenarios at Vesuvius. J. Volcanol. Geotherm. Res. 178 (3), 416–453.

Chapter 13

Extreme Volcanic Risks 1: Mexico City

Hugo Delgado Granados[1] and Susanna Jenkins[2]

[1] *Departamento de Vulcanología, Instituto de Geofísica, Universidad Nacional Autónoma de México, México,* [2] *Department of Earth Sciences, University of Bristol, Wills Memorial Building, Queens Road, Bristol, UK*

ABSTRACT

The Mexico City (MC) area lacks an integrated volcanic hazard assessment. Here, we discuss the volcanic setting of MC and previous studies to quantify volcanic hazard in the MC area. In identifying possible sources for future volcanic ash falls in MC, we consider the eruptive activity of 13 central volcanoes and two monogenetic volcanic fields. The prevalent wind regimes in the region are accounted for using a multisource probabilistic approach. This assessment shows the probability of ashfall distribution in MC that may affect critical infrastructure (i.e., hospitals, airport). We also discuss false alarm reports (based on events like heating and deformation of the ground, earthquake swarms, and anthropogenic lava flows) related to the possible appearance of a volcano in southern MC. Because MC is a place where more than 20 million inhabitants are exposed to volcanic hazards, it is mandatory to pursue more stratigraphic studies in MC in order to improve probabilistic evaluations of vent-opening location and occurrence. Local authorities should implement a large education program to prepare the population against eruption contingencies.

13.1 MEXICO CITY AND THE METROPOLITAN ZONE OF THE VALLEY OF MEXICO

Mexico City (MC) is located in the central part of Mexico (Figure 13.1). It is also known as *Distrito Federal* (DF; Federal District) and is identified as the capital of the country with about 8,851,080 inhabitants (INEGI, 2011), thus considered the third largest human settlement worldwide. DF is not a part of any of the 31 states that constitute the country (*Estados Unidos Mexicanos* or Mexico), but it is the largest urban area and the principal political, academic, economic, financial, business, and cultural center. Even though the boundaries of DF are well defined, due to its location within a basin, where it shares the

Volcanic Hazards, Risks, and Disasters. http://dx.doi.org/10.1016/B978-0-12-396453-3.00013-7

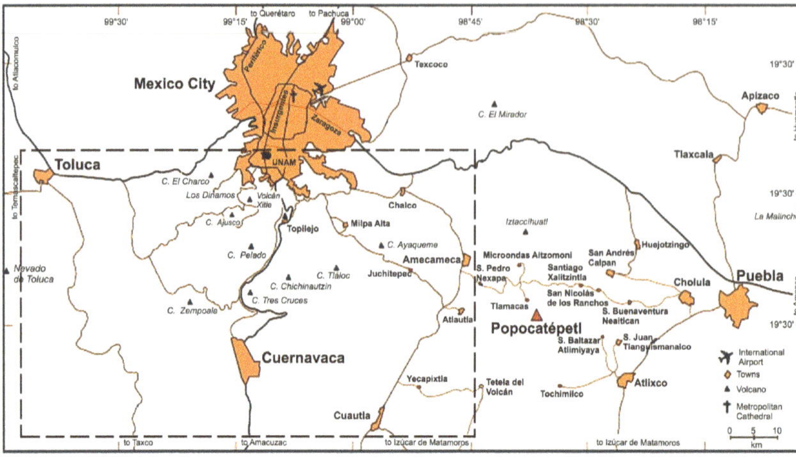

FIGURE 13.1 Mexico City surrounded by volcanoes. Key infrastructure such as the MC airport and the National University can be seen. Dashed line shows the location of map shown in Figure 13.3. Red triangle is Popocatépetl volcano.

geology and the typical problems of a densely populated area, the metropolitan area of MC extends its limits beyond DF to include 60 municipalities of the *Estado de México* (State of Mexico) and one of the state of Hidalgo. This large human conglomerate is called the *Zona Metropolitana del Valle de México* (ZMVM; Metropolitan Zone of the Valley of Mexico) and has about 20,116,842 inhabitants (INEGI, 2011). Because geological hazards and, particularly, volcanic hazards may have an impact in the entire ZMVM, from now on, when referring to MC, the reference actually will be given to the entire ZMVM.

MC is located in the central part of the Trans-Mexican Volcanic Belt (TMVB; Figure 13.2). The origin of the structure and volcanism in this part of the province is linked to the subduction of the Cocos plate beneath the North American plate; and the geodynamic parameters resulting from this relationship indicate an extensional tectonic regime (Gómez-Tuena et al., 2005). The basin of Mexico is a tectonic depression greater than 2 km deep (Alaniz Alvarez and Nieto Samaniego, 2005) limited at the south by La Pera fault system (LPFS) (Delgado Granados et al., 1995), which controls in part the distribution of volcanism of the Chichinautzin monogenetic volcanic field.

MC is bounded by several volcanic ranges: the Sierra de Las Cruces to the west, Sierra Chichinautzin to the south, Sierra Nevada to the east, and Sierra de Pachuca to the north, although the Sierra de Guadalupe rises at the northern limits of Distrito Federal.

Several volcanoes in the region have had eruptive activity in the short- (years to tens of years), mid- (tens to hundreds of years), and long-term (hundreds to thousands of years). The geological evidence indicates that some stratovolcanoes have deposited ash in MC at several times during the

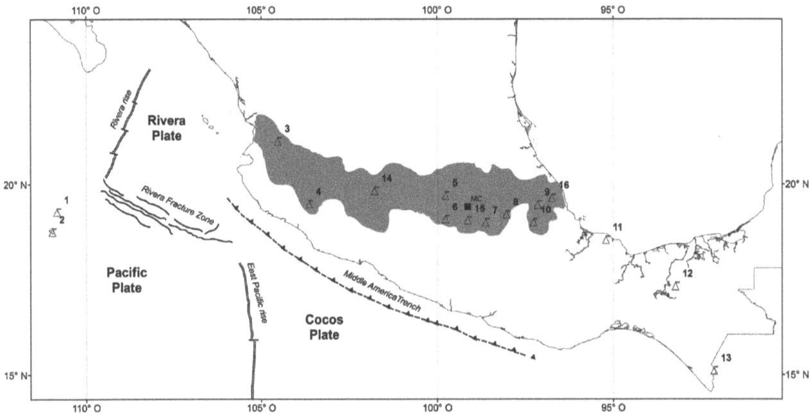

FIGURE 13.2 Location map of Mexico City (MC) in the context of the Tran-Mexican Volcanic Belt and regional tectonics. Volcanoes: (1) Bárcena, (2) Evermann, (3) Ceboruco, (4) Volcán de Fuego de Colima, (5) Jocotitlán, (6) Nevado de Toluca, (7) Popocatépetl, (8) La Malinche, (9) Cofre de Perote, (10) Pico de Orizaba, (11) San Martín Tuxtla, (12) El Chichonal, and (13) Tacaná. Monogenetic volcanic fields: (14) Michoacán-Guanjuato, (15) Chichinautzin, and (16) Xalapa.

Holocene. Also, the short-lived small monogenetic volcanoes have dispersed ash into MC at various times during the same period. Because of the large concentration of population, infrastructure, and relevance of MC as the capital of the country, it is very important and timely to have a look into the state of the art in volcanic hazards assessment for MC.

Several cities in the world have performed a variety of hazards assessment in order to cope with the possibility of a disaster due to the high population concentrations. For instance, Yamamoto and Nakada (2014) evaluated the volcanic hazard in the Tokyo Bay area in terms of sector collapse and ashfall. The latter can be of large social and economic impact due to the expected thicknesses in the Tokyo metropolitan area. One of the main risks is mainly related to the electric power outage. On the other hand, the Auckland Volcanic Field, a young monogenetic area (<250,000 years) with 49 basaltic centers where the last eruption occurred 600 years ago, underlies the urban area of the largest city of New Zealand. It has been the subject of several estimates of hazards estimation and eruption forecasting using the Bayesian Event Tree (Lindsay et al., 2010; Bebbington and Cronin, 2011). So, because MC is also a densely populated area, it is crucial to revise the volcanic hazards in the region.

This chapter builds on hazard assessments made previously by discussing the eruptive activity of active volcanoes whose products are capable of reaching MC and providing a preliminary evaluation on the probability for those volcanoes that may produce a deposit of any kind in MC. Finally, a report of false alarms related to the appearance of a monogenetic volcano in southern MC is presented, as well as the methodology applied to tackle them.

13.2 VOLCANIC HAZARD ASSESSMENTS FOR MC

No integrated volcanic hazard assessment exists for MC so far. Several attempts have, however, been made to forecast the possible occurrence or location of a new eruptive event in the region using probabilistic, geological, and gas emission approaches. In the following, we summarise these attempts.

13.2.1 Early Probabilistic Approaches

13.2.1.1 A Preliminary Evaluation of the Eruption Rate in the Southern Valley of Mexico

Scandone (1979) recognized the difficulty in making hazard assessments at monogenetic volcanic fields as compared with central volcanoes where the location of the eruption center is known. However, an estimate of the probability of eruption in southern MC is approached considering the methods proposed by Wickmann (1966), taking into account the presence of Popocatépetl volcano and the eruption of Xitle volcano in southern MC. The eruption rate for the MC's surroundings was calculated with different methods: (1) using historical records of two volcanoes, (2) using geological records, and (3) calculating the eruption rate by dividing the total volume of products emitted by a single volcanic center by the number of centers in the Chichinautzin Volcanic Field (CVF). The mean eruption rate obtained by Scandone (1979) was of 10^{-3} eruptions/year (or 0.83×10^{-3} eruptions/year, the resulting average of the three methods applied). The author states that the probability of eruption is distributed over a large area of 1,900 km^2, where the highest probability of volcanic eruption in the southern Valley of Mexico lies in an E–W band through the Sierra Chichinautzin. The eruption rate calculated for Colima and Popocatépetl volcanoes was: 5×10^{-2} eruptions/year and 2.4×10^{-2} eruptions/year, respectively.

13.2.1.2 A Model for the Transportation and Deposition of Volcanic Ash in the Valley of Mexico

Sánchez-Bibriesca et al. (1989) proposed a two-dimensional mathematical model to describe the behaviour of volcanic ash in the atmosphere from a vertical column, and the concentration, dispersion and precipitation of ash from the top and bottom of the umbrella region by wind and gravitational settling. The method was applied to two hypothetical eruptions of volcanoes in the vicinity of MC in order to estimate the thickness of the resulting ash deposit. The first case was for Popocatépetl volcano, and the second for a small emerging volcano similar to Xitle volcano in the CVF.

Probabilistic methods were employed by Sánchez-Bibriesca et al. (1989) to estimate the probability of exceeding a certain thickness of ash in the MC region. The ash dispersed from Popocatépetl to MC required wind directions blowing towards the northwest and north-northwest, whereas ash in MC from a volcano such as Xitle required wind transport directed from the southwest and south-southwest. For ash coming from Popocatépetl volcano reaching MC and depositing a thickness of 0.01–1.0 m, the exceedance probability was of 75.7–0.07 × 10^{-5} within a return period of 13.2–15,297 kyr. In the case of a "new Xitle," the results were of 5.9–0.05 × 10^{-5} within a return period of 168–21,373 kyr.

According to Sánchez-Bibriesca et al. (1989), the return period for a new eruption of Popocatépetl that deposits 0.01 m (1 cm) of ash in MC is larger than 5,200 years. An eruption of a "new Xitle" however, may deposit 0.2 m or more ash in southern MC with a return period of more than 100,000 years. Based on these results the authors estimated preliminarily a remote probability for the deposit of ash in MC.

13.2.2 Geological Hazards Assessments

Scandone (1979), based on descriptions by Mooser (1969) for the Tlapacoya archaeological site in southeastern MC (who counted 12 volcanic events during the last 23,000 years), obtained an eruption rate of 0.53 × 10^{-3} eruptions/year. Similarly, Scandone (1979) used the morphological data and dating information published by Bloomfield (1975) indicating that 41 volcanic cones were formed in CVF in the last 40,000 years. This resulted in a calculated eruption rate of 1.03 × 10^{-3} eruptions/year. Taken together and considering 1,000 events in the CVF, and an age of 0.7 Myr for the volcanism in this field, Scandone (1979) estimated an eruption rate for CVF of 1.43 × 10^{-3} eruptions/year.

According to Martin del Pozzo et al. (1997), 50 volcanoes have erupted in the vicinity of MC during the Holocene. The most recent eruptions were at two vents: Chichinautzin and Xitle volcanoes, the latter being the last eruption in the region.

Siebe et al. (2004) mentioned that the recurrence interval of eruptions in the CVF is less than 1,700 years and average eruption rate for the region is 0.6 km^3/1,000 years. Delgado Granados and Villalpando Cortés (2008) considered that, if the recurrence interval is less than 1,700 years and the last eruption occurred 1670 to 1977 years ago (according to the estimates by Delgado Granados et al., 1998; Siebe, 2000), the region of the CVF would be very close to witnessing the birth of a new volcano.

Espinasa-Pereña and Delgado Granados (2013) considered that hazards in monogenetic fields such as the CVF are not restricted to the processes and events occurring during an eruption. Events may occur before and after an eruption, and they might be considered as hazardous not only to the people but also to the environment. Building expansion in inhabited areas on volcanic deposits, particularly on top of lava flows where hidden lava tubes exist, represents a hazard for the people who build or live in those places. Also, people living in these areas need an easy solution for sewage but the impact of waste disposal through lava-tube systems sets up a long-term hazard in the form of environmental pollution. The expansion of the cities toward the cave systems also affects the surrounding fauna, threatening the preservation of endangered species.

13.2.3 CO_2 Flux from Soils as an Indicator of Hazardous Zones

Forecasting future eruptions is still a hard-to-tackle issue, temporally and spatially; estimating the location for a new volcano to be born is a complex task. In contrast to large volcanoes, for which the location of their craters is known, in the case of active volcanic fields it is difficult to forecast where the next eruption may take place. Monogenetic volcanic fields are regions whose geologic setting allows the birth of new volcanoes according to an unveiled periodicity. Volcanoes are born at the intersection of faults and fractures used by the ascending magmas. At monogenetic volcanic fields, eruptive products from previous volcanoes cover the traces of faults and fractures so that their identification is difficult. Southern MC is a region likely to witness the birth of a new eruptive vent in the future, and its probability to occur increases with time. So, it is necessary to explore ways to identify the most likely locations of the next eruption for planning and prevention purposes to mitigate the effects of an eruption in the MC area. Delgado Granados and Villalpando Cortés (2008) proposed a method to identify the most tectonically active zones in the CVF that might be used by ascending magmas, through the distribution patterns of the CO_2 emission rates from soils of the region. This method has also been used in New Zealand to evaluate the activity of the monogenetic Auckland Volcanic Field (Mazot et al., 2013).

The first measurement of CO_2 in the MC region was carried out in April 2005, and two zones with high emission of CO_2 were reported (Figure 13.3). Interestingly, the zone with the highest value of CO_2 emission was a place where 6 months later the epicenter of a shallow earthquake occurred (Carlos

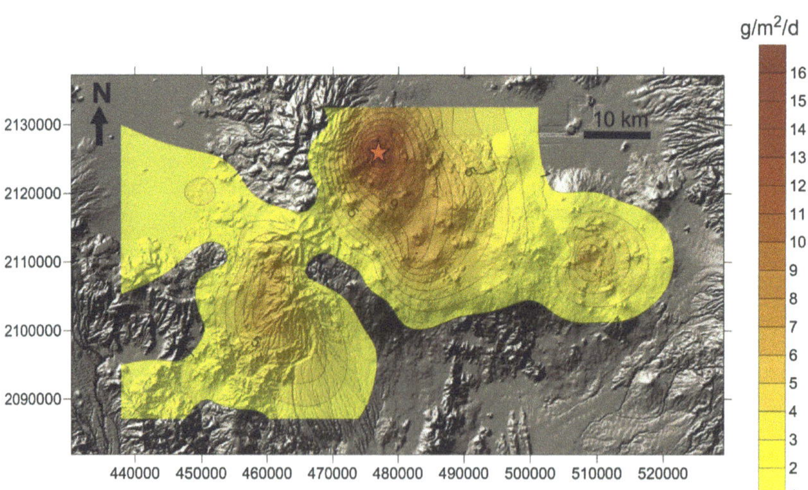

FIGURE 13.3 Contour map of the CO_2 emissions from soils in the Chichinautzin Volcanic Field in April 2005. The star shows the epicenter of an earthquake that occurred in October 2005. Units of the contours are in $g/m^2/day$.

Valdés, personal communication). These high-emission zones were associated with active tectonic features, which might act as conduits for ascending magmas, although no certainty can be claimed so far. These zones represent elements to be taken into account for mitigation planning.

13.3 POSSIBLE SOURCES FOR ASHFALL IN MC

Here, we identify polygenetic and monogenetic volcanoes that may disperse volcanic ash in MC in the long-term. This includes volcanoes that lie at distance, but upwind, from MC and those that erupt infrequently or typically with small magnitude. We consider it important to model all potential sources of ash and use a numerical model to quantify the likely range of possible deposits, and their associated probabilities, in MC. Potential sources are defined as volcanoes that have been active in the Holocene and that lie within 1,000 km of downtown MC. The very low probabilities associated with any unusual ash deposits in MC will mean that their contribution to the overall hazard is negligible but quantifying this is an important step in assessing the ashfall hazard in MC. Location of these volcanoes and the tectonic setting of TMVB can be seen in Figure 13.2 and we discuss each in turn here, distinguishing between polygenetic and monogenetic volcanoes.

Bárcena and Evermann volcanoes are located at the northern edge of the Mathematicians Ridge, an abandoned segment of the East Pacific Rise (EPR) c.3.15 Ma BP (Sclater et al., 1971; Anderson and Davis, 1973). Even though they are on an abandoned sector of the EPR, they are still active volcanoes. Ceboruco and Fuego de Colima volcanoes are the result of the subduction of the Rivera plate beneath the North American plate. Jocotitlán, Nevado de Toluca, Popocatépetl, La Malinche, Cofre de Perote, Pico de Orizaba, San Martin Tuxtla, El Chichonal, and Tacaná volcanoes, as well as the monogenetic fields of Michoacán-Guanajuato, Chichinautzin and Xalapa, are related to the subduction of Cocos plate beneath the North American plate. As seen in Figure 13.2, the TMVB is parallel to the trench in the westernmost part of the volcanic province. However, eastward from the Michoacán-Guanajuato Volcanic Field (MGVF), the TMVB lacks parallelism with the trench until the appearance of El Chichonal and Tacaná volcanoes that are parallel to the trench as well as the Central American volcanic chain. This is explained in terms of the inclination of the subducting plates (Pardo and Suárez, 1995).

13.3.1 Polygenetic Volcanoes

13.3.1.1 Bárcena (19.30 N, 110.82 W; 332 m)

San Benedicto Island is located more than 400 miles south of Cabo San Lucas, Baja California, and more than 600 km west of Manzanillo, Colima, in the Pacific Ocean. The island is approximately at the same latitude as Fuego de Colima, Parícutin, and Popocatépetl volcanoes. St Benedict is a volcanic island originated at the intersection of the Clarion and Mathematicians Ridge fracture zones and is the north-easternmost island of the Revillagigedo Archipelago. It comprises a series of Pleistocene trachytic lava domes and a cinder cone whose age is probably some hundreds of years. The birth and eruptive activity of Bárcena volcano occurred abruptly on August 1952. The eruptive activity lasted 7 months and became an advent cone of the volcanic edifice of San Benedicto Island. The 1952–1953 eruptions were characterized by strong explosions and pyroclastic flows, forming a cone greater than 100 m high that covered in part the cinder cone. The eruption ended in 1953 with the emission of two small lava domes in the crater and extrusion of a lava flow that formed a delta southeast of the cone (Richards, 1959, 1965).

13.3.1.2 Evermann (18.78 N, 110.95 W; 1,050 m)

Socorro Island is the south-easternmost island of Revillagigedo Archipelago in the Pacific Ocean. It occupies the top of a predominantly submarine basaltic shield volcano that measures 4.5×3.8 km. The origin of Socorro Island dates back to 540,000 years ago (Bohrson et al., 1996). It was built during the first phase of activity by a succession of lava flows whose accumulation ended 370,000 years ago, when large explosive activity resulted in a crater caldera. The formation of Mount Evermann, the main volcano, started 180,000 years ago and is formed by a cone and several domes. There are rhyolitic lava domes on the north, west, and southeast flanks. The volcano's growth rate has been estimated as 7×10^{-4} km^3/year. During 1848, 1896, and 1951, several eruptions were reported. Due to the remoteness of this island, the explosive activity is not described in detail. In 1951, a small phreatic eruption formed a column of blocks and steam 1,200 m altitude. In 1993–1994, it had an underwater eruption from a vent 3 miles west of the island, large blocks of more than 5 m floated at sea without associated explosive activity (Siebe et al., 1995a).

13.3.1.3 Ceboruco (21.13 N, 104.51 W; 2,280 m)

Ceboruco volcano is a ~60 km^3 volcano located in the western part of the TMVB, in a rift valley known as the Tepic graben. The eruptive history of Ceboruco is divided into three stages: (1) characterized by effusive andesitic eruptions, a Plinian eruption destroyed the volcano's summit and formed the first caldera ~1,000 years ago; (2) the second stage ended with another large Plinian eruption that formed a second caldera; and (3) emplacement of an andesitic dome in the center of the second caldera. In the last 1,000 years, eight eruptions have occurred allowing to consider an average eruption recurrence of every 125 years. The 1542 and 1567 eruptions are not well documented as in the case of the 1870–1872 explosive eruption, which produced 1.1 km^3 of dacitic lava, blanketing a large area of the western flank (Frey et al., 2004; Sieron and Siebe, 2008).

13.3.1.4 Volcán De Fuego De Colima (19.51 N 103.62 W; 3,850 m)

Volcán de Fuego de Colima is regarded as the most active volcano of Mexico. It grew within the caldera of the collapsed Nevado de Colima volcano and on its slopes there are the lava domes known as El Volcancito and Los Hijos del Volcán with heights of 120 and 300 m (Lugo Hubp et al., 1993). A historical record exists of intense eruptive activity at the volcano. This is due to the generation of short-length lava flows ejected from the main crater, and in 1869 El Volcancito (located on the northeast side of the volcano) produced block and ash pyroclastic flows, surges, ashfalls, lahars, and debris avalanches. Frequent historical eruptions have been documented. Over the past 500 years, the volcano has a frequency of explosive type activity with a number that exceeds 30 eruptions among which are those of 1585, 1606, 1622, 1690, 1818, 1869, 1890, 1903, 1913, and the most recent, which began in 1961 and continues to date (Martin del Pozzo et al., 1995; De la Cruz-Reyna, 1993).

13.3.1.5 Jocotitlán (19.72 N, 99.76 W; 3,900 m)

The Jocotitlán volcano is a dome complex built during the Pleistocene with andesitic to dacitic lava flows. The collapse of the eastern flank, ~10,000 years ago, was due to the tectonic features of the region and the steepness of its slopes. An earthquake may have triggered the collapse, because several large-scale seismic events have occurred in the region such as the Acambay earthquake in 1912. The collapsed mass covered an area of 80 km^2 and reached a distance of 12 km. The collapse was followed by an explosive event. After these events, lava domes filled the crater left by both processes. The last explosive eruption occurred ~700–900 years BP, producing block and ash pyroclastic flows and surges (Siebe et al., 1992).

13.3.1.6 Nevado De Toluca (19.11 N, 99.76 W; 4,680 m)

Nevado de Toluca is a late Pleistocene–Holocene calc-alkaline composite volcano consisting of andesites and dacites (Bloomfield and Valastro, 1974; Cantagrel et al., 1981; García- Palomo et al., 2002; Macías, 2005). The volcano experienced several sector collapses (Macías et al., 1997; Capra and Macías, 2000; Macías, 2005) and a very intense eruptive activity during the last 50,000 years. The most extensive deposits of up to 30 m thick are those of 37,000 (Garcia Palomo et al., 2002), ~24,500 (Lower Toluca Pumice; Bloomfield and Valastro, 1974, 1977), and 10,500 years BP (Upper Toluca Pumice; UTP). Of special interest is the eruption that generated the UTP deposit from four Plinian columns; two of them covered 2,000 km^2; an area currently occupied by the cities of Toluca and Mexico with a volume of 14 km^3. A dacite dome was emplaced in the crater of the volcano ~9,100 years BP (Arce et al., 2003). The last eruption took place approximately 3,300 years ago in the central crater (Macías et al., 1997), producing a pyroclastic flow and pyroclastic surge.

13.3.1.7 Popocatépetl (19.02 N, 98.62 W; 5,426 m)

At least three previous major cones to the current Popocatépetl were destroyed in the Pleistocene, producing massive debris avalanche deposits covering broad areas south of the volcano. Nexpayantla's lava flows have uniformly normal magnetic polarity (Delgado Granados et al., 1988; Conte et al., 2004), suggesting that they were erupted within the last 780,000 years. Throughout the lifetime of Nexpayantla, an estimated 500–600 km^3 of magma was extruded as lava and ash flows (Robin and Boudal, 1987). The catastrophic eruption that destroyed Nexpayantla distributed a 30 km^3 debris avalanche to the south and southeast of Popocatépetl and was accompanied by a >1 km^3 Plinian eruption (Robin and Boudal, 1987). The total mass of mobilized rock associated with this eruption was 10 times that of the 1980 Mount St Helens eruption. Remnants of the caldera wall that resulted from the catastrophic eruption of Nexpayantla are visible to the north of Popocatépetl (Espinasa-Pereña and Martin-Del Pozzo, 2006). After some time, the growth of Ventorrillo commenced within the caldera walls of Nexpayantla. Throughout its lifetime, an estimated 40 km^3 of magma was extruded as lava and ash flows (Robin and Boudal, 1987). The catastrophic eruption that destroyed Ventorrillo distributed ~9 km^3 of material as a debris avalanche to the south and southwest of Popocatépetl and a widely dispersed Plinian ash fall (Macías, 2007). Remnants of the caldera wall that resulted from the catastrophic eruption of Ventorrillo are visible to the west-northwest of Popocatépetl (Espinasa-Pereña and Martin-Del Pozzo, 2006). ^{14}C dating indicates the major, cone-destroying eruption occurred about 23,000 years ago (Siebe et al., 1995b). The current edifice has had at least seven major Plinian eruptions in this time frame (Schaaf et al., 2005). In MC, 10–20 cm isopach of fall deposits from the ~14,100 years BP Plinian eruption have been reported (Sosa-Ceballos et al., 2012). Frequent historical eruptions 1354, 1363, 1509, 1512, 1519–1530, 1539–1540, 1548, 1562–1570, 1571, 1592–1594, 1642, 1663–1665, 1697, 1720–1919, 1920–1927, December 21, 1994 began a new phase that continues to date.

13.3.1.8 La Malinche (19.23 N, 98.03 W; 4,503 m)

Much of this volcano was built during the Pleistocene. Pre-Malinche comprises most of the eruptive history, but the products of La Malinche stage cover the associated deposits. This stage started 45,000 years BP consisting of pumice fall and pyroclastic deposits, followed by the emplacement of several lava domes (Castro and Siebe, 2007). Then, widespread pumiceous fallout covered the slopes of the volcano at 21,500 years BP, and related pyroclastic flows and lahars reached large distances. A partial sector collapse occurred at 20,900 years BP, and at 15,900 years BP a pumice-and-ash flow occurred. The last eruptive period started between 12,000 and 9,000 years BP, and the most intensive episode occurred at 7,500 years BP. After a long period of quiescence, an eruption of ash fallout accompanied by small pyroclastic flows occurred 3,100 years BP.

13.3.1.9 Cofre De Perote (19.49 N, 97.15 W; 4,282 m)

This volcano is a basaltic andesite, andesitic–trachyandesitic to dacitic, massive low-angle Quaternary shieldlike structure, located at the northern part of the Cofre de Perote-chain Pico de Orizaba. This 1.3 Ma volcano was built in five stages. During the fourth stage, edifice collapses occurred at 40,000 and 11,000–13,000 years BP. During the Pleistocene-to-Holocene, in the last stage, nearly simultaneous eruption of monogenetic cones of alkaline and calc-alkaline nature occurred. A group of cinder cones located on the northeast flank and numerous lava flows reached the vicinity of the city of Xalapa, the youngest lava (El Volcancillo volcano) was dated at 870 years BP (Siebert and Carrasco-Nuñez, 2002; Carrasco-Nuñez et al., 2009). This date represents the age of a monogenetic cone though, rather than the age of lavas from the large stratovolcano, which is likely not to be an active volcano.

13.3.1.10 Pico De Orizaba (Citlaltépetl) (19.03 N, 97.27 W; 5,675 m)

Citlaltépetl (Pico de Orizaba) is an active volcano, currently in a dormant state. It was formed in three stages during the Middle Pleistocene. Construction of the Torrecillas and Espolón de Oro was contemporary with activity of Sierra Negra volcano, followed by a collapse that produced voluminous debris avalanches and lahars (Carrasco-Núñez, 2006). Modern Citlaltépetl emitted andesitic to dacitic lavas during the late Pleistocene–Holocene. During early Holocene (9,000–8,500 years BP), repetitive explosive eruptions deposited an intercalation of pumice fallout and scoria and pumice-rich pyroclastic flow deposits accompanied by the development of an outpouring of lava to produce a dome (Rossotti et al., 2006). Historical eruptions consist of moderate explosive activity and outpouring of dacitic lava flows. The most recent eruption was reported in the seventeenth century. The eruption probabilities for Citlaltépetl have been established for eruptions of volcanic explosivity index (VEI) 2 (1.28×10^{-2} eruptions/year), VEI 3 (1.72×10^{-3} eruptions/year), VEI 4 (2.3×10^{-4} eruptions/year), and VEI 5 (3×10^{-5} eruptions/year) (De la Cruz-Reyna and Carrasco-Nuñez, 2002).

13.3.1.11 San Martin Tuxtla (18.30 N, 94.73 W; 1,250 m)

More than 250 pyroclastic cones and maars of the Los Tuxtlas volcanic field surround this volcano. It is an alkalic basaltic volcano, elongated in a NW–SE direction. Two large historic eruptions were recorded in the years 1664 and 1793–1796. The eruption that started in 1793 originated in two cinder cones at the summit crater and produced a wide dispersion of ash and lava flows that spread 3.5 km on the northeast flank (Espindola et al., 2007, 2010). This eruption started as a phreatomagmatic eruption followed by Strombolian activity (Espindola et al., 2010).

13.3.1.12 El Chichonal (17.36 N, 93.23 W; 1,150 m)

El Chichonal is built on evaporites and limestones of Jurassic-Early Cretaceous, Early to Mid-Cretaceous dolomitic limestones, limestones, Tertiary sandstones, and shales (Canul and Rocha, 1981; Canul-Dzul et al., 1983; Duffield et al., 1984). The largest cone was built 220,000 years ago. It has two large concentric craters; an older dome is located to the northwest. Eruptions occurred at 550, 900, 1250, 1500, 1600, 1900, 2000, 2500, 3100, 3700, 7700 years BP, and in 1982 (Espíndola et al., 2000; Macias et al., 2008). The explosive eruption of 1982 destroyed the dome top and was accompanied by pyroclastic flows that devastated an area that stretches for 8 km. These eruptions created the new crater 1 km wide and 300 m deep.

13.3.1.13 Tacaná (15.13 N, 92.11 W; 4,060 m)

The Tacaná volcanic complex started its formation in the late Pleistocene nested within the preexisting San Rafael caldera. This volcano is composed of volcanic structures: (1) Chichuj, an andesitic volcano with pyroclastic deposits destroyed by sector collapse; (2) Tacaná, made of basaltic-andesite lava flows, andesitic and dacitic lava domes (38,000–16,000 years BP); (3) Plan de las Ardillas volcano consisted of an andesitic dome and two lava flows ~30,000 years BP; and (4) San Antonio volcano, made of lava flows, andesitic and dacitic domes destroyed by a Peléan-type eruption (1,950 years BP). The summit of Tacaná volcanic complex is elongated with an NE–SW direction, which is the direction of the Tacaná graben (García-Palomo et al., 2006). Volcanism has migrated to the southwest and a small lava dome in the crater of the San Antonio volcano is the youngest unit located on the southwest flank. Numerous viscous lava flows occur on the northern and southern flanks. The historical activity has consisted of small phreatic eruptions in 1855, 1878, 1949, 1986, and strong explosive activity in the presence of pyroclastic flows 1,950 years ago (Macías et al., 2000; Mora et al., 2004).

13.3.2 Monogenetic Volcanic Fields

13.3.2.1 Michoacán-Guanajuato Volcanic Field (19.48 N, 102.25 W; 3,860 m)

This field is a region of active monogenetic volcanism comprising an area of 40,000 km^2 (Hasenaka and Carmichael, 1985a), consisting mainly of cinder cones, shield volcanoes, stratovolcanoes, lava domes, and maars. The latest historical eruptions were Jorullo in 1759–1774 and the Paricutin in 1943–1952. This volcanism is intimately associated with the subduction of the Cocos plate beneath the North American plate, and the volcanoes are located at a distance of 190–440 km from the Middle America Trench. 75 percent of the volcanoes, however, are at 200–300 km distance, and the highest density of volcanoes is located 250 km away from the trench. The average density of volcanoes per unit area is 2.5 cones per 100 km^2 and the average spacing between the cones is 2 km (Hasenaka and Carmichael, 1985b). Between 1,000 and 2,000 Quaternary volcanoes exist in the MGVF (Connor, 1987). At least 16 volcanoes are less than 10,000 years old and around 78 are under 40,000 years old (Hasenaka and Carmichael, 1985b). The volume of these volcanoes is 0.021 km^3 on average.

In the MGVF exist several shield, and composite volcanoes as well as stratovolcanoes. Culiacán and Cerro Grande shield volcanoes yield ages of 2.1 and 2.3 Ma, respectively; the latter has a volume of 54 km^3. The shield volcanoes of this region are very different from other parts of the world due to their chemical composition, size, and origin. Paracho volcano is an example of a composite volcano ~60,000 years old, and Tancítaro volcano is an example of a stratovolcano, whose approximate age is 530,000 years BP but with Holocene eruptive activity (Ban et al., 1992; Hasenaka et al., 1994; Ownby et al., 2007; Ownby et al., 2011). The eruptive activity in this field is recurrent and further eruptions in the region are expected in the future, either by the emergence of new cinder cone, from a shield or composite volcanoes and stratovolcanoes.

13.3.2.2 Chichinautzin Volcanic Field (19.08 N, 99.13 W; 3,930 m)

The CVF is a monogenetic volcanic terrain above 3,000 m, elongated in an E–W direction, extending from the flanks of Popocatépetl volcano in the east to the slopes of Nevado de Toluca in the west. Its northern boundary encompasses parts of MC and in the south to the north of the State of Morelos. The CVF has one of the highest densities of monogenetic volcanoes of the TMVB (i.e., number of volcanoes per unit area). It incorporates at least 220 structures including Quaternary volcanic shield volcanoes, lava flows, lava and cinder cones, whose products range from basalts to dacites (Martin del Pozzo, 1982; Meriggi et al., 2008) and covers an area of approximately 2,500 km^2. Bloomfield (1975) reported a density of 0.1/km^2 cinder cones. Most of the research conducted in the CVF focused on the Xitle volcano and associated flows found in the southern part of MC (Velasco-Tapia and Verma, 2001). Xitle volcano has been studied in detail because its lavas destroyed and buried the ancient pre-Hispanic city of Cuicuilco, located 7 km NE of the cone. Here Libby (1955) dated coal found under the Xitle flows for the first time, and found that it belonged to an age of 2,422 ± 250 years BP. However, a recent compilation of radiometric data indicates that Xitle erupted about 1,670 ± 35 years BP and geological and mapping studies at Xitle reported an age of 1,977 ± 47 years BP (Delgado Granados et al., 1998; Siebe, 2000). Cuicuilco was probably abandoned as a direct result of this eruption with varied estimated ages (Siebe, 2000).

Recent studies by Siebe et al. (2004a, 2004b), Meriggi et al. (2008), Guilbaud et al. (2009) in the CVF provide a better understanding of volcanism in the region. More research is still needed to recognize the eruptive evolution of CVF and refine the eruption recurrence times.

13.3.2.3 Xalapa Volcanic Field (19.67 N, 96.75 W; 2,000 m)

Rodríguez et al. (2009) defined the Xalapa Monogenetic Volcanic Field in order to include a series of monogenetic eruptive centers included in the "Cofre de Perote Vent Cluster" (CPVC) and the "Naolinco Volcanic Field" (NVC) described by Siebert and Carrasco-Núñez (2002) as well as more than 50 scoria cones located in the surroundings of the city of Xalapa. This field extends over 2,400 km^2 and includes more than 50 Late Pleistocene scoria cones, shield volcanoes, and tuff rings of basaltic, basaltic andesitic, and andesitic nature with a calc-alkaline signature. The volcano ages are grouped in: (1) >2.0 Ma, (2) 400,000–250,000 years BP, and (3) <100,000 years BP. The youngest volcanic feature in the NVC is Coacotzintla lava flow (~2,980 years BP), and comes from the cinder cone Rincón de Chapultepec. The age of El Volcancillo at CPVC is younger though (870 years BP; Carrasco-Nuñez et al., 2005) and can be included in the Xalapa Volcanic Field (XVF) as explained above.

13.4 A MULTISOURCE, PROBABILISTIC APPROACH FOR HAZARDS ASSESSMENT

Ashfall hazard modeling typically considers a single explosive source. In the case of hazards assessment for a region such as MC, it is necessary to provide an in-depth assessment and comparison of the ash fall hazard from more than one source. A probabilistic modeling methodology developed by Jenkins et al. (2012a,b) is applied here and represents a method for the preliminary assessment of multivolcano ash fall hazard in MC. As a first draft, the method sources eruption information from the Smithsonian Institution Global Volcanism Program catalog of Holocene volcanism. As such, inaccuracies present within the database are carried through. Therefore, future studies will look to incorporating local data and expertise into modeling. We present the results here as a first-order estimate and methodology, which highlights volcanoes that may pose a threat to MC. Ashfall hazard, in terms of probable ash fall thickness and event duration, was assessed for areas surrounding the MC.

13.4.1 Wind Field

Wind patterns play an important role in the distribution of ash. Information for 20 years on wind direction and speed from the surface to >20,000 m, obtained from the Servicio Meteorológico Nacional (National Weather Service), was organized and analyzed by layers (low- and upper-altitude winds; Figures 13.4

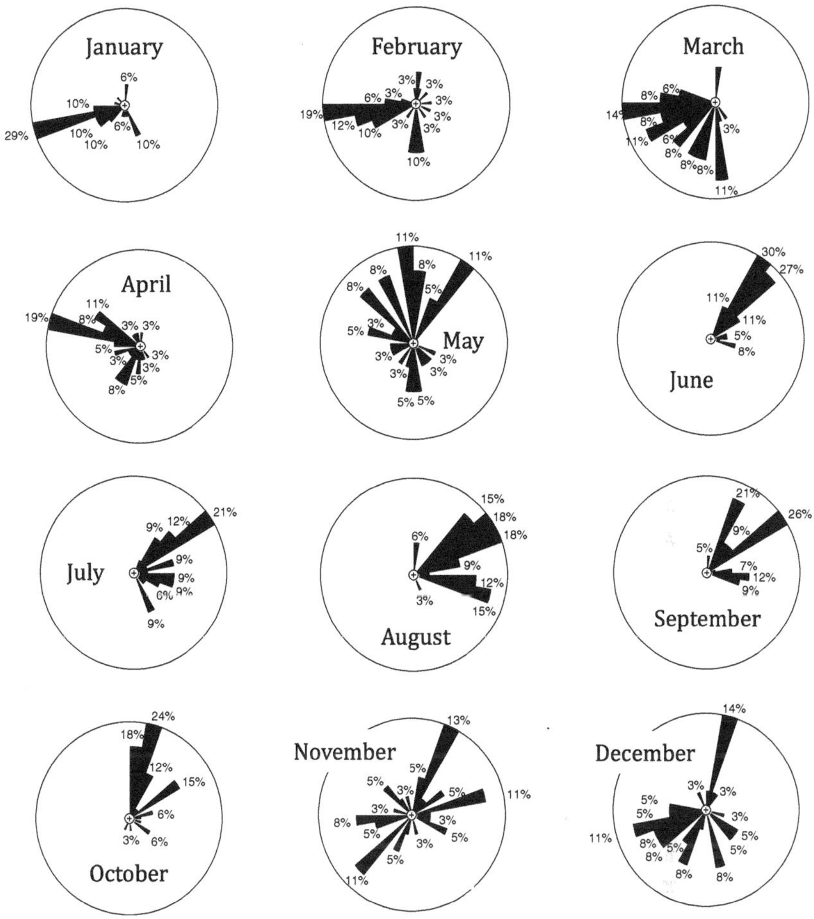

FIGURE 13.4 Lower wind patterns in the MC. Histograms show monthly winds below 5,500 m asl, indicating the frequency of occurrence at every direction in percent, each bar represents 10°.

and 13.5), strata (eight; Figure 13.6), and levels (six levels selected; Figure 13.7) (Delgado Granados, 2002) for the MC. Surface winds (below 3,000 m) have a strong N–S tendency and the average speed does not exceed 5 m/s. In some months, winds between 3,000 and 5,800 m show the same trends as the upper winds (>5,800 m), whereas other months have a scattered distribution but are transitional between surface winds and upper wind with speeds between 5 and 10 m/s. The upper winds (up to 20,000 m) between November and April blow from the west and southwest, and between June and September from the east (east-northeast, east, east-southeast). May and October are transitional months. Between April and May, winds keep the same west and southwest pattern up to 16,500 m and from this altitude and up to

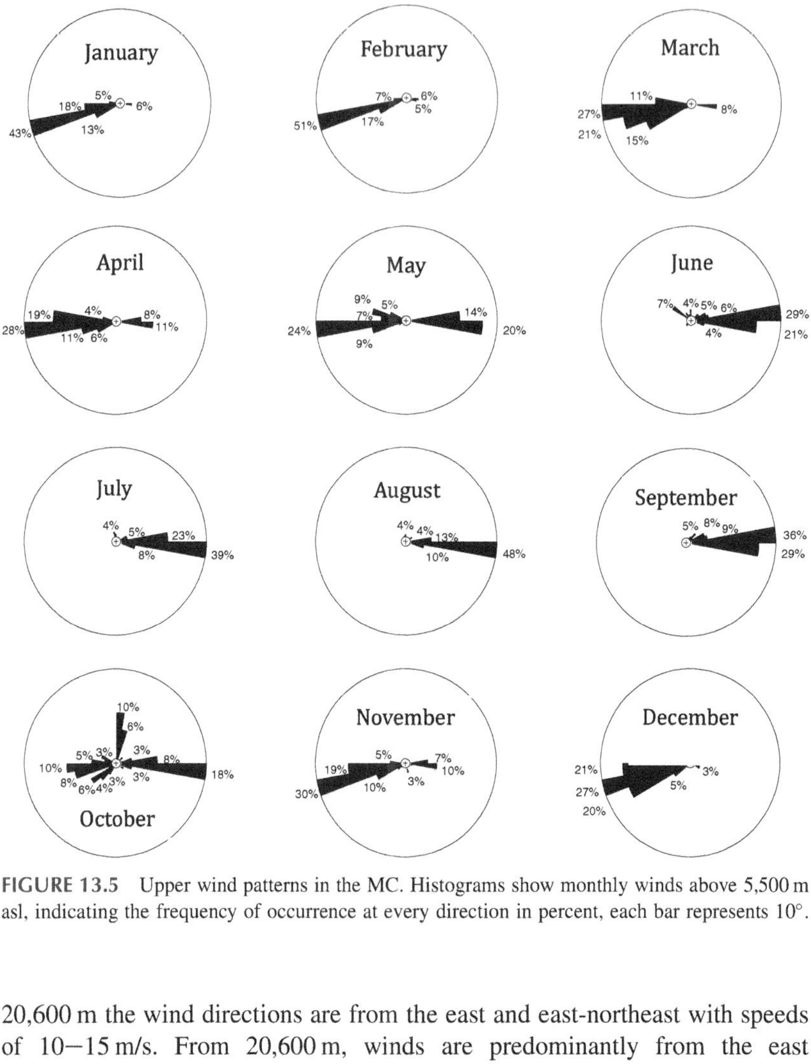

FIGURE 13.5 Upper wind patterns in the MC. Histograms show monthly winds above 5,500 m asl, indicating the frequency of occurrence at every direction in percent, each bar represents 10°.

20,600 m the wind directions are from the east and east-northeast with speeds of 10–15 m/s. From 20,600 m, winds are predominantly from the east (80–100°) between April and November. Between December and March, winds from the west are also observed and can be up to 20–30 m/s, particularly between July and September.

Available information indicates that the topography has a great influence on winds less than 5,800 m; winds above this level show more defined patterns.

13.4.2 Probabilistic Approach

The identification of defined patterns that characterize the behavior of winds in the atmosphere is of great importance to predict whether those

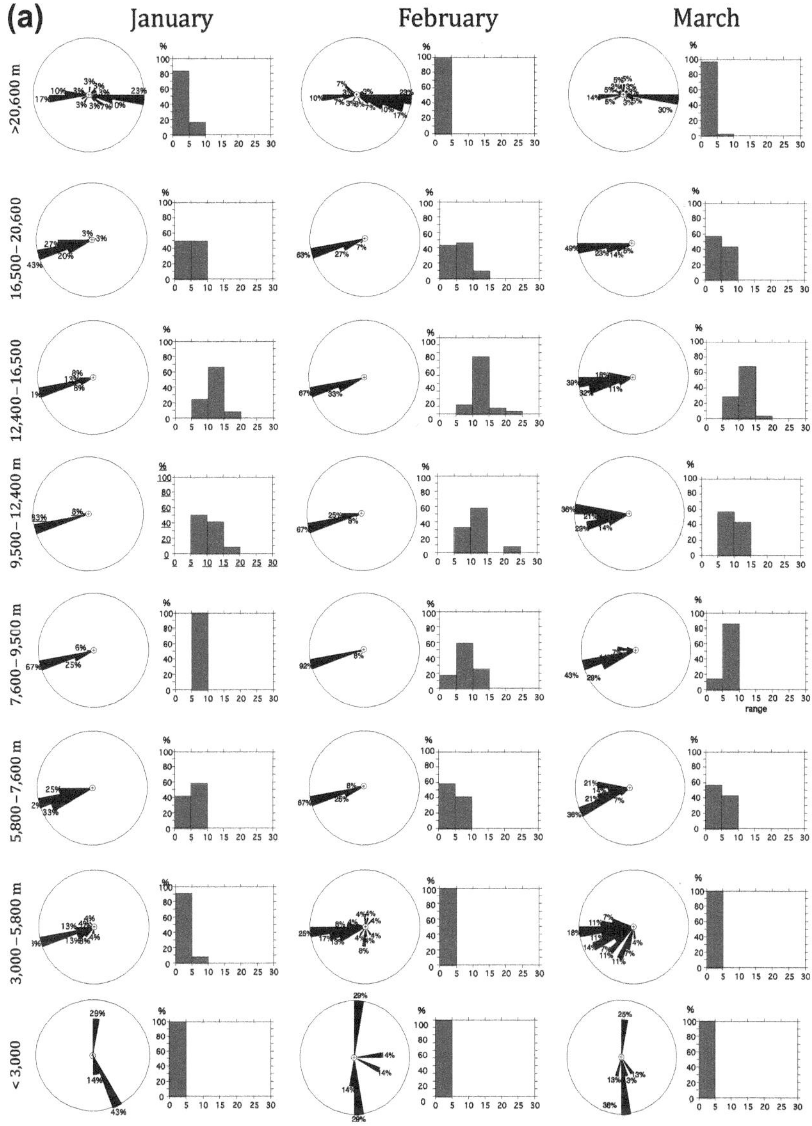

FIGURE 13.6 Stratified wind patterns in the MC. Circular histograms show wind direction indicating frequency in percent and rectangular histograms show the frequency of wind speed in m/s for eight strata. (a) Histograms from January to March, (b) April–June, (c) July–September, and (d) October–December.

likely trends continue in the future. It is possible to apply these results in evaluating hazards associated with a future eruption in the MC region; however, we assume that wind patterns will not be modified by outside atmospheric disturbances, which occur on an occasional basis. For

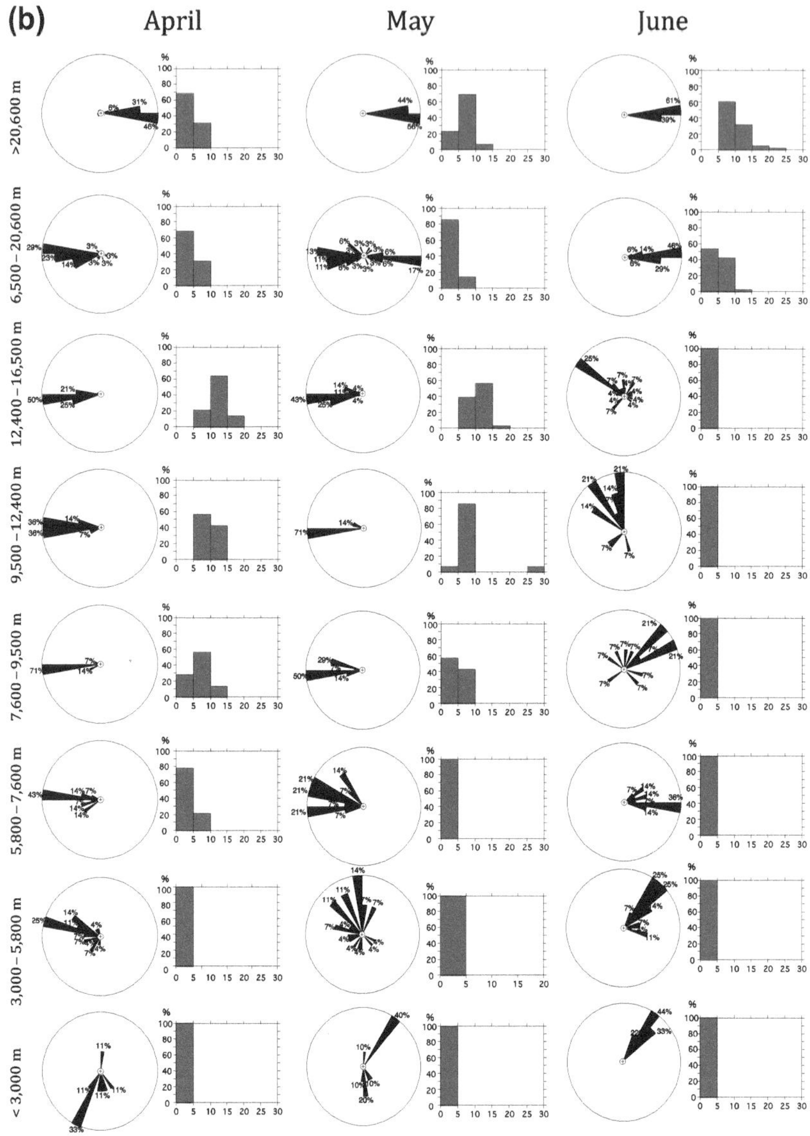

FIGURE 13.6 Continued

example, the forecast presented here may vary significantly according to changes in wind patterns due to events such as "El Niño" or "La Niña." Although we have not separated them out and may have independent assessments, they likely have been incorporated as we consider decades of wind data.

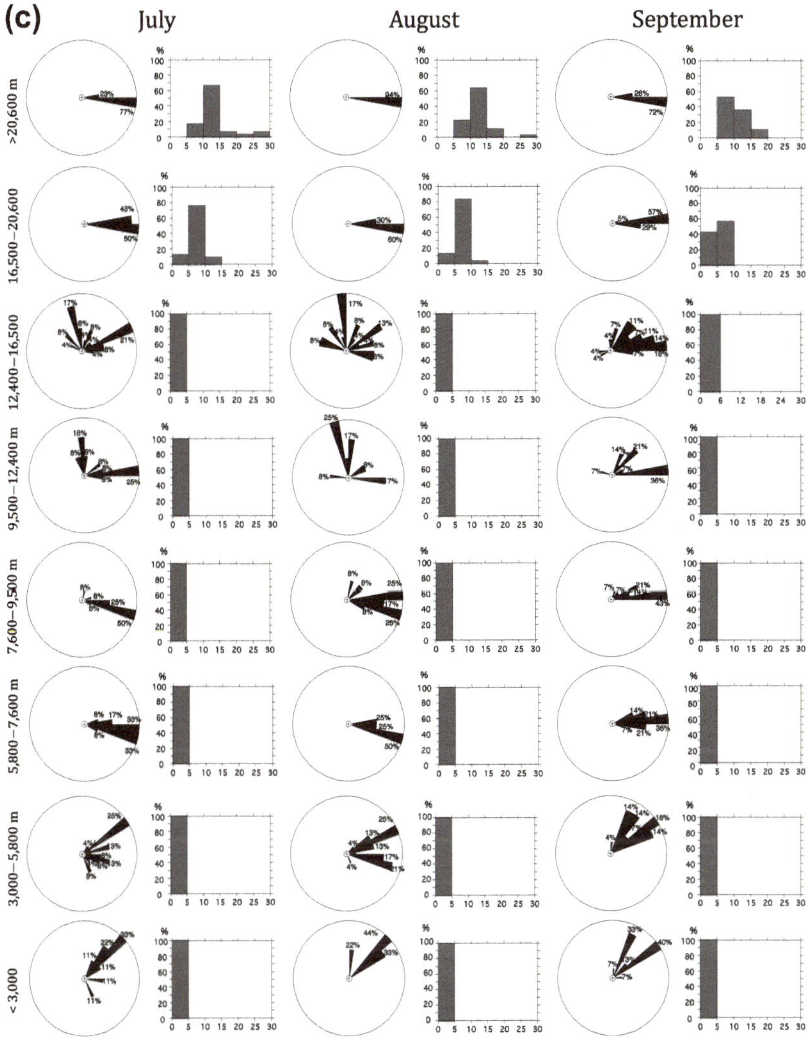

FIGURE 13.6 Continued

Ashfall hazard in MC has been estimated following the methodology laid out in Jenkins et al. (2012a). Ash hazard from large-magnitude (VEI ≥4) eruptions at multiple volcanoes has been simulated using a fully probabilistic framework to account for uncertainty in wind conditions and eruption variables. The ash dispersal model ASHFALL (Hurst, 1994) has been used to simulate more than 180,000 ashfall "footprints" from the 16 volcanoes lying within 1,000 km of MC (Table 13.1). As mentioned previously, those volcanoes that contribute little to ash hazard in MC

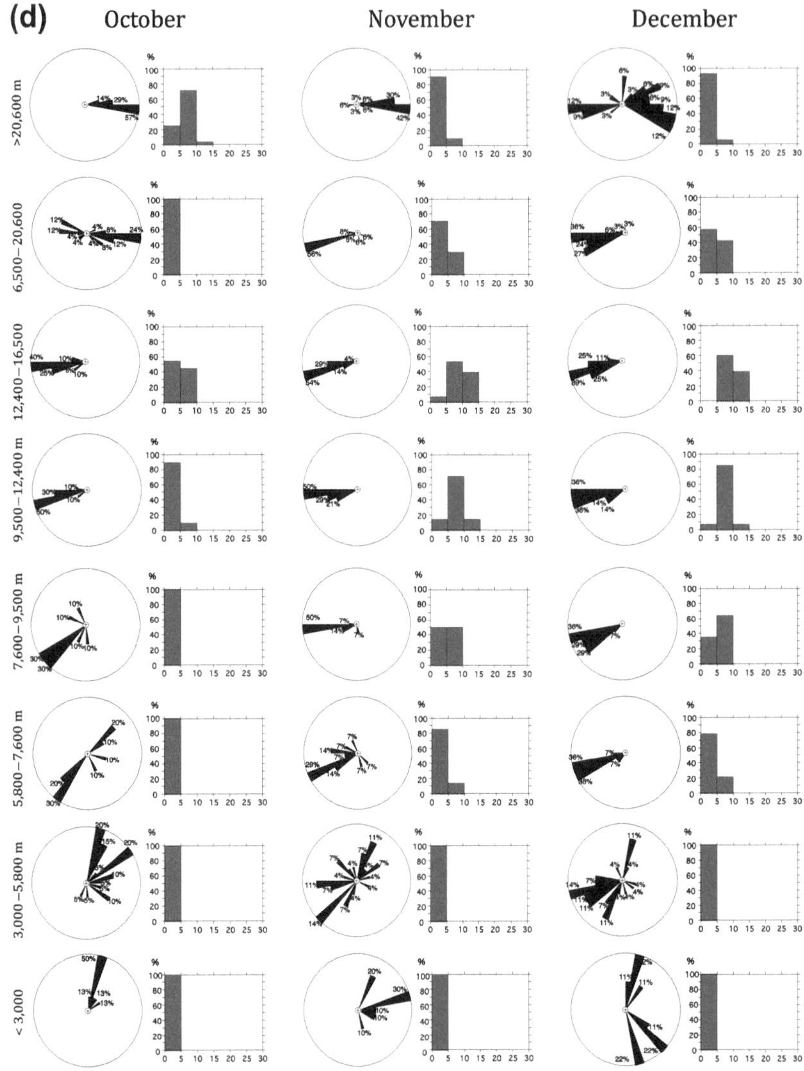

FIGURE 13.6 Continued

because of likely eruption style or distance produce low or no probability of impact in the region; however, it is important to quantify this contribution and to consider all sources preventing any subjective bias that may be present in choosing potential ash sources. Eruptions were simulated into 45 years of local wind data from the ECMWF ERA-40 dataset to incorporate the range of likely wind directions and speeds, with height above the volcano, in a future eruption. Eruption frequency–magnitude relationships

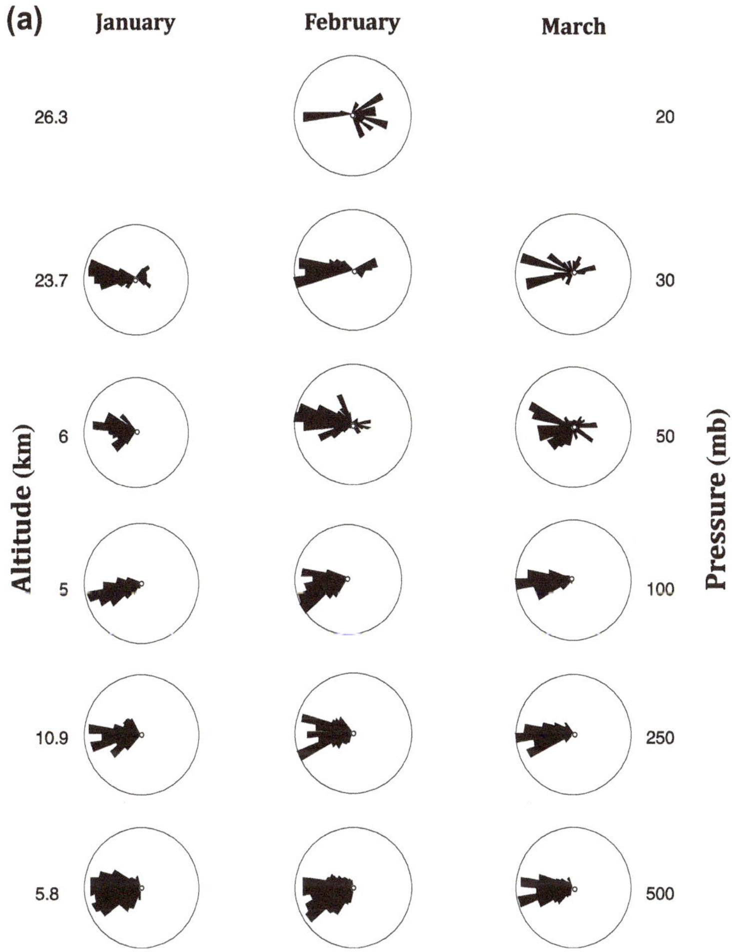

FIGURE 13.7 Wind patterns for selected levels. Histograms indicate provenance of the winds for a given level. (a) January–March, (b) April–June, (c) July–September, and (d) October–December.

(Table 13.1) have been derived from the analysis of the Smithsonian Institution VOTW 4.0 catalog (Siebert et al., 2010), after first accounting for data completeness of the eruption records (see Jenkins et al., 2012a). A limitation of this approach is the reliance upon the Global Volcanism Program catalog of Holocene activity (Siebert et al., 2010), which omits some data that are available locally. Future iterations should therefore look to revise the frequency–magnitude estimates made in Table 13.1 and remodel accordingly.

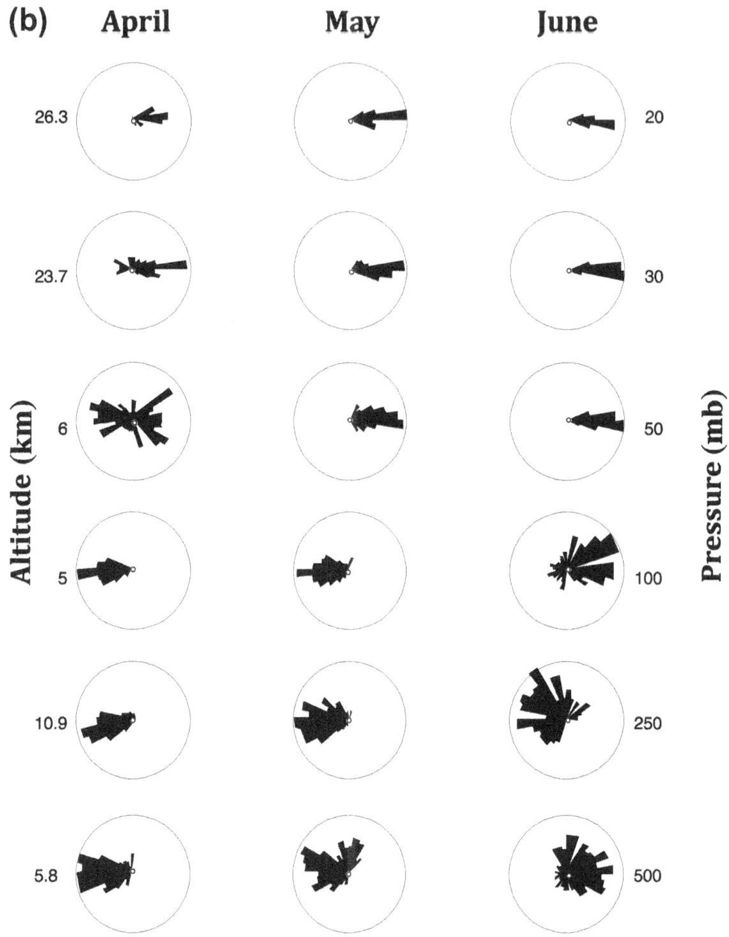

FIGURE 13.7 Continued

13.4.3 Preliminary Assessment

Results show that ash thicknesses of 1 mm or greater—enough to disrupt critical infrastructure such as power, water supply, and aviation (Wilson et al., 2012)—may be expected on average every 1,000 to 1,500 years in MC (Figure 13.8(a)). Larger thicknesses of 100 mm, which may begin to cause structural damage to weak roofs and will likely destroy some crops and infrastructure (Jenkins et al., 2014) in MC, may be expected rarely, on average every 10,000 years (Figure 13.8(b)). This does not include the contribution from smaller magnitude eruptions (VEI $\leq$3) and as such may be conservative.

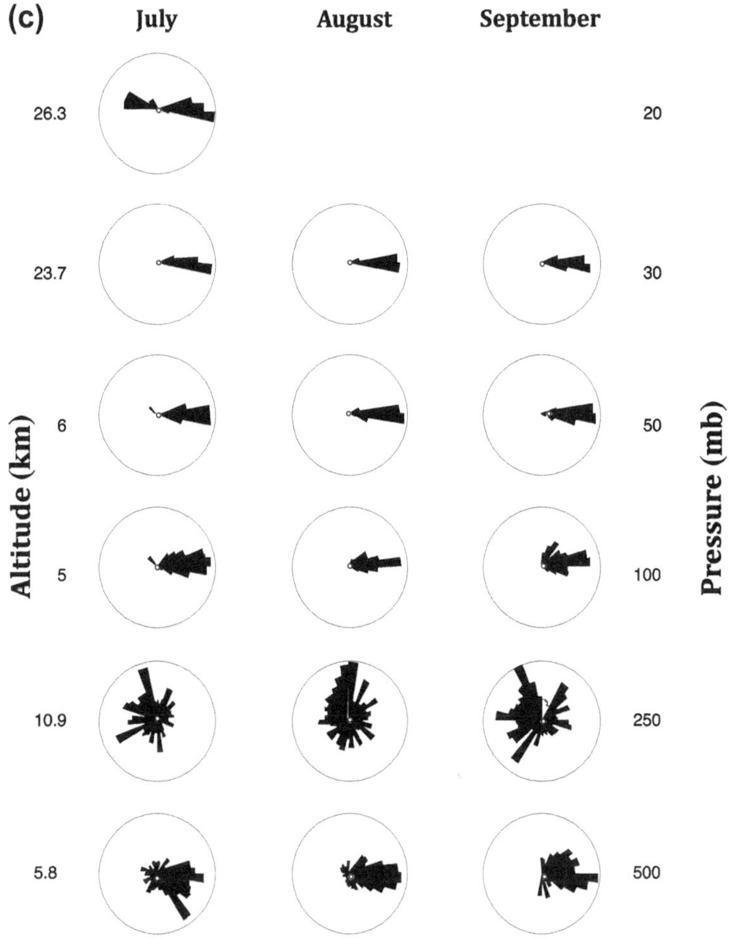

FIGURE 13.7 Continued

We considered the long-term hazard at key sites that are important when planning for a volcanic crisis (Figure 13.9(a)). The resultant ash hazard curves can be used to draw approximate return periods for increasing ash depths. The disaggregated curves (Figure 13.9(b)) show the contribution of all volcanoes impacting the site and show that, taking the eruption record provided by the Smithsonian, Popocatépetl dominates the ash hazard in MC because of its relatively active eruption history, its location southeast of the city and the predominant east-west winds in the area.

These results may be considered first approximations of the ashfall hazard in MC and highlight the importance of considering the hazard from multiple volcanoes: further, more detailed analysis is required for use in emergency management or planning purposes.

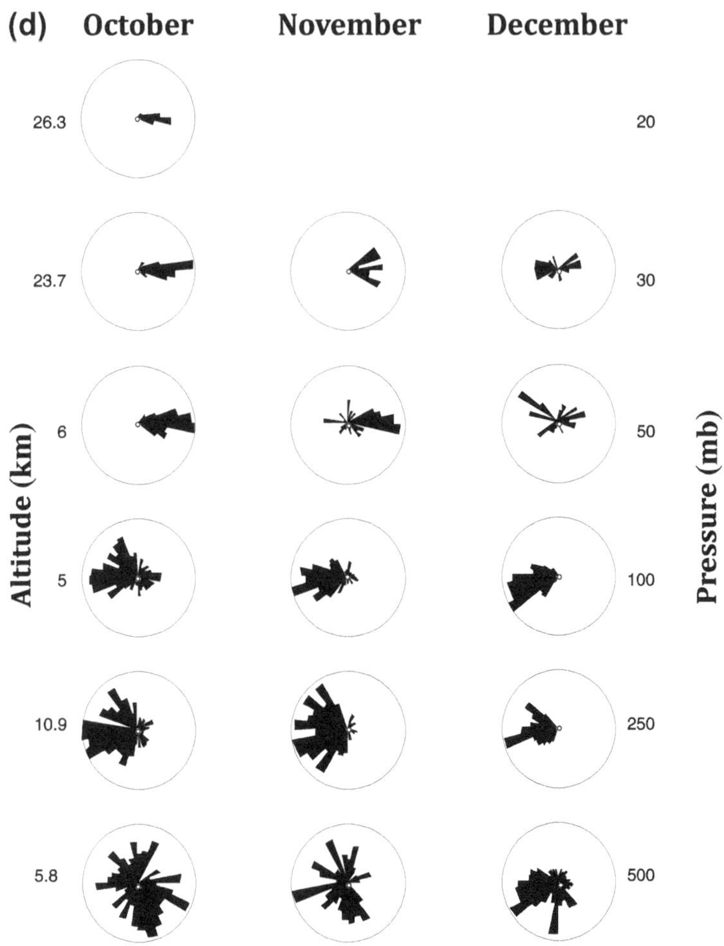

FIGURE 13.7 Continued

13.5 LIVING WITH THE EVERLASTING POSSIBILITY OF THE FORMATION OF A NEW VOLCANO IN THE VICINITY OF MC: DEALING WITH FALSE ALARMS

When evaluating hazards in a monogenetic volcanic field (i.e., CVF, MGVF, XVF), it is important to consider not only the processes occurring during an eruption but also those that can happen before an eruption occurs. Although many alerts in the MC region turn out to be false alarms, it is important to treat each one as a potential volcanic precursor and deal with them in a systematic way. False alarms are of a diverse nature and should be tackled as proposed by Espinasa-Pereña and Delgado Granados (2013). Paying attention to false alarms is important but they have a cost that needs to be evaluated too.

TABLE 13.1 Eruption Probabilities sourced from the Smithsonian Institution Global Volcanism Program for Each Volcano Considered Within the Ash Hazard Assessment. Note that Only Large Magnitude Eruptions (VEI ≥4) are Simulated Because of Their Potential for Producing Ashfalls that Extend Over Greater Distances and Produce Larger Ash Thicknesses. Values with Dashes are Considered of Negligible Probability and as Such are Not Modeled. The Average Return Period of Small-Magnitude (VEI ≤3) Eruptions is Included here for Completeness

			Average Return Period (Years)				
Volcano	**Longitude**	**Latitude**	**VEI ≤3**	**VEI 4**	**VEI 5**	**VEI 6**	**VEI 7**
1) Bárcena	19.3	−110.82	141	6,900	–	–	–
2) Socorro	18.78	−110.95	143	17,730	–	–	–
3) Ceboruco	21.125	−104.508	127	5,760	54,420	248,140	1.2 million
4) Colima	19.514	−103.62	6	256	2,420	11,050	55,260
5) Jocotitlán	19.73	−99.758	4,881	220,870	2.1million	9.5 million	48 million
6) Nevado de Toluca	19.108	−99.758	3,417	154,610	1.5 million	6.7 million	33 million
7) Popocatépetl	19.023	−98.622	15	656	6,200	28,300	141,480
8) La Malinche	19.231	−98.032	1,298	58,710	555,190	2.5 million	13 million
9) Cofre de Perote	19.492	−97.15	869	107,760	–	–	–
10) Pico de Orizaba	19.03	−97.268	10,251	463,820	4.4 million	20 million	100 million
11) San Martin Tuxtla	18.57	−95.2	424	52,520	–	–	–
12) El Chichonal	17.36	−93.228	257	3,680	16,780	–	–
13) Tacaná	15.13	−92.112	156	6,650	62,840	286,530	1.4 million
14) Michoacán-Guanajuato	19.85	−101.75	518	25,380	–	–	–
15) Chichinautzin	19.08	−99.13	7,849	384,620	–	–	–
16) Xalapa	19.67	−96.75	3,292	161,290	–	–	–

VEI, volcanic explosivity index.

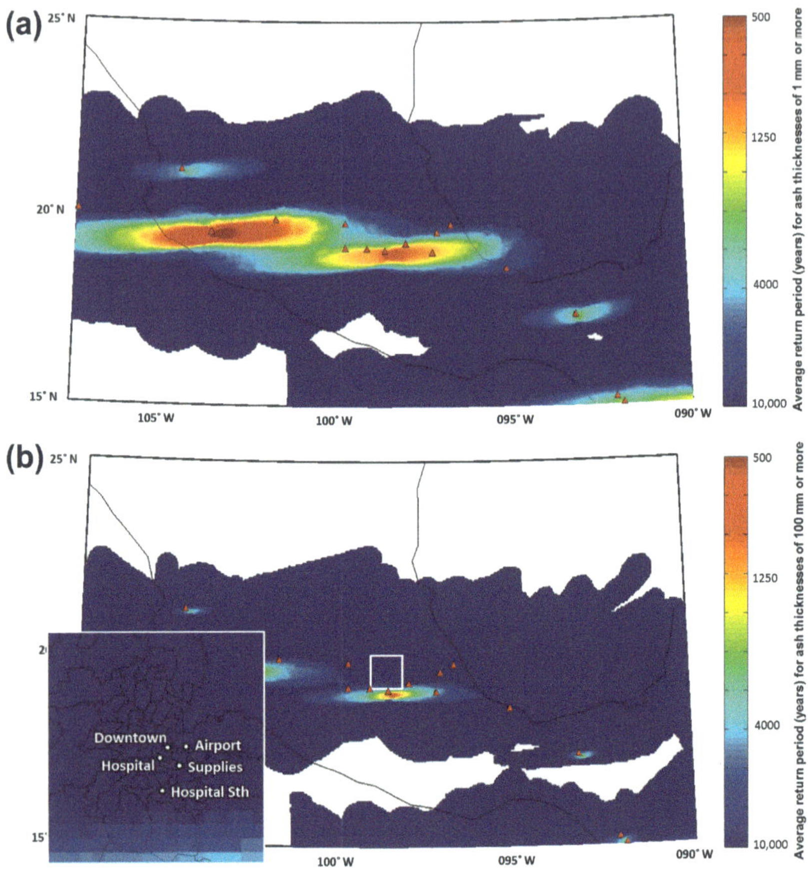

FIGURE 13.8 Average return period (in years) for ashfall thicknesses exceeding (a) 1 mm and (b) 100 mm in Mexico. Note that this does not include the contribution of eruptions of volcanic explosivity index (VEI) 3 or smaller and has a log base 10 scale. Inset box in (b) shows the location of some key facilities within Mexico City. "Supplies" refers to the very important Center for Supplies (so-called "Central de Abasto") of MC. "Hospital Sth" refers to the zone where several health facilities are clustered in southern MC whereas "Hospital" refers to the so-called "Siglo XXI" large hospital complex in MC.

13.5.1 Heating Related to Decomposition of Organic Matter at Lacustrine Basins and Landfills

Organic matter decomposes in such a way that, several gases are released in the process. Among the released gases is methane (CH_4), a colorless, odorless (Hensher and Button, 2003), highly flammable, and explosive gas, when there is 10–15 percent concentration in air (Demirbas, 2010), which means that this can be a hazard in volcanic areas. Several volcanic fields around the world are settled in intermontane valleys. In Mexico, all monogenetic volcanic fields are

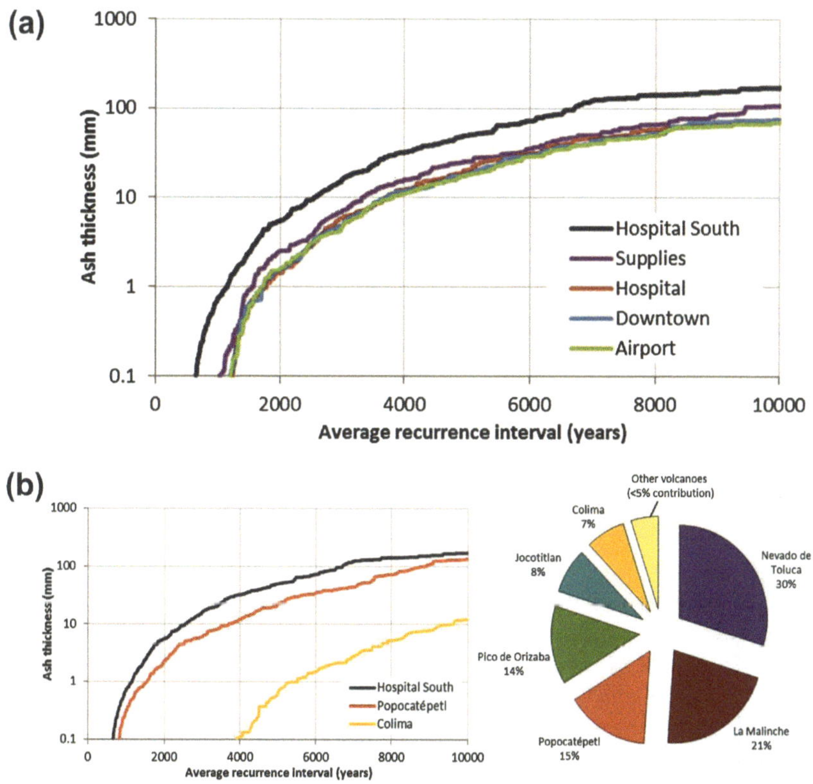

FIGURE 13.9 (a) Ashfall hazard curves for key sites in Mexico City (MC) (shown in Figure 13.8(b)). The curves show the average return period for ashfalls of increasing thickness. Ash hazard at downtown, hospital, and airport are approximately similar because of their location to the north of the city. (b) The disaggregated curves for a major hospital in the south, which shows the greatest hazard because of its proximity to active volcanoes in the south of the city. The contribution of assessed volcanoes to the ash hazard in MC by way of the proportion of simulated eruptions that impact the City is shown in the pie chart. These show that for the hospital in the south, simulations from Nevado de Toluca most frequently produced ashfalls in MC but Popocatépetl dominates the ash hazard, because of its relatively high probability of eruption.

in close relationship with lacustrine basins active during the Pleistocene, and impacted by the Quaternary climate changes. These basins have been the location of repeated events of inundation and drying (Lozano García et al., 1993; Caballero Miranda, 1996, 1997a,b; Caballero Miranda and Ortega Guerrero, 1998; Caballero Miranda et al., 1999; Caballero Miranda et al., 2010), resulting in the burial of large amounts of organic matter. At places like the vicinity of Lake Xochimilco, where large amounts of organic matter have been piled up together with volcanic ashes and surrounded by lava flows of various ages, underground fires are reported when the release of methane starts to be ignited. This is a process not only active in the basin of Mexico, reports are from places where volcanism is considered extinct (such as Jerez, Zacatecas) but people have established a relationship between the occurrence of these underground fires and the possibility to witness the birth of a new volcano, as in the case of Parícutin in 1943.

On the other hand, landfills are sites where all sorts of waste is dumped; some landfills are legal while others are not. At landfills, there is also a strong release of methane and, for this reason, several dumping sites require ventilation pipes to be built for this gas to escape safely. However, this is not the case in many landfills around the world, and particularly at illegal dumping sites, where after use, they are buried, thus allowing the decomposition to occur without allowing the produced gases to escape. In these situations, underground fires are also generated by ignition of methane. People report these fires believing that they are related to a possible volcano-birth event (Espinasa-Pereña and Delgado Granados, 2013).

13.5.2 Heating Related to Underground Fires

In this case, different from the one described above, the fires are not related to methane but with combustion of organic matter (such as tree leaves) and other flammable material (i.e., paper, cardboard, etc.) in small-scale lava tubes (on the order of tens of centimeters to a few meters wide), with different lengths (tens of meters long), relatively isolated from the surface (but with enough connection to allow the entrance of air). The ignition of flammable materials starts at the beginning of the rainy season in the MC region. This process seems to be related with the presence of flammable grass on the surface ignited by the sun and small water drops acting as a magnifying glass. The ignited materials burn flammable material inside the lava tubes and the heat is transported by the heated air through the cracks and fractures in the lavas.

Espinasa-Pereña and Delgado Granados (2013) describe in detail a case of this type. People living in monogenetic volcanic fields know the possibility of having a new volcanic event in the area and so they report these events.

13.5.3 Deformation Related to Landslides

Some landslides cause surficial deformation provoking fracturing and bulging of the terrain. The occurrence of this deformation at volcanic areas (such as the MGVF) can be mistakenly considered as a volcanic process. We portray an example of this in the following lines.

On May 22, 1992, a series of fractures caused alarm in the population of El Calvario, Michoacán (Mexico). The fractures described a cirque of about 500 m in diameter, at the skirts of a hill with a slope of $\sim 22°$. The fractures were part of an east-northeast–west-southwest fault running for more than 10 km at the base of the Mesa La Cantera-Cerro La Mesa-Los Potrerillos Casahuatera mountain range, and with displacements of the order of meters in the vertical and horizontal direction. Measurements indicated changes in the ground level of up to 1.2 m within a month. Fractures affected an irrigation channel (which was previously reinforced and embankments constructed with packages of sand) and water deposits. At the water tanks, cracks related to the fracturing process were evident, although some tanks did not show cracking but subsidence of the ground was evident on the surface and formed upheaved and subsided blocks. These small-scale horst and graben structures were formed in a short time (~ 1 day) and in this time frame, the cracks enlarged and widen. At lower altitudes, small hills or crests emerged in small mounds or ridges (~ 1 m above the previous level) that emerged within a month. A nearby house, however, was not affected at all. The owner did not hear or feel anything strange that might be associated with the appearance of fractures or the crests. The area affected by the fracturing coincided with a landslide occurred about 21 years ago, according to locals.

This is a simple description of a landslide but countryside people do not have geological background and collective memory is too short to remember that these processes are common in this tectonically active area.

13.5.4 Deformation due to Underground Gas Pressure

During the decomposition of organic matter, gases are released as mentioned above, but when ventilation to the atmosphere is not allowed, the gases accumulate underground. Accumulation of gases can produce bulging of the surface and eventually provoke explosions. An example of this kind of event occurred in June 2010. An explosion occurred in Chimalhuacán, State of Mexico (Mexico), after a bulging process that produced fractures of tens of meters (Figure 13.10). Measurements made at this site during this

FIGURE 13.10 Fractures along the landfill of Chimalhuacán, Estado de Mexico. After bulging of the ground, the gases produced an explosion at the site. (http://www.jornada.unam.mx/2010/06/16/estados/033n1est).

TABLE 13.2 Site Location and Gas Emissions from Soils Measured at Every Site

Site	Longitude	Latitude	CH_4 (g/m²/d)	H_2S (g/m²/d)	CO_2 (g/m²/d)
3	−98.962905	19.449185	89.948304	0.006816	189.39888
3a	−98.962905	19.449153	237.136584	0.0020448	389.08848
3b	−98.962817	19.449139	38.2948416	0.0006816	100.40008
3c	−98.962819	19.449142	0	0.003408	206.3644
3d	−98.962818	19.449144	0	0.0074976	20.2576
3e	−98.962578	19.449184	18.152712	0.0047712	27.29496
3f	−98.962471	19.449154	6.007488	0.0061344	6.20928
2	−98.962346	19.449051	89.9367264	0.0013632	251.28048
2a	−98.962078	19.448982	135.6756432	0.0109056	171.13712
2b	−98.962079	19.448983	13.0399152	0.0006816	14.8808
1	−98.962082	19.448983	4.6072416	0.0013632	15.70888
1a	−98.962091	19.448977	94.2101472	0.0074976	322.6476
1b	−98.962172	19.448899	24.9397584	0.003408	79.82744
1c	−98.962189	19.448902	1567.056782	0.2406048	2813.99888
1d	−98.962173	19.448901	220.1123664	−0.0047712	448.16112
5	−98.962279	19.449526	3.0272208	0.0020448	8.81496
5a	−98.962251	19.449745	4.2669888	0.0047712	1.7952
5b	−98.962503	19.449721	2.3647248	0.0020448	15.0348
5c	−98.962503	19.44964	76.21116	0.0040896	222.44904

emergency indicated emissions of CO_2, H_2S, and CH_4 as high as 2,810, 0.24, and 1,570 $g/m^2/d^1$, respectively (see Table 13.2).

The people, fearing for a new volcano being born at their vicinity, reported these events. A way to cope with false alarms related to swelling of the ground and explosions, which are extremely hazardous to life and property, is by carrying out soil gas emission measurements, particularly of carbon dioxide and methane (Delgado Granados et al., 2009; Espinasa-Pereña and Delgado Granados, 2013).

13.5.5 Earthquake Swarms

MC is the location for the occurrence of earthquakes associated with tectonic features and dormant volcanoes. The LPFS is a major tectonic feature in southern MC (Delgado Granados et al., 1995, 1997; Lermo Samaniego et al., 1995; Campos Enríquez et al., 1999). The occurrence of earthquakes and seismic swarms in southern MC are signs of tectonic activity, whereas microseismicity is associated with volcanic edifices and monogenetic volcanic fields, which are instrumentally recorded. People report to the authorities the seismicity felt in a region like MC because sometimes they fear that the seismicity is related to the birth of a new volcano.

Although shallow seismic swarms are frequent, for instance, in places like the CVF, the truth is that little is known about how to distinguish purely tectonic events related to rupture at active faults from tectonic events related to magma ascent.

13.5.6 Anthropogenic Lava Flows

Espinasa-Pereña and Delgado Granados (2013) give an account of the formation of a very small-size, short-lived anthropogenic lava flow within MC. This is a hilarious account of thieves stealing metallic manholes and melting them for resale that resulted in the melting of the concrete-based furnace, which eventually reached the street as a "lava flow." This calls the attention to the fact that, even the most experienced volcanologists might be challenged by these accounts seeming so natural that they can be taken as real, as can be seen in Figure 13.11. People, authorities, and scientists are taken by surprise by these events but the reports are normally issued by the people as clear signs of a volcano-birth event in a region prone to witnessing such events.

13.5.7 A Methodology to Cope with False Alarms at Monogenetic Volcanic Fields

Espinasa-Pereña and Delgado Granados (2013) proposed a methodology to deal with false alarms depicted in Figure 13.12. After manifestation of the event and people's call to the authorities and scientists, the most important issue is to distinguish between the anthropogenic and natural character of the event. This can be accomplished by using electrochemical sensors, IR instruments, regular chemical analyses or rocks, seismometers and/or GPS, and electronic distance-meters. Applying this methodology, the underground fires should not show related seismicity, at least no volcano-tectonic events should be present, but the deformation related to landslides could show some tremor and even, some tectonic activity related to the movement of tectonic features. However, no volcano-related gas emissions such as SO_2 would be expected, although other gases could be expected. During deformation related to underground gas-pressure at landfills, CO_2, H_2S, and/or CH_4 could also be expected; however, no SO_2 should be present, and fracturing/tectonic activity (deeper than 50 m) would be absent. The earthquakes are always present at

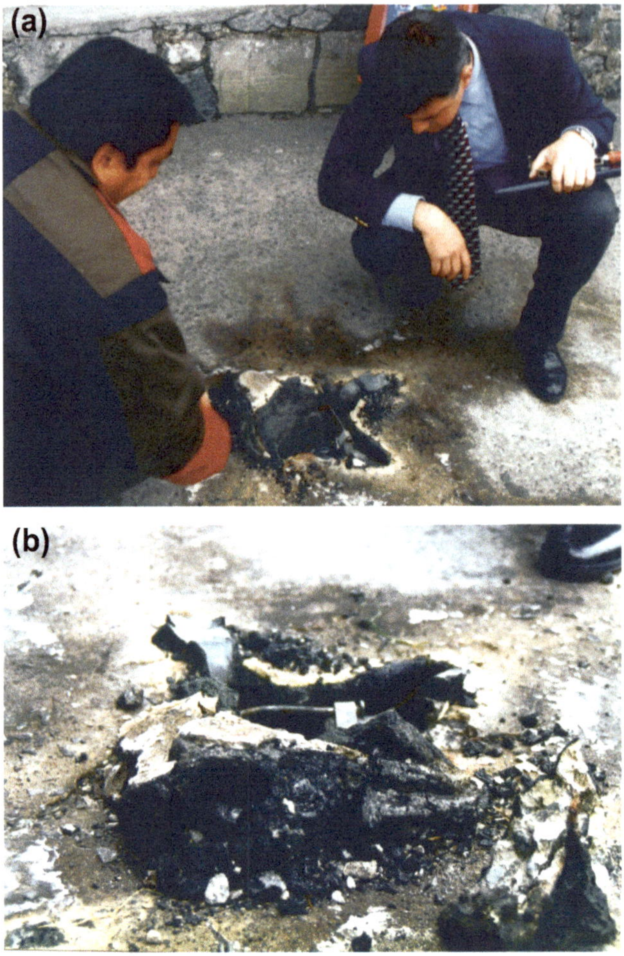

FIGURE 13.11 View of the anthropogenic "lava flow". (a) "Lava flow" emitted through a telephone pipeline. (b) Detail of the reported lava flow.

monogenetic volcanic fields and changes through time in CO_2 emissions from soils can be expected; however, no volcano-related gases would be present (i.e. SO_2) unless a volcanic event is underway and then, these gases can be present and CO_2 emissions from soils may show changes of an order or orders of magnitude. Some of the most intriguing events would be the anthropogenic lava flows. Even the chemical analyses may reveal a normal volcanic nature of the rocks as in the case of Iztapalapa, Mexico (an andesite), as mentioned by Espinasa-Pereña and Delgado Granados (2013). In this case, no volcano-tectonic events would be present, and no volcano-related gases (CO_2, SO_2) would be released.

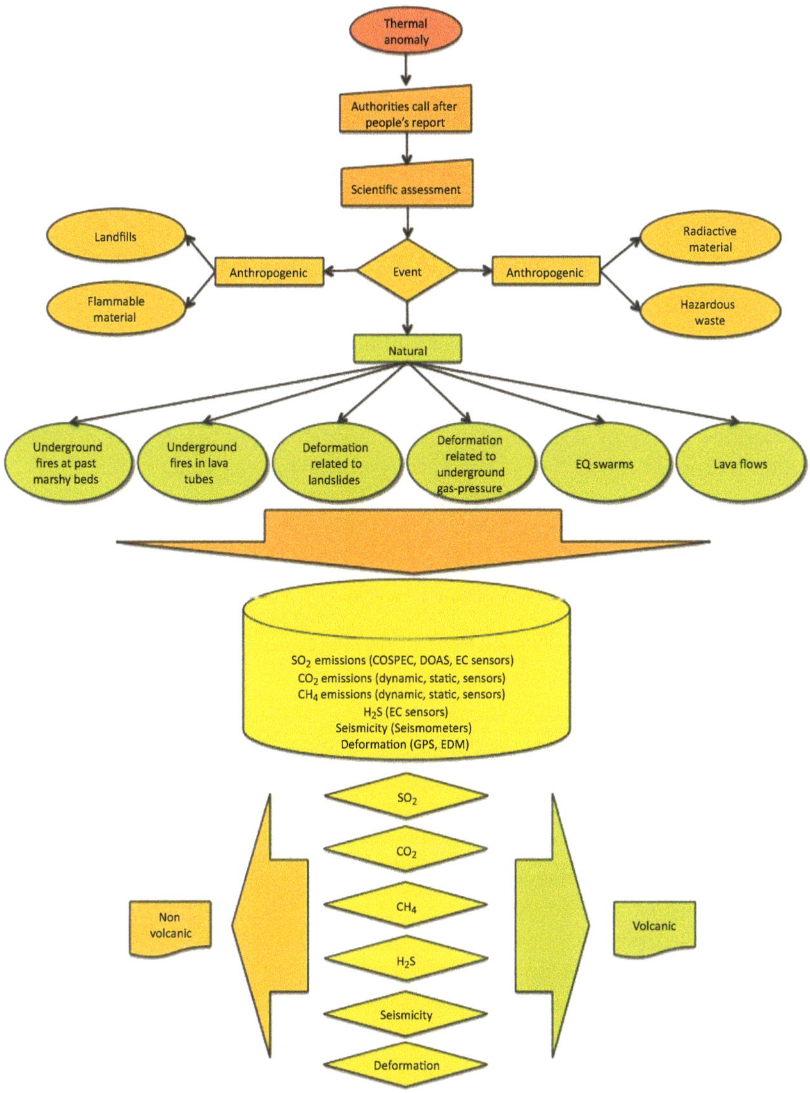

FIGURE 13.12 Methodology proposed for the attention of alerts in monogenetic volcanic fields in order to elucidate the nature of the cause of alert.

13.6 FUTURE PERSPECTIVES

Volcanic hazard assessment at largely populated areas such as MC is a very complex task. It should involve the study of geological features in the greatest detail possible. In the case of CVF, better detail of the previous eruptions is needed, including more and better dating of the last eruptive events. This includes, of course, more extensive geologic and stratigraphic work in the area.

This is true not only for the monogenetic volcanism but also for the large volcanic centers such as Popocatépetl and Nevado de Toluca. Interestingly, despite the importance of Popocatépetl, it still lacks a geologic map. In addition to providing a better understanding of the volcanoes that could impact MC, these studies can help support the development of more robust and applicable frequency–magnitude relationships for use in probabilistic ash fall modeling for the MC region.

Another important task is a widespread education program for the people to recognize the events leading up to a new volcanic event. It is important to ensure that people and authorities can notice the differences between volcanic activities related to large central volcanoes and the activity related to a small monogenetic volcano, and the signals that might be associated with each. The accounts of previous experiences should be emphasized to the population and authorities, for them to know what has been experienced before and be able to more accurately report the events, during a near-future event.

To the authorities, it is important to make clear that attendance of all sorts of alerts, even if they are false alerts, provides a very important opportunity to gather experience that can be applied during the next odd report, in spite of how hilarious the report might be.

13.7 SUMMARY

In this chapter, we have discussed the main volcanic hazard threatening MC and how previous authors have assessed the hazard. Multiple volcanoes, both monogenetic volcanic fields and large volcanoes that surround the region, particularly to the south, threaten MC. We present a new probabilistic ash hazard assessment methodology for ashfall in MC that accounts for all possible volcanic sources. Volcanic hazard assessments for MC require more geological assessments in order to better refine frequency–magnitude relationships and to develop a better understanding of the previous activity, which may inform future activity. Here, we have discussed what is currently known about volcanoes that may produce ash in MC. We also provide an in-depth assessment of wind conditions at MC, which gives an indication of how ash may most likely be dispersed in a future event.

Finally, the importance of false alarms, where suspected volcanic activity or precursors are reported, is considered. The many different types of report highlight that the population are aware that volcanic activity is possible but also that further education is needed to help distinguish between anthropogenic, volcanic, and other natural activity. Future studies will be important in refining volcanic hazard assessments, emergency management planning, and in educating the population.

ACKNOWLEDGMENTS

The first author acknowledges the partial financial support of the Instituto de Geofísica, UNAM, and Direción General de Asuntos del Personal Académico for additional financial support (grant IN108209). The second author acknowledges the support of an AXA research fellowship. Discussions with Servando de la Cruz (Instituto de Geofísica, UNAM) and Eduardo Reynoso (Instituto de Ingeniería, UNAM) are appreciated. Patricia Jácome Paz carried out the measurements at the Chimalhuacán landfill.

REFERENCES

Alaniz-Alvarez, S.A., Nieto-Samaniego, A.F., 2005. El sistema de fallas Taxco-San Miguel de Allende y la Faja Volcánica Transmexicana, dos fronteras tectónicas del centro de México activas durante el Cenozoico. Bol. conmemorativo del centen. Soc. Geol. Mex. T LVII (1), 65–82.

Anderson, R., Davis, E., 1973. A topographic interpretation of the mathematician ridge, clipperton ridge, east pacific rise system. Nature 241, 191–193.

Arce, J.L., Macías, J.L., Vázquez, S.L., 2003. The 10.5 KA Plinian eruption of Nevado de Toluca, México: stratigraphy and hazard implications. Geol. Soc. Am. Bull. 115 (2), 230–248.

Bloomfield, K., 1975. A late quaternary volcano field in central Mexico. Geol. Rundsch. 64, 476–497.

Bloomfield, K., Valastro, S., 1974. Late Pleistocene eruptive history of Nevado de Toluca Volcano, Central México. Geol. Soc. Am. Bull. 85, 901–906.

Bloomfield, K., Sanchez, G.R., Wilson, L., 1977. Plinian eruptions of Nevado de Toluca volcano, México. Geol. Rundsch. 66, 120–146.

Ban, M., Hasenaka, T., Delgado Granados, H., Takaoka, N., 1992. K-Ar ages of lavas from shield volcanoes in Michoacan–Guanajuato, Mexico. Geofís. Int. 31 (4), 467–473.

Bebbington, M., Cronin, S., 2011. Spatio-temporal hazard estimation in the Auckland Volcanic Field New Zealand with a new event-order model. Bull. Volcanol. 73, 55–72. http://dx.doi.org/10.1007/s00445-010-0403-6.

Bohrson, W.A., Reid, M.R., Grunder, A.L., Heizler, M.T., Harrison, T.M., Lee, J., 1996. Prolonged history of silicic peralkaline volcanism in the eastern Pacific Ocean. JGR 101 (B5), 11457–11474. http://dx.doi.org/10.1029/96JB00329.

Caballero Miranda, M., 1996. The diatom flora of two acid lakes in Central Mexico. Diatom Res. 11, 227–240.

Caballero Miranda, M., 1997a. The last glacial maximum in the Basin of México: the diatom record between 34,000 and 15,000 years BP from Lake Chalco. Quat. Int. 43/44, 125–136.

Caballero Miranda, M., 1997b. Reconstrucción paleolimnológica del Lago de Chalco durante el último máximo glaciar, el registro de diatomeas entre 34,000 y 15,000 años A.P. Revista Mex. Ciencias Geol. 14, 91–100.

Caballero Miranda, M., Ortega Guerrero, B., 1998. Lake levels since about 40,000 years ago at lake Chalco, near Mexico City. Quat. Res. 50, 69–79.

Caballero Miranda, M., Lozano, S., Ortega Guerrero, B., Urrutia Fucugauchi, J., Macias, J.L., 1999. Environmental characteristics of lake Tecocomulco, northern basin of Mexico, for the last ca. 50,000 years. J. Paleolimnol. 22, 399–411.

Caballero Miranda, M., Lozano-García, S., Vázquez-Selem, L., Ortega, B., 2010. Evidencias de cambio climático y ambiental en registros glaciales yen cuencas lacustres del centro de México durante el último máximo glacial. Bol. Soc. Geol. Mex. 62, 359–377.

Campos Enríquez, J.O., Morales Rodríguez, H.F.y., Delgado Granados, H., 1999. Evidencia gravimétrica del sistema de fallas La Pera. In: Simposio Internacional Volcán Popocatépetl, Secretaría de Gobernación, CENAPRED, México D. F., Marzo 22–24, p. 6.

Enríquez, Campos, 1999. J. O., Morales Rodríguez, H. F. y Delgado Granados, H. Evidencia gravimétrica del sistema de fallas La Pera. Simposio Internacional Volcán Popocatépetl, Secretaría de Gobernación, CENAPRED, México D. F., 6. Marzo 22-24.

Cantagrel, J.M., Robin, C., Vincent, P., 1981. Les grandes etapes d'evolution d'un volcan andesitique composite: exemple du Nevado de Toluca. Bull. Volcanol. 44, 177–188.

Canul, R.F., Rocha, V.L., 1981. Informe geológico de la zona geotérmica de El Chichonal, Chiapas, México: México, D.F. Comisión Federal de Electricidad, Informe 32–81, 38 pp.

Canul-Dzul, R.F., Razo-Montiel, A., Rocha-López, V., 1983. Geología e historia volcanológica del volcán Chichonal, estado de Chiapas en Sociedad Geológica Mexicana, VI Convención Geológica Nacional. El Volcán Chichonal: México, D F. Universidad Nacional Autónoma de México, Instituto de Geología, 3–22.

Capra, L., Macías, J.L., 2000. Pleistocene cohesive debris fl ows at Nevado de Toluca volcano, central Mexico. J. Volcanol. Geotherm. Res. 102 (1–2), 149–167. http://dx.doi.org/10.1016/S0377-0273(00)00186-4.

Carrasco-Núñez, G., Righter, K., Chesley, J., Siebert, L., Aranda-Gómez, J.J., 2005. Contemporaneous eruption of calc-alkaline and alkaline lavas in a continental arc (Eastern Mexican Volcanic Belt): chemically heterogeneous but isotopically homogeneous source. Contrib. Mineral. Petrol. 150, 423–440.

Carrasco-Núñez, G., Díaz-Castellón, R., Siebert, L., Hubbard, B., Sheridan, M.F., Rodríguez, S.R., 2006. Multiple edifice-collapse events in the Eastern Mexican Volcanic Belt: the role of sloping substrate and implications for hazard assessment. In: Tibaldi, A., Lagmay, A. (Eds.), The Effects of Basement Structural and Stratigraphic Heritages on Volcano Behavior: J. Volcanol. Geotherm. Res. 158, 151–176.

Carrasco-Núñez, G., Siebert, L., Díaz-Castellón, R., Vázquez-Selem, L., Capra, L., 2009. Evolution and hazards of a long-quiescent compound shield-like volcano: cofre de Perote, Eastern Trans-Mexican Volcanic Belt. J. Volcanol. Geothermal Res. 197 (1–4), 209–224.

Castro-Govea, R., Siebe, C., 2007. Late Pleistocene–Holocene stratigraphy and radiocarbon dating of La malinche volcano, Central Mexico. J. Volcanol. Geother. Res. 162, 20–42.

Connor, C.B., 1987. Structure of the Michoacán-Guanajuato volcanic field, Mexico. J. Volcanol. Geother. Res. 33 (1–3), 191–200.

Conte, G., Urrutua-Fugugauchi, A., Goguitchaichvili, A., Soler-Arechalde, A.M., Morton-Bremea, O., 2004. Paleomagnetic study of lavas from the Popocatépetl volcanic region, central Mexico. Int. Geol. Rev. 46, 210–225.

De la Cruz-Reyna, S., 1993. Random patterns of occurrence of volcanic eruptions at Colima volcano, Mexico. J. Volcanol. Geotherm. Res. 55, 51–68.

De la Cruz-Reyna S, Carrasco-Núñez, G., 2002. Probabilistic hazard analysis of Citlaltépetl (Pico de Orizaba) Volcano, Eastern Mexican Volcanic Belt. J. Volcanol. Geother. Res. 113, 307–318.

Delgado Granados, H., 2002. Patrones de viento en las cercanías del volcán Popocatépetl, los productos balísticos y sus efectos en: las cenizas volcánicas del Popocatépetl y sus efectos para la Aeronavegación e infraestructura Aeroportuaria. Segob-cenapred, 51–81.

Delgado Granados, H., Carrasco Nuñez, G., Urrutia Fucugauchi, J., Casanova Becerra, J.M., 1988. Analysis of the eruptive records of the Popocatépetl volcano, México. In: Kagoshima International Conference on volcanoes, Kagoshima, Japan, proceedings, pp. 510–513.

Delgado Granados, H., Nieto Obregón, J., Silva Romo, G., Mendoza Rosales, C., Arellano Gil, J., Lermo Samaniego, J.F., Rodríguez González, M., 1995. La Pera Detachment Fault System: Active Faulting South of México City (II): Geological Evidence, Reunión Anual de la Unión Geofísica Mexicana, Puerto Vallarta, Jalisco, del 13–17 de noviembre 1995, GEOS, 15, 2, 64.

Delgado Granados, H., Nieto Obregón, J., Lermo Samaniego, J., Silva Romo, G., Mendoza Rosales, C., Campos Enriquez, J.O., 1997. La Pera Fault System: A Major Active Structure in Central Mexico. Annual Meeting of the American Geophysical Union. San Francisco, California, E.U.A., p. F823.

Delgado Granados, H., Villalpando Cortés, R.E., 2008. Método para pronosticar la localización de un nuevo volcán al sur de la Ciudad de México. Revista TIP, Revista Especializada en Ciencias Químico-Biológicas 11 (1), 5–16.

Delgado Granados, H., Molinero, R., Cervantes, P., Nieto-Obregón, J., Lozano-Santa-Cruz, R., Macías-González, H.L., Mendoza-Rosales, C., Silva-Romo, G., 1999 (1998). Geology of Xitle volcano in southern Mexico City – a 2,000-year-old monogenetic volcano in an urban area. Revista Mexicana de Geociencias 15 (2), 115–131.

Demirbas, A., 2010. Methane Gas Hydrate. Springer-Verlag, 102 pp.

Duffield, W.A., Tilling, R.I., Canul, R., 1984. Geology of el Chichón volcano, Chiapas, México. J. Volcanol. Geothermal Res. 20, 117–132.

Espinasa-Pareña, R., Martín-Del Pozzo, A.L., 2006. Morphostratigraphic evolution of Popocatépetl volcano, México. Geol. Soc. Am. Spec. Pap. 402, 115–137.

Espinasa-Pereña, R., Delgado Granados, H., 2013. Hazards related to lava tubes and caves in the Sierra Chichinautzin monogenetic volcanic field (México). In: Rose, W.I., Palma, J.L., Delgado Granados, H., Varley, N. (Eds.), Understanding Open-vent Volcanism and Related Hazards. Geological Society of America. Special Paper 498. http://dx.doi.org/10.1130/2013.2498(10).

Espíndola, J.M., Macías, J.L., Tilling, R.I., Sheridan, M.F., 2000. Volcanic history of El Chichón volcano (Chiapas, México) during the Holocene, and its impact on human activity. Bull. Volcanol. 62, 90–104.

Espíndola, J.M., Zamora-Camacho, A., Godinez, M.L., Rodriguez-Elizarraras, S., 2007. The 1793 Eruption of San Martin Volcano (Los Tuxtlas, Veracruz, Mexico). American Geophysical Union, Fall Meeting 2007, abstract #V41D-0802.

Espíndola, J.M., Zamora-Camacho, A., Godinez, M.L., Schaaf, P., Rodríguez, S.R., 2010. The 1793 eruption of san martín tuxtla volcano, Veracruz, Mexico. J. Volcanol. Geotherm. Res. 197, 188–208.

Frey, H., Lange, R.A., Hall, C.M., Delgado Granados, H., 2004. Magma eruption rates constrained by $^{40}Ar/^{39}Ar$ chronology and GIS for the Ceboruco-San Pedro volcanic field, western Mexico. GSA Bull. 116 (3/4), 259–276. http://dx.doi.org/10.1130/B25321.1.

García-Palomo, A., Macías, J.L., Arce, J.L., Capra, L., Garduño, V.H., Espíndola, J.M., 2002. Geology of Nevado de Toluca volcano and surroundings areas, Central México: Boulder, Colorado. Geological Society of America. Map and Chart Series, MCH080, 1 mapa, 26 pp. texto.

García-Palomo, A., Macías, J.L., Arce, J.L., Mora, J.C., Hughes, S., Saucedo, R., Espindola, J.M., Escobar, R., Layer, P., 2006. Geological evolution of the tacaná volcanic complex, México-Guatemala. Special Paper. In: Rose, et al. (Eds.), Natural Hazards in Central America: Boulder, Colorado. Geological Society of America, pp. 39–57.

Gómez-Tuena, A., Orozco-Esquivel, T., Ferrari, L., Petrogénesis Ígnea de la Faja Volcánica Transmexicana, 2005. Volumen Conmemorativo del Centenario, Temas Selectos de la Geología Mexicana, LVII Boletín de la Sociedad Geológica Mexicana 3, 227–285.

Guilbaud, M.-N., Siebe, C., Agustin-Flores, J., 2009. Eruption dynamics of the young high-Mg basaltic-andesite Pelagatos scoria cone southeast of México City. Bull. Volcanol. 71, 859–880.

Hasenaka, T., Carmichael, I.S.E., 1985a. A compilation of location size and geomorfologial parameters of volcanoes of the Michoacan-Guanajuato volcanic field, Central Mexico. Geofís. Int. 24, 577–607.

Hasenaka, T., Carmichael, I.S.E., 1985b. The cinder cones of Michoacan–Guanajuato, Central Mexico-Their age, volume and distribution and magma discharge rate. J. Volcanol. Geotherm. Res. 25, 105–124.

Hasenaka, T., Ban, M., Delgado Granados, H., 1994. Contrasting volcanism in the Michoacan–Guanajuato volcanic field: shield volcanoes vs. cinder cones. Geofís. Int. 33 (1), 125–138.

Hensher, D.A., Button, K.J., 2003. Handbook of Transport and the Environment. Emerald Group Publishing, 168 pp.

Hurst, A.W., 1994. ASHFALL, a Computer Program for Estimating Volcanic Ash Fallout: Report and Users Guide. GNS Science.

INEGI (Instituto Nacional de Estadística y Geografía), 2011. Panorama sociodemográfico del Distrito Federal. Censo de Población y Vivienda 2010, ISBN 978-607-494-205-7, 50 pp.

Jenkins, S., Magill, C., McAneney, J., Blong, R., 2012a. Regional ash fall hazard I: a probabilistic assessment methodology. Bull. Volcanol. 74 (7), 1699–1712.

Jenkins, S., McAneney, J., Magill, C., Blong, R., 2012b. Regional ash fall hazard II: Asia-Pacific modelling results and implications. Bull. Volcanol. 74 (7), 1713–1727.

Jenkins, S.F., Spence, R.J.S., Fonseca, J.F.B.D., Solidum, R.U., Wilson, T.M., 2014. Volcanic risk assessment: Quantifying physical vulnerability in the built environment. J. Volcanol. Geotherm Res. 276 (0), 105–120.

Lermo Samaniego, J.F., Nieto Obregón, J., Delgado Granados, H., Rodríguez González, M., Huidrobo González, A., Silva Romo, G., Mendoza Rosales, C., Arellano Gil, J., 1995. La Pera Detachment Fault System: Active Faulting South of México City (I): Seismological Evidence. Reunión Anual de la Unión Geofísica Mexicana, Puerto Vallarta, Jalisco, del 13 al 17 de noviembre de 1995, resúmen GEOS 15, 2, 67.

Libby, F.W., 1955. Radiocarbon Dating. University Chicago Press, Chicago, 124.

Lindsay, J., Marzocchi, W., Jolly, G., Constantinescu, R., Jacopo, S., Sandri, L., 2010. Towards real-time eruption forecasting in the Auckland Volcanic Field: application of BET_EF during the New Zealand neational disaster xercise "Ruaumoko". Bull. Volc. 72 (2), 185–204. http://dx.doi.org/10.1007/s00445-009-0311-9.

Lozano García, S., Ortega Guerrero, B., Caballero Miranda, M., Urrutia Fucugauchi, J., 1993. Late pleistocene and holocene Palaeoenvironments of Chalco Lake, Central Mexico. Quat. Res. 40, 332–342.

Lugo Hubp, J., Martín del Pozzo, A.L., Vazquez Selem, L., 1993. Estudio geomorfológico del complejo volcánico de Colima. Geofís Int. 32 (4), 633–641.

Macías, J.L., 2005. Geología e historia eruptive de lagunos de los grendes volcanes de México. Bol. Soc. Geol. Mex. Volumen Conmemorativo del Centenario Temas Selectos Geol. Mex. LVII (3), 379–424.

Macías, J.L., 2007. Geology and eruptive history of some active volcanoes of México. Geol. Soc. Am. Spec. Pap. 422, 184–332.

Macías, J.L., Arce, J.L., García, P.A., Siebe, C., Espíndola, J.M., Komorowski, J.C., Scott, K., 1997. Late Pleistocene-Holocene cataclysmic eruptions at Nevado de Toluca and Jocotitlan volcanoes, Central México. In: Link, K.P., Kowallis, B.J. (Eds.), Proterozoic to recent stratigraphy, tectonics, and volcanology, Utah, Nevada, southern Idaho and Central México: Bringham Young Univesity, Geology Studies, 42, 1, 493–528.

Macías, J.L., Espíndola, J.M., García-Palomo, A., Scott, K.M., Hughes, S., Mora, J.C., 2000. Late Holocene Peléan style eruption at Tacaná Volcano, México–Guatemala: past, present, and future hazards. Geol. Soc. Am. Bull. 112 (8), 1234–1249.

Macías, J.L., Capra, L., Arce, J.L., Espíndola, J.M., García Palomo, A., Sheridan, M.F., 2008. Hazard map of El Chichón volcano, Chiapas, México: constraints posed by eruptive history and computer simulations. J. Volcanol. Geotherm. Res. 175 (4), 444–458.

Martín Del Pozo, A.L., 1982. Monogenetic volcanism in sierra chichinautzin, Mexico. Bull. Vulcanol. 45 (1), 9–24.

Martin del Pozzo, A.L., Sheridan, M., Barrera, D., Hubp, J.L., Vázquez, L., 1995. Mapa de peligros del volcán de Colima. Instituto de Geofisica UNAM, Mexico D,F.

Mazot, A., Smid, E.R., Schwendenmann, L., Delgado Granados, H., Lindsay, J., 2013. Soil CO_2 flux baseline in an urban monogenetic volcanic field: the Auckland Volcanic Field, New Zealand. Bull. Volcanol. 75, 757. http://dx.doi.org/10.1007/s00445-013-0757-7.

Martin del Pozzo, A.L., Córdoba, C., López, J., 1997. Volcanic impact on the southern Basin of Mexico during the Holocene. Quat. Int. 43/44, 181–190.

Meriggi, L., Macías, J.L., Tommasini, S., Capra, L., Conticelli, S., 2008. Heterogeneous magmas of the Quaternary Sierra Chichinautzin volcanic field (central Mexico): the role of an amphibole-bearing mantle and magmatic evolution processes. Revista Mex. Ciencias Geol. 25 (2), 197–216.

Mooser, F., 1969. Tefracronología de la Cuenca de México para los últimos treinta mil años, Arquelogía. UNAM, México, 12–15.

Mora, J.C., Macías, J.L., García-Palomo, A., Arce, J.L., Espíndola, J.M., Manetti, P., Vaselli, O., Sánchez, J.M., 2004. Petrology and geochemistry of the Tacaná volcanic complex, Mexico-Guatemala: evidence for the last 40,000 yr of activity. Geofis. Int. 43 (3), 331–359.

Ownby, S., Delgado Granados, H., Lange, R.A., Hall, C.M., 2007. Volcán Tancítaro, Michoacán, Mexico 40Ar/39Ar constraints on its history of sector collapse. J. Volcanol. Geotherm. Res. 161, 1–14. http://dx.doi.org/10.1016/j.jvolgeores.2006.10.009.

Ownby, S., Lange, R.A., Hall, C.M., Delgado Granados, H., 2011. Origin of andesite in the deep crust and eruption rates in the Tancítaro-Nueva Italia Region of the Central Mexican arc. Geol. Soc. Am. Bull. 123 (1–2), 274–294. http://dx.doi.org/10.1130/B30124.1.

Pardo, M., Suárez, G., 1995. Shape of the subducted rivera and cocos plates in southern Mexico: seismic and tectonic implications. JGR 100 (B7), 12357–12373.

Richards, A.F., 1959. Geology of the Islas revillagigedo, méxico. Brith and development of volcán bárcena, Isla san benedicto. Bull. Volcanol. 2 (22), 73–123.

Richards, A.F., 1965. Geology of the Islas revillagigedo. 3. Effects of erosion on Isla san Bendicto 1952–61, on Isla san benedicto 1952–61. Following the birth of volcan Barcena. Bull. Volcanol. 28, 381–403.

Robin, C., Boudal, C., 1987. A gigantic Bezymianny-type event a the beginning of modern Volcán Popocatépetl. J. Volcanol. Geotherm. Res. 31, 115–130.

Rodríguez, S.R., Morales-Barrera, W., Layer, P., González-Mercado, E., 2009. A Quaternary monogenetic volcanic field in the Xalapa region, eastern Trans-Mexican Volcanic Belt: geology, distribution and morphology of the volcanic vents. J. Volcanol. Geotherm. Res. 197, 149–166. http://dx.doi.org/10.1016/j.jvolgeores.2009.08.003.

Rossotti, A., Carrasco-Núñez, G., Rosi, M., Di Muro, A., 2006. Eruptive dynamics of the "Citlaltépetl pumice" at citlaltépetl volcano, eastern Mexico. J. Volcanol. Geotherm. Res. 158, 401–429.

Sánchez Bribiesca, J.L., Fuentes Mariles, O., Esteva, M., 1989. Desarrollo de un modelo predictivo del transporte y depósito de cenizas volcánicas para el Valle de México. Instituto de Ingeniería, UNAM, México. Serie Azul No. 525, 117pp.

Scandone, R., 1979. Preliminary evaluation of the volcanic hazard in the southern valley of Mexico. Geofís. Int. 18, 21–35.

Schaaf, P., Stimac, J., Siebe, C., Macías, J.L., 2005. Geochemical evidence for mantle origin and crustal processes in volcanic rocks from Popocatépetl and surrounding monogenetic volcanoes, central Mexico. J. Petrol. 46, 1243–1282.

Sclater, J.G., Anderson, R.N., Bell, M.L., 1971. Elevation of ridges and evolution of the central eastern Pacific. J. Geophys. Res. 76, 7888–7915.

Siebe, C., 2000. Age and archaeological implications of Xitle volcano, southwestern Basin of Mexico-City. J. Volcanol. Geotherm. Res. 104, 45–64.

Siebe, C., Komorowski, J.C., Sheridan, M.F., 1992. Morphology and emplacement of an unusual debris-avalanche deposit at Jocotitlán volcano, Central Mexico. Bull. Volcanol. 54 (7), 573–589.

Siebe, C., Komorowski, J.C., Navarro, C., McHone, J., Delgado Granados, H., Cortés, A., 1995a. Submarine eruption near Socorro Island, Mexico: geochemistry and scanning electron microscopy studies of floating scoria and reticulite. J. Volcanol. Geotherm. Res. 68, 239–270.

Siebe, C., Macías, J.L., Abrams, M., Rodríguez, S., Castro, R., Delgado Granados, y H., 1995b. Explosive Volcanism in Central-Eastern México. Geological Society of America (Field Guide Book).

Siebe, C., Rodríguez-Lara, V., Schaaf, P., Abrams, M., 2004a. Radiocarbon ages of Holocene Pelado, Guespalapa, and Chichinautzin scoria cones, south of Mexico City: implications for archaeology and future hazards. Bull. Volcanol. 66, 203–225.

Siebe, C., Arana Salinas, L., Abrams, M., 2004b. Geology and radiocarbon ages of Tláloc, Tlacotenco, Cuauhtzin, Hijo del Cuauhtzin, Teuhtli, and Ocusacayo monogenetic volcanoes in the central part of the Sierra Chichinautzin, México. J. Volcanol. Geotherm. Res. 141, 225–243.

Siebert, L., Carrasco-Núñez, G., 2002. Late Pleistocene to precolumbian behind-the-arc mafic volcanism in the eastern Mexican Volcanic Belt; implications for future hazards. J. Volcanol. Geotherm. Res. 115, 179–205.

Siebert, L., Simkin, T., Kimberly, P., 2010. Volcanoes of the World, third ed. University of California Press, Berkeley. 568 pp.

Sieron, K., Siebe, C., 2008. Revised stratigraphy and eruption rates of Ceboruco stratovolcano and surrounding monogenetic vents (Nayarit, Mexico) from historical documents and new radiocarbon dates. J. Volcanol. Geotherm. Res. 68, 241–264.

Sosa-Ceballos, G., Gardner, J.E., Siebe, C., Macías, J.L., 2012. A caldera-forming eruption ~14,100 14C yr BP at Popocatépetl volcano, México: Insights from eruption dynamics and magma mixing. J. Geotherm. Res. 213-214, 27–40.

Velasco Tapia, F., Verma, S.P., 2001. First partial melting inversion model for a rift-related origin of the Sierra de Chichinautzin Volcanic Field, central Mexican Volcanic Belt. Int. Geol. Rev. 43, 788–817.

Wickmann, F.E., 1966. Repose period patterns of volcanoes. Arck. Miner. Geol. 4, 291–367.

Wilson, T.M., Stewart, C., Sword-Daniels, V., Leonard, G.S., Johnston, D.M., Cole, J.W., Wardman, J., Wilson, G., Barnard, S.T., 2012. Volcanic ash impacts on critical infrastructure. Phys. Chem. Earth Pt. A/B/C 45-46, 5–23.

Yamamoto, T., Nakada, S., 2014. Extreme Volcanic Risks 4: Mount Fuji. In: Volume on Volcanic Hazards, Risks and Disasters. Elsevier (This volume).

Chapter 14

Extreme Volcanic Risks 2: Mount Fuji

Takahiro Yamamoto[1] and Setsuya Nakada[2]

[1] *Geological Survey of Japan, AIST, Tsukuba, Ibaraki, Japan,* [2] *Volcano Research Center, Earthquake Research Institute, University of Tokyo, Bunkyo-ku, Tokyo, Japan*

ABSTRACT

Fuji Volcano is the largest and closest active volcano to the Tokyo metropolitan area. The 1707 Hoei eruption was the last of this volcano. Fuji Volcano has two extreme risks inherent to its geological and geographical features. The first is a result of its unique tectonic setting; the volcano is located on the collision boundary between the Northeast Japan and Izu-Mariana arcs. Large earthquakes have occurred repeatedly in this zone triggering sector collapses and ensuing debris avalanches, including the Gotemba debris avalanche at 900 BC. The Fujikawa-kako faults cutting the volcano is the most active fault system in Japan, and a forecast indicates that the scale of the next earthquake would be *M 8* and that it is likely to occur some time in the near future within several hundred years. The second risk is related to damage of electric networks in the Tokyo metropolitan area by explosive eruptions. It was forecasted that the fallout thickness by the Hoei-type Plinian eruptions would be ten to several centimeters around the Tokyo Bay area. The electricity in this area highly depends on thermal power plants clustered around the Tokyo Bay. Furthermore, half of the thermal power plants have gas turbines. Although there is an air filter system in the thermal power plants, it may be clogged up by ash and partially or totally block the airflows into the gas turbines. To protect the thermal power plants from the ashfall hazard, the electric company must stop the plant, remove the fallout, and replace the air filters or other broken parts. As a result, massive outages may disrupt social and economic activities in the metropolitan area.

14.1 INTRODUCTION

Fuji Volcano (3,776 m in altitude) is the highest mountain in the country and the symbol of Japan. The volcano has been dormant since the 1707 Hoei eruption, but is ~95 km west-southwest from the Tokyo metropolitan area (Figure 14.1) and has a large potential to damage the country. As a response to

Volcanic Hazards, Risks, and Disasters. http://dx.doi.org/10.1016/B978-0-12-396453-3.00014-9

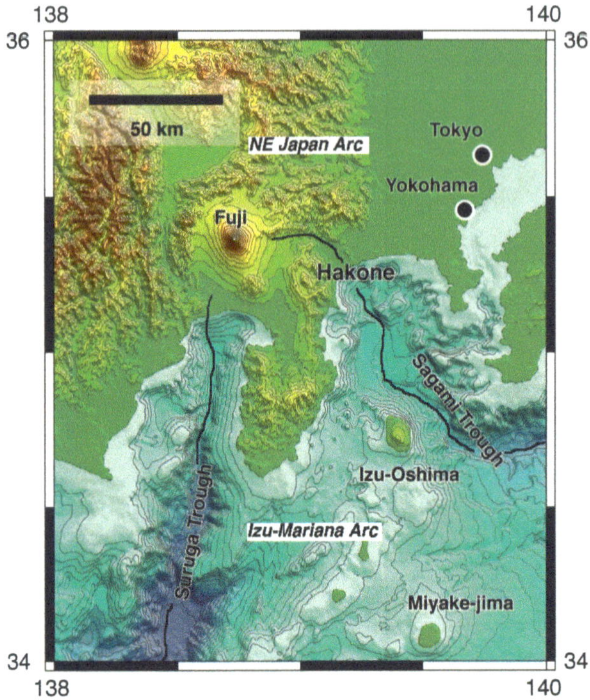

FIGURE 14.1 Index map of Fuji Volcano. Hakone, Izu-Oshima, and Miyake-jima are also active volcanoes. The Suruga and Sagami troughs are subduction boundaries between the Eurasian and Philippine Sea plates. The Izu-Mariana arc on the Philippine Sea plate is colliding with the Northeast Japan arc on the Eurasian plate beneath Fuji Volcano. *After Takada et al. (2013).*

the deep low-frequency (DLF) earthquake activities beneath the volcano from 2000 to 2001 (Ukawa, 2005), the Japanese Government set up the Review Committee of Volcanic Hazard Mitigation of Mt. Fuji (RC) and conducted its first comprehensive research to make new hazard maps for future eruptions of Fuji Volcano (Cabinet Office, Government of Japan, 2004; hereafter referred as CO, 2004). Furthermore, crustal deformation generated by the M_w *9.0* 2011 off the Pacific coast of Tohoku earthquake triggered the M_w *5.9* East Shizuoka earthquake beneath the southern flank, just above the magma system of Fuji Volcano, 4 days after the 2011 main shock (Fujita et al., 2013). This event aroused great public attention about the activity of Fuji Volcano, and the Japanese Government proclaimed the Proposal for Large-scale Volcano Disaster Countermeasures on May 2013 (CO, 2013).

Fuji Volcano is situated in a special position, geologically and geographically, imposing a destructive instability of the edifice and, as a consequence, the possibility of a serious social impact upon the country.

We describe here the eruptive history of Fuji Volcano and two extremely hazardous phenomena: a sector collapse of the volcano caused by a large-magnitude earthquake and ashfall damages on electricity in the Tokyo metropolitan area.

14.2 CHARACTERISTICS OF FUJI VOLCANO

Fuji Volcano has grown in the collision zone between the Northeast Japan and Izu-Mariana arcs. The Sagami and Suruga troughs (Figure 14.1) are subduction boundaries between the Philippine Sea plate (above which is the Izu-Mariana arc) and the Eurasian plate that forms the Northeast Japan arc. The magmatic activity of Fuji Volcano is caused by the subduction of the Pacific plate beneath the Philippine Sea plate and the Eurasian plate. So, crusts are duplicated under the volcano. The southwestern sector of Fuji Volcano has been deformed by the active Fujikawa-kako faults (black solid lines F in Figure 14.2), which are northern branches of the Suruga Trough and the most active faults in Japan. These are reverse faults with western upheaval. The long-term average slip rate is estimated to be ~10 m/kyr (Yamazaki, 1992). Six rupture events have taken place in the last 10 kyr, and individual magnitudes of earthquakes were presumably *M 8* (Figure 14.3; Headquarters for Earthquake Research Promotion, 2010; hereafter referred as HERP, 2010). The last rupture event occurred in the seventh or eighth century.

Fuji Volcano has issued mostly basaltic magmas since the beginning. Its basalt has been evolved with FeO*/MgO ratio larger than 1.6 and shows large variation of incompatible elements concentration without changing in silica content (Takahashi et al., 1991; Togashi et al., 1991; Fujii, 2007). This trend is quite different from other volcanoes of the Izu-Mariana arc (e.g., Izu-Oshima, Miyake-jima in Figure 14.1), and indicates that the magma reservoir beneath Fuji Volcano is deeper than other volcanoes of the Izu-Mariana arc (Fujii, 2007; Kaneko et al., 2010). The magma chamber may have been formed in 15- to 20-km-deep granitic crust of the Philippine Sea plate beneath the Eurasian plate.

Fuji Volcano is characterized by (1) size as the largest basaltic volcano in Japan, (2) the variation of erupted volume and eruption intervals, and (3) the variation of eruption style (Takada et al., 2013). The volume of the edifice is 400–500 km^3. Eruptions with an erupted volume of more than 1 km^3 occurred several times during the last 15,000 years; on the other hand, low volcanic activity periods occurred between 6,000–3,600 BC, and during the last 900 years, not considering the 1707 Hoei eruption. This eruptive pattern makes more difficult to predict future activity (Yamamoto et al., 2005a). Various types of eruptions occur at Fuji Volcano, such as Plinian, lava flow, and pyroclastic flow eruptions. This makes it more difficult to prepare a hazards map (Aramaki, 2007).

FIGURE 14.2 Geological map of Fuji volcano (Takada et al., 2014). Units in Red = lava flows younger than 2.3 ka, Blue–green–gray units = lava flows from 17 to 2.3 ka, yellowish units = volcaniclastic fan deposits. A = Aokigahara lava flow, M = Mishima lava flow, Ob = Obuchi lava flow, S = Saruhashi lava flow, G = Gotemba debris avalanche deposits, T = Tanukiko debris avalanche deposit, U = Umafusegawa debris avalanche deposit, H = Hoei crater, O = Osawa Kuzure. Black solid lines of F are Fujikawa-kako faults.

14.3 ERUPTIVE HISTORY OF FUJI VOLCANO

The eruptive history of Fuji Volcano is summarized in Figure 14.3. Tephra studies suggest that the older activities of Fuji Volcano started at ~100 ka (Machida, 1964, 2007). Fuji Volcano is divided into Older Fuji Volcano and Younger Fuji Volcano (Tsuya, 1968; Miyaji, 1988). According to a detailed ^{14}C study (Yamamoto et al., 2005a), the volcanic activity of Older Fuji

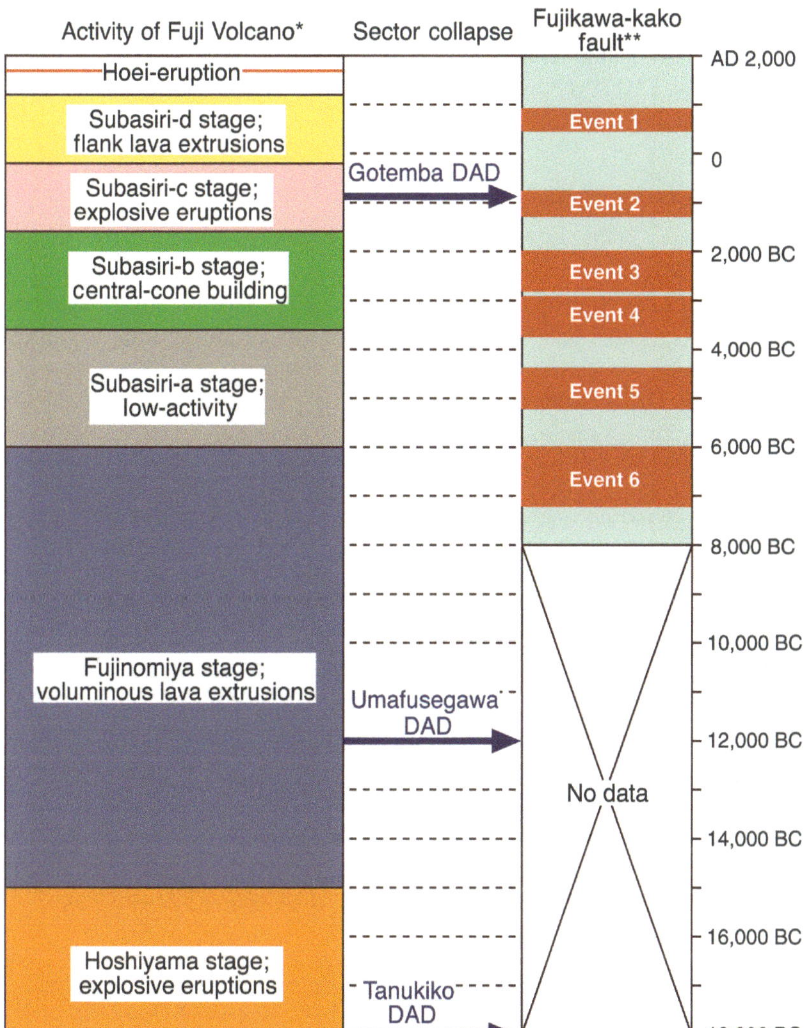

FIGURE 14.3 Eruptive history of Fuji Volcano. Blue arrows indicate the age of debris avalanche deposits (DAD) from Fuji Volcano. In the right column, rupture events of the Fujikawa-kako faults are shown (in red) and the thickness of bars indicates the approximate age range of rupture. * = Yamamoto et al. (2005a, 2007); ** = Headquarters for Earthquake Research Promotion (2010).

Volcano is known as the Hoshiyama stage (older than Cal 15,000 BC); and Younger Fuji Volcano is composed of the Fujinomiya stage (Cal 15,000 BC to Cal 6,000 BC), and the Subashiri stage (younger than Cal 6,000 BC) (Yamamoto et al., 2007). The long-term eruption rate of Fuji Volcano is around 4–5 km^3/kyr for an edifice volume of 400–500 km^3 and an age of 100 kyr.

Miyaji (2007) estimated the long-term eruption rate since the last 11 kyr in detail: 17−3 km^3 DRE (Dense Rock Equivalent)/kyr for 11−8 ka, 3.3−0.4 km^3 DRE/kyr for 5.6−3.7 ka, 2.4 km^3 DRE/kyr for 3.7−2.3 ka, 1.3 km^3 DRE/kyr for 2.3−0.3 ka (Figure 14.4).

14.3.1 100 ka−17 ka (15,000 BC)

The volcanic activity of Older Fuji Volcano (Hoshiyama stage) is characterized by Plinian or sub-Plinian explosive eruptions of basaltic magma. During this stage, the volcano ejected more than 140 tephra units toward the east (Uesugi, 1990) and formed about 10 m of eolian deposits (so-called "Kanto Loam") on plateaus around Tokyo, Yokohama, and Chiba (Machida, 1964, 2007). Distal facies of numerous lahar deposits are commonly making up volcanic fans at the volcano's skirt. The volcano may have been covered

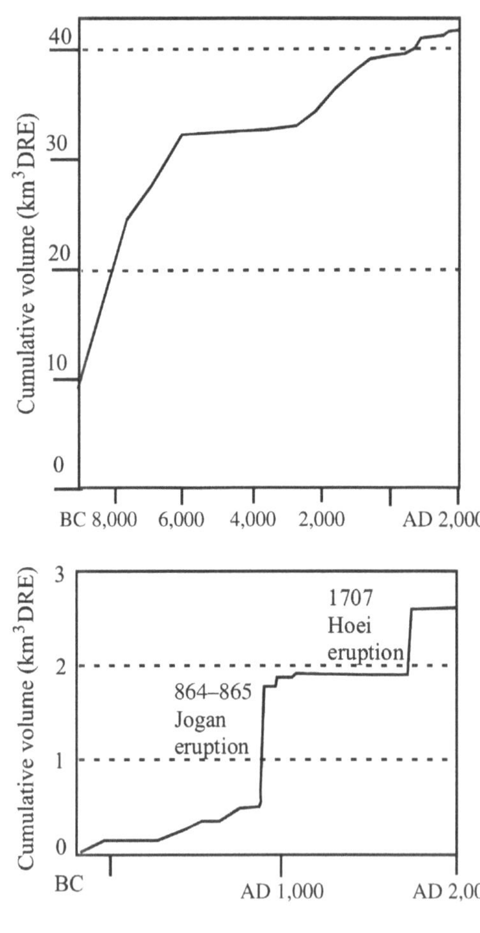

FIGURE 14.4 Cumulative volume for total products erupted by Fuji Volcano during the last 11,000 years (upper) and 2,200 years (lower). Eruption rates of Fuji Volcano have varied with its activity stages (upper). The 864−865 Jogan and 1707 Hoei eruptions were extraordinary large in volume during historic time (lower). However, both eruptions were quite different in style; the Jogan was an effusive eruption of lava flows, and the Hoei was a Plinian eruption. *After Miyaji (2007).*

with ice during the last glaciation from 100 to 20 ka. The main part of Older Fuji Volcano collapsed toward the southwest at Cal 18,000 BC and east at Cal 12,000 BC to form the Tanukiko and Umafusegawa debris avalanche deposits (T and U units in Figure 14.2, respectively) (Yamamoto et al., 2007).

14.3.2 15,000 BC–6,000 BC

The volcanic edifice was built up again to construct Younger Fuji Volcano (Fujinomiya stage) within the Tanukiko avalanche scar. Many voluminous basaltic lava flows in the foot of the volcano were erupted from the central portion. The erupted volumes of the Mishima lava flow, the Obuchi lava flow, and the Saruhashi lava flow (M, Ob, and S units in Figure 14.2) are estimated to be 4, 3, and 1.1 km^3, respectively (Togashi et al., 1991), and the longest runout of the Saruhashi lava flow is the longest (>40 km from the summit).

14.3.3 6,000 BC–3,600 BC

The volcanic activity strongly decreased during this period (Figure 14.3; Subashiri-a stage), and Fuji black soil was formed around the volcano, marking the boundary of older and younger tephra groups of Fuji Volcano (Machida, 1964).

14.3.4 3,600 BC–1,700 BC

The eruptive activity increased again to build the present volcanic cone (Figure 14.3; Subashiri-b stage), which is made up of many thin basaltic lava flows from the central conduit, upon the Fujinomiya-stage edifice (Takada et al., 2007; Ishizuka et al., 2007). The summit region had been formed until Cal 2,100 BC (Yamamoto et al., 2005a). Flank lava extrusions also occurred during this period.

14.3.5 1,700 BC–300 BC

The explosive basaltic Plinian or sub-Plinian eruptions at the summit during Cal 1,700 BC–1,300 BC were followed by flank explosive eruptions forming large scoria cones on the northwestern and southeastern flanks during Cal 1,300 BC–1,000 BC (Figure 14.3; Takada et al., 2007; Ishizuka et al., 2007). The volcano of the eastern flank had collapsed around at Cal 900 BC to form Gotemba debris avalanche (G unit in Figure 14.2; Miyaji et al., 2004). The collapsed volume is estimated to be ~1 km^3 and the collapse involved the remains of the eastern flank of the Hoshiyama-stage edifice. After the collapse, the summit basaltic explosive eruptions became dominant. These eruption products cover the uppermost part of the summit crater rim and the eastern avalanche scar. Some bomb and scoria fall deposits emplaced on the steep western upper flank generated pyroclastic flows toward the west (Yamamoto et al., 2005b).

14.3.6 300 BC–AD 1,100

The Cal 300 BC Yufune-2 eruption was the last one from the summit crater and followed only by basaltic flank eruptions (Figure 14.3; Subashiri-d stage). The volcanic activity on the eastern-northeastern flank was relatively higher than that on the other flanks before Cal AD 700. The frequency of flank eruption became high during Cal AD 700–1,000 in any direction from the summit; the eruptions took place at 20- to 30-year intervals in this period (Takada et al., 2007; Yamamoto et al., 2011). Their fissure eruption sites are restricted within 13.5 km from the summit. The 864–866 Jogan eruption, one of the huge eruptions in Fuji Volcano, occurred on the lower northwestern flank to effuse the Aokigahara lava flow (A unit in Figure 14.2), with a total erupted volume of 1.4 km^3 (Takahashi et al., 2007). The lava flow entered into a paleolake, called "Seno-umi," on the northwestern foot to split into Lake Shoji, and Lake Sai. The volcanic activity became low during AD 1,100–1,700. Some historical documents reported the existence of smoking at the summit during this period (Koyama, 2007).

14.3.7 The 1707 Hoei Eruption

The last eruption, the 1707 Hoei eruption, was Plinian in nature and occurred at the three craters of the southeastern flank (Figure 14.5). The erupted volume was estimated at 0.7 km^3 DRE (Miyaji, 1988), and the ashfall reached distal areas such as Tokyo and Yokohama, approximately 100 km east of the volcano (Figure 14.6). The eruption started on December 16, 1707 and completed on January 1, 1708. The eruption continued for 15 days. Miyaji et al. (2011) have divided the deposit into three stages on the basis of the patterns of the eruptive pulses, whose characteristics are described as follows: Stage I had two energetic eruptive pulses with at least 20 km high columns, each showing an intense initial outburst, followed by a decrease in intensity. The eruption sequence indicates ruptures of highly overpressured dacite and andesite magma chambers. The initial silicic eruption was followed by basaltic magma withdrawal from a deep and voluminous magma chamber. Stage II consisted of discrete sub-Plinian pulses of relatively degassed basaltic magma. Stage III was principally characterized by activity producing a sustained column without a clear repose time. During this stage, the column height appears to have been more than 16 km. The Hoei eruption was followed by the generation of a lot of lahars. About 100 years were necessary to recover from the disaster (Inoue, 2007).

FIGURE 14.5 Southeastern view of Fuji Volcano. The 1707 Hoei eruption generated three craters (H) on the southeastern flank of the volcano. PL is the power line between the Sakuma Frequency Convert Station and the Higashi-Fuji Substation (Figure 14.9). This power line passes through the flank fissure zone of Fuji Volcano.

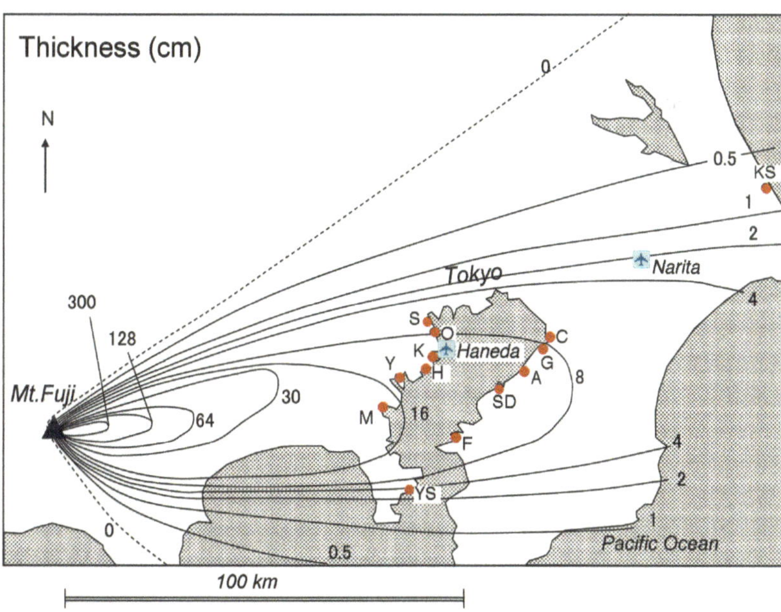

FIGURE 14.6 Isopach map of the 1707 Hoei fall deposit. Numerals are thickness of the fallout in centimeters. Red circles show the thermal power stations (A = Anegasaki, C = Chiba, F = Futtu, G = Goi, H = Higashi-Ogishima, K = Kawasaki, KS = Kashima, M = Minami-Yokohama, O = Ooi, S = Shinagawa, SD = Sodegaura, Y = Yokohama, YS = Yokosuka, Table 14.2). Haneda and Narita are international airports. *Modified from Miyaji and Koyama (2007).*

14.4 GEOPHYSICAL MONITORING

Although surface volcanic activity of Fuji Volcano has been quiet in recent years, the DLF earthquakes have been identified in the mid-crustal depth beneath the summit area since the start of monitoring in the early 1980s by using high-quality seismic observations. The predominant frequencies of most events range from 1 to 3 Hz, far lower than those of tectonic earthquakes of the same magnitude range. The DLF earthquake activity at Fuji Volcano is characterized by successive occurrence of small low-frequency earthquakes for several minutes to a few tenths of minutes. Such events had occurred about 15 times in a year on average, before the year 2000.

The DLF activity increased extremely in the period from October 2000 to May 2001 (Ukawa, 2005). The activity returned to the usual level in June 2001. The DLF earthquakes are clustered around an area centered at 2–4 km northeast from the summit. The principal depths of the DLF events range from 10 to 15 km. During the high-activity period, no significant depth change was recognized either in epicenter or focal depth, indicating no evidence for magma ascent (Nakamichi et al., 2005; Ukawa, 2005). Ground tilt observations also indicated no abnormal changes relating to volcanic activity.

This unusual seismic activity, however, prompted the government to improve seismic and crustal deformation observations, especially in the summit and in the high-altitude area, partly because of the high degree of public interest in volcanic activity of Fuji Volcano. The geophysical monitoring of Fuji Volcano is performed by several institutes including National Research Institute for Earth Science and Disaster Prevention, Earthquake Research Institute of Tokyo University, Japan Meteorological Agency (JMA), Geographical Survey Institute, and Yamanashi Institute of Environmental Sciences. Most of the data are transmitted to JMA for real-time monitoring of Fuji Volcano.

14.5 SECTOR COLLAPSE OF FUJI VOLCANO

14.5.1 Frequency of Debris Avalanches

Alongside pyroclastic flows, debris avalanches stemming from sector collapses rank as one of the most dangerous types of volcanic processes that might produce a disaster. A debris avalanche is a large-scale landslide that travels at high speed of up to 100 m/s, and can flow for distances of 10 km or more (Siebert et al., 1987; Shea and van Wyk de Vries, 2008). They occur when a section of a steep-sided edifice, such as a stratovolcano or lava dome, breaks loose after being triggered by an earthquake or volcanic eruption. Their deposits partially retain the structure of the original edifice, but they are generally fragmented and consist of nonadhesive clasts. Moreover, deposits are thin and spread out, and the ratio of drop height to runout length (H/L = apparent friction coefficient) decreases as the volume of debris increases (Figure 14.7).

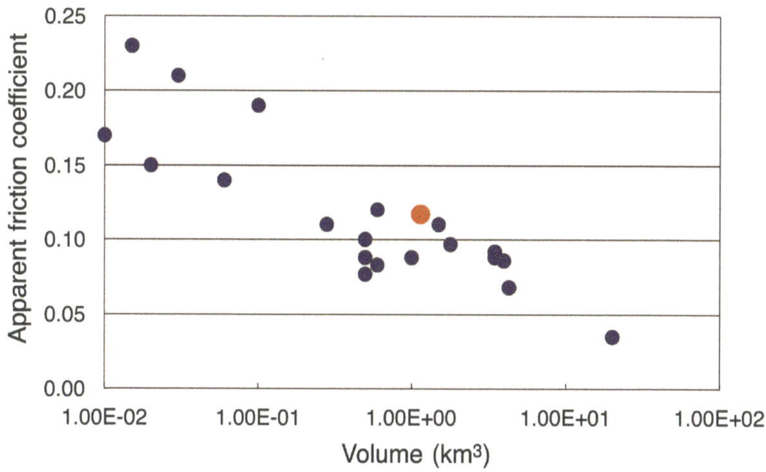

FIGURE 14.7 Volume versus apparent friction coefficient of debris avalanche deposits. Blue circle = debris avalanches of 10 stratovolcanoes in the Northeast Japan arc, red circle = the Gotemba debris avalanche. *After Yamamoto et al. (2002).*

As shown clearly through numerical experiments, these characteristics are able to explain a debris avalanche fairly well provided that the avalanche travels as a particle flow (Campbell et al., 1995).

In the case of Fuji Volcano, it is well known that there are three debris avalanche deposits with a volume of $>10^{-1}$ km^3 in the past 20,000 years (Figure 14.3). This frequency is clearly an order of magnitude larger than the average frequency of 10 stratovolcanoes in the Northeast Japan arc that was several collapses with $>10^{-1}$ km^3 volume per 100,000 years (Yamamoto et al., 2002). Since the eruption rate of Fuji Volcano is greater than that of these volcanoes, naturally its sectors will recover more quickly following an edifice collapse. However, it is certain that there is some reason for the triggering frequency of edifice collapses being greater than that for volcanoes in other regions.

14.5.2 Trigger of the Sector Collapse

The Gotemba debris avalanche deposit (G unit in Figure 14.2) has been dated at about Cal 900 BC (Miyaji et al., 2004). This collapse age is quite consistent with the rupture event 2 of the Fujikawa-kako faults that has been shown by a trench excavation survey (Figure 14.3; HERP, 2010). Eruptions and earthquakes are the typical causes of debris avalanches. In the case of the famous Mount St Helens eruption in 1980, the edifice deformed and destabilized, and then collapsed because magma had intruded into the edifice (Lipman and Mullineaux, 1981). However, in this case as well, the direct trigger was an earthquake. Also, an *M* 5 earthquake just below the edifice triggered the 1888 phreatic eruption and sector collapse of Bandai Volcano (Yamamoto et al.,

1999). In the case of the Gotemba debris avalanche deposit, the evidence for a magmatic eruption has not been found so far. However, the close temporal proximity between the Gotemba debris avalanche and the fault dislocation (Figure 14.3) points to the possibility that the large earthquake on the Fujikawa-kako faults triggered the sector collapse that produced the Gotemba debris avalanche deposit.

The Fujikawa-kako faults have caused six large earthquakes since 10,000 years ago, but the rupture event 2 only triggered the sector collapse (Figure 14.3). The event 2 was presumably an interlocked earthquake of the Sagami and Suruga troughs and directly dislocated the edifice between the two troughs, because a rupture along the Kozu fault, which is an extension of the Sagami Trough (Figure 14.1), took place between 2,800 and 3,200 years ago (HERP, 2009). The Cal 18,000 BC Tanukiko and Cal 12,000 BC Umafuse-gawa debris avalanches also lack magmatic products accompanying the eruption. Although their triggers are uncertain, the high frequency of sector collapses of Fuji Volcano is probably a consequence of the high tectonic activity level of the Fujikawa-kako faults.

14.5.3 Risk of the Sector Collapse

A forecast indicates that the scale of the next earthquake along the Fujikawa-kako faults would be M 8 and that it is likely to occur some time in the near future within several hundred years; the long-term probability of earthquake occurrence is 2–18 percent for the next 30 years (HERP, 2010). Although a large sector collapse (>0.1 km^3) is more rare than a rupture event, it is important to consider that the present summit region of Fuji Volcano is gravitationally unstable. The western slope of the summit is steeper than 34°, a slope beyond the angle of repose of most pyroclastic materials (Figure 14.8; Yamamoto et al., 2005b). Osawa Kuzure, a deep gully within the western slope (O in Figure 14.2), appeared 1,000 years ago immediately after rupture event 1 of the Fujikawa-kako faults, and has discharged voluminous debris flows ever since (Figure 14.8; Hanaoka et al., 2007). On the other hand, Fuji Volcano has no summit eruptions since 300 BC and undergoes erosion by the expansion of Osawa Kuzure. There is a horizontal distance of 12 km at the closest distance from the current summit of Fuji Volcano to the communities at its western base and a vertical drop of ~3200 m. Therefore, if a debris avalanche occurs with an H/L of less than 0.25, it will reach the communities (Figure 14.8). If the amount of collapsed material required for a debris avalanche to reach communities were on the order of 10^{-2} km^3 (Figure 14.7), an event 2 orders of magnitude smaller than the Gotemba debris avalanche could be the direct cause of a disaster. Predicting the occurrence of debris avalanches brought on by earthquakes is difficult without the ability to predict earthquakes. However, the next large earthquake is expected to put portions of the edifice, including the peak, at high risk for small-scale collapse.

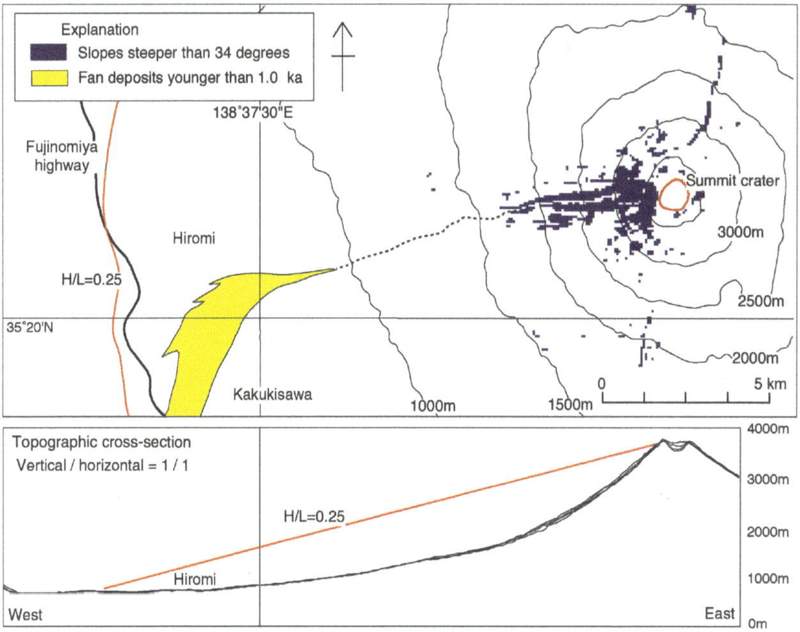

FIGURE 14.8 Distribution of steep slopes (>34°) and west-east topographic cross-section through Fuji Volcano. H/L = height difference/horizontal distance. *Modified from Yamamoto et al. (2005b).*

14.6 ASHFALL DAMAGE ON ELECTRICITY IN THE TOKYO METROPOLITAN AREA

14.6.1 Ashfall Hazard Due to the Hoei-Type Eruption

The 1707 Hoei eruption occurred after a 2,000-year interval without Plinian-type explosions from Fuji Volcano. Although Fuji Volcano shed frequently lava flows from flank fissures without voluminous tephra dispersion during the Subashiri-d stage (300 BC to AD 1,100), the activity of Fuji Volcano has presumably shifted to a new stage since AD 1,100 (Yamamoto et al., 2005a). Plinian-type explosive eruptions were common rather than rare in 10,000-year lifetime of Fuji Volcano (Machida, 1964, 2007). So, the Hoei-like explosive eruption is the most likely style for future activity of Fuji Volcano. RC estimated the maximum possible area of ashfall in cases similar to the Hoei eruption, showing 10-cm- and more than 2-cm-thick fallout in Yokohama and Tokyo, respectively (CO, 2004). Three hundred years ago, already about one million people were in Edo, which is the name for Tokyo until AD 1,868. The impacts in Edo from the Hoei eruption (Figure 14.6) were limited to respiratory problems suffered by many residents, probably caused by inhalation of fine ash (Koyama, 2009). However, the present more densely populated and modern cities within the ashfall hazard area, such as Tokyo, Yokohama, and Chiba, would be much more vulnerable to a similar event. It is critical to delineate the worst-case scenario for future eruptions of Fuji Volcano in order to plan mitigation measures in case of a Hoei-type explosive eruption.

Although ashfalls rarely endanger human life directly, threats to public health, disruption to critical infrastructure services, aviation, buildings, and primary production can lead to significant societal and economic impacts (e.g., Blong, 1984; Wardman et al., 2012b). Economical loss during a future Hoei-type explosive eruption was estimated to be 1.2–1.6 trillion Yen in fair days, 1.8–2.2 trillion Yen in rainy days, and 2.1–2.5 trillion Yen in rainy seasons, including damages of houses, lifelines, roads, railways, airways, various manufacturing industries, tourism, and so on (CO, 2004). However, RC's accounts may underestimate damages to electricity networks, which are very vulnerable to disruption from ashfalls (Wardman et al., 2012a,b). RC assumed no ashfall damage in fair-day cases and 18 percent insulator flashover in rainy-day cases with more than 1-cm-thick ashfall using data of outage accidents around active Sakurajima Volcano, Japan (CO, 2004). Recent studies show that the flashover risk is much higher than RC's prerequisite (Wardman et al., 2012a; Wilson et al., 2012). Furthermore, the RC's estimation lacked more severe problems in electricity that may cause large-scale blackouts in the Tokyo metropolitan area as follows.

14.6.2 Electric Power Supply to the Tokyo Metropolitan Area

The Tokyo Electric Power Company (TEPCO) supplies the Tokyo metropolitan area representing 10 percent of Japan's territory with 35 percent of the nation's population with electric power (Table 14.1). The peak demand of this area is ~50 GW, which is 34 percent of the total for the entire nation. TEPCO owns163 hydroelectric, 25 thermal, and three nuclear (Fukushima Daiichi, Fukushima Daini, and Kashiwazaki-Kariwa; Figure 14.9) power stations and has a 66 GW generation capacity in total as of the end of FY2011 (Table 14.1). However, after the Fukushima nuclear accident triggered by the devastating 2011 tsunami (IAEA, 2011), the Fukushima Daiichi Nuclear Power Station was decommissioned, and other two nuclear power stations have been suspended since the accidents. Currently, TEPCO generates most of the electricity by thermal power stations, and somehow maintains a 50 GW capacity. The thermal power generation has continued to play a central role in supplying electric power in Japan, because reoperation of nuclear power stations must be inspected by the New Regulatory Requirements for Commercial Nuclear Power Reactors since July 2013 (Nuclear Regulation Authority, 2013). TEPCO has been building additional thermal generators in the station since the Fukushima accident (Table 14.2).

TABLE 14.1 Service Area and Generation Capacity of the Tokyo Electric Power Company (TEPCO) as of the End of FY2011

	Area (km^2)	Popula-tion (million)	Peak Demand (GW)	Generation Capacity (GW)				
				Hydro	Thermal	Nuclear	New Energy	Total
TEPCO	39,509	44.73	49.99	8.982	40.148	17.308	0.034	66.472
Total for entire nation	377,955	127.73	156.6	35.602	125.483	46.343	0.580	208.008

Data from TEPCO http://www.tepco.co.jp/en/corpinfo/overview/pdf-4/04-e.pdf. New energy consists of wind, photovoltaic, waste, geothermal, and biomass power generation.

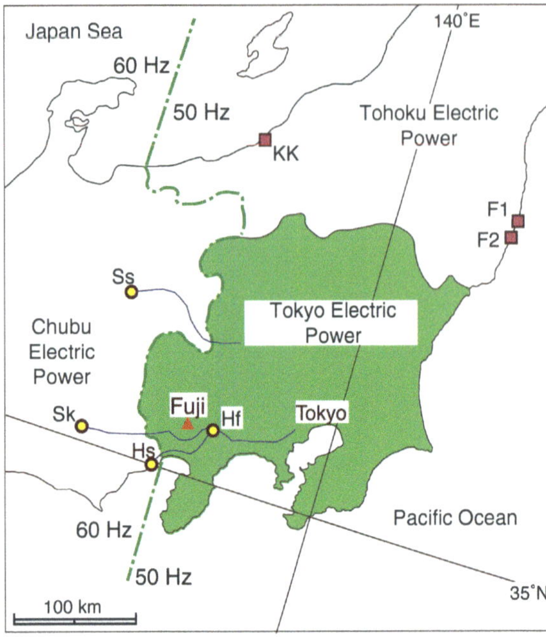

FIGURE 14.9 Service areas of the electric power companies around Tokyo. Pink squares and yellow circles are the nuclear power stations of the Tokyo Electric Power Company (F1 = Fukushima Daiichi, F2 = Fukushima Daini, K = Kashiwazaki-Kariwa) and the frequency convert stations (Hf = Higashi-Fuji, Hs = Higashi-Shimizu, Sk = Sakuma, Ss = Shin-Shinano) between 50 and 60 Hz electricity (dotted line). The Fukushima Daiichi Nuclear Power Station was decommissioned due to the nuclear accident caused by the 2011 tsunami, and other two nuclear power stations have been suspended since the accident.

14.6.3 Vulnerability of Electricity in the Tokyo Metropolitan Area

The first problem in electricity is that the TEPCO's thermal power stations are concentrated around the Tokyo Bay, except three stations (Figure 14.5). Namely, these power stations are located within the more than 4-cm-thick ashfall hazard area of the Hoei eruption, and they can be affected at once by a single explosive eruption. Not only volcanic but also seismic hazards, including liquefaction of reclaimed land and tsunami, can cause heavy damages in electricity due to this geographical problem. Although such clustering of the plants is efficient for fueling, power supply, and maintenance, it has a serious vulnerability against various geological hazards.

The second problem is that TEPCO employs gas turbines for a half of the thermal power generation (Table 14.2). TEPCO has replaced old steam power plants with combined cycle (CC), advanced combined cycle (ACC), and more advanced combined cycle (MACC) plants, as occasion demands; these

TABLE 14.2 List of the Thermal Power Stations of Tokyo Electric Power Company (TEPCO) as of August 2014

Ppwer Station	Max Output (kW)	Start of Operation	Type	Fuel
Chiba	4,380,000	2000	ACC	LNG
		2014	MACC	
Goi	1,886,000	1963	ST	LNG
		1993	GT	
Anegasaki	3,600,000	1967	ST	LNG, HO
	[5,600]	[2011]	[DE]	[LO]
Sodegaura	3,600,000	1974	ST	LNG
Futtu	5,040,000	1986	CC	LNG
		2003	ACC	
		2010	MACC	
Yokosuka	2,274,000	1964	ST	HO
		1971	GT	LO
Kawasaki	2,000,000	2009	MACC	LNG
Yokohama	3,325,000	1964	ST	HO
		1998	ACC	LNG
Minami-Yokohama	1,150,000	1970	ST	LNG
Higashi-Ogishima	2,000,000	1987	ST	LNG
Kashima	5,660,000	1971	ST	HO
		2014	ACC	CG
Ooi	1,050,000	1971	ST	HO
	[81,000]	[2011]	[GT]	[CG]
Hirono	4,400,000	1980	ST	HO
Shinagawa	1,140,000	2003	ACC	CG
Hitachinaka	2,000,000	2003	ST	Coal

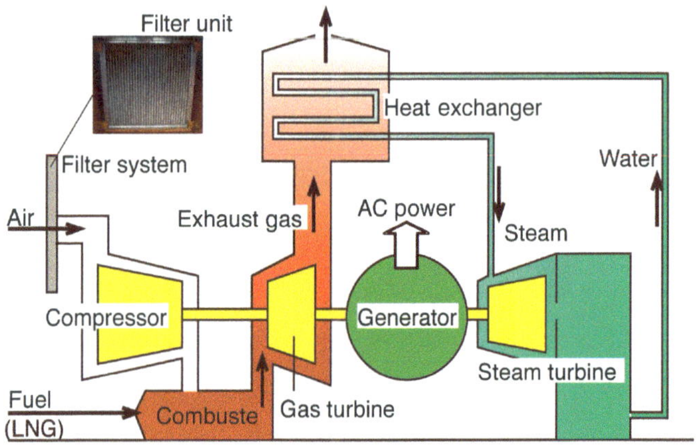

FIGURE 14.10 Schematic diagram for the combined cycle (CC) power generator. CC is a method of generating electric power that combines gas turbine power generation with steam turbine power generation. The air filter system is fitted with more than 100 filter units (60 cm by 60 cm; Photo by the Shinwa Corporation Ltd.). *Modified from TEPCO http://www.tepco.co.jp/en/challenge/energy/thermal/power-g-e.html.*

CC systems are designed for maximum efficiency in which the hot exhaust gases from the gas turbine are used to raise steam to power a steam turbine with both turbines being connected to electricity generators (Figure 14.10). That is why the TEPCO's thermal power stations are weak in ashfall hazard similar to jet engines of aircrafts (Casadevall, 1994; Miller and Casadevall, 2000). In the gas turbine, fresh atmospheric air flows through a compressor that brings it to higher pressure. Energy is then added by spraying fuel into the air and igniting it so the combustion generates a high-temperature flow. This high-temperature high-pressure gas enters a turbine, where it expands down to the exhaust pressure, producing a shaft work driving compressor and electric generator (Figure 14.10). Inlet gas temperatures of the turbines are 1,100, 1,300, and 1,500 °C in CC, ACC, and MACC, respectively; they are sufficiently higher than the melting point of ash. Although there is an air filter system in the thermal power plants, it may be clogged up by ash and partially or totally block the airflows into the gas turbines. Furthermore, inhalation of ash into the gas turbine may be manifested in one or more of the following ways: (1) deposition of remelted ash on hot-section components, (2) erosion of compressor blades, (3) partial or complete blocking of fuel nozzles similar to jet engines (Dunn and Wade, 1994). An incident of the ashfall over thermal power plants has never taken place in Japan so far; therefore the plants are been operated without special protection. The damage that the gas turbine would experience depends upon the air filter system. However, the filter system was not tested when operating in the ashfall, according to its maker (Shinwa Corporation Ltd., personal communication).

The third problem is that TEPCO has not yet showed a plan for countermeasures against the ashfall damages, including insulator flashover, transformer damage, and line collapse. TEPCO must suspend the plant, remove the fallout, and replace the air filter units (Figure 14.10) or other broken parts in order to protect the thermal power plants from the ashfall hazard. However, the electric power supply to the Tokyo metropolitan area depends strongly on the TEPCO's thermal generation in the ashfall hazard zone. So, massive controlled outage caused by the ashfall may disrupt social and economic activities in the whole metropolitan area over the ashfall hazard zone. Transmission of electricity from the Chubu Electric Power Company also has a problem of difference in electric frequency; 50 Hz in TEPCO and 60 Hz in the western power companies (Figure 14.9). Although there are three stations converting the frequency and transmitting 600 MW in total, the power lines from two conversion stations to Tokyo transit the eastern foot of Fuji Volcano (Figure 14.5 and 14.9) and can be collapsed by thick scoria fall during a Hoei-type eruption. On the other hand, the northern Tohoku Electric Power Company, which was damaged by the 2011 tsunami, does not have much capacity transmitting to TEPCO. The Japanese Government proclaimed the Proposal for Large-scale Volcano Disaster Countermeasures on May 2013 and stressed the necessity for synthetic research of the ashfall hazard on electric facilities (CO, 2013). TEPCO should draw up a disaster response plan and establish a backup system to avoid the blackout on the future explosive eruption of Fuji Volcano.

14.7 CONCLUSION

Although Fuji Volcano has been superficially inactive since the last 1707 Hoei eruption, it has a large hazardous potential due to various volcanic phenomena. Among all of them, we focused on sector collapses of the edifice and ashfall hazards on the Tokyo metropolitan area. In general, sector collapses are low-probability phenomena; at the moment, they fall largely outside the realm of disaster prediction. However, given the unique tectonic setting of Fuji Volcano, which is located on the collision boundary between the Northeast Japan and Izu-Mariana arcs, huge earthquakes have repeatedly occurred in this zone thus triggering sector collapses and the ensuing debris avalanches. So, the possibility of a debris avalanche occurring in the near future must not be ignored. Even a small-volume (10^{-2} km^3) collapse including the summit area can produce a debris avalanche destroying the communities on the foot. On the other hand, Plinian or sub-Plinian explosions such as the 1707 Hoei eruption would cause ashfall hazards in the Tokyo metropolitan area. It was forecasted that the fallout thickness during Hoei-type eruptions would be ten to several centimeters in the area around the Tokyo Bay area. Ashfalls covering the metropolitan area can lead to significant societal impacts. Especially, electricity networks in the Tokyo metropolitan area are very vulnerable to disruption due to ashfalls, because the electricity depends mainly on thermal power plants clustered around the Tokyo Bay. Furthermore, half of the thermal power plants use gas turbines, which can easily be damaged by inhalation of ash, as in the case of jet engines of aircraft. Although there is an air filter system in the thermal power plants, they may be clogged up by ash and partially or totally block the airflows into the gas turbines. In order to protect the thermal power plants from the ashfall hazard, the electric company must suspend the plant, remove the fallout, and replace the air filters or other broken parts. As a result, massive outages during interruption of electric supply may disrupt social and economical activities in the whole metropolitan area over the ashfall hazard zone.

REFERENCES

Aramaki, S., 2007. Volcanic disaster mitigation maps of Fuji volcano and overview of mitigation programs. In: Aramaki, S., Fujii, T., Nakada, S., Miyaji, N. (Eds.), Fuji Volcano. Yamanashi Instit Environ Sci, pp. 451–475 (in Japanese with English abstract).

Blong, R.J., 1984. Volcanic Hazards. A Sourcebook on the Effects of Eruption. Academic Press, 424 p.

Cabinet Office, Government of Japan (CO), 2004. Report of the Review Committee of Volcanic Hazard Mitigation of Mt. Fuji. (in Japanese) http://www.bousai.go.jp/kazan/fujisan-kyougikai/report/index.html.

Cabinet Office, Government of Japan (CO), 2013. Proposal for large-scale volcano disaster countermeasures. (in Japanese) http://www.bousai.go.jp/kazan/kouikibousai/pdf/20130516_teigen.pdf.

Campbell, C.S., Cleary, P.W., Hopkins, M., 1995. Large-scale landslide simulations: global deformation, velocities and basal friction. J. Geophys. Res. 100 (B5), 8267–8283.

Casadevall, T.J., 1994. The 1989–1990 eruption of Redoubt volcano, Alaska: impacts on aircraft operations. J. Volcanol. Geotherm. Res. 62, 301–316.

Dunn, M.G., Wade, D.P., 1994. Influence of volcanic ash clouds on gas turbine engines. In: Casadevall, T.J. (Ed.), Volcanic Ash and Aviation Safety: Proceedings of the First International Symposium on Volcanic Ash and Aviation Safety, vol. 2047. US Geol Surv Bull, pp. 107–117.

Fujii, T., 2007. Magmatology of Fuji volcano. In: Aramaki, S., Fujii, T., Nakada, S., Miyaji, N. (Eds.), Fuji Volcano. Yamanashi Instit Environ Sci, pp. 233–244 (in Japanese with English abstract).

Fujita, E., Kozono, T., Ueda, H., Kohno, Y., Yoshioka, S., Toda, N., Kikuchi, A., Ida, Y., 2013. Stress field change around the Mount Fuji volcano magma system caused by the Tohoku megathrust earthquake, Japan. Bull. Volcanol. 75, 679–682.

Hanaoka, M., Tomita, Y., Ito, M., 2007. Osawa Kuzure (failure) and debris flows in Mt. Fuji. In: Aramaki, S., Fujii, T., Nakada, S., Miyaji, N. (Eds.), Fuji Volcano. Yamanashi Instit Environ Sci, pp. 407–425 (in Japanese with English abstract).

Headquarters for Earthquake Research Promotion (HERP), 2010. Long-term evaluation of the Fujikawa-kako faults (in Japanese). http://www.jishin.go.jp/main/chousa/10oct_fujikawa/index.htm.

Headquarters for Earthquake Research Promotion (HERP), 2009. Long-term Evaluation of the Kannawa and Kozu-Matsuda Faults (in Japanese). http://www.jishin.go.jp/main/chousa/09jun_kannawa/index.htm.

IAEA, 2011. Report of Japanese Government to IAEA Ministerial Conference on Nuclear Safety – Accident at TEPCO's Fukushima Nuclear Power Stations. http://www.iaea.org/newscenter/focus/fukushima/japan-report/.

Inoue, K., 2007. Distribution of sediment-related disasters after the Hoei eruption of Fuji volcano in 1707, based on historical documents. In: Aramaki, S., Fujii, T., Nakada, S., Miyaji, N. (Eds.), Fuji Volcano. Yamanashi Instit Environ Sci, pp. 427–439 (in Japanese with English abstract).

Ishizuka, Y., Takada, A., Suzuki, Y., Kobayashi, M., Nakano, S., 2007. Eruption ages and whole-rock chemistries of scoria cones on the northern to western slope of Fuji volcano based on trenching surveys. Bull. Geol. Surv. Jpn. 57, 357–376 (in Japanese with English abstract).

Kaneko, T., Yasuda, A., Fujii, T., Yoshimoto, M., 2010. Crypto-magma chambers beneath Mt. Fuji. J. Volcanol. Geotherm. Res. 193, 161–170.

Koyama, M., 2007. Database of eruptions and other activities of Fuji volcano, Japan, based on historical records since AD 781, Fuji volcano. In: Aramaki, S., Fujii, T., Nakada, S., Miyaji, N. (Eds.), Fuji Volcano. Yamanashi Instit Environ Sci, pp. 119–136 (in Japanese with English abstract).

Koyama, M., 2009. Eruption of Fuji Volcano and Hazard Map. Kokon Shoin, Tokyo, 174p. (in Japanese).

Lipman, P.W., Mullineaux, D.R. (Eds.), 1981. The 1980 Eruptions of Mount St. Helens, vol. 1250. US Geol Surv Prof Pap, Washington, p. 844.

Machida, H., 1964. Tephrochronological study of volcano Fuji and adjacent areas. J. Geography 73, 293–308, 337–350. (in Japanese with English abstract).

Machida, H., 2007. Development of Fuji volcano: a review from Quaternary tephrochronology. In: Aramaki, S., Fujii, T., Nakada, S., Miyaji, N. (Eds.), Fuji Volcano. Yamanashi Instit Environ Sci, pp. 29–44 (in Japanese with English abstract).

Miller, T.P., Casadevall, T.J., 2000. Volcanic ash hazards to aviation. In: Sigurdsson, H., Miller, T.P., Casadevall, T.J. (Eds.), Encyclopedia of Volcanoes. Elsevier, New York, pp. 915–930.

Miyaji, N., 1988. History of younger Fuji volcano. J. Geol. Soc. Jpn. 94, 433–452 (in Japanese with English abstract).

Miyaji, N., 2007. Eruptive history, eruption rate and scale of eruptions for the Fuji volcano during the last 11,000 years, Fuji volcano. In: Aramaki, S., Fujii, T., Nakada, S., Miyaji, N. (Eds.), Fuji Volcano. Yamanashi Instit Environ Sci, pp. 79–95 (in Japanese with English abstract).

Miyaji, N., Koyama, M., 2007. Recent studies on the 1707 (Hoei) eruption of Fuji volcano. In: Aramaki, S., Fujii, T., Nakada, S., Miyaji, N. (Eds.), Fuji Volcano. Yamanashi Instit Environ Sci, pp. 339–348 (in Japanese with English abstract).

Miyaji, N., Kan'no, A., Kanamaru, T., Mannen, K., 2011. High-resolution reconstruction of the Hoei eruption (AD 1707) of Fuji volcano, Japan. J. Volcanol. Geotherm. Res. 207, 113–129.

Miyaji, N., Togashi, S., Chiba, T., 2004. A large-scale collapse event at the eastern slope of Fuji volcano about 2900 years ago. Bull. Volcanol. Soc. Jpn. 49, 237–248 (in Japanese with English abstract).

Nakamichi, H., Ukawa, M., Sakai, S., 2005. Precise hypocenter locations of midcrustal low-frequency earthquakes beneath Mt. Fuji, Japan. Earth Planets Space 56, e37–e40.

Nuclear Regulation Authority, 2013. Enforcement of the New Requirements for Commercial Nuclear Power Reactors. NRA, Japan http://www.nsr.go.jp/english/e_news/data/13/0912.pdf.

Shea, T., van Wyk de Vries, B., 2008. Structural analysis and analogue modeling of the kinematics and dynamics of rockslide avalanches. Geosphere 4, 657.

Siebert, L., Glicken, H., Ui, T., 1987. Volcanic hazards from Bezymianny- and Bandai-type eruptions. Bull. Volcanol. 49, 435–459.

Takada, A., Mannen, K., Yamamoto, T., 2013. B03: Fuji and Hakone volcanoes: typical strato-volcanoes in Japan, IAVCEI 2013 field trip guide. Bull. Volcanol. Soc. Jpn. 58. B3-1-26.

Takada, A., Yamamoto, T., Ishizuka, Y., Nakano, S., 2014. Geological map of Fuji Volcano, 2nd Edition (Ver.1). GSJ Open-File Report, no.592. Geol. Sur. Jpn. AIST. https://www.gsj.jp/researches/openfile/openfile2013/openfile0592.html.

Takada, A., Ishizuka, Y., Nakano, S., Yamamoto, T., Kobayashi, M., Suzuki, Y., 2007. Characteristic and evolution inferred from eruptive fissures of Fuji volcano, Japan. In: Aramaki, S., Fujii, T., Nakada, S., Miyaji, N. (Eds.), Fuji Volcano. Yamanashi Instit Environ Sci, pp. 183–202 (in Japanese with English abstract).

Takahashi, M., Hasegawa, H., Tsukui, M., Nemoto, Y., 1991. Evolution of magma-plumbing system beneath Fuji volcano: on the view point of whole-rock chemistry. Bull. Volcanol. Soc. Jpn. 36, 281–296 (in Japanese with English abstract).

Takahashi, M., Matsuda, F., Yasui, M., Chiba, T., Miyaji, N., 2007. Geology of the Aokigahara lava flow field in Fuji volcano and its implications in the process of Jogan eruption. In: Aramaki, S., Fujii, T., Nakada, S., Miyaji, N. (Eds.), Fuji Volcano. Yamanashi Instit Environ Sci, pp. 303–338 (in Japanese with English abstract).

Togashi, S., Miyaji, N., Yamazaki, H., 1991. Fractional crystallization in a large tholeiitic magma chamber during the early stage of the younger Fuji volcano, Japan. Bull. Volcanol. Soc. Jpn. 36, 269–280 (in Japanese with English abstract).

Tsuya, H., 1968. Geologic map of Mt. Fuji, 1:50, 000. Miscellaneous map series 12. Geol Sur. Jpn.

Uesugi, Y., 1990. Standard tephra columns for the eastward area of volcano Fuji. Part 1: S-25 ~ Y-141. Bull. Assoc. Kanto Quat. Res. 16, 3–28 (in Japanese).

Ukawa, M., 2005. Deep low-frequency earthquake swarm in the mid-crust beneath Mount Fuji (Japan) in 2000 and 2001. Bull. Volcanol. 68, 47–56.

Wardman, J.B., Wilson, T.M., Bodger, P.S., Cole, J.W., Johnston, D.M., 2012a. Investigating the electrical conductivity of volcanic ash and its effect on HV power systems. Phys. Chem. Earth 45–46 (Parts A/B/C), 128–145.

Wardman, J.B., Wilson, T.M., Bodger, P.S., Cole, J.W., Stewart, C., 2012b. Potential impacts from tephra fall to electric power systems: a review and mitigation strategies. Bull. Volcanol. 74, 2221–2241.

Wilson, T.M., Stewart, C., Sword-Daniels, V., Leonard, G.S., Johnston, D.M., Cole, J.W., Wardman, J., Wilson, G., Barnard, S.T., 2012. Volcanic ash impacts on critical infrastructure. Phys. Chem. Earth 45–46 (Parts A/B/C), 5–23.

Yamamoto, T., Nakamura, Y., Glicken, H., 1999. Pyroclastic density current from the 1888 phreatic eruption of Bandai volcano, NE Japan. J. Volcanol. Geotherm. Res. 90, 191–207.

Yamamoto, T., Takada, A., Shimokawa, K., 2002. Debris avalanches of Fuji voclano. Chikyu Mon. 24, 640–644 (in Japanese).

Yamamoto, T., Takada, A., Ishizuka, Y., Nakano, S., 2005a. Chronology of the products of Fuji volcano based on new radiometric carbon ages. Bull. Volcanol. Soc. Jpn. 50, 53–70 (in Japanese with English abstract).

Yamamoto, T., Takada, A., Ishizuka, Y., Miyaji, N., Tajima, Y., 2005b. Basaltic pyroclastic flows of Fuji volcano, Japan: characteristics of the deposits and their origin. Bull. Volcanol. 67, 622–633.

Yamamoto, T., Ishizuka, Y., Takada, A., 2007. Surface and subsurface geology at the southwestern foot of Fuji volcano, Japan: new stratigraphy and chemical variations of the products. In: Aramaki, S., Fujii, T., Nakada, S., Miyaji, N. (Eds.), Fuji Volcano. Yamanashi Instit Environ Sci, pp. 97–118 (in Japanese with English abstract).

Yamamoto, T., Nakano, S., Takada, A., Kobayashi, M., 2011. New stratigraphy of the recent ejecta on the eastern flank of Fuji volcano, Japan. Bull. Geol. Surv. Jpn. 62, 405–424 (in Japanese with English abstract).

Yamazaki, H., 1992. Tectonics of a plate collision along the northern margin of Izu peninsula, central Japan. Bull. Geol. Surv. Jpn. 43, 603–657.

Chapter 15

Volcanic Gas and Aerosol Hazards from a Future Laki-Type Eruption in Iceland

Anja Schmidt
School of Earth and Environment, University of Leeds, Leeds, United Kingdom

ABSTRACT

The eruptions of Eyjafjallajökull in 2010 and Grímsvötn in 2011 in Iceland were stark reminders that society is very vulnerable to volcanic hazards. This chapter assesses the potential impacts of a future Icelandic Laki-type eruption on air quality, human health, and aviation by combining atmospheric modeling, volcanological data sets, and epidemiological evidence. Depending on the duration of such an eruption, air quality could be degraded for weeks or months to a degree that visibility across Europe may be noticeably reduced and public health is at risk. In excess of 100,000 fatalities could occur in Europe due to such an eruption, depending on the length of exposure of individuals to fine particles. In contrast to the health effects caused by exposure to fine particles, sulfur dioxide (gas) concentrations at the surface away from Iceland appear to be a minor hazard. At flight altitude, a hazard from sulfur dioxide to aircraft passenger health and comfort is, however, implied.

15.1 INTRODUCTION

Iceland has one of the highest volcanic eruption frequencies in the world with one eruption every 5 years on average (Thordarson and Larsen, 2007). The eruptions of Eyjafjallajökull in 2010 and of Grímsvötn in 2011 in Iceland were stark reminders that society is increasingly vulnerable to volcanic hazards and that the potential of disruption to everyday functions in society is high (e.g., Harris et al., 2012). Volcanic ash clouds from Eyjafjallajökull in April/May 2010 repeatedly resulted in restrictions of controlled European airspace for commercial aviation for up to 9 days at the time (e.g., Budd et al., 2010; Gudmundsson et al., 2012). These restrictions subsequently affected several industry sectors other than the aviation industry and resulted in an economic

Volcanic Hazards, Risks, and Disasters. http://dx.doi.org/10.1016/B978-0-12-396453-3.00015-0

loss of at least 1.7 billion US dollars (Budd et al., 2010; IATA, 2010; Mazzocchi et al., 2010). However, the hazards from volcanic eruptions are not restricted to volcanic ash clouds (Chapter 4 in this volume deals with volcanic ash and aviation hazards). In the following, this chapter discusses a less prominent, but equally hazardous phenomenon: volcanic gas and aerosol plumes from long-lasting flood lava eruptions in Iceland.

It is well known that sulfur-bearing species released by volcanic eruptions can degrade air quality and affect human health (e.g., Baxter, 2000; Hansell and Oppenheimer, 2004; Hansell et al., 2006; Longo et al., 2008, 2010). The AD 1783–1784 Laki eruption in Iceland is the best-studied example of a prolonged flood lava eruption. Laki released 120 Megatons (Mt) of sulfur dioxide (SO_2) into the atmosphere over the course of 8 months (Thordarson and Self, 1993; Thordarson et al., 1996; Thordarson and Self, 2003), which is about the same as the current annual anthropogenic SO_2 flux to the atmosphere of about 116 Mt (Smith et al., 2011).

Contemporary records of the 1780s imply that the Laki eruption induced severe environmental stress well beyond the shores of Iceland (Grattan and Pyatt, 1994; Stothers, 1996; Grattan, 1998; Jacoby et al., 1999; Thordarson and Self, 2003). The volcanic gases and aerosol particles from the eruption might also have contributed to increased morbidity and mortality rates across Europe (e.g., Durand and Grattan, 1999; Grattan et al., 2003a; Thordarson and Self, 2003; Grattan et al., 2005); although, other factors like disease cannot be excluded (Witham and Oppenheimer, 2005). Although the recurrence frequency of a Laki-type eruption is two orders of magnitude lower than the average eruption frequency in Iceland (or one eruption every 200–500 years as opposed to one every 5 years), it is useful to quantify how society might be impacted if such an eruption happened today.

By combining volcanological data sets, epidemiological evidence, and atmospheric modeling, this chapter describes the potential impacts of a future Laki-type eruption on air quality, human health, and aviation. Sections 15.2 and 15.3 discuss volcanological aspects of the Laki eruption and the recurrence frequencies of volcanic eruptions in Iceland. Sections 15.4 and 15.5 describe the gas and aerosol hazards to humans that could arise across Europe as well as potential hazards to aviation in European and North Atlantic airspace.

15.2 THE AD 1783–1784 LAKI ERUPTION

In Iceland, the Laki eruption is known as "Skaftáreldar," which translates to "Skaftá Fires" with "Skaftá" denoting the fact that the Laki lava flooded into the gorge of the river Skaftá (Thordarson and Self, 1993). The 27-km-long Laki fissure system is located in the Eastern Volcanic Zone (EVZ in Figure 15.1(a)) and erupted in the ice-free part of the Grímsvötn volcanic system. In total there are 30 active volcanic systems in Iceland. Each volcanic

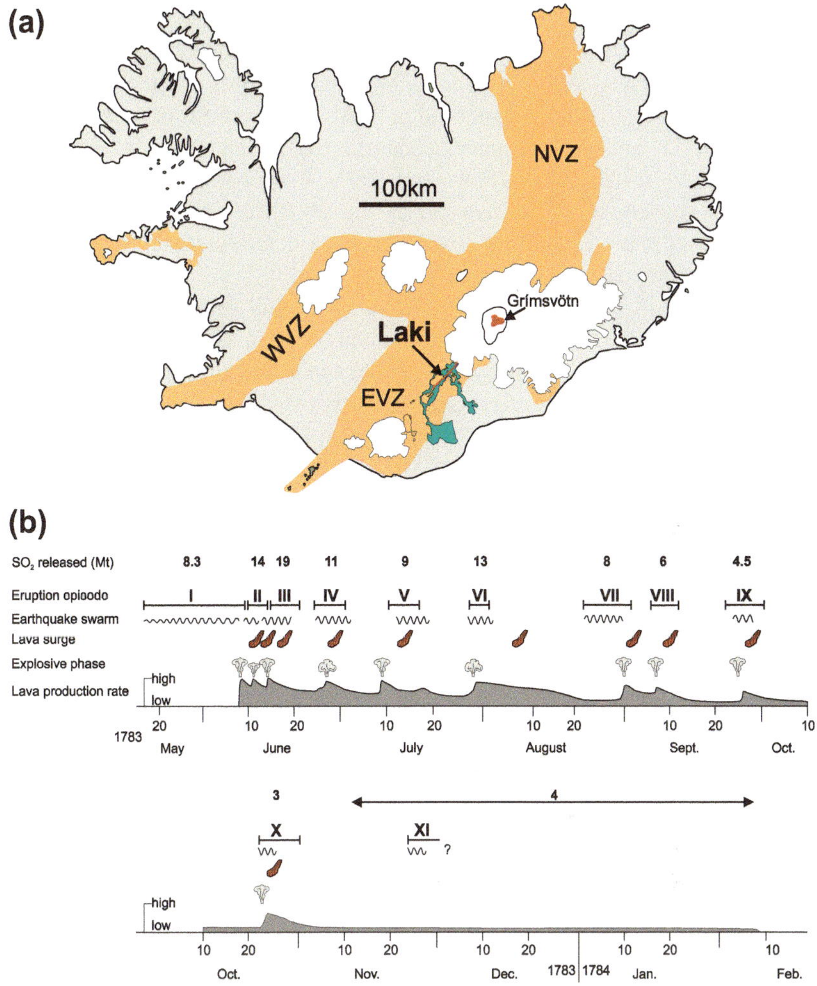

FIGURE 15.1 (a) Geological setting of the AD 1783–1784 Laki fissures (red line) and lava flow (blue) with active volcanic zones in Iceland shown in orange (WVZ = Western Volcanic Zone, EVZ = Eastern Volcanic Zone, NVZ = Northern Volcanic Zone). (b) Sequence of events during the Laki eruption with eruption episodes indicated by the numerals I through X including the mass of SO_2 released (in Megatons, Mt) from the vents. The eruption plumes indicate the occurrence of explosive phases at the vents, and the lava production rate is represented by the gray curve. *Figures courtesy of Thorvaldur Thordarson (University of Iceland).*

system is characterized by a conspicuous tectonic architecture and distinct magma chemistry and typically has a lifetime of between 0.5 and 1.5 million years (Thordarson and Larsen, 2007). The Grímsvötn volcanic system has been active since the Holocene and has since yielded a total eruptive magma volume of around 55 km^3 (Thordarson and Larsen, 2007). Over the course of

8 months, Laki produced an eruptive volume of 15.1 km^3 (Thordarson and Self, 1993), which is about 30 percent of the total erupted magma volume during the Holocene. For comparison, 30 years of almost constant activity at Kīlauea volcano (Hawai'i) produced a total lava volume of about 4 km^3 (USGS, 2013), and two rifting episodes in the Krafla volcanic system in Iceland (1724–1729 Mývatn Fires and 1975–1984 Krafla Fires) produced about 0.5 km^3 of lava (Thordarson and Larsen, 2007). In terms of composition, Laki produced basaltic lavas and its tephra is a quartz-tholeiite (Sigmarsson et al., 1991; Thordarson et al., 1996).

The course of events during the Laki eruption has been reconstructed by Thordarson and Self (1993,2003) using a combination of historical accounts, tephra stratigraphy, and detailed analyses of the volcanic deposits. The diary written by the Icelandic priest Jón Steingrímsson (Steingrímsson, 1788; 1998) offers a unique insight into the course of the eruption as well as the nature of the activity that typified it. Based on this and other chronicles, we know that the eruption commenced on June 8, 1783 and lasted until February 7, 1784. Fire fountains reached 1.5 km height during periods of peak discharge, as opposed to 200–600 m during periods of lower discharge (Thordarson and Self, 1993; Thordarson et al., 1996). The eruption featured a total of 10 eruption episodes (indicated by the numerals I to X in Figure 15.1(b)). Each eruption episode was associated with the opening of a new fissure segment, each extending the length of the vent system by several kilometers. The opening of a new fissure segment was accompanied by a short-lived (between 0.5 and 4 days) explosive phase followed by a longer-lived lava-producing phase. As illustrated in Figure 15.1(b), the eruption was most vigorous during the first 1.5 months producing around 60 percent of the total lava volume by mid-July (eruption episode V). By October 1783, around 93 percent of the total lava volume had been expelled (eruption episode X). The remaining 3 months of the activity were characterized by passive vent degassing.

By analyzing the major element and volatile composition of glass inclusions and groundmass glass in the quenched products from the Laki eruption, Thordarson et al. (1996) estimated the total mass of volatile species released during Laki: ~122 Tg of SO_2, ~235 Tg of H_2O, ~15 Tg of Cl, and ~7 Tg of HF. Degassing in fissure eruptions such as Laki occurs in two stages: by explosive activity or lava fountaining at the vents, and from the lava as it flows away from the vents. Thordarson et al. (1996) estimated that more than 80 percent of the volatile species (by mass) that escaped from the magma were released at the vents and the remainder from the lava upon emplacement (Table 15.1). At the time of peak activity (eruption episodes I to III = 8 days), a total of 40 Tg of SO_2 was released from the vents (Figure 15.1(b)). This is equivalent to the total SO_2 release by two 1991 Mt Pinatubo-sized eruptions. In the last 3 months of the Laki eruption, a total of about 4 Tg of SO_2 was released from the vents (Figure 15.1(b)).

TABLE 15.1 Overview of AD 1783–1784 Laki Eruption Episodes Including the Duration of Each Episode, the Timing of Peak SO_2 Emissions, the Total Mass of SO_2 Releases in Megatons (Mt), and the Total Mass of SO_2 Released from the Laki Vents after Thordarson et al. (1996). Table 15.1 lists the Best Estimate of SO_2 Mass Fluxes Available for the Purpose of a Modeling Study. However, Measurement Uncertainties of ±20 percent are Commonly Associated with Each Estimate

Eruption Episode	Episode Duration	Timing of Peak SO_2 Emissions	Total SO_2 Released, Mt	SO_2 Released at Vents, Mt
I	8–10 Jun	8 Jun	10.3	8.3
II	11–13 Jun	11–12 Jun	16.7	13.5
III	14–21 Jun	14–15 Jun	23.2	18.7
IV	25 Jun–1 Jul	26–28 Jun	13.4	10.8
V	9–21 Jul	9–10 Jul	11.0	8.9
VI	29 Jul–9 Aug	29–30 Jul	16.3	13.2
VII	31 Aug–4 Sep	31 Aug?	9.6	7.7
VIII	7–14 Sep	7 Sep	7.3	5.9
IX	24–29 Sep	24–25 Sep	5.4	4.4
X	25–30 Oct	25 Oct	3.6	2.9
Last 3 months	1 Nov 1783–7 Feb 1784	–	5.1	4.1
Total			121.9	98.4

Source: Table modified from Thordarson and Self (2003).

The magnitude and duration of the climatic and environmental impact of an eruption are governed by its eruption style and any temporal changes in eruption style including the injection height and the mass flux of volcanic volatile species (e.g., Robock, 2000; Schmidt et al., 2010). For Laki, eyewitness records indicate that the eruption columns above the vents reached heights greater than 9 km, and plume-rise model calculations (Stothers et al., 1986; Woods, 1993) indicate heights up to 13 km during the time of peak activity (Thordarson and Self, 1993,2003). Thus, volatile species released during explosive phases of activity reached the lower stratosphere above Iceland (the tropopause in Iceland is at 9–11 km in summer). In contrast, volatile species from passive vent degassing were predominantly emitted into the atmospheric boundary layer (Thordarson and Self, 1993,2003).

15.3 FREQUENCY OF ICELANDIC ERUPTIONS AND LIKELIHOOD OF A LAKI-TYPE ERUPTION

Iceland is one of the volcanically most active places in the world. By combining mapping of volcanic deposits, historical chronicles, and ice core data of the last 1,130 years, Thordarson and Larsen (2007) suggested that, on average, there is one eruption every 5 years in Iceland. This mean frequency is further corroborated by analyses of the Holocene eruption record (Óladóttir et al., 2008,2011; Thordarson and Höskuldsson, 2008). Given the 5-year eruption frequency, the total number of eruptions that may have taken place in Iceland over the last 24 million years is on the order of 5 million. Between the years 1990 and 2013 there have been seven eruptions (1991 Hekla; 1996 Gjálp; 1998 Grímsvötn; 2000 Hekla; 2004 Grímsvötn; 2010 Eyjafjallajökull; and 2011 Grímsvötn), which corresponds to an average eruption frequency over this period of just over 3 years.

Thordarson and Larsen (2007) identified at least four volcanic eruptions in the last 1,130 years that are characterized as flood lava eruptions; that is, they produced magma volumes significantly greater than 1 km^3. These are (1) Eldgjá AD 934–938 (Katla volcanic system, EVZ); (2) Hallmundarhraun ~AD 900 (Langjökull volcanic system, Western Volcanic Zone); (3) twelfth or thirteenth century Frambruni (Bárðarbunga-Veiðivötn volcanic system, Northern Volcanic Zone); and (4) Laki AD 1783–1784 (Grímsvötn volcanic system, EVZ). Therefore, on average one flood lava eruption takes place in Iceland every 280 years. Statistically, this recurrence interval is based on a short period of time of 1,130 years and only four events, hence substantial uncertainty is associated with this estimate. Based on fissure eruptions of any erupted volume in the last 1,130 years, the average recurrence rate is 88 years (Thordarson et al., 2003; Thordarson and Larsen, 2007). Taking into account all uncertainties involved, one Laki-type eruption every 200–500 years seems to be an appropriate figure.

15.4 VOLCANIC GAS AND AEROSOL HAZARDS FROM A FUTURE LAKI-TYPE ERUPTION

Based on the historical records describing the aftermath of the Laki eruption (e.g., Steingrímsson, 1788,1998; Grattan and Pyatt, 1994; Thordarson, 1995; Grattan, 1998,2005;Durand and Grattan, 1999; Grattan et al., 2003b,2005; Thordarson and Self, 2003; Witham and Oppenheimer, 2005), it seems appropriate to ask: What hazards could society face if a Laki-type eruption were to happen today? Since several factors may have contributed to the above-average mortality in 1783/84 (e.g., Witham and Oppenheimer, 2005) and because the vulnerability of society has changed significantly since the 1780s, one cannot simply scale the impacts described in the historical records in order to assess how a Laki-type eruption would affect modern society. Therefore, in this chapter, atmospheric modeling, epidemiological evidence, and volcanological data sets are combined to simulate the dispersal and the concentrations of volcanic SO_2 and sulfuric acid aerosol particles for a Laki-type eruption in a present-day context. By using a global aerosol model (GLOMAP; Mann et al., 2010) that tracks the number and the size of sulfuric acid aerosol particles, the degradation of air quality over Europe resulting from such an eruption can be quantified. Based on these model simulations, the likely scale of pollution-induced mortality in Europe today can be assessed using epidemiology. The model simulations also allow for an assessment of SO_2 concentrations at flight altitude and ground level, which can then be put into context with current air quality standards to assess whether SO_2 concentrations could reach hazardous levels affecting human health and aircraft operations in general.

15.4.1 Impact on Air Quality in Europe

The mass concentration of particles with diameters less than 2.5 μm (PM2.5) is a common metric for setting air quality standards to protect public health (WHO, 2008). Using the GLOMAP model, PM2.5 mass concentrations (in units of microgram per cubic meter; $\mu g/m^3$) at ground level are calculated based on two model simulations: (1) a baseline simulation that includes natural and anthropogenic emissions representative of the year 2000 and (2) a volcanically perturbed simulation, which additionally to the emissions in (1) includes the Laki SO_2 emissions by mimicking the course and SO_2 mass flux of the Laki eruption in 1783. The difference in PM2.5 mass concentrations between these two simulations allows one to quantify the contribution of a Laki-type eruption to air pollution. The model was run using reanalyzed meteorological fields from the European Centre for Medium-Range Weather Forecasts (Dee et al., 2011). To better understand the sensitivity of the results to meteorological variability, the model was run with meteorological fields for the years 2003 and 2005. The reader is referred to Schmidt et al., (2011) for further details of the model and its setup.

In the baseline model simulation, representative of anthropogenic pollution in the year 2000, annual European-mean PM2.5 mass concentrations of about 12 $\mu g/m^3$ are simulated with peak concentrations reaching about 29 $\mu g/m^3$ in the most polluted areas of Europe (Figure 15.2(a)). In Schmidt et al. (2011, Figure S2) it is also shown that GLOMAP simulates baseline PM2.5 mass concentrations in good agreement with PM2.5 measurements throughout Europe. The European Environment Agency sets an annual mean air quality standard of 25 $\mu g/m^3$ for PM2.5 (EEA, 2013) and the World Health Organization sets the same threshold, but for an averaging period of 24 h (WHO, 2000). For a Laki-type eruption, it is useful to first consider the degradation of air quality during the most vigorous phases at the beginning of the eruption. Averaging PM2.5 mass concentrations over the first 3 months of the eruption, changes in European-mean PM2.5 concentrations equate to

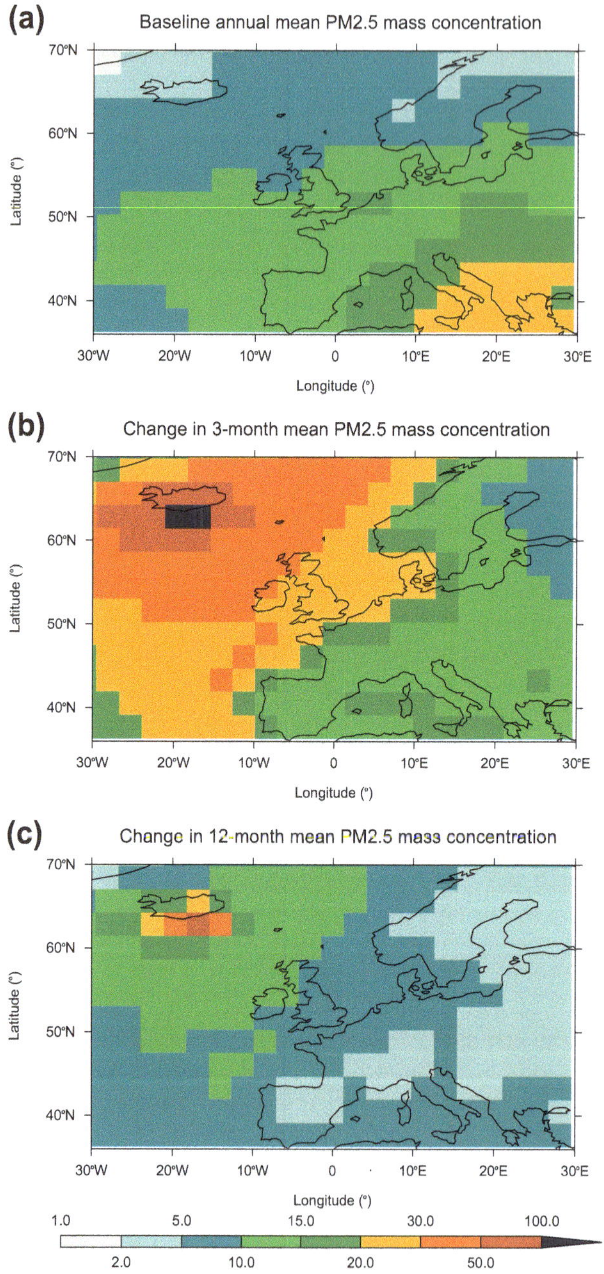

FIGURE 15.2 Simulated PM2.5 mass concentrations at the surface using the GLOMAP model. Annual mean surface PM2.5 mass concentrations (µg/m^3) in (a) the baseline simulation without the Laki emissions. Panel (b) shows the absolute change in PM2.5 mass concentrations from the

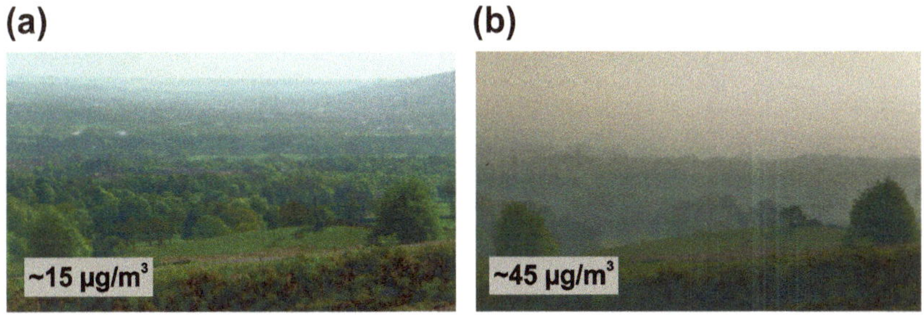

FIGURE 15.3 Illustrative comparison of pollution-induced visibility reductions with (a) showing typical particulate matter concentrations (PM2.5) on a clean day and (b) on a polluted day in April 2011. For the Laki model simulation, daily mean particulate matter concentrations across Northern Europe during the first 3 months are of about the same magnitude as in (b). Note that the comparison is for illustrative purposes only because visibility reductions for the same change in particulate matter concentrations will depend on relative humidity and chemical composition of the aerosol particles. *Photographs were taken by K.S. Carslaw (University of Leeds) in April 2011 near Burley in Wharfedale (United Kingdom).*

about 23 $\mu g/m^3$ (Figure 15.2(b)), which is equivalent to a 120 percent increase from the baseline. Locally, PM2.5 concentrations increase by up to 320 percent over northern Europe and up to 60 percent over southern Europe due to the eruption. When averaged over a longer period of 12 months, European-mean PM2.5 concentration changes equate to 8 $\mu g/m^3$ due to the eruption (Figure 15.2(c)). Annual European-mean concentration changes of 8 $\mu g/m^3$ are a factor of about 4 greater than the peak annual mean concentration changes attributable to shipping emissions (Corbett et al., 2007). Shipping-induced PM2.5 in turn is associated with more than 60,000 premature deaths worldwide per year (Corbett et al., 2007).

In Schmidt et al. (2011), daily mean PM2.5 mass concentration changes have been analyzed for the same model simulations. There it is shown that daily European-mean PM2.5 concentrations could exceed 25 $\mu g/m^3$ throughout the first 2 months of the eruption. Over land areas in Europe (excluding Iceland), daily mean concentrations of 25 $\mu g/m^3$, are on average, exceeded on almost twice as many days due to the eruption compared with the baseline simulation. While recognizing that PM2.5 concentration changes will vary locally on a daily basis, Figures 15.3(a) and (b) illustrate that visibility could be significantly reduced due to such a change in PM2.5 mass concentrations.

baseline simulation due to the Laki eruption averaged over the first 3 months of the eruption. Panel (c) shows the absolute change in PM2.5 concentrations from the baseline simulation due to the Laki eruption averaged over the first year following the onset of the Laki eruption in June. The data are shown as the mean concentrations from two simulations using 2003 and 2005 meteorological fields. *Figure adapted from Schmidt et al. (2011).*

15.4.2 Effects of Exposure to Fine Particulate Matter Pollution on Mortality Rates across Europe

Given the increase in fine particulate matter air pollution, the likely effects on mortality rates across Europe can be quantified by applying current epidemiological evidence. Epidemiological cohort studies have found a statistically significant link between both short-term (i.e., days) and long-term (i.e., months to years) exposure to PM2.5 and associated increases in morbidity and premature mortality risks (Ostro, 1984; Dockery et al., 1993; Pope and Dockery, 2006; Brook et al., 2010). Several epidemiological studies also found strong associations between sulfate-dominated air pollution and premature mortality (e.g., Dockery et al., 1992; Pope and Dockery, 2006; Longo et al., 2008, 2010).

To quantify the effects of a Laki-type eruption on human health, epidemiological concentration–response functions are applied using the model output of PM2.5 mass concentrations, population data, and baseline mortality rates for the year 2004 (see also Schmidt et al., 2011). Most simply, concentration–response functions link changes in PM2.5 concentrations to changes in the risk of premature mortality (Ostro, 2004). Although ever more evidence exists that volcanic air pollution is associated with adverse health effects (e.g., Longo et al., 2008, 2010), there are no concentration–response functions that are derived specifically for volcanic air pollution. However, given the strength of the evidence for PM2.5-induced mortality relative to the main constituent (sulfate) occurring in a volcanic plume, concentration–response functions recommended for PM2.5 are applied here. Specifically, concentration–response functions for two different health risks are considered: adult cardiopulmonary mortality associated with long-term exposure to PM2.5 and mortality (all ages) associated with short-term exposure to PM2.5 (see also Schmidt et al., 2011). The mortality rates calculated for long-term and short-term exposure should not be added together because the long-term exposure effects are likely to already include those from short-term exposure. Overall, these two averaging times represent alternative assumptions about the biological mechanisms associated with exposures to volcanic air pollution (Ostro, 2004). The most conservative approach of estimating mortality rates would be to assume that short-term exposure is most relevant during a Laki-type eruption. The functions applied only address a subset of the adverse health effects associated with exposure to PM2.5, and any potential health effects from exposure to volcanic ash or volcanic SO_2 are neglected. Whether or not exposure to volcanic ash causes adverse health effects remains equivocal (e.g., Baxter et al., 1981; Hansell et al., 2006; Horwell and Baxter, 2006; Durant et al., 2010).

Based on the model simulations, between 27,500 (95 percent confidence interval 22,600–32,200) (using 2003 meteorology) and 30,100 (95 percent confidence interval 24,800–35,300) (using 2005 meteorology), additional fatalities could occur from short-term exposure to PM2.5 during the first

8 months of the eruption. In Europe, daily mortality rates from short-term exposure remain in excess of 100 deaths per day until mid-September, which coincides with the most vigorous phases of the eruption (Schmidt et al., 2011). When assuming that long-term exposure is relevant, the total number of fatalities is much greater: ~139,000 (95 percent confidence interval 51,000–224,000) excess cardiopulmonary deaths using 2003 meteorology and ~144,000 (95 percent confidence interval 53,000–232,000) using 2005 meteorology. Both short-term and long-term mortality rates vary spatially depending on pollutant transport and the number of people exposed. The greatest number of excess cardiopulmonary fatalities (Figure 15.4(a)) per year occurs in the Benelux (the Netherlands, Belgium, and Luxembourg) with 8,400 and 8,800 fatalities for 2003 and 2005, respectively. Averaged for 2003 and 2005, a total of 142,000 (95 percent confidence interval 52,000–228,000) excess fatalities could occur in Europe with 7,400 fatalities in the southeast of the United Kingdom (total of 21,200 across the United Kingdom), 19,000 in Germany, and 13,000 in France (Figure 15.4(a)). Figure 15.4(b) shows cardiopulmonary mortality per 1,000 people per gridbox, which can be thought of as the risk of premature mortality in each model gridbox. Clearly, the risk of premature mortality is greater closer to the volcanic vent (Figure 15.4(a)), whereas the total number of fatalities in each model gridbox is a function of population density and pollutant transport (Figure 15.4(b)).

15.4.3 Sulfur Dioxide Concentrations at Flight Altitude and Ground Level

Aside from the potential health effects of its oxidation products, SO_2 is toxic, and atmospheric concentrations may be sufficiently high to cause concern about public health both at ground level and flight altitude (UK National Risk Register, 2012). Using the GLOMAP model, daily mean SO_2 concentrations for the period June 8 to July 31 were analyzed for the two periods of meteorology used in the model (i.e., years 2003 and 2005). Note that daily mean concentrations will not capture much shorter-term peak concentrations. For simplicity, one flight level (FL330) corresponding to about 10 km altitude above sea level in the model has been analyzed. The model results are put into context with air quality legislation devised for SO_2 concentrations at ground level to protect public health. In the United Kingdom, 15-min mean SO_2 mass concentrations greater than ~266 $\mu g/m^3$ at ground level are considered moderate air pollution, which would trigger a public health warning (Connolly et al., 2013). On a 24-h mean basis, the European Commission sets a threshold for SO_2 at 125 $\mu g/m^3$ to protect public health (EEA, 2013). In the following, this threshold set by the European Commission is used to analyze the model output.

For the Laki simulation, at FL330, daily mean SO_2 concentrations are greater than 125 $\mu g/m^3$ up to 14 days out of the 54 days consider (Figure 15.5(a)). Even at distances greater than 1,000 km from the vent, daily

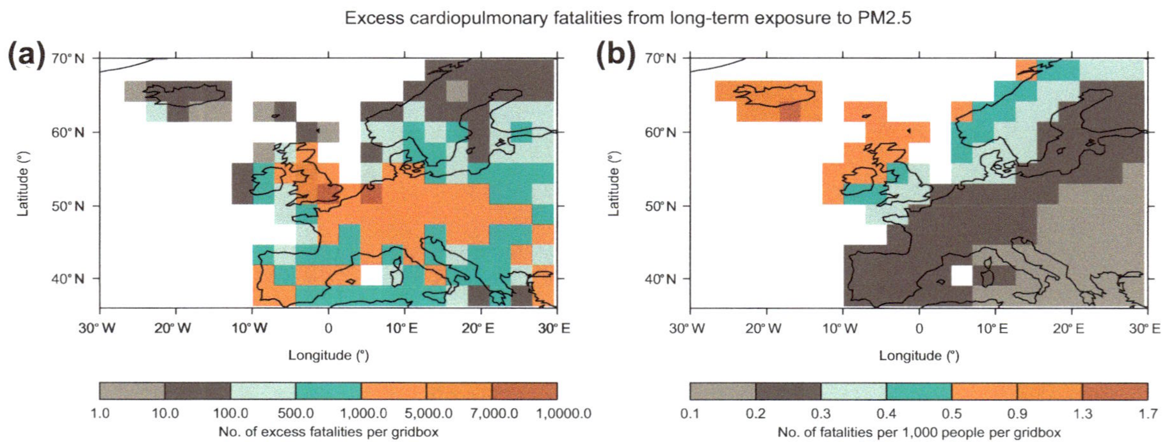

FIGURE 15.4 Number of excess cardiopulmonary fatalities in Europe resulting from long-term exposure to PM2.5 in the year following the onset of a Laki-type eruption. (a) Number of excess cardiopulmonary fatalities per model gridbox, and (b) number of fatalities per 1,000 people per model gridbox. The data are shown as the mean for the simulations using 2003 and 2005 meteorological fields. *Figure based on data in Schmidt et al. (2011).*

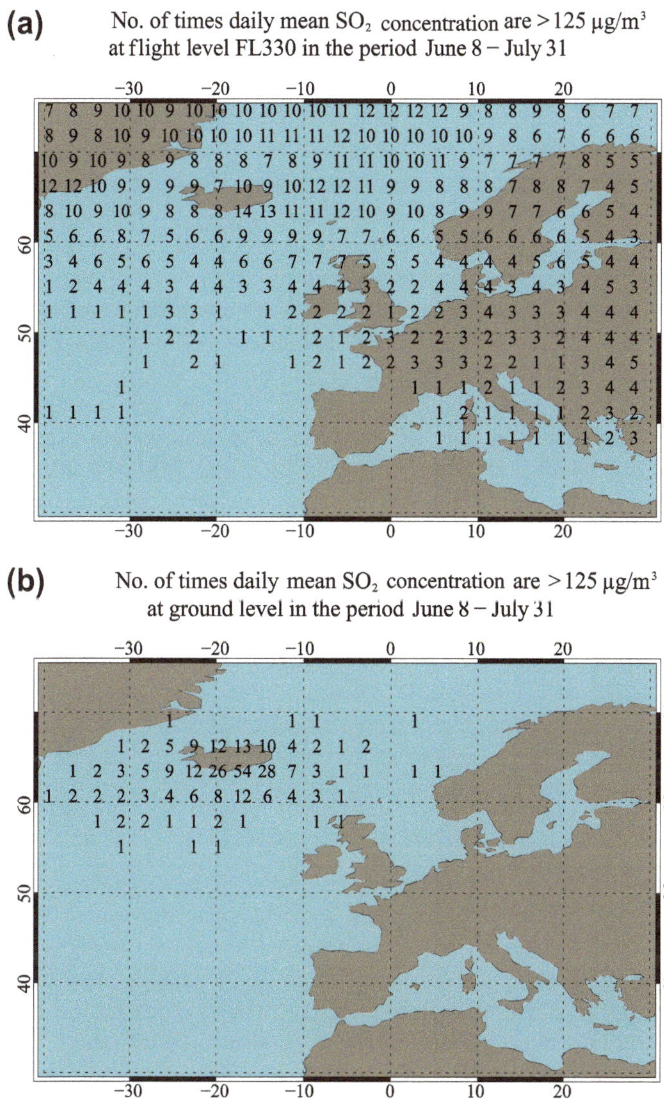

FIGURE 15.5 Number of times simulated daily mean sulfur dioxide (SO_2) mass concentrations (μg/m^3) are greater than 125 μg/m^3 in the period June 8 to July 31 (= 54 days) due to the Laki eruption at (a) flight level FL330 (about 10 km above sea level) and at (b) ground level. The data are shown as the mean for the simulations using 2003 and 2005 meteorological fields.

mean concentrations exceed 125 μg/m^3 but on fewer occasions (Figure 15.5(a)). In the near field of the volcanic vent, daily mean SO_2 concentrations greater than 1,000 μg/m^3 occur at FL330 mainly as a result of reduced conversion of SO_2 to sulfuric acid aerosol particles (due to a lack of oxidants). Overall, these modeling results imply that SO_2 concentrations at flight altitude may pose a risk to aircraft passenger health if current air quality standards devised for ground-level pollution are considered relevant also at flight altitude. If the evidence base (i.e., threshold concentrations that are considered hazardous to aircraft and/or aircraft passengers) changes in the future, this conclusion may not hold and the risk level would need to be revised. A full interpretation of the results is generally hampered by the lack of evidence as to whether or not short-term exposure to SO_2 concentrations greater than 125 μg/m^3 poses a health risk to aircraft passengers. Elevated SO_2 and sulfuric acid aerosol particle concentrations at flight altitude could also result in increased maintenance frequencies for aircraft (increasing the total cost of ownership) as a result of sulfidation (e.g., Bernard and Rose, 1990; Casadevall, 1994a,b; Casadevall et al., 1996; Eliaz et al., 2002; Fisher, 2013), but little is known about aircraft exposure durations and exposure frequencies.

In the immediate vicinity of the Laki vents (South East Iceland, Fire District), at ground level, the modeled daily mean SO_2 concentrations are greater than 1,000 μg/m^3. This is considered to be severe pollution and would result in public health warnings based on current air quality legislation in the United Kingdom (Connolly et al., 2013). In the far field, ground-level SO_2 concentrations remain well below the current European daily mean SO_2 air quality threshold of 125 μg/m^3 in most parts of Europe (Figure 15.5(b)). Based on these modeling results, which sampled only very limited meteorological variability and used one specific eruption source term, the increase in fine particulate matter concentrations due to oxidation of SO_2 forming sulfuric acid aerosol particles poses a greater public health risk at ground level than gas-phase SO_2 at ground level in Western and Central Europe (see Section 4.2).

15.5 DISCUSSION

In January 2012, Laki-type eruptions were added to the United Kingdom's National Risk Register for Civil Emergencies as one of the highest priority risks with potentially "widespread impacts on health, agriculture and transport" (UK National Risk Register, 2012). The evidence for the decision to add a Laki-type eruption to the Risk Register mainly stems from the contemporary records of the 1780s, but further evidence has been gained from atmospheric modeling of the likely impacts on air quality and human health (Schmidt et al., 2011).

For eruptions in Iceland, it is worth reminding oneself that a recurrence of a Laki-type flood lava eruption is much less probable (one event every 200–500 years) than that of any other type of eruption (one eruption every 5 years on average). The main hazards from explosive volcanic eruptions are dispersal of volcanic ash into European and North Atlantic airspace and proximal tephra fallout (e.g., Gudmundsson et al., 2012; Stevenson et al., 2012), aside from local hazards due to the resuspension of volcanic ash (Leadbetter et al., 2012). Leadbetter and Hort (2011) used a dispersion model to show that there is about a one-in-three chance of ash from a short-duration eruption of Hekla volcano in Iceland being dispersed into European airspace. Volcanic ash from small-magnitude eruptions such as Eyjafjallajökull 2010 can be transported over long distances affecting regions several thousands of kilometers away from the vent (Stevenson et al., 2012). Swindles et al. (2011) estimated that there is a 16 percent probability of ash from Icelandic eruptions reaching Europe in any given decade on the basis of cryptotephra occurrences across Europe in the past 7,000 years. Explosive volcanic eruptions in Iceland are usually of a short duration (hours to weeks) (Thordarson and Larsen, 2007); therefore, the potential hazards are likely to be of short duration, too. It is mainly the long eruption duration, on the order of weeks to months or even years, together with the sustained gas release that makes Icelandic flood lava eruptions a potential high-impact hazard for Europe.

Risk is more readily perceived when put into context. Based on the GLOMAP model simulation, it is estimated that a Laki-type eruption could cause more than 25,000 additional deaths due to short-term exposure to fine particles. If these estimates are based on the likelihood of long-term exposure, then more than 100,000 additional deaths could occur throughout Europe due to such an eruption. For context, Cohen et al. (2005) estimated that globally 800,000 deaths in the year 2000 could be attributed to urban outdoor air pollution, measured as PM2.5. Of this, 23,000 deaths (with a 95 percent confidence interval of 10,000–43,000) would occur in Western and Central Europe. Thus, even 25,000 additional deaths in Europe due to a Laki-type eruption would be significant. In context with the death toll from volcanic activity, the 1902 eruption of Pelée killed about 29,000 people and about 23,000 people were killed during the 1985 Nevado del Ruiz eruption (Witham, 2005). During the twentieth century, a total of about 93,000 people lost their lives in 260 volcanic events (Witham, 2005). Given that the long-term exposure mortality estimates are of the same order of magnitude as the total death toll from volcanic activity in the twentieth century, a future Laki-type eruption has a very high health hazard potential.

In the future, a full probabilistic risk assessment accounting for variability in the eruption source term parameters including the mass flux of volatile species and their injection height is required to fully characterize the hazards from a Laki-type eruption (Loughlin et al., 2012). The eruptions of Eyjafjallajökull and Grímsvötn in 2010 and 2011 demonstrated that society is increasingly vulnerable to volcanic hazards and that substantial economic and societal disruption can result even from a small-magnitude eruption (Budd et al., 2010; IATA, 2010; Mazzocchi et al., 2010; Gudmundsson et al., 2012; Harris et al., 2012). If a Laki-type eruption happens tomorrow, some of the hazards, in particular the health hazards at ground level, may be mitigated through effective monitoring and forecasting of SO_2 and particulate matter concentrations. Early warning systems together with public health warnings and advice in the event of threshold exceedances could effectively facilitate hazard mitigation. Similar practices are already in place on the Island of Hawai'i to protect public health from volcanic activity at Kīlauea volcano (e.g., Vog Measurement and Prediction Project). Currently, Volcanic Ash Advisory Centers are not required to monitor and/or forecast volcanic SO_2 and sulfuric acid aerosol particle dispersion and/or concentrations at flight altitude. In future, these requirements might change if the aviation industry deems such forecasts as valuable additional information for aviation safety.

15.6 SUMMARY

The awareness that short-term and long-term exposure to air pollution is associated with adverse health effects led to the introduction of air quality legislation to protect public health in Europe and other parts of the world. In this chapter, it has been shown that a future Laki-type eruption in Iceland could significantly degrade air quality in Western and Central Europe due to an increase in fine particulate matter concentrations (Figure 15.2) resulting from the oxidation of volcanic sulfur dioxide. Depending on the duration of such an eruption, air quality could be degraded for weeks or months to a degree that visibility may be noticeably reduced (Figure 15.3) and public health is at risk. Depending on the length of exposure of individuals to volcanic air pollution, in excess of 100,000 fatalities could occur throughout Europe due to such an eruption (Figure 15.4).

In contrast to the aerosol particle hazards to human health, sulfur dioxide gas concentrations at ground level away from Iceland appear to be a minor hazard, whereas at flight altitude a hazard to passenger health from sulfur dioxide is implied (Figure 15.5). A full judgment whether or not sulfur dioxide concentrations as simulated here pose a hazard to aircraft passengers and/or aircraft/avionics is not possible. This is mainly due to the limited knowledge of the consequences associated with aircraft exposure to sulfur dioxide. If the evidence base (i.e., threshold concentrations, durations, and exposure frequencies that are considered hazardous) changes in future, a more complete hazard assessment is possible. At flight altitude, the presence of sulfuric acid aerosol particles can result in airframe damage and abrasion and etching of acrylic windows (Bernard and Rose, 1990; Casadevall et al., 1996; Miller and Casadevall, 2000), and sulfurous odors inside an aircraft cabin can cause distress to pilots and passengers (Guffanti et al., 2008; see supporting material in Guffanti et al., 2010; Harris et al., 2012 and references therein; Phys.org, 2012). Therefore, monitoring and forecasting of volcanic sulfur species may provide additional safety information for the aviation industry (e.g., Carn et al., 2009).

Monitoring of volcanic unrest and monitoring of volcanic gas and aerosol concentrations are key requirements for increasing our capabilities to mitigate gas and aerosol hazards in the event of a Laki-type eruption. It is also likely that hazards directly affecting, for instance, Central Europe have knock-on effects that might reach a global scale, which requires developing and strengthening partnerships between scientists, forecasters, and emergency agencies on an international level.

ACKNOWLEDGMENTS

This chapter greatly benefited from a constructive and helpful review by Andy Harris (Université Blaise Pascal-Université Clermont II, France). I am also very grateful to Thorvaldur Thordarson (University of Iceland) and Claire S. Witham (Met Office, United Kingdom) who provided helpful comments on an earlier version of this chapter. I also acknowledge funding through an Academic Research Fellowship from the School of Earth and Environment, University of Leeds, UK, and NERC grant NE/I015612/1, as well as usage of the NERC UK High End Computing Service.

REFERENCES

Baxter, P.J., 2000. Impacts of eruptions on human health. In: Sigurdsson, H., Houghton, B.F., McNutt, S.R., Rymer, H., Stix, J. (Eds.), Encyclopedia of Volcanoes. Academic Press, pp. 1035–1043.

Baxter, P.J., Ing, R., Falk, H., French, J., Stein, G.F., Bernstein, R.S., Merchant, J.A., Allard, J., 1981. Mount St Helens eruptions, May 18 to June 12, 1980: an overview of the acute health impact. JAMA 246 (22), 2585–2589.

Bernard, A., Rose Jr, W.I., 1990. The injection of sulfuric acid aerosols in the stratosphere by the El Chichón volcano and its related hazards to the international air traffic. Nat. Hazards 3 (1), 59–67. http://dx.doi.org/10.1007/bf00144974.

Brook, R.D., et al., 2010. Particulate matter air pollution and cardiovascular disease: an update to the scientific statement from the American Heart Association. Circulation 121 (21), 2331–2378.

Budd, L., Griggs, S., Howarth, D., Ison, S., 2010. A Fiasco of volcanic proportions? Eyjafjallajökull and the closure of European airspace. Mobilities 6 (1), 31–40. http://dx.doi.org/10.1080/17450101.2011.532650.

Carn, S., Krueger, A., Krotkov, N., Yang, K., Evans, K., 2009. Tracking volcanic sulfur dioxide clouds for aviation hazard mitigation. Nat. Hazards 51 (2), 325–343. http://dx.doi.org/10.1007/s11069-008-9228-4.

Casadevall, T.J., 1994a. The 1989–1990 eruption of Redoubt Volcano, Alaska: impacts on aircraft operations. J. Volcanol. Geotherm. Res. 62 (1–4), 301–316. http://dx.doi.org/10.1016/0377-0273(94)90038-8.

Casadevall, T.J., 1994b. Volcanic Ash and Aviation Safety: Proceedings of the First International Symposium on Volcanic Ash and Aviation Safety, 450 pp. Seattle, Washington.

Casadevall, T.J., Delos Reyes, P.J., Schneider, D.J., 1996. The 1991 Pinatubo eruptions and their effects on aircraft operations. In: Newhall, C.G., Punongbayan, R.S. (Eds.), Fire and Mud: Eruptions and Lahars of Mount Pinatubo, Philippines. University of Washington Press, Philippines Institute of Volcanology and Seismology, Quezon City/Seattle, pp. 625–636.

Cohen, A.J., et al., 2005. The global burden of disease due to outdoor air pollution. J. Toxicol. Environ. Health Part A 68 (13–14), 1301–1307.

Connolly, E., Fuller, G., Baker, T., Willis, P., 2013. Update on Implementation of the Daily Air Quality Index, 1–11 pp. Department for Environment Food and Rural Affairs.

Corbett, J.J., Winebrake, J.J., Green, E.H., Kasibhatla, P., Eyring, V., Lauer, A., 2007. Mortality from ship emissions: a global assessment. Environ. Sci. Technol. 41 (24), 8512–8518. http://dx.doi.org/10.1021/es071686z.

Dee, D.P., et al., 2011. The ERA-Interim reanalysis: configuration and performance of the data assimilation system. Q.J. Roy. Meteorol. Soc. 137 (656), 553–597.

Dockery, D.W., Pope, C.A., Xu, X.P., Spengler, J.D., Ware, J.H., Fay, M.E., Ferris, B.G., Speizer, F.E., 1993. An association between air-pollution and mortality in 6 United-States cities. N. Engl. J. Med. 329 (24), 1753–1759.

Dockery, D.W., Schwartz, J., Spengler, J.D., 1992. Air pollution and daily mortality: associations with particulates and acid aerosols. Environ. Res. 59 (2), 362–373.

Durand, M., Grattan, J., 1999. Extensive respiratory health effects of volcanogenic dry fog in 1783 inferred from European documentary sources. Environ. Geochem. Health 21, 371–376. http://dx.doi.org/10.1023/A:1006700921208.

Durant, A.J., Bonadonna, C., Horwell, C.J., 2010. Atmospheric and environmental impacts of volcanic particulates. Elements 6 (4), 235–240.

EEA, 2013. Air quality in Europe—2013 report. Copenhagen, Denmark, 107 pp.

Eliaz, N., Shemesh, G., Latanision, R.M., 2002. Hot corrosion in gas turbine components. Engineering Failure Analysis 9 (1), 31–43. http://dx.doi.org/10.1016/S1350-6307(00)00035-2.

Fisher, S.M., 2013. Sulfidation: Turbine Blade Corrosion (edited).

Grattan, J., 1998. The distal impact of Icelandic volcanic gases and aerosols in Europe: a review of the 1783 Laki Fissure eruption and environmental vulnerability in the late 20th century. Geological Society, London Eng. Geol. Spec. Publ. 15 (1), 97–103.

Grattan, J., 2005. Pollution and paradigms: lessons from Icelandic volcanism for continental flood basalt studies. Lithos 79 (3–4), 343–353.

Grattan, J., Durand, M., Taylor, S., 2003a. Illness and elevated human mortality in Europe coincident with the Laki Fissure eruption. In: Oppenheimer, C., Pyle, D.M., Barclay, J. (Eds.), Volcanic Degassing, vol. 213. Geological Society, London, UK, Special Publications, pp. 401–414.

Grattan, J., Durand, M., Taylor, S., 2003b. Illness and elevated human mortality in Europe coincident with the Laki Fissure eruption. In: Oppenheimer, C., Pyle, D.M., Barclay, J. (Eds.), Volcanic Degassing. Geological Soc Publishing House, Bath, pp. 401–414.

Grattan, J., Rabartin, R., Self, S., Thordarson, T., 2005. Volcanic air pollution and mortality in France 1783-1784. C. R. Geosci. 337 (7), 641–651.

Grattan, J.P., Pyatt, F.B., 1994. Acid damage to vegetation following the Laki Fissure eruption in 1783-an historical review. Sci. Total Environ. 151 (3), 241–247.

Gudmundsson, M.T., et al., 2012. Ash generation and distribution from the April-May 2010 eruption of Eyjafjallajökull, Iceland. Sci. Rep. 2.

Guffanti, M., Casadevall, T.J., Budding, K., 2010. In: Encounters of Aircraft with Volcanic Ash Clouds: A Compilation of Known Incidents, 1953–2009. U.S. Geological Survey Data Series 545, ver. 1.0, 12.

Guffanti, M., Schneider, D.J., Ewert, J.W., Targosz, S., 2008. Impact on Aviation Operations of Volcanic Gas and Ash Clouds from the 2008 Eruptions of Okmok and Kasatochi, Alaska. American Geophysical Union, Fall Meeting 2008. *Abstract #A53B-0277.*

Hansell, A., Oppenheimer, C., 2004. Health hazards from volcanic gases: a systematic literature review. Arch. Environ. Health 59 (12), 628–639.

Hansell, A.L., Horwell, C.J., Oppenheimer, C., 2006. The health hazards of volcanoes and geothermal areas. Occup. Environ. Med. 63 (2), 149–156.

Harris, A.J.L., Gurioli, L., Hughes, E.E., Lagreulet, S., 2012. Impact of the Eyjafjallajökull ash cloud: a newspaper perspective. J. Geophys. Res.: Solid Earth 117 (B9), B00C08. http://dx.doi.org/10.1029/2011jb008735.

Horwell, C., Baxter, P., 2006. The respiratory health hazards of volcanic ash: a review for volcanic risk mitigation. Bull. Volcanol. 69 (1), 1–24. http://dx.doi.org/10.1007/s00445-006-0052-y.

IATA, 2010. IATA Economics Briefing: The Impact of Eyjafjallajokull's Volcanic Ash Plume.

Jacoby, G.C., Workman, K.W., D'Arrigo, R.D., 1999. Laki eruption of 1783, tree rings, and disaster for northwest Alaska Inuit. Quat. Sci. Rev. 18 (12), 1365–1371.

Leadbetter, S.J., Hort, M.C., 2011. Volcanic ash hazard climatology for an eruption of Hekla volcano, Iceland. J. Volcanol. Geotherm. Res. 199 (3–4), 230–241.

Leadbetter, S.J., Hort, M.C., von Löwis, S., Weber, K., Witham, C.S., 2012. Modelling the resuspension of ash deposited during the eruption of Eyjafjallajökull in spring 2010. J. Geophys. Res.: Atmos. 117 (D20). http://dx.doi.org/10.1029/2011jd016802. D00U10.

Longo, B.M., Rossignol, A., Green, J.B., 2008. Cardiorespiratory health effects associated with sulphurous volcanic air pollution. Public Health 122 (8), 809–820. http://dx.doi.org/10.1016/j.puhe.2007.09.017.

Longo, B.M., Yang, W., Green, J.B., Crosby, F.L., Crosby, V.L., 2010. Acute health effects associated with exposure to volcanic air pollution (vog) from increased activity at Kilauea volcano in 2008. J. Toxicol. Environ. Health, Part A 73 (20), 1370–1381. http://dx.doi.org/10.1080/15287394.2010.497440.

Loughlin, S.C., Aspinall, W.P.A., Vye-Brown, C., Baxter, P.J., Braban, C., Hort, M., Schmidt, A., Thordarson, T., Witham, C.S., 2012. Large-magnitude Fissure Eruptions in Iceland: Source Characterisation.

Mann, G.W., Carslaw, K.S., Spracklen, D.V., Ridley, D.A., Manktelow, P.T., Chipperfield, M.P., Pickering, S.J., Johnson, C.E., 2010. Description and evaluation of GLOMAP-mode: a modal global aerosol microphysics model for the UKCA composition-climate model. Geosci. Model Dev. 3 (2), 519–551.

Mazzocchi, M., Hansstein, F., Ragona, M., 2010. The 2010 volcanic ash cloud and its financial impact on the European aSirline Industry. CESifo Forum 11 (2), 92–100.

Miller, T.P., Casadevall, T.J., 2000. Volcanic ash hazards to aviation. In: Sigurdsson, H. (Ed.), Encyclopedia of Volcanoes. Academic Press, San Diego, CA, pp. 915–930.

Óladóttir, B., Larsen, G., Sigmarsson, O., 2011. Holocene volcanic activity at Grímsvötn, Bárdarbunga and Kverkfjöll subglacial centres beneath Vatnajökull, Iceland. Bull. Volcanol. 73 (9), 1187–1208. http://dx.doi.org/10.1007/s00445-011-0461-4.

Óladóttir, B., Sigmarsson, O., Larsen, G., Thordarson, T., 2008. Katla volcano, Iceland: magma composition, dynamics and eruption frequency as recorded by Holocene tephra layers. Bull. Volcanol. 70 (4), 475–493. http://dx.doi.org/10.1007/s00445-007-0150-5.

Ostro, B., 1984. A search for a threshold in the relationship of air-pollution to mortality - a reanalysis of data on London winters. Environ. Health Perspect. 58 (DEC), 397–399.

Ostro, B., 2004. Outdoor Air Pollution: Assessing the Environmental Burden of Disease at National and Local Levels. World Health Organization, Geneva, WHO Environmental Burden of Disease Series, No. 5.

Phys.org, 2012. Smelling Grímsvotn: Signal of a volcanic eruption, 2014 (last accessed 03 February). http://phys.org/news/2012-09-grmsvotn-volcanic-eruption.html.

Pope, C.A., Dockery, D.W., 2006. Health effects of fine particulate air pollution: Lines that connect. J. Air Waste Manage. Assoc. 56 (6), 709–742.

Robock, A., 2000. Volcanic eruptions and climate. Rev. Geophys. 38 (2), 191–219. http://dx.doi.org/10.1029/1998RG000054.

Schmidt, A., Carslaw, K.S., Mann, G.W., Wilson, M., Breider, T.J., Pickering, S.J., Thordarson, T., 2010. The impact of the 1783-1784 AD Laki eruption on global aerosol formation processes and cloud condensation nuclei. Atmos. Chem. Phys. 10 (13), 6025–6041.

Schmidt, A., Ostro, B., Carslaw, K.S., Wilson, M., Thordarson, T., Mann, G.W., Simmons, A.J., 2011. Excess mortality in Europe following a future Laki-style Icelandic eruption. Proc. Natl. Acad. Sci. 108 (38), 15710–15715. http://dx.doi.org/10.1073/pnas.1108569108.

Sigmarsson, O., Condomines, M., Grönvold, K., Thordarson, T., 1991. Extreme magma homogeneity in the 1783–84 Lakagigar Eruption: origin of a large volume of evolved basalt in Iceland. Geophys. Res. Lett. 18 (12), 2229–2232.

Smith, S.J., van Aardenne, J., Klimont, Z., Andres, R.J., Volke, A., Delgado Arias, S., 2011. Anthropogenic sulfur dioxide emissions: 1850–2005. Atmos. Chem. Phys. 11 (3), 1101–1116. http://dx.doi.org/10.5194/acp-11-1101-2011.

Steingrímsson, J., 1788. Fullkomid skrif um Sídueld (A complete description on the Sida volcanic fire). In: Safn til Sögu íslands, pp. 1–57. Copenhagen, 1907–1915.

Steingrímsson, J., 1998. Fires of the Earth. The Laki Eruption 1783-1784. Nordic Volcanological Institute and University of Iceland Press, Reykjavík.

Stevenson, J.A., et al., 2012. Distal deposition of tephra from the Eyjafjallajökull 2010 summit eruption. J. Geophys. Res.: Solid Earth 117 (B9). http://dx.doi.org/10.1029/2011jb008904. B00C10.

Stothers, R.B., 1996. The great dry fog of 1783. Clim. Change 32 (1), 79–89.

Stothers, R.B., Wolff, J.A., Self, S., Rampino, M.R., 1986. Basaltic fissure eruptions, plume heights, and atmospheric aerosols. Geophys. Res. Lett. 13 (8), 725–728. http://dx.doi.org/10.1029/GL013i008p00725.

Swindles, G.T., Lawson, I.T., Savov, I.P., Connor, C.B., Plunkett, G., 2011. A 7000 yr perspective on volcanic ash clouds affecting northern Europe. Geology 39 (9), 887–890. http://dx.doi.org/10.1130/g32146.1.

Thordarson, T., 1995. Volatile Release and Atmospheric Effects of Basaltic Fissure Eruptions, 570 pp. University of Hawaii, Honolulu.

Thordarson, T., Höskuldsson, Á., 2008. Postglacial volcanism in Iceland. Jökull 58, 197–228.

Thordarson, T., Larsen, G., 2007. Volcanism in Iceland in historical time: volcano types, eruption styles and eruptive history. J. Geodyn. 43 (1), 118–152.

Thordarson, T., Self, S., 1993. The laki (Skaftar fires) and Grimsvötn eruptions in 1783-1785. Bull. Volcanol. 55 (4), 233–263.

Thordarson, T., Self, S., 2003. Atmospheric and environmental effects of the 1783-1784 Laki eruption: a review and reassessment. J. Geophys. Res.-Atmos. 108 (D1). http://dx.doi.org/10.1029/2001JD002042, 4011.

Thordarson, T., Self, S., Miller, D.J., Larsen, G., Vilmundardóttir, E.G., 2003. Sulphur release from flood lava eruptions in the Veidivötn, Grímsvötn and Katla volcanic systems, Iceland. Geol. Soc. Lond. Spec. Publ. 213 (1), 103–121. http://dx.doi.org/10.1144/gsl.sp.2003.213.01.07.

Thordarson, T., Self, S., Oskarsson, N., Hulsebosch, T., 1996. Sulfur, chlorine, and fluorine degassing and atmospheric loading by the 1783-1784 AD Laki (Skaftar fires) eruption in Iceland. Bull. Volcanol. 58 (2–3), 205–222.

UK National Risk Register, 2012. National Risk Register of Civil Emergencies, 1–55 pp.

USGS, 2013. The Ongoing Pu'u 'Ō'ō Eruption of Kīlauea Volcano, Hawai'i—30 Years of Eruptive Activity, 1–6 pp. U.S. Geological Survey.

WHO, 2000. Air Quality Guidelines for Europe. World Health Organization, Copenhagen.

WHO, 2008. The Global Burden of Disease: 2004 Update. edited. World Health Organization.

Witham, C.S., 2005. Volcanic disasters and incidents: a new database. J. Volcanol. Geotherm. Res. 148 (3–4), 191–233. http://dx.doi.org/10.1016/j.jvolgeores.2005.04.017.

Witham, C.S., Oppenheimer, C., 2005. Mortality in England during the 1783-4 Laki Craters eruption. Bull. Volcanol. 67 (1), 15–26. http://dx.doi.org/10.1007/s00445-004-0357-7.

Woods, A.W., 1993. A model of the plumes above basaltic fissure eruptions. Geophys. Res. Lett. 20, 1115–1118. http://dx.doi.org/10.1029/93GL01215.

Chapter 16

Explosive Super-Eruptions and Potential Global Impacts

Stephen Self[1,2]

[1] *Volcano Dynamics Group, Department of Environment, Earth and Ecosystems, The Open University, Walton Hall, Milton Keynes, UK,* [2] *Department of Earth and Planetary Science, University of California, Berkeley, CA, USA*

ABSTRACT

Rare, but extremely large, explosive super-eruptions lead to the catastrophic formation of huge calderas, devastation of substantial regions by pyroclastic density currents or flows and their deposits, and ashfalls that cover continent-sized areas. The effects of future super-eruptions will be felt globally or at least by a whole hemisphere. The most widespread effects are likely to derive from the volcanic gases released, particularly sulfur gases that are converted into sulfuric acid aerosols in the stratosphere. These will remain for several years, promoting changes in atmospheric circulation, causing surface temperatures to fall in many regions, leading to short-term, temporary reductions in light levels, and severe and unseasonable weather ("volcanic winter"), but are not expected to be severe enough to cause major climatic changes. Major disruptions to global societal infrastructure can be expected for periods of months to years, and the cost to global financial markets will be high and sustained.

16.1 INTRODUCTION

The geologic record contains ample evidence of giant explosive volcanic eruptions, preserved as pyroclastic deposits (Mason et al., 2004), which were mainly from volcanoes that erupted evolved silicic magma (generally rhyolite or rhyodacite). Evidence suggests that these rare, explosive super-eruptions covered whole continents and/or ocean basins with volcanic ash. Super-eruptions have been defined as those yielding more than 10^{15} kg of magma (>450 km^3 of dense, nonvesicular rhyolitic magma; Sparks et al., 2005; Self, 2006; or magnitude (M) 8), broadly equivalent to ~1,000 km^3 of ash deposits. Extremely rare examples have produced vast amounts of rhyolitic magma (>1 × 10^{16} kg, or >4,500 km^3), equal to nine on the eruption magnitude (M)

Volcanic Hazards, Risks, and Disasters. http://dx.doi.org/10.1016/B978-0-12-396453-3.00016-2

scale (Lipman, 1997; Pyle, 2000; Mason et al., 2004). The Volcanic Explosivity Index (Newhall and Self, 1982) equivalent is VEI 8–9.

This chapter reviews the nature of explosive super-eruptions and summarizes some of the environmental effects and hazards that a super-eruption may cause. It presents an assessment of the effects of very large explosive eruptions that could have consequences well beyond those associated with historic volcanic activity. We begin with our understanding of the frequency, recurrence, and duration, then turn to brief descriptions of the eruption style and deposits that would be produced, and the gases that are released. The atmospheric impact and potential influence on global temperatures and weather of super-eruptions is a topic in which opinions are changing, and this is discussed. Finally, issues facing society after a super-eruption, including the potential socioeconomic impacts, are presented.

16.2 SUPERSIZED ERUPTIONS

An eruption magnitude scale based on magma mass must also include basaltic magmas, and large igneous provinces contain the deposits of both basaltic and silicic super-eruptions (Bryan et al., 2010; Bryan and Ferrari, 2013). The basaltic variety has been erupted in huge volumes, forming extensive sheet lava flows and constructing, by repeated eruptions, continental flood basalt provinces. These constitute another type of super-eruption, but because Earth is not presently in a phase that produces flood basalts (Self, 2006), and these eruptions have modest VEIs (VEI 4–5; Self et al., in press), we will not consider them further here as a source of future hazard. This book is more concerned with eruptions that may impact Earth in the near future, from a human lifetime perspective. Moreover, all M 9 eruptions discovered to date are from ancient silicic or basaltic large igneous provinces (Bryan et al., 2010; Lipman, 1997) and their relevance to volcanism in the more recent past (e.g., over the Quaternary period) is uncertain.

Explosive super-eruptions of compositionally evolved magmas lead to caldera collapse and the formation of huge sheets of pyroclastic density current (or flow) flow deposits (ignimbrites), as well as very extensive co-ignimbrite ashfall layers (e.g., Matthews et al., 2012), and are accompanied by injection of noxious gases into the atmosphere. The effects are therefore expected to be more violent and damaging than those of the considerably smaller eruptions that society has experienced in historic times (Self, 2006).

Unlike some other extreme natural hazards, some degree of detectable warning should occur for an impending super-eruption (Lowenstern and Hurwitz, 2008). Furthermore, super-eruptions are potentially long-lasting (continuing at least for many days, and in some cases intermittently for weeks to perhaps many hundreds of years; Wilson, 2008) compared to a brief but intense major earthquake or tsunami. The volcanic ash and gases released high into the atmosphere might have severe worldwide effects on climate and weather. Thus, huge explosive eruptions are one of the few natural phenomena that can produce truly global catastrophic effects. They may also differ from other hazards (except meteorite impacts; Rampino, 2002) in that, for several years after the eruption, they will produce persistent atmospheric effects.

16.3 THE NEXT SUPER-ERUPTION?

Questions that arise are "When will the next super-eruption happen?, "Where will the next super-eruption happen?", and "What environmental effects be caused?" The first two can only be answered by using the geological record as a guide, but the evidence is difficult to interpret. In fact, it is obvious that the present record of known super-eruptions is woefully incomplete. Because M 8 eruptions are so rare (the youngest super-eruption is the 26.5 ka, ~530 km^3, Oruanui event from Taupo Caldera, New Zealand; Wilson, 2001), our knowledge of the recent geological record is unable to capture any meaningful periodicity for such events (Deligne et al., 2010), if such a periodicity exists.

An initial survey of the largest magnitude explosive eruptions (Mason et al., 2004) recognized only 8 in the Quaternary period (last ~2.5 Ma) and suggested that super-eruptions have been very rare, with an average frequency of only one every 100,000–200,000 years. Thus, our incomplete knowledge base makes prediction of recurrence rates difficult; the LaMEVE database, part of the VOGRIPA initiative (Crosweller et al., 2012) lists 26 M 8 eruptions in the Quaternary, suggesting a slightly higher recurrence interval of 1 every 96,000 years. Moreover, long-term averages may be misleading as the current eruption record suggests that super-eruptions are clustered temporally. Underreporting is severe for records of even large eruptions in a geologically well-documented country like Japan (Kiyosugi et al., 2010) and all records have more events in the most recent periods of geologic time (last 50,000, or last 100,000 years, for example) than in previous periods. Thus our understanding of past frequency makes it very difficult to predict when the next super-eruption might happen; this is especially relevant from our human experience perspective. From a future hazards point of view, it is more sensible to concentrate on the effects of M 7 eruptions, which may occur every ~780 years (Deligne et al., 2010), and which have occurred in historic times (e.g., Tambora AD 1815, M 6.9—almost a 7 at 30–33 km^3 dense rock equivalent; Self et al., 2004; or the AD 1257 Samalas eruption; Lavigne et al., 2013). Such eruptions will certainly happen again in the relatively near future.

The question of where the next super-eruption might happen is also interesting. It might come from a volcano that has already produced a super-eruption (i.e., from a known super-volcano), or it might come from a volcanic system that has not yet attained the capability to erupt a sufficiently huge mass of magma. Globally, at least six volcanic systems are known to have produced a super-eruption in the last few million years or so, and which might therefore be seen as capable of another super-eruption: Yellowstone and Long Valley (USA), the Altiplano-Puna and Cerro Galan complexes (Central Andes in Argentina, Bolivia, and Chile), Toba (Indonesia), and Taupo Volcanic Zone (New Zealand). Others may have occurred but the true sizes of the eruptions have not been fully established. As stated above, the most recent well-documented super-eruption took place ~26,000 years ago at Taupo caldera and the most recent eruption approaching the extreme end of the size spectrum is the Young Toba Tuff (YTT) event 74,000 years ago (~2,800 km^3; Rose and Chesner, 1987; Chesner, 2012; Storey et al., 2012). [Note added in press: Costa et al., 2014 make a re-assessment of the volume of the YTT eruption products using an advanced ash dispersal model; it shows wider and thicker dispersal of the ash fall deposit. The deposits of this remarkable eruption could have been as large as 5,300 cubic km (M 9.1). Effects on the atmosphere and land surface would have been concomitantly increased.]

Most of the recognized super-volcanoes have produced more than one super-eruption, but they also produce much smaller eruptions. Yellowstone, Long Valley, Toba, the Altiplano-Puna complexes, and the Taupo Volcanic Zone have all had several to many relatively much smaller eruptions since their last super-eruptions. The sizes of their next eruptions will depend on the rate of magma production within these volcanic systems and the strength of the rocks containing the magma chambers.

The above factors are influenced by the rate at which basaltic magma enters the crustal, silicic magma system from the underlying mantle. Basalt provides mass and heat to generate large volumes of rhyolite by fractional crystallization and melting of continental crust, and can also maintain huge volumes of magma in an eruptible state. Prolonged heating over millions of years weakens the crust, as proposed by de Silva et al. (2006) for the Altiplano-Puna province. Crustal stretching in the Taupo Volcanic Zone aids basalt production, crustal heating, and rifting, which allows a high, long-term rate of rhyolite magma eruption, perhaps the world's highest (Houghton et al., 1995).

16.4 PRODUCTS OF SUPER-ERUPTIONS

Huge explosive eruptions of silicic magmas lead to caldera collapse and the formation of vast sheets of pyroclastic flow deposits (ignimbrites), as well as very extensive ashfall layers and injection of noxious gases into the atmosphere (Self and Blake, 2008; Wilson, 2008). The effects of VEI/M 8 super-eruptions will therefore be almost unimaginably more damaging than those of the considerably smaller eruptions (VEI 5–6) that society has experienced in recent historic times, such as Laki (Iceland) 1784–1784 (see Chapter 15), Tambora, or Pinatubo (Philippines) in 1991.

The largest known Quaternary super-eruptions produced 2,000–3,000 km^3 of magma ($4–7 \times 10^{15}$ kg, or M 8.6–8.8 (Pyle, 2000), equivalent to

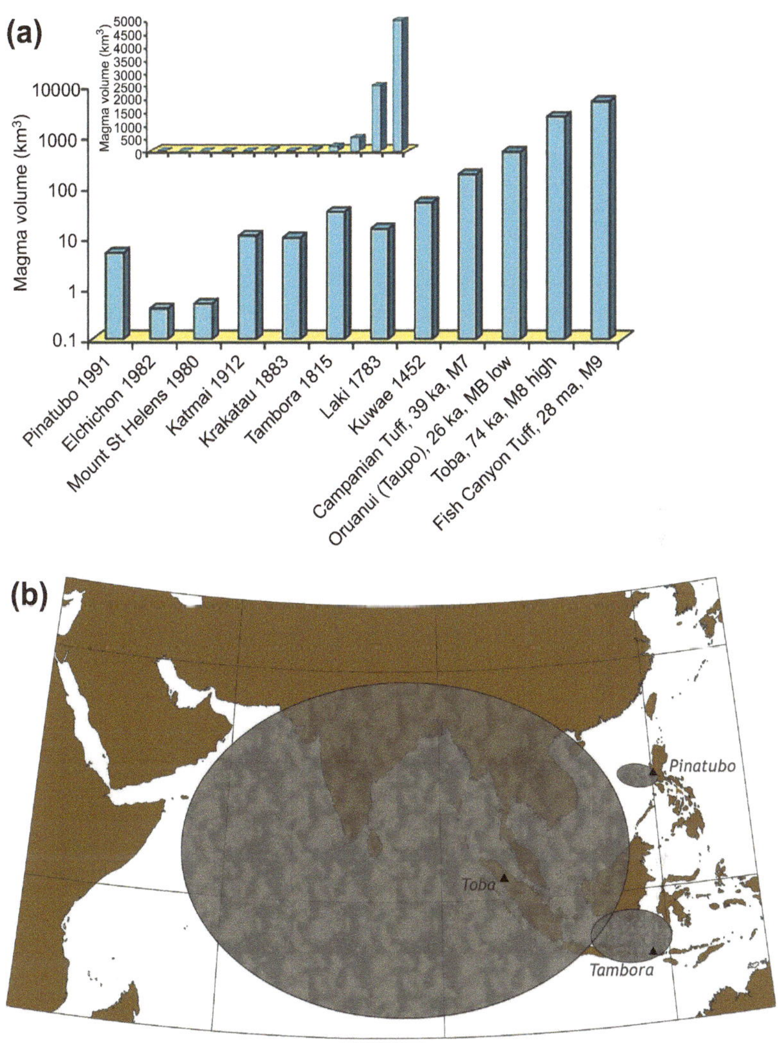

FIGURE 16.1 (a). Histograms showing the log of volumes of magma produced in selected eruptions, as mentioned in text. Inset shows same data on linear scale. (b). Map showing areas covered by volcanic ash (~5 cm isopach) from eruptions of Pinatubo (Luzon, Philippines) in 1991, Tambora (Sumbawa, Indonesia) in 1815, and Toba (Sumatra, Indonesia) ~74,000 years ago. Ashfall volumes from historic eruptions such as Pinatubo 1991 (~2 km^3 DRE, second biggest in twentieth century) and Tambora in 1815 (~25 km^3, one of the biggest in the past 1,000 years) are dwarfed by those of super-eruptions (Toba ~1,000 km^3).

5,000–8,000 km^3 of volcanic ash deposits), and include the great eruptions from Yellowstone and Toba super-volcanoes (Figure 16.1(a)). All super-eruptions are associated with formation of a caldera at least 100 km^2 in area. In the case of Toba, the present lake-filled caldera formed incrementally by three super-eruptions measures 100 × 40 km and covers ~3,000 km^2. The Japanese calderas of Aso, Kikai, and Aira are large enough to have been formed by super-eruptions, but it is possible that the eruptions that formed these calderas may have been smaller, particularly for the case where the present caldera is the result of a series of eruptions (e.g., Aso).

16.4.1 Solid Products

Explosive super-eruptions produce extensive ignimbrite and ashfall deposits. Those from the Young Toba eruption cover more than 20 million km^2, or 4 percent of Earth's surface area, with the ashfall extending across the Indian Ocean and South China Sea (Matthews et al., 2012, Figure 16.1(b)), and covering the Indian peninsula and South Asia (Gatti et al., 2013). The ashfall is produced by high-altitude ash- and gas-laden atmospheric eruption columns that penetrate through the troposphere and into the stratosphere where they spread in all directions and rotate, irrespective of stratospheric winds (Baines and Sparks, 2005). The ashfalls are the most widespread solid eruptive products (Figure 16.2).

The most immediately destructive, widespread eruptive phenomenon in super-eruptions is the pyroclastic flows, or pyroclastic density currents. These emanate from collapsing eruptive columns above the volcanic vents (see Chapter 5). They are extremely hot (hundreds of degrees Celsius, °C), rapidly moving mixtures of ash and gas that hug the ground and flow for many tens of

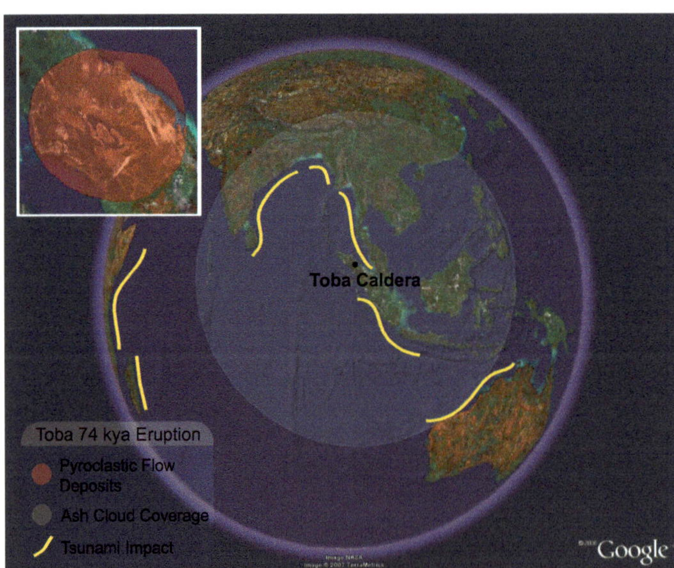

FIGURE 16.2 Cartoon of Toba YTT products showing, inset caldera and pyroclastic flow distribution, and ashfall distribution with possible coasts affected by tsunami caused by pyroclastic flows entering the sea. YTT, Young Toba Tuff. *Figure by S. Self, after by Miller and Wark (2008).*

FIGURE 16.3 Photo of Toba lake (caldera) with ignimbrite plateau shown in background. *Photo courtesy of S. Blake.*

kilometers, or in extreme cases for as much as 200 km, laying down ignimbrite deposits (Figure 16.3). Major pyroclastic flow systems such as those that formed Toba's ignimbrite deposit can cover up to 20,000 km^2 around the source volcano and bury the landscape with a layer of ash and pumice fragments up to 200 m thick. During emplacement, they will obliterate everything in their path.

16.4.2 Gaseous Products

The large eruption rates and extreme explosiveness of super-eruptions are driven by the high volatile (gas) contents of the magma. The gases are released as the magma rises from the crustal magma chamber to the surface, in a similar manner to gas released when a bottle of carbonated beverage is opened. The most abundant constituents of the gas are water (~5 percent or more of the magma by mass) and carbon dioxide (up to 0.1 percent), but eruptive emissions of these gases will make little difference to the high concentrations already in the atmosphere (e.g., for CO_2, see Gerlach, 2011). Much moist tropospheric air is also taken up in the eruption columns and transferred to the relatively dry stratosphere. More importantly, sulfur (S) and halogens (Cl, F, and perhaps Br) released from the magma cause significant changes to the normal atmospheric concentrations of these gases or of the acid aerosols H_2SO_4, HCl, and HF (Timmreck, 2012). This is especially so in the stratosphere where such species normally occur in extremely low abundance. The sulfur content is particularly important because S gas yields stratospheric aerosols that influence atmospheric opacity and hence have strong effects on climate.

Pre-eruptive (nondegassed) concentrations of S (and other volatile constituents) in magmatic liquid can be estimated by analyzing tiny (100 μm) inclusions of glass (quenched magmatic liquid) found in crystals that grew within the magma chamber, trapping the liquid and sealing it before it could degas. However, studies of glass inclusions from recent explosive eruptions such as El Chichón (1982) and Pinatubo (1991) have found that the S contents determined this way are insufficient, by one to two orders of magnitude, to supply the large amounts of SO_2 (7 and 20×10^9 kg SO_2, respectively) that Total Ozone Mapping Spectrometer (TOMS) satellite instruments detected in the stratosphere after these eruptions (Bluth et al., 1997). The abundant "excess sulfur" is accounted for by an S-rich gas phase having been present in the pre-eruptive magma. Estimating the amount of S released from a super-eruption such as Toba requires a way of estimating the amount of gas that harbors the excess sulfur and the concentration of S in this gas, and we must ask whether such a sequestered phase existed before eruption (Chesner and Luhr, 2010). In terms of whether a super-eruption causes a significant atmospheric aerosol load, the question may not be that important in the light of recent modeling of large sulfur emission from volcanoes (see subsections on atmospheric impact via aerosols and modeling).

There are two ways in which the S yield of super-eruptions can be constrained. The first is to consider super-eruptions (none of which has ever been witnessed by scientists) as simply larger versions of smaller eruptions (witnessed and studied). For historic eruptions, TOMS (and its successors) data show a positive correlation between the mass of SO_2 and the mass of magma erupted (Blake, 2003). Extrapolating this trend, the amounts of SO_2 that could be potentially released by a 2,000–3,000 km^3 super-eruption are about 3,500 to $20,000 \times 10^9$ kg. However, this approach may be misleading.

The second method is to quantify the S concentrations in the pre-eruptive magmatic liquid and gas, and then use this, together with the mass fraction of the gas phase, to calculate a total S content for the magma. In using this method, it is important to recognize that the partitioning of S between the gas phase and the magmatic liquid phase is highly dependent on the oxidation state of the magma. In reduced magmas, the partition coefficient (mass of S in gas)/(mass of S in magmatic liquid) is about 1, but it is about 1,000 in more oxidized magmas. This means that most of the S in an oxidized magma can reside in the gas phase. Using this approach, and assuming 5 wt. percent gas is present in the magma, Scaillet et al. (2004) estimated that the Toba magma had an S content of 0.022 wt. percent, therefore indicating a release of 1.4×10^{12} kg S ($2,800 \times 10^9$ kg SO_2) from the 2,800 km^3 of magma. By contrast, the rhyolite magmas erupted from Yellowstone are considerably more reduced and can therefore be expected to have lower S yields. For Yellowstone's 2.0 M Huckleberry Ridge or 0.6 M Lava Creek eruptions ($\sim 7 \times 10^{15}$ kg magma each; Mason et al., 2004), this approach predicts only 2×10^{11} kg S ($= 400 \times 10^9$ kg SO_2) despite having unleashed roughly the same amount of magma as Toba. This is "merely" 20 times the amount of SO_2 produced by Pinatubo in 1991 ($\sim$20 Tg, or 20×10^9 kg) of SO_2 from $\sim$5 km^3 of dacite magma; Self et al., 1996), or about only three to four times as much as that produced by the 1783 eruption of basaltic lava at Laki (Iceland) or the 1815 climactic explosive eruption of Tambora.

The question raised by Oppenheimer (2002,2011)—"was the Young Toba Tuff eruption an ash giant but a sulphur wimp?" is still not solved. No report has been made to date of more mafic magma erupted along with the Toba rhyolite magma, which could have increased the magma S content by late intrusion into the magma system. Although considerable uncertainties exist in the calculation of S emissions from ancient eruptions, the analysis above indicates that a smaller volume of oxidized magma can release more S gas than a larger volume of reduced magma.

16.5 EFFECTS OF SUPER-ERUPTIONS

16.5.1 Near- to Mid-field: Impact of Caldera and Pyroclastic Deposit Formation

The effects of very large eruptions, beginning at the local scale and extending outward from source, are now considered (Table 16.1). Caldera collapse may enlarge, or be encompassed within, a previous collapse structure depending on the prior history of thc volcano. Most super eruption calderas are partly backfilled with pyroclastic deposits (intracaldera ignimbrite). Caldera formation, which takes place catastrophically and engulfs everything on the surface, would destroy any artificial structures in the region over hundreds to thousands of squares kilometers in the region of the vent(s). If it was an island volcano, and/or the coast was within the reach of pyroclastic flows, the collapse could also generate tsunami (e.g., Krakatau (Sunda Straits, Indonesia), 1883; Carey et al., 1996; Miller and Wark, 2008; Figure 16.2).

Pyroclastic flows are lethal and highly destructive to buildings and other structures (Valentine, 1998; see Chapter 5). Even areas covered by parts of a pyroclastic flow that leave thin deposits would be obliterated, as borne out by the tragic effects of the small-scale flows that went through St Pierre, Martinique, on May 8, 1902. Only three people out of an estimated 30,000 survived, yet the deposits in the town are only centimeters to a few meters thick. After a very large eruption, the widespread pyroclastic flow deposits would have to be contended with. These would be many tens, up to several hundreds, of meters thick over areas of thousands to tens of thousands of squares kilometers, thicker in valleys and on plains, and thinner over hills. All buildings and artificial structures, indeed, all infrastructure, will be buried in thick pumice and ash deposits and subjected to high temperatures (400–800 °C), and acidic gases and leachates.

Rain falling onto fresh volcanic deposits can transform them into mudflows (lahars) in rivers and stream catchments, leading to possible damming of rivers, with ensuing breakout floods. Lahars caused the highest proportions of death and destruction associated with the Pinatubo eruption, but this flood-flow activity occurred in the 4–5 years after the activity died down (Torres et al., 2004).

TABLE 16.1 List of Phenomena and Hazards Associated With a Toba-size Super-eruption Based on Effects of Various Eruptive Phenomena and Deposits that Would be Formed (after Self, 2006)

Ash fallout and deposit
Roof collapse in built-up areas—a local effect out to distances where ashfall is a few centimeters thick (hundreds of kilometers from the volcano). Exacerbated if rain occurs or ashfall is wet. *Agriculture*—disruption for at least a growing season over most of the area receiving ash fallout. Longer-term beneficial changes to soil composition. *Drinking water, waste disposal*—potential for both chemical and filtration/blockage problems associated with water supply. Blockage of sewage treatment processes. *Aviation*—risk to flying aircraft while ash still airborne (days to weeks); problems with landing and takeoff until airports cleared. *Power generation*—effects of ash on hydroelectric and nuclear power plants due to ash in water intakes. *Power distribution*—electric pylons and power lines might be susceptible to ash loading and associated electrostatic effects, possibly exacerbated if ashfall is wet. *Health* (see gas and aerosols).
Pyroclastic flows and deposits
Burial of all objects on ground and *fires* on a local scale, up to perhaps 50–80 km from source volcano. *Refugees*—if pyroclastic flows were predicted, widespread evacuation would be required, which, depending on the area, could lead to a large number of persons requiring relocation. *Tsunami*—if the volcano is near the coast, pyroclastic flows entering the sea could cause tsunami. In the region around the island volcano of Krakatau in 1883, most of 30,000 casualties were due to tsunami. *Secondary mudflows (lahars)* in rivers and stream catchments after rainfall on fresh volcanic deposits. Possible damming of rivers, with ensuing breakout floods. (Lahars caused the highest proportions of death and destruction associated with the Pinatubo eruption, but in the years after the activity died down.)
Gas and aerosols
Climate change—dominantly cooler temperatures for a few years after the eruption might change agricultural yields. Some areas may undergo warming, and there might be short-term, very warm spells that could also affect growing crops. Changes in rainfall patterns may influence liability to flooding in certain areas. *Dry-fog* and *acid aerosol air pollution*—a Laki-type dry fog in the lower atmosphere (composed of sulfur dioxide gas and sulfuric acid aerosols) could induce respiratory illness, as could fine ash ($<10\ \mu m$) and other minerals in the ash. Such clouds can attain complete global coverage within a hemisphere. Chemical etching effects of *aerosol particles on aircraft* engines and instrumentation. *Ozone depletion*—stratospheric aerosols will serve to catalyze ozone loss, permitting more UV-B flux to the ground in high- and midlatitude regions, the effect lasting a few years after the eruption.

TABLE 16.1 List of Phenomena and Hazards Associated With a Toba-size Super-eruption Based on Effects of Various Eruptive Phenomena and Deposits that Would be Formed (after Self, 2006)—Cont'd

General
Disruption of national and international relief efforts and cooperation, and of some communications (satellite-borne instruments may not be able to perform normally due to ash and/or aerosols situated in atmosphere between sensor and Earth's surface). *Effects of all the above on world financial markets.*
Severity of the effects of deposits would depend on distance from source volcano. Impact of atmospheric aerosols in any region will depend on the location of volcano and season of eruption.

The low bulk density of parts of most pyroclastic flows enables them to cross water. During the 1883 eruption of Krakatau, pyroclastic flows crossed 40 km of the sea surface to hit the coast of southern Sumatra (Carey et al., 1996; cf. Allen and Cas, 2001). However, most of the deaths associated with the Krakatau eruption were apparently caused by tsunami generated when the pyroclastic flows entered the sea (Maeno and Imamura, 2011; Simkin and Fiske, 1983; Paris et al., 2014).

16.5.2 Atmospheric and Environmental Effects

16.5.2.1 Ashfall

Airborne ash is resident for only a few hours to days, unless replenished by continued activity, during which it is an extreme hazard to anything flying, including aircraft (Casadevall, 1993). Aircraft transport would be halted, or severely hampered over a wide region of a hemisphere, affecting transport on a global scale (see Chapter 4). Quite thin ashfall layers (from a few cm to 10–15 cm) can cause roof collapse (depending on construction style and standards, and wetting by rainfall), and as little as 1 cm of ash can cause severe disruption to agricultural production, if it falls in the growing season. Satellite communication and electrical grid systems may well be highly vulnerable to interruption by ash clouds. Hydroelectric and nuclear powers stations could be immobilized due to ash in water intakes, and power lines brought down by ash loading and associated electrostatic effects (Wardman et al., 2012; see also Chapter 3). Widespread ash on the land surface (e.g., Costa et al., 2012) would cause immeasurable suffering and damage, if it occurred today (e.g., Ayris and Delmelle, 2012b).

16.5.2.2 Atmospheric Impact via Aerosols

Explosive volcanic eruptions can produce global aerosol clouds in the stratosphere, and historic explosive events such as Tambora in 1815 (Indonesia) have had significant short-term climatic effects on global temperatures (Robock, 2000; Timmreck, 2012; Schmidt and Robock, submitted). The volcanic SO_2 released into the stratosphere is oxidized to form aerosol droplets of partly frozen H_2SO_4 that are in the size range (0.1−1 μm) to effectively backscatter and absorb incoming solar radiation (Robock, 2000; Self, 2005). Therefore, the net effect at the Earth's surface and in the lowest atmosphere is usually cooling. Although it produced two orders of magnitude less magma than a small super-eruption, the Pinatubo eruption of 1991 illustrates these effects. The Pinatubo aerosol cloud encircled the Earth in less than 2 weeks, and by 3 months had covered much of the global atmosphere (Self et al., 1996). Global temperatures were 0.5 °C below normal for 2 years after the eruption, slightly offsetting global warming (Hansen et al., 1996).

The largest and youngest explosive eruption of the last few 100,000°years was that of the Young Toba Tuff, which occurred during a relatively rapid period of interglacial−glacial cooling and sea-level fall. It was once thought that the eruption might have played a role, via stratospheric dust and aerosols, that could have accelerated the cooling by causing a "volcanic winter" with global temperature decreases of 3−5 °C and coolings of 10 °C during the growing season at mid- to high latitudes for a year or more (G. Jones et al., 2005), with possible longer-term climatic effects (Rampino et al., 1988). This interpretation has, however, been called into question (Oppenheimer, 2002; Robock et al., 2009), along with much of the perceived climatic impact of the eruption. It is now fairly certain that the atmospheric impact of Toba was less than originally anticipated, and that it did not initiate a glaciation or global cooling trend.

16.5.2.3 Modeling Studies of Toba's Aerosol Impact

Initial assessments using global circulation models of the climate impact of a Toba-like eruption assumed that there would be a large S gas emission and that a dense aerosol cloud would be formed. Some of the first attempts were made to assess the effects of a large stratospheric burden of sulfate aerosols. Timmreck and Graf (2005) used an atmospheric model with an SO_2 gas input of 100 times that of the Pinatubo eruption from a source at 45° N, similar to the location of Yellowstone volcano. They found that the season of injection is important, with almost global spread of aerosols from an eruption in the midlatitude Northern Hemisphere in summer, but with the aerosol cloud more restricted to the Northern Hemisphere if the eruption occurred in winter. The results from this study support those of other climate model runs that simulate the response to a huge eruption like Toba (e.g., Robock et al., 2009). Both studies indicated severe, short-term cooling with regional temperatures plummeting by as much as 10 °C for a few years, followed by a longer-term (up to 10 years) recovery period. Such an impact, combined with much reduced rainfall predicted by the climate models, could potentially kill off tropical rain forests. However, these studies may have overestimated the aerosol effects.

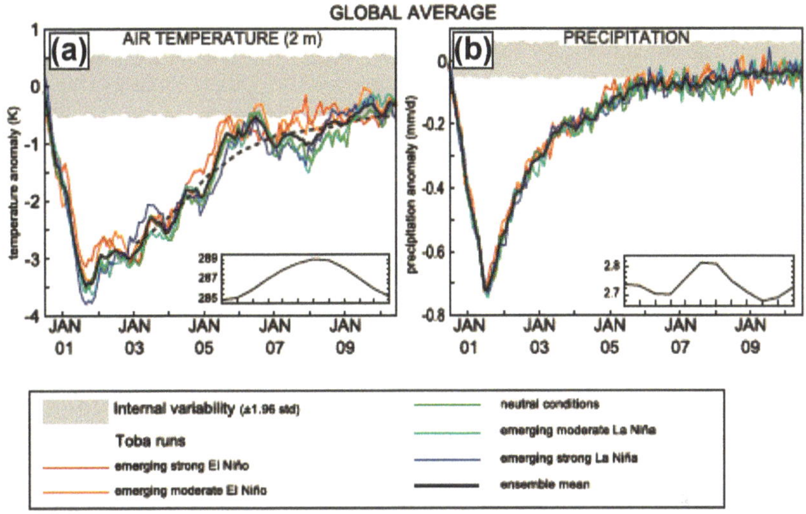

FIGURE 16.4 Plot from Timmreck et al. (2010) showing global (a) temperature and (b) precipitation anomalies model results for a "Toba" run of an Earth system model. The dashed line in (a) depicts a 3 year running average.

The much larger S gas releases predicted for super-eruptions may imply greater masses of aerosols formed; these would also attain hemispheric or global atmospheric coverage rapidly (Baines and Sparks, 2005). But other studies suggest that aerosol droplets formed after massive SO_2 emissions would become larger more quickly, and less radiatively effective, and would settle out of the stratosphere faster, thus self-limiting the duration and magnitude of the effects (Pinto et al., 1989; Timmreck et al., 2009, 2010).

Future modeling of the effects of large eruptions must include microphysical processes such as coagulation, which cause particles to grow to large sizes and have a lower optical depth per unit mass. Timmreck et al. (2010) and English et al. (2013) simulated these effects using three-dimensional microphysics models and found, for model runs intended to represent the Youngest Toba Tuff eruption, that global mean temperature response was about three times weaker (change of about −3 °C) than previously estimated using climate models without microphysics (Figure 16.4). Other work has suggested that the chemical reactions forming the aerosols from huge SO_2 clouds would initially dehydrate the stratosphere, thus prolonging the time over which aerosols form (Bekki, 1995), but Robock et al. (2009) suggest this would be a small effect. It is not yet possible to predict with confidence the particle concentration of an aerosol cloud, or its longevity at various concentrations, after a very large eruption, but most recent model runs show a muted response when compared with original estimates.

Stratospheric aerosols will serve to catalyze ozone loss (Solomon, 1999), increasing the ground-level UV-B flux in high- and midlatitude regions, the effect lasting a few years after the eruption ends. Further, the extensive ashfall deposits generated by a super-eruption may have global consequences. Covering a continent-sized area with white rhyolitic pumice and ash will increase the surface albedo, changing the land–atmosphere energy exchange. A blanketing ash layer will also kill off vegetation and alter sources and sinks involved in land–atmosphere gas exchange (M. Jones et al., 2007). Further, if large masses of ash fall over large areas of ocean, a potential may exist for significant ocean-surface fertilization, due in part to iron (Fe) in the ash, causing CO_2 drawdown from the atmosphere (e.g., Cather et al., 2009; Ayris and Delmelle, 2012a).

16.5.2.4 Implications

Botanical and palynological studies suggest that very cold-sensitive tropical vegetation would be affected disastrously by abrupt low-latitude cooling from Toba (Williams, 2012). Furthermore, computer model simulations show that an ~10 °C reduction in temperatures during the early growing season can kill about 50 percent of temperate to subarctic evergreen forests and severely damage the surviving trees. Temperate deciduous trees and other less cold-hardy vegetation would be expected to fare even worse. These results suggest that a global ecological crisis, with reductions in the amounts of plants and animals, might be caused by an eruption such as Toba. However, no hard evidence exists that the Toba eruption produced huge amounts of aerosols. Evidence for such abrupt environmental changes in the aftermath of an explosive super-eruption might be detectable in high-resolution records (e.g., lake and bog sediments and ice cores), but these records have not been adequately explored for the period around the Toba event to rigorously find evidence for or against a major environmental impact.

Although no major extinctions have been discovered at the time of Toba's blast, human genetic studies may indicate that prior to about 50,000 years ago, the population of humans suffered a severe bottleneck, followed by rapid population increases, technological innovations, and the spread out of Africa (Ambrose, 1998; Rampino and Ambrose, 2000). The predicted environmental and ecological effects of the Toba eruption lend support to a possible connection between that volcanic event and the human population bottleneck, and other faunal and floral changes have also been postulated to have occurred (Williams, 2012). Again, such interpretations are controversial, and no consensus of opinion exists about the climatic and biotic impact of Earth's youngest and biggest volcanic outburst (Lane et al., 2013).

The enormous S releases predicted, but probably not realistically, for super-eruptions may not thus imply aerosol clouds of unprecedented opacity to incoming radiation. Stratospheric aerosols also serve to catalyze ozone loss, increasing the ground-level UV-B flux in high- and midlatitude regions for a few years after the eruption. Further halogens can be released by large-scale silicic volcanism, potentially leading to effects on ozone and UV radiation

(Kutterolf et al., 2013). More studies with general circulation atmospheric models are required to further evaluate the potential climatic effects and other feedback within global environmental cycles.

16.6 SOCIETAL IMPACTS OF SUPER-ERUPTIONS

We now live in a complex, globally interrelated society that is highly vulnerable to disruption. Major historic eruptions have caused mass starvation in vulnerable regions (Auker et al., 2013), but the vulnerability of agriculture to volcanic phenomena may not now be as great as it was before the twentieth century. However, almost 90 percent of the world's population lives in the Northern Hemisphere where most food production is concentrated, thus it is especially vulnerable to the effects from major eruptions. Today's globalized society is heavily dependent on communications and other technologies that are highly susceptible to disruption, and many areas are now densely populated. The predicted effects and consequences of a large explosive eruption are almost entirely detrimental to mankind in the short term. These range from roof collapse due to the weight of an ash layer to the possible disruptive effects of ash and aerosol clouds on satellite communications. Airborne ash presents problems to aviation and electricity transmission. Ash will block roads, railways, and airport runways, causing widespread problems. Ash will also be a health hazard to the eyes and lungs; for example, airborne fine volcanic particulates can contain respirable particles of cristobalite (a form of silica) that can cause silicosis (Horwell et al., 2003; Baxter, 2005). There are also bound to be effects that we cannot yet anticipate.

The consequences of future large-scale eruptions are in detail very challenging to predict because of the variability of the volcanic phenomena and the complex interlinking of important societal necessities. Telecommunications and air transport are likely to be extremely vulnerable, significantly compromising the effectiveness of disaster response in the wake of the large-scale eruptions. The cost of all natural disasters is rising steeply (Smolka, 2006; see also Chapter 12) and effects on world financial markets are likely to be acute.

16.7 SUMMARY AND FUTURE CONCERNS

Wherever on Earth they occur, a very small, but real, threat exists from very large explosive eruptions up to, and including, super-eruptions. Eruptions are unstoppable, and super-eruptions are potentially long-lasting. One well-studied case (Wilson and Hildreth, 1997) is the ~600 km^3 Bishop Tuff (Long Valley caldera, California, USA), which is estimated to have lasted about 6 days with only one short pause. Other super-eruptions may have included several high-intensity eruptive phases spread over weeks to even years with periods of quiescence in between (Wilson, 2008). A new study has suggested even longer hiatuses (10^3 years) in one of the giant Yellowstone eruptions (Ellis and Mark, 2013), underlining that what we map as, and assume is, the product of a single giant eruption may have a more complex emplacement history, which is important knowledge for mankind in case of a repeat event.

The detailed consequences of future large-scale eruptions are in detail very challenging to predict. The effects of each explosive phase will be immediate (widespread ashfall, pyroclastic flows) and persistent, presenting problems for years after the last major explosive phase due to atmospheric aerosols and debris flows (lahars) from reworked unstable, fresh deposits. Catastrophic inundation of wide areas by ignimbrites, suspension of fine ash in the atmosphere, and global dispersal of aerosols are likely to seriously damage human health; terrestrial and marine ecosystems; food, water, and energy supplies; and communication and transport networks (Baxter, 2005; Self, 2005). The rising cost of all natural disasters (Smolka, 2006) and the economic cost of recovery from any future large-scale eruptions will be a major burden on society. Telecommunications and air transport will be extremely vulnerable. Predicting all effects of super-eruptions is problematic because many may be outside modern experience.

Much exists still to learn about super-eruptions. For example, we do not know over what duration, or range of durations, an eruption of this size builds up, and whether the magnitude and course of events in the impending eruption can be predicted (Wilson, 2008). At present exists no scientific or evidence-based way of determining whether precursor eruptions might be heralding a very large outburst rather than a smaller magnitude eruption. Of concern is a situation where a volcano that is presently unrecognized as a potential large-eruption site, or that is evolving toward its first super-eruption, or that is not monitored, becomes restless. In these cases, scientists and/or the government concerned may not recognize the signs of what is forthcoming. This uncertainty underlines the fact that we need to know much more about past super-eruptions: which volcanoes have produced them, what the ages are, what the true size was, and what types of deposits were produced. We also need atmospheric scientists to further model the effects of super-eruptions on the weather and climate systems.

ACKNOWLEDGMENTS

I thank Paolo Papale for inviting me to contribute this chapter, and Sarah Brown and Sian Crosweller for supplying me with updates on the LaMEVE database. Andrew Tupper is thanked for a helpful review of the draft manuscript.

REFERENCES

Allen, S.R., Cas, R.A.F., 2001. Transport of pyroclastic flows across the sea during the explosive, rhyolitic eruption of the Kos Plateau Tuff, Greece. Bull. Volcanol. 62, 441–456.

Ambrose, S.H., 1998. Late Pleistocene human population bottlenecks, volcanic winter, and differentiation of modern humans. J. Hum. Evol. 34 (6), 623–651.

Auker, M.R., Sparks, R.S.J., Siebert, L., Crosweller, H.S., Ewart, J., 2013. A statistical analysis of the global historical volcanic fatalities record. J. Appl. Volcanol. 2 http://dx.doi.org/10.1186/2191-5040-2-2.

Ayris, P., Delmelle, P., 2012a. Volcanic and atmospheric controls on ash iron solubility: a review. Phys. Chem. Earth 45–46, 103–112. JPCE-D-11-00132.

Ayris, P.M., Delmelle, P., 2012b. The immediate environmental effects of tephra emission. Bull. Volcanol. 74 (9), 1905–1936.

Baines, P.G., Sparks, R.S.J., 2005. Dynamics of giant ash clouds from supervolcanic eruptions. Geophys. Res. Lett. 32 (L24808) http://dx.doi.org/10.1029/2005GL024597.

Baxter, P.J., 2005. Human effects of volcanism. In: Ernst, G.G.J., Marti, J. (Eds.), Environmental Effects of Volcanic Activity. Cambridge University Press, pp. 271–303.

Bekki, S., 1995. Oxidation of volcanic SO_2: a sink for stratospheric OH and H_2O. Geophys. Res. Lett. 22, 913–916.

Blake, S., 2003. Correlations between eruption magnitude, SO_2 yield, and surface cooling. In: Oppenheimer, C., Pyle, D.M., Barclay, J. (Eds.), Volcanic Degassing, vol. 213. Geological Society of London Special Publication, pp. 371–380.

Bluth, G.J.S., Rose, W.I., Sprod, I.E., Krueger, A., 1997. Stratospheric loading of sulfur from explosive volcanic eruptions. J. Geol. 105, 671–683.

Bryan, S.E., Ukstins Peate, I.A., Self, S., Peate, D., Jerram, D.A., Mawby, M.R., Miller, J., Marsh, J.S., 2010. The largest volcanic eruptions on earth. Earth-Sci. Rev. 102, 207–229.

Bryan, S.E., Ferrari, L., 2013. Large igneous provinces and silicic large igneous provinces: progress in our understanding over the past 25 years. Geol. Soc. Am. Bull v. 125, 1053–1078. http://dx.doi.org/10.1130/B30820.1.

Carey, S., Sigurdsson, H., Mandeville, C., Bronto, S., 1996. Pyroclastic flows and surges over water: an example from the 1883 Krakatau eruption. Bull. Volcanol. 57, 493–511.

Casadevall, T., 1993. Volcanic hazards and aviation safety: lessons of the past decade. FAA Aviat. Saf. J. 2, 3–11.

Cather, S.M., Dunbar, N.W., McDowell, F.W., McIntosh, W.C., Scholle, P.A., 2009. Climate forcing by iron fertilization from repeated ignimbrite eruptions: the icehouse–silicic large igneous province (SLIP) hypothesis. Geosphere 5, 315–324.

Chesner, C.A., 2012. The Toba Caldera complex. Quat. Int. 258, 5–18. http://dx.doi.org/10.1016/j.quaint.2011.09.025.

Chesner, C., Luhr, J., 2010. A melt inclusion study of the Toba Tuffs, Sumatra, Indonesia. J. Volcanol. Geotherm. Res. 197, 259–278.

Costa, A., Folch, A., Macedonio, G., Giaccio, B., Isaia, R., Smith, V.C., 2012. Quantifying volcanic ash dispersal and impact of the Campanian Ignimbrite super-eruption. Geophys. Res. Lett. 39, L10310. http://dx.doi.org/10.1029/2012GL051605.

Costa, A., Smith, V.C., Macedonio, G., Matthews, N.E., 2014. The magnitude and impact of the Youngest Toba Tuff super-eruption. Frontiers in Earth Science 2, Art. 16, 1–8. http://dx.doi.org/10.1186/2191-5040-1-4.

Crosweller, H.S., Arora, B., Brown, S.K., 12 others, 2012. Global database on large magnitude explosive volcanic eruptions (LaMEVE). J. Appl. Volcanol. 1. http://dx.doi.org/10.1186/2191-5040-1-4.

Deligne, N.I., Coles, S.G., Sparks, R.S.J., 2010. Recurrence rates of large explosive volcanic eruptions. J. Geophys. Res. 115, B06203. http://dx.doi.org/10.1029/2009JB006554.

de Silva, S., Zandt, G., Trumbell, R., Viramonte, J., 2006. Large-scale silicic volcanism – the result of thermal maturation of the crust. Adv. Geosci. 1, 215–230.

Ellis, B., Mark, D.F., 2013. Super-eruptions and silicic volcanism from the Yellowstone volcanic field. Geol. Today 29, 133–137.

English, J.M., Toon, O.B., Mills, M.J., 2013. Microphysical simulations of large volcanic eruptions: Pinatubo and Toba. J. Geophys. Res., Atmos. 118, 1880–1895.

Gatti, E., Saidin, M., Talib, K., Rashidi, N.-A., Gibbard, P., Oppenheimer, C., 2013. Depositional processes of reworked tephra from the Late Pleistocene Youngest Toba Tuff deposits in the Lenggong Valley, Malaysia. Quat. Res. 79, 228–241.

Gerlach, T., 2011. Volcanic versus anthropogenic carbon dioxide. Eos Trans. AGU 92 (24), 201–203.

Hansen, J., Sato, M., Ruedy, R., 1996. A Pinatubo climate modelling investigation. In: The Mount Pinatubo Eruption: Effects on the Atmosphere and Climate. NATO Advanced Study Institute Series 14. Springer-Verlag, Heidelberg, pp. 233–272.

Horwell, C.J., Sparks, R.S.J., Brewer, T.S., et al., 2003. Characterization of respirable volcanic ash from the Soufriere Hills volcano, Montserrat, with implications for human health hazards. Bull. Volcanol. 65, 346–362.

Houghton, B.F., Wilson, C.J.N., McWilliams, M.O., Lanphere, M.A., Weaver, S.D., Briggs, R.M., Pringle, M.S., 1995. Chronology and dynamics of a large silicic magma system: central Taupo Volcanic Zone, New Zealand. Geology 23, 13–16.

Jones, G.S., Gregory, J.M., Stott, P.A., Tett, S.F.B., Thorpe, R.B., 2005. An AOGCM simulation of the climate response to a volcanic super-eruption. Clim. Dyn. 45. http://dx.doi.org/10.007/s00382-005-0066-8.

Jones, M.R., Sparks, Valdes, P., 2007. The climatic impact of supervolcanic ash blankets. Clim. Dyn. 29 (6), 553–564.

Kiyosugi, K., Connor, C., Sparks, R.S.J., Crosweller, H.S., Siebert, L., Takarada, S., December 2010. How many explosive eruptions are missing from the geologic record? Analysis of the quaternary record of large magnitude explosive eruptions in Japan. In: AGU Fall Meeting Abstracts.

Kutterolf, S., Hansteen, T., Appel, K., Freundt, A., Krüger, K., Pérez, W., Wehrmann, H., 2013. Combined bromine and chlorine release from large explosive volcanic eruptions: a threat to stratospheric ozone? Geology 41, 707–710.

Lane, C.S., Chorn, B.T., Johnson, T.C., 2013. Ash from Toba supereruption in Lake Malawi shows no sign of volcanic winter in East Africa at 75 ka. Proc. Natl. Acad. Sci. 110, 8025–8029.

Lavigne, F., Degeai, J.-P., Komorowski, J.-C., 13 others, 2013. Source of the great A.D. 1257 mystery eruption unveiled, Samalas volcano, Rinjani Volcanic Complex, Indonesia. Proc. Natl. Acad. Sci. 110, 16,742–16,747.

Lipman, P.W., 1997. Chasing the volcano. Earth, 33–39 (December).

Lowenstern, J.B., Hurwitz, S., 2008. Monitoring a supervolcano in repose: heat and volatile flux at the Yellowstone caldera. Elements 4, 35–40.

Maeno, F., Imamura, F., 2011. Tsunami generation by a rapid entrance of pyroclastic flow into the sea during the 1883 Krakatau eruption, Indonesia. J. Geophys. Res. 116, B09205.

Mason, B.G., Pyle, D.M., Oppenheimer, C., 2004. The size and frequency of the largest explosive eruptions on Earth. Bull. Volcanol. 66, 735–748.

Matthews, N.E., Smith, V.C., Antonio Costa, A., Durant, A.J., Pyle, D.M., Pearce, N.J.G., 2012. Ultra-distal tephra deposits from super-eruptions: examples from Toba, Indonesia and Taupo Volcanic Zone, New Zealand. Quat. Int. 258, 57–79.

Miller, C.F., Wark, D.A., 2008. Supervolcanoes and their explosive supereruptions. Elements 4, 11–16.

Newhall, C.G., Self, S., 1982. The volcanic explosivity Index (VEI): an estimate of explosive magnitude for historical volcanism. J. Geophys. Res. 87, 1231–1238.

Oppenheimer, C., 2002. Limited global change due to the largest known quaternary eruption, Toba ~ 74 kyr BP? Quat. Sci. Rev. 21, 1593–1609.

Oppenheimer, C., 2011. Eruptions that Shook the World. Cambridge University Press, Cambridge, pp. 181–207.

Paris, R., Switzer, A.D., Belousova, M., Belousov, A., Ontowirjo, B., Whelley, P.L., Ulvrova, M., 2014. Volcanic tsunami: a review of source mechanisms, past events and hazards in Southeast Asia (Indonesia, Philippines, Papua New Guinea). Nat. Hazards 70 (1), 447–470.

Pinto, J.P., Turco, R.P., Toon, O.B., 1989. Self-limiting physical and chemical effects in volcanic eruption clouds. J. Geophys. Res. 94 (D8), 11165–11174.

Pyle, D.M., 2000. The sizes of volcanic eruptions. In: Sigurdsson, H., et al. (Eds.), The Encyclopedia of Volcanoes. Academic Press, London, pp. 263–269.

Rampino, M.R., 2002. Supereruptions as a threat to civilizations on Earth-like planets. Icarus 156, 562–569.

Rampino, M.R., Ambrose, S.H., 2000. Volcanic winter in the Garden of Eden: the Toba super-eruption and the late Pleistocene human population crash. In: McCoy, F.W., et al. (Eds.), Volcanic Hazards and Disasters in Human Antiquity, vol. 345. GSA Special Paper, pp. 71–82.

Rampino, M.R., Self, S., Stothers, R.B., 1988. Volcanic Winters. Annu. Rev. Earth Planet. Sci. 16, 73–99.

Robock, A., 2000. Volcanic eruptions and climate. Rev. Geophys. 38, 191–207.

Robock, A., Ammann, C.M., Oman, L., et al., 2009. Did the Toba volcanic eruption of ~ 74ka B.P. produce widespread glaciation? J. Geophys. Res. 114, D10107. http://dx.doi.org/10.1029/2008JD011652.

Rose, W.I., Chesner, C.A., 1987. Dispersal of ash in the great Toba eruption, 75 ka. Geology 15, 913–917.

Scaillet, B., Luhr, J.F., Carroll, M.R., 2004. Petrological and volcanological constraints on volcanic sulfur emissions to the atmosphere. In: Robock, A., Oppenheimer, C. (Eds.), Volcanism and the Earth's Atmosphere, vol. 139. American Geophysical Union Geophysical Monograph, pp. 11–40.

Schmidt, A., Robock, A. Volcanism, the atmosphere, and climate through time. In: Schmidt, A., Fristad, K., Elkins-Tanton, L., in press, Volcanism and Global Environmental Change, Cambridge University Press, 310 pp.

Self, S., 2005. Effects of volcanic eruptions on the atmosphere and climate. In: Ernst, G.G.J., Marti, J. (Eds.), Environmental Effects of Volcanic Activity. Cambridge University Press, pp. 152–174.

Self, S., 2006. The effects and consequences of very large explosive volcanic eruptions,. Philos. Trans. R. Soc. A 364, 2073–2097. http://dx.doi.org/10.1098/rsta.2006.1814.

Self, S., Blake, S., 2008. Consequences of explosive super-eruptions. Elements 4, 41–46.

Self, S., Gertisser, R., Thordarson, Th, Rampino, M.R., Wolff, J.A., 2004. Magma volume, volatile emissions, and stratospheric aerosols from the 1815 eruption of Tambora. Geophys. Res. Lett. 31, L20608. http://dx.doi.org/10.1029/2004GL020925.

Self, S., Schmidt, A., Mather, T.A. Emplacement characteristics, timescales, and volatile release rates of continental flood basalt eruptions on Earth. In: Keller, G. (Ed.), GSA Special Paper 505, Flood Basalts, Impacts and Extinctions, in press.

Self, S., Zhao, J.-X., Holasek, R.E., Torres, R.C., King, A.J., 1996. The atmospheric impact of the Mount Pinatubo eruption. In: Newhall, C.G., Punongbayan, R.S. (Eds.), Fire and Mud: Eruptions and Lahars of Mount Pinatubo, Philippines. Philippine Institute of Volcanology and Seismology, Quezon City, and University of Washington Press, Seattle, pp. 1089–1115.

Simkin, T., Fiske, R.S., 1983. Krakatau 1883: The Volcanic Eruption and its Effects. Smithsonian Institution Press, Washington DC, 464 pp.

Smolka, A., 2006. Natural disasters and the challenge of extreme events: risk management from an insurance perspective. Philos. Trans. R. Soc., A 364, 2147–2165.

Solomon, S., 1999. Stratospheric ozone depletion: a review of concepts and history. Rev. Geophys. 37 (3), 275–316.

Sparks, R.S.J., Self, S., Grattan, J.P., Oppenheimer, C., Pyle, D.M., Rymer, H., 2005. Super-Eruptions: Global Effects and Future Threats: Report of a Geological Society of London Working Group. The Geological Society, London, 24 pp. http://www.geolsoc.org.uk/template.cfm?name=Super1.

Storey, M., Roberts, R.G., Saidin, M., 2012. Astronomically calibrated 40Ar/39Ar age for the Toba supereruption and global synchronization of late quaternary records. Proc Natl Acad Sci USA. 109 (46), 18684–18688.

Timmreck, C., 2012. Modeling the climatic effects of large explosive volcanic eruptions, Wiley Interdisciplinary Reviews. Clim. Change 3 (6), 545–564.

Timmreck, C., Graf, H.-F., 2005. The initial dispersal and radiative forcing of a mid-latitude super-eruption: a Northern Hemisphere case study. Atmos. Chem. Phys. Discussions 5, 7283–7308.

Timmreck, C., Graf, H.-F., Lorenz, S.J., Niemeier, U., Zanchettin, D., Matei, D., Jungclaus, J.H., Crowley, T.J., 2010. Aerosol size confines climate response to volcanic supereruptions. Geophys. Res. Lett. 5, L24705. http://dx.doi.org/10.1029/2010GL045464.

Timmreck, C., Lorenz, S.J., Crowley, T.J., Kinne, S., Raddatz, T.M., Thomas, M.A., Jungclaus, J.H., 2009. Limited temperature response to the very large AD 1258 volcanic eruption. Geophys. Res. Lett. 36, L21708. http://dx.doi.org/10.1029/2009GL040083.

Torres, R.C., Mouginis-Mark, P.J., Self, S., 3 others, 2004. Monitoring the evolution of the Pasig-Potrero alluvial fan at Pinatubo volcano using a decade-long remote sensing data set. J. Volcanol. Geotherm. Res. 138, 371–392.

Valentine, G.A., 1998. Damage to structures by pyroclastic flows and surges, inferred from nuclear weapons effects. J. Volcanol. Geotherm. Res. 87, 117–140.

Wardman, J.B., Wilson, T.M., Bodger, P.S., Cole, J.W., Stewart, C., 2012. Potential impacts from tephra fall to electric power systems: a review and mitigation strategies. Bull. Volcanol. 74 (10), 2221–2241.

Williams, M., 2012. Did the 73 ka Toba super-eruption have an enduring effect? Insights from genetics, prehistoric archaeology, pollen analysis, stable isotope geochemistry, geomorphology, ice cores, and climate models. Quat. Int. 269, 87–93.

Wilson, C.J.N., 2001. The 26.5 ka Oruanui eruption, New Zealand: an introduction and overview. J. Volcanol. Geotherm. Res. 112, 133–174.

Wilson, C.J.N., 2008. Physical volcanology of supereruptions and supervolcanoes. Elements 4, 29–34.

Wilson, C.J.N., Hildreth, W., 1997. The Bishop Tuff: new insights from eruptive stratigraphy. J. Geol. 105, 407–439.

Chapter 17

Integration of European Volcano Infrastructures

Giuseppe Puglisi[1], Kristín S. Vogfjörd[2], Patrick Bachelery[3] and Teresa Ferreira[4]

[1] *Istituto Nazionale di Geofisica e Vulcanologia, Sezione di Catania – Osservatorio Etneo, Catania, Italy,* [2] *Icelandic Meteorological Office, Reykjavík, Iceland,* [3] *Lab. Magmas et Volcans, Observatoire de Physique du Globe de Clermont-Ferrand, Blaise Pascal University – CNRS–IRD, Clermont-Ferrand, France,* [4] *Universidade dos Açores, Centro de Vulcanologia e Avaliação de Riscos Geológicos, Ponta Delgada, Portugal*

ABSTRACT

Volcano Observatories, as well as some laboratories of scientific institutions, can be considered as research infrastructures (RIs) that foster cutting-edge research studies and novel technological developments for volcano monitoring and understanding of volcanic processes. The volcanological observation systems are typically multidisciplinary and multiparametric, and can be carried out from both ground and space. In this chapter, we provide an overview of the current European research infrastructures and their ability in tracking the volcano-related processes. Despite the high level of the European volcanological research, an analysis of the main drawbacks highlighted heterogeneities in the geographic distribution and quality of the monitoring systems and in the governance structures, as well as poor collaboration and data sharing among the RIs. In this view, we here propose a vision and a possible plan for the implementation of an integrated RI, focused on the volcano observations that join together and coordinate the single-sited European research infrastructures. Such an integrated RI will have a noteworthy impact both on science and society in terms of understanding of volcanic processes and related hazards, and improvements of the relationship with the decision-makers.

17.1 RATIONALE

Research infrastructure (RI) is a term that includes "facilities, resources and related services used by the scientific community to conduct top-level research in their respective fields" (from European Strategy Forum on Research Infrastructures, ESFRI; http://ec.europa.eu/research/infrastructures/index_en.

Volcanic Hazards, Risks, and Disasters. http://dx.doi.org/10.1016/B978-0-12-396453-3.00017-4

cfm?pg=home). This definition links the concept of RI with that of research studies that move the frontiers of the knowledge forward.

Current examples of RIs include top-class laboratories, such as CERN in Geneva, or large databases, such as that on the Survey of Health, Ageing and Retirement in Europe (http://www.share-project.org/home0.html), or massive computing resources, such as the grid/cloud computing facilities implemented in neuGRID to develop drugs (http://www.neugrid.eu/pagine/home.php).

From these few examples it is evident that at the base of the concept of an RI, there is the possibility to collect, save, share, and analyze large amounts of qualified data. Furthermore, the larger the scientific community involved in the RI activity, the more relevant is the impact of the RI and the more crucial the role of cutting-edge technologies (e.g., informatics, electronics, nanotechnologies, etc.). In some cases, the impact of RIs on the research is priceless! For instance, considering only the above examples, even to unskilled people it is clear that experiments that led to a Noble Prize or the modeling of new molecules would simply have been impossible without them. Another consideration resulting from the current example of RIs is that they could be "single-sited" (a single resource at a single location), "distributed" (a network of distributed resources), or "virtual" (the service is provided electronically).

As far as volcanology is concerned, Volcano Observatories (VOs) are the main RIs since they represent the basic resource for research in volcanology and provide the usual facilities for the design, implementation, and testing of monitoring techniques. The aims of the VOs and laboratories are: (1) to implement, upgrade, and maintain the observation systems; (2) to timely and accurately inform and alert the authorities/media/communities living in or close to volcanoes about impending or current volcanic activity; (3) to implement and maintain databases of volcanological data for successive studies or educational aims; and (4) to support civil protection authorities in defining and implementing initiatives aimed at mitigating volcanic risk. The commitments to the decision-makers involved in risk management vary from one country to another.

Furthermore, many research and academic institutions carrying out experiments, modeling, and scientific studies on volcanic areas and products can be considered as RIs in volcanology. As a whole they can be referred to as Volcanological Research Institutions (VRIs), and their role is mainly in the education sector or in top-level research. Usually both VOs and VRIs are "single-sited" and both have greatly benefitted from the fast technological evolution.

The last decades have seen a continuous progress of volcanological monitoring networks on active volcanoes worldwide. The progress is not only in the proliferation of automatic stations installed on volcanoes and sites where physical or chemical parameters are intermittently measured, but also in the types of parameters measured and the accuracy of the measurements. This progress is the result of technological development as well as the positive feedback between the monitoring and research activities, and it produces a large amount of data suitable for deeply improving the knowledge of volcanic processes.

Furthermore, the evolution of computational resources, both hardware and software, can now enable analysis and modeling of vast amounts of data allowing investigation of surface phenomena (e.g., DInSAR maps), internal structure of volcanoes (e.g., seismic tomography), and volcanic processes (e.g., numerical modeling of magmatic sources), which were simply impossible to envisage 20 or 30 years ago.

This rough overview of the main characteristics of the current volcanological RIs shows that they fit the main features of the RIs mentioned above, in that: they collect, analyze, and model large amount of data on volcanological processes, and their evolution largely depends on the growth of the applicable technologies (Saccorotti et al., this Volume). The evolution of these RIs has accompanied the growth of the volcanological community over the last decades and has enabled deepening of the knowledge of volcanic processes.

Nevertheless, the current impact of the volcanological RIs on the volcanological community seems limited due to shortcomings in data sharing and in providing access to laboratories and computing resources. This observation, which will be analyzed in more detail in the following subchapter, supports the creation of a network of RIs to allow evolution from the current configuration of "single-sited" volcanological RIs toward an integrated volcanological "distributed" RI.

17.2 STATE OF THE ART OF THE EUROPEAN VOLCANOLOGICAL RIS

Active volcanoes in Europe constitute a direct threat for millions of European citizens in both on the continent and on European islands. In fact, the ash cloud from the recent 2010 Eyjafjallajökull eruption in Iceland highlighted that even sporadic and remote volcanic eruptions can indirectly affect the lives of hundreds of millions of people living in continental Europe (Schmidt, this Volume). Moreover, due to the existence of European territories also in the West Indies and the Indian and Pacific oceans, eruptions of "European" volcanoes can have a much wider impact outside Europe.

Our ability to understand volcanic unrest and forecast eruptions depends on both the capability of the monitoring systems to detect and quantify effects of the rising magma and on the scientific knowledge necessary to interpret them. These premises make evident the strategic importance of an integrated RI focused on volcanological observations for both science and society.

Most of the European volcanological RIs, either VOs or VRIs, are located close to active volcanic areas. This implies that the geographical distribution of these RIs is determined by the location of the volcanoes that are not uniformly distributed in European continent and furthermore are sometimes located in overseas territories of the European countries (Figure 17.1).

The active volcanic areas on the European continent are located along the Eurasian plate boundaries (Figure 17.1 (a)).

Along the southern boundary, volcanoes located in the Greek islands and in southern Italy are the most studied. They can be considered the cradle of volcanology. The activity of these volcanoes is well known over the last few

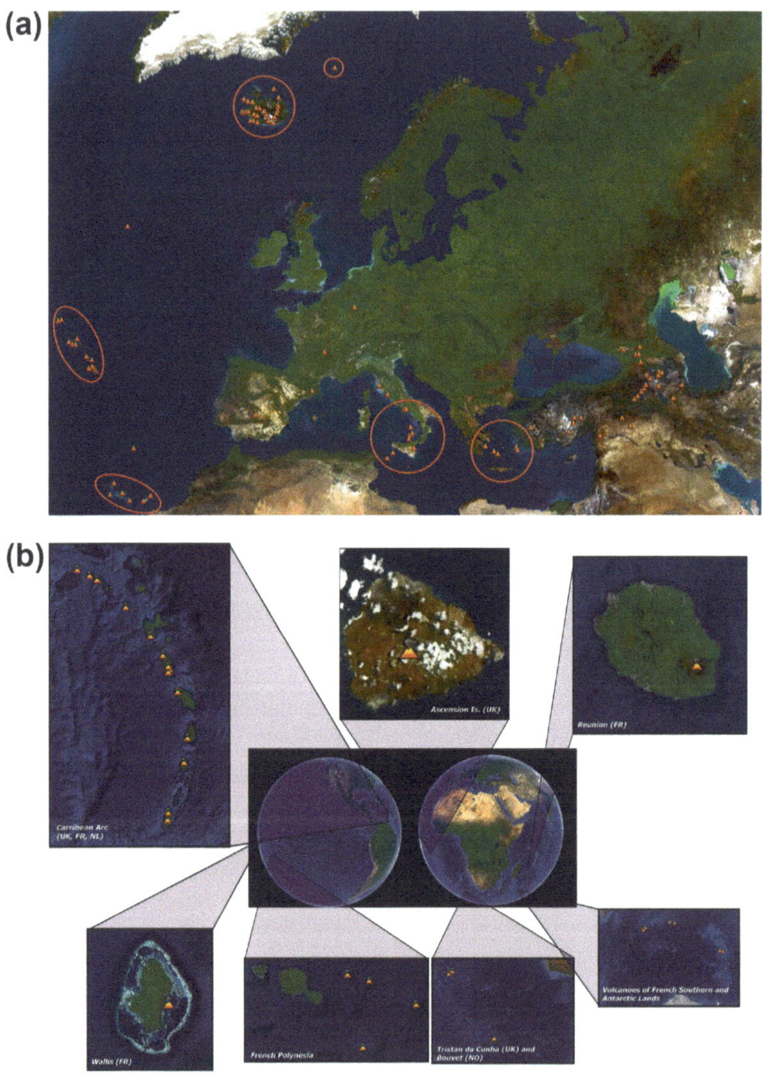

FIGURE 17.1 Geographical distribution of European volcanoes (red triangles): (a) on the European continents, plus Canary Islands and (b) in overseas countries.

thousand years and their catastrophic effects have conditioned the evolution and history of the ancient cultures at the center of the Mediterranean Sea (e.g., Santorini eruption about 2000 BCE and Vesuvius eruption CE 79), as well as influenced their mythology and religions (e.g., Hephaestus or Vulcan gods in Greek and Latin religion, respectively) (Druitt et al., 1999; Oppenheimer, 2011). Since the eighteenth and nineteenth centuries, these volcanoes have been the natural laboratories for scientists coming from all over Europe. In particular, these early studies were most frequent on the volcanoes of South Italy (Etna and Vesuvius) thanks to the frequent eruptive events and the wide spectrum of eruptive phenomena, ranging from fumarolic activity to plinian explosions (e.g., Goete, 1816; Spallanzani, 1792). These specific conditions favored the birth of modern volcanology, specifically in southern Italy, thanks to the implementation of the first monitoring networks. The first active volcanological observatory, with a clear commitment to volcano monitoring, was the Osservatorio Vesuviano (OV), founded in 1841. Later on, at the beginning of the twentieth century, infrastructures for monitoring volcanic activity were installed on Etna (e.g., the volcanological Observatory of Catania University was equipped in 1926, at 2,800 m altitude) and initiatives to implement and organize volcano monitoring at a national level were updated during the 1980s and 1990s (e.g., the setup of the National Group of Volcanology—GNV), leading to the establishment of the INGV in 2000. Institutes which have merged the various institutions involved in geophysical and volcanological research in Italy, including OV, are the Istituto Nazionale di Geofisica (ING) and the Institutes of Consiglio Nazionale delle Ricerca (CNR) located in Catania, Milano, Palermo, and Pisa. INGV has a clear mandate for carrying out research and monitoring activities on Italian volcanoes, as well as to contribute to the scientific management of surveillance activities through an agreement with the Department of Civil Protection.

Currently, volcanological monitoring networks of southern Italian volcanoes are comprised of hundreds of seismic, geodetic, and geochemical permanent sensors deployed mainly on Mt Etna, Vesuvius, Campi Flegrei, and the Aeolian islands (Stromboli, Lipari, and Vulcano), and less abundant on Pantelleria island, in the Sicilian channel, and Albanian Hills, close to Rome. Furthermore, periodic geodetic (Aloisi et al., 2003; D'Auria et al., 2012; Greco et al., 2012; La Spina et al., 2013), geochemical (Chiodini et al., 2007; Salerno et al., 2009), geophysical (Giudicepietro et al., 2010; Patané et al., 2004; Aiuppa et al., 2010; Napoli et al., 2008; Muré et al., 2013), and geological surveys (Calvari et al., 2011) are also carried out at thousands of benchmarks and measurement points, and on networks of video, IR, and thermal cameras. Finally, analytical laboratories and facilities for analyzing satellite data are also included in the monitoring activities (Corsaro et al., 2011).

Probably due to the quiescent state of activity at volcanoes on the Greek islands (Santorini and Nisyros), the implementation of monitoring systems on Greek volcanoes has not developed as necessary. Monitoring is mainly based

FIGURE 17.2 A permanent GPS station close to the summit craters of Mt Etna.

on a few seismic, geodetic, and geochemical stations and periodic geophysical and geochemical surveys (Figure 17.2).

On the western boundary of the Eurasian plate, along the mid-Atlantic ridge, are located two of the main volcanic areas in Europe: the Azores islands in the south and Iceland in the north.

The Azores archipelago is a Portuguese archipelago comprising nine volcanic islands, about 1,600 km from Europe, displaying a northwest alignment that extends for more than 600 km, crossing the mid-Atlantic Ridge. It is located in a region where a very particular geodynamic setting exists, resulting from the Azores triple junction, formed by the North America, Eurasia, and Nubia lithospheric plates, combined with a large area of anomalous magmatic production attributed to the existence of a deep mantle plume. Significant seismic and volcanic activity is mainly associated with the plate boundaries. The active volcanic systems display a wide variety of eruptive styles, from submarine to subaerial, from basaltic and effusive to trachytic and highly explosive, magmatic, and hydromagmatic. The first noticed Azores historical eruption was coincident with the settlement of São Miguel Island, some time between AD 1439 and 1443, and took place at Furnas volcano. Since then, at least 28 volcanic eruptions occurred distributed in almost all of the islands and submarine areas located along Eurasia and Nubia boundary. The most recent eruption, located at the Serreta submarine ridge, spanned between 1998 and 2001 exhibiting unusual volcanic products that were named lava balloons. Seismic swarms are frequent at some volcanic systems, particularly at Fogo volcano where during recent years unrest episodes have being recorded. The VO is conducting a multiparametric monitoring program at a regional and volcanic system scale, based on permanent geophysical, geodetic, geochemical, environmental, and meteorological networks, complemented with discrete data measurements and sampling campaigns, aiming a better

FIGURE 17.3 A GNSS monitoring station at the Azores Islands.

understanding of the relationships between the regional tectonic regimes, magma rise, and volcanic processes (Figure 17.3). Also, in a multihazard assessment perspective, the monitoring networks provide critical information for a better advisory of civil protection authorities of all the hazards that can be triggered during unrest or eruptive episodes (Gaspar et al., 2011). The Azores archipelago represents one of the most challenging regions in the world for volcano monitoring, once both tectonic and volcanic processes are acting and overlapping in the same narrow strip of the lithosphere increasing the difficulty of precursory events of volcanic activity to be disclosed.

Iceland is a volcanic island on the mid-Atlantic ridge in the North Atlantic Ocean. The ridge comes on land in southwest Iceland, but interaction with a hot spot located under southeast Iceland, shifts the rift zone 100 km eastward as it crosses the island, generating two separate rift zones in the south, one in the north and two transform zones offsetting the volcanic zones in the north and south (Einarsson, 2008). Interaction with the plume and a 19 mm/year spreading rate on the rift (Sella et al., 2002; Árnadóttir et al., 2008) leads to active volcanism in the volcanic zones, with over 20 eruptions per century (Thordarson and Höskuldsson, 2008), and a high seismicity rate, with repeated earthquakes of magnitudes up to M7 in the two transform zones offsetting the volcanic zones (Einarsson, 2008; Stefánsson et al., 2008).

The volcanic zones are characterized by en-echelon volcanic systems consisting of a central volcano and an associated fissure swarm extending up to several tens of kilometers from the volcano. Volcanism ranges from effusive basaltic fissure eruptions to explosive silicic eruptions in the central volcanoes. The most active volcanoes are Grímsvötn, Hekla, Katla, and Bárdarbunga, all of which, except Hekla, are subglacial (Thordarson and Larsen, 2007).

The magma–ice interaction leads to phreatomagmatic explosive eruptions that can generate volcanic plumes of up to10–20 km height and catastrophic subglacial floods (jökulhlaups) from the glacier terminus. Major explosive eruptive rifting episodes occur in the fissure swarms, some during historical time (Eldgjá 934, Laki 1783). Submarine eruptions can occur offshore, the most recent one at Surtsey in 1963. Over 30 volcanoes have erupted in the Holocene.

Activity and unrest in the volcanoes as well as in the seismic zones are monitored by nationwide and regional, permanent, multidisciplinary systems: seismic, geodetic, hydrologic, meteorologic, remote sensing, and other campaign measurements (geochemical, petrological mapping of ashfall and lava flows). The designated state VO is located at the country's meteorological office, but operation of the GPS network is in collaboration with other VRIs. Rock laboratories and many of the geological observations and analyses are also operated and performed at other national VRIs. Collaboration between the RIs at a national level has resulted in a comprehensive monitoring of Icelandic volcanoes. Recent developments in the FUTUREVOLC supersite project have enabled extension of the networks into the Vatnajökull glacier, closer to some of the most active volcanoes, as well as extending the collaboration to European VRIs (infrasound, seismic arrays, remote sensing, lightning).

Placing the VO at a meteorological office is highly advantageous considering the frequency of explosive eruptions generating high-rising, ash-laden volcanic plumes that can be carried by winds over the European continent to cause massive disturbance to aviation, like the disorder caused by the Eyjafjallajökull 2010 plume. This colocation allows close and expedient interaction with Volcanic Ash Advisory Centers modeling the ash dispersion over the continent (Figure 17.4).

FIGURE 17.4 New permanent seismic station on a nunatak in the Vatnajökull ice cap.

In the Norwegian Sea, 650 km north of Iceland and further along the plate boundary, volcanic activity also occurs on the Norwegian island Jan Mayen. The island is 373 km^2 in area and at its northern end sits the ice-covered, basaltic stratovolcano Beerenberg. This volcano has erupted six times since the eighteenth century, with the last eruption occurring in 1985.

The closest volcanic area which does not belong to the Eurasian plate is the Canary Islands, located in the Nubia plate. This Spanish archipelago comprises seven main islands that form an active volcanic chain extending 500 km across the eastern Atlantic, with its eastern island only 100 km from the northwest African coast. In this oceanic place, the Nubia plate is almost stationary and volcanic activity is present since the late Cretaceous, developing in an intraplate geodynamic setting, at a passive continental margin. Although historical eruptions covering the last 500 years have been mainly basaltic effusive, strong explosive eruptions mark the history of some volcanoes, such as Las Cañadas edifice in Tenerife Island. Volcano monitoring at the Canary Islands has seen supported by regional networks but since the unrest period of El Hierro in 2011 that culminated with a submarine volcanic eruption, a strong monitoring effort was done with the deployment of dense seismic and geodetic monitoring and other monitoring equipment. This submarine eruption also produced lava balloons, but the most remarkable aspect is the several ground deformations, posteruptive episodes that El Hierro is still undergoing.

Active volcanoes in overseas European territories are in the Caribbean Arc (La Soufriere and Montagne Pelée in the French Lesser Antilles, Montserrat, etc.), Pacific Ocean (French Polynesia, Wallis), Indian Ocean (Piton de la Fournaise in La Réunion, Crozet, Saint-Paul, Amsterdam) and South Atlantic (Tristan da Cunha, Ascension, Bouvet) (Figure 17.1 (b)). These volcanoes represent a wide range of volcanic activities, from basaltic and mainly effusive volcanoes to very explosive and dangerous andesitic stratovolcanoes, as evidenced by the eruption of Mt Pelée in 1902 or the ongoing eruption of Soufriere Hills in Montserrat.

Subduction zone volcanoes in the Lesser Antilles, such as La Soufrière in Guadeloupe, Montagne Pelée in Martinique or Soufriere Hills in Montserrat, are known for their potentially very large volcanic hazards (Boudon et al., 2007; Komorowski et al., 2008). Nearly all the islands along the arc are the result of subduction-related volcanism, with andesitic stratovolcanoes producing explosive eruptions, coupled with the growth of lava domes generating pyroclastic flows. The first observatory was built in 1903 in Martinique after the devastating eruption of 1902 (Lacroix, 1904). This observatory functioned until 1925, when the volcano no longer showed signs of activity. Then the observations were stopped, only 4 years before a new big eruption of Montagne Pelée in 1929–1932. This is a good example of the need for long-term observations on volcanoes.

The current observatories in Martinique and Guadeloupe were created in 1935 and 1950, respectively. Their networks, mainly based on seismological,

FIGURE 17.5 Monitoring station near the top of La Soufrière in Guadeloupe Island (Lesser Antilles). Photo: OVSG/IPGP.

geodetic, and geochemical methods, are focused on monitoring the volcanic activity (Figure 17.5). In addition, VOs in Guadeloupe and Martinique are also operating large monitoring networks on a regional scale in the Lesser Antilles arc, for the purpose of monitoring tectonic seismicity and tsunamis.

In La Réunion Island, Piton de la Fournaise is one of the most active hot spot volcanoes in the world (Lénat et al., 2012). Since the 1980s, the mean eruption rate is more than two per year, giving fantastic opportunities to observe the developing of eruptions in such a context (Peltier et al., 2009; Massin et al., 2011; Roult et al., 2012). The Piton de la Fournaise Volcano Observatory was created in 1980. The current monitoring network is focused on the activity of the Piton de la Fournaise volcano. Particular attention is given to identify eruption precursors for predicting lava flow parameters (time, amplitude, lava flow emission point, etc.). More recently, the monitored area was extended to the whole island in order to better understand the deep regional seismicity and magma injections.

For French volcanoes, the monitoring networks are mainly based on seismic, geodetic, and geochemical methods. Permanent sensors are largely used by the VOs, but the observatory staff members also perform periodic measurements and sampling. Other studies (petrology, geochemistry, etc.) and remote sensing surveys from satellite and ground sensors are also realized by VRIs, as part of the French National Service of Observation in Volcanology.

In addition to the above listed VOs, several additional VRIs hold volcanological databases or laboratories of relevance to pan-European volcanological RIs. A rough survey of the state of the art in volcanology allows the recognition of dozens of VRIs across Europe. These belong to universities, public or private research centers, and space agencies where data collected on active volcanoes are stored, analyzed, and modeled for research or teaching purposes. These facilities are not listed here, as any such list runs the risk of being incomplete. However, since the overall focus of these facilities is the same as that of the VOs, i.e., geological, geophysical, or geochemical observations related to volcanoes, the natural conclusion is that the integrated European volcanological infrastructure should encompass all RIs, focusing on "volcano observations."

17.3 GAP ANALYSIS AND SOCIAL OR SCIENTIFIC NEEDS

The European volcanological community as a whole is, by far, one of the most evolved and scientifically advanced worldwide. Many monitoring systems as well as modeling approaches are continuously tested and implemented in the European VOs and VRIs, also profiting from the possibility of using the European volcanoes as natural laboratories (e.g., Mt Etna, Campi Flegrei, Iceland, Piton de la Fournaise, etc.).

Nevertheless, this community still presents some nonhomogeneous features in the collection and sharing of data and in the organization of monitoring and research activities that could limit its evolution.

The main problems are related to the heterogeneity of the organization of the VOs and laboratories. Aside from the few cases where the volcanoes are objectively undermonitored (which are specific problems of organization), on the whole, heterogeneity of an organization produces difficulties in the definition of a common data policy to foster science, in the capacity to create large and strong networks to sustain the cooperation among VOs and in guaranteeing common standards of intervention in all of Europe in the event of a transnational volcanic crisis.

A short analysis of the main limiting features is reported below.

17.3.1 The Availability and Quality of Instrumentation

ven in a preliminary overview, the distribution and quality of the volcanic observation systems seem heterogeneous among the different European volcanic areas. In some countries (e.g., Italy, France or Iceland), the distribution and organization (e.g., link with governmental authorities) of the VOs allow a sufficient amount of instrumentation to be deployed on the active volcanoes. Conversely, in other countries (e.g., Greece or Norway), the active volcanoes are undermonitored (e.g., in number of involved personnel, in number or quality of instruments). In between, there are other cases, where the good quality of observations and/or organization perhaps suffers from a limited number of observations.

17.3.2 Standardization in Data Collection, Processing, and Archival

Volcanology is an intrinsic multidisciplinary science. The measurable phenomena of internal or external volcanic processes are both physical and chemical in nature. Thus the observation systems are aimed at measuring all spectra of ground movements (from seismic ground motion to slow ground deformations), variations in gravity or magnetic fields, variations in flux/temperatures/chemical compositions of volcanic gases (either in the atmosphere, soils, groundwater, or rocks), and dimensions and nature of volcanic products (e.g., lava flows, pyroclastic deposits, rock chemistry), and so on.

In some cases, the physical/chemical parameters are measured and processed by using standardized approaches (e.g., seismic and GPS signals or rock chemistry) inherited from other communities (seismology, geodesy, petrology). In other cases, the observations require the design and implementation of specific techniques, which—obviously—might not be standardized (this is the case, for instance, of several gas geochemistry measurements). Due to these circumstances, the World Organization of Volcano Observatories (WOVO) has started to define standard formats in its database (WOVOdat). These formats are continuously evolving and have not been adopted worldwide until now, due to the heterogeneities in current volcanic observing systems (see the point above). However, WOVOdat represents a valued starting point for the implementation of a shared practice for archival volcanological data.

17.3.3 Data Openness and Dissemination; Cooperation

Except for very few cases (e.g., some seismic stations, GPS data, or some thermal and video cameras), access to raw data is quite difficult. Conversely, parametric event data (e.g., earthquake locations or time series of deformations) or results from monitoring activities (e.g., maps of lava flows) are often available through web portals, frequently related to outreach activities organized to inform the society about the current volcanic activity. However, in general, data access is unevenly regulated due to the heterogeneity of the organizational operating of the VOs and laboratories.

Until now, the cooperation among VO or laboratories is not well structured and is limited to sporadic episodes during, for instance, time-limited R&T projects aimed at studying specific themes of volcanology (e.g., modeling of volcanic processes) or implementing new technologies (e.g., the use of Earth observation system in volcano monitoring).

17.3.4 Governance and Legal Structures

In general, the VOs and laboratories are governmental agencies or research institutions, but occasionally universities implement and maintain specific networks or laboratories. Furthermore, in some cases, observatories are managed in cooperation between university or research institutions and local administrations, under specific legal agreements.

This wide variety of legal status derives from the different history of the VOs or VRIs (which in some cases are more than a century old) and the type of links between them and the funding agencies, which generally are the government and/or civil protection authorities, and produces differences in their organizations and aims. For instance, in some cases, VOs are mainly technical structures aimed only at implementing and maintaining network and observing systems, while in other cases there is an intrinsic structural link between monitoring and research activities in the same observatory. Obviously this implies wide differences in the governance among the European VOs or VRIs, which makes it difficult to identify common practices of governance in the observatories in Europe.

The previous analysis of the criticalities existing in the current organization of the European volcanological monitoring and research activities highlights the needs to design and implement an integrated RI on volcano observations to create a network linking all the existing RIs.

17.4 PRINCIPLES OF THE VOLCANO OBSERVATION RI

Whichever the optimum design of the European integrated RI of volcano observations will be, it will have to take into account that the evolution of the volcanological knowledge has been based on the positive feedback between research and monitoring activities. The integrated RI should keep this peculiarity and eventually foster all these initiatives aimed at strengthening the positive feedback mentioned above.

Among the many examples that we can use to show the positive feedback between the monitoring activities and the research, even not strictly related to the volcanological studies, we can cite the use of SAR interferometry in volcanology. In the early 1990s, the remote sensing scientific community designed, implemented, and tested new airborne and satellite, side-looking radar sensors to perform topographic surveys by using the interferometric analysis of the signals. Volcanoes were one of the first natural objects surveyed due the good radar reflectance of their surfaces and the wide variety of geological features. In a few years, a large amount of SAR images become available on active volcanic areas, as Mt Etna and Hawaii, putting in evidence that the applications of SAR interferometry might be beyond the simple topographic mapping, moving to the measurement of the ground deformations due to the movement of magma within the volcanoes. The positive applications of this new technique to measure the ground deformation at the scale of the whole volcano allowed the volcanologists to have a "holistic" approach for studying the surface dynamics (instead of considering, for example, the dynamics of each fault/rift/vent as isolated from the context) and to study volcanic processes occurring at greater depths than has been investigated before. Furthermore, starting from the consideration that ground deformations are often a clear precursor of an impending eruption, these early applications also suggested the possibility of using large data sets of SAR images to continuously monitor the evolution of the ground deformations. This idea fostered in the remote sensing community the notion to implement new techniques to manage long time series of SAR images. This feedback brought about the implementation of PS or SBAS techniques, which are at the base of the many approaches used today to monitor the evolution of ground-deformation patterns on several volcanoes worldwide.

This example shows that an effective distributed RI in volcanology has to be implemented according to models that take into account both the need to have a network among VOs and VRIs and the link with research communities not even strictly related to the studies on volcanoes. This is a real challenge, since this idea involves institutions spread over all of Europe, with different organizations, mandates, technologies or structures (i.e., the VO and VRI), and scientific communities only sporadically working together (see the above example of volcanological and remote sensing communities).

To successfully face this challenge, two issues seem the most critical to be coped with: the overall architecture of this RI and the policy of the management of the information. Indeed, the standardization and interoperability of the databases, as well as the implementation of e-infrastructures and the definition of a common "data policy," will improve the knowledge of the volcanic processes and the implementation of the monitoring systems through joint research activities, by reinforcing the virtuous feedback between fundamental and applied research on volcanoes.

Data generated by volcano monitoring are as heterogeneous as the complexity of the volcanic processes themselves. They include geological, geochemical, geophysical, meteorological, and remote sensing data sets to monitor active volcanoes. These data range from the physical sample of data (e.g., sample of rocks or volcanic fluids) to digital products (i.e., files), which require different database structures and quality control procedures to be defined and implemented.

A nonexhaustive list of the data currently acquired on the active volcanoes includes:

- Seismic waveforms (earthquakes, tremor), earthquake parameters (locations, magnitudes, source mechanisms), and other derived parameters (wave number and back azimuth of phases), infrasound waveforms and derived parameters (wave number and back azimuth of events),
- Geodetic time series (strain, GNSS (Global Navigation Satellite System)) and derived velocities,
- SAR surface observations and derived InSAR images,
- Gravity, magnetic, electromagnetic, thermal data,
- Geochemical data of volcanic fluids (samples and composition of volcanic gases in atmosphere, fumaroles, soils, and water),
- Ash deposits (componentry, grain-size distribution, chemical composition),
- Maps of the distribution of eruptive fissures, vents, deposits of volcanoes (lava flows, pyroclastic falls, flows, etc.),

- Plume observations (plume height at the source, radar images, optical and IR video images), including meteorological data,
- Radiosonde observations (meteorological: pressure, temperature, wind velocity; chemistry: volatiles),
- IR and UV absorption spectra of plumes, lightning in volcanic plumes,
- Rock samples as well as physical (pressure, temperature, density, viscosity) and chemical (composition of rocks/minerals) parameters derived from petrological analyses and rock laboratory measurements (variations in physical and chemical properties with pressure and temperature).

The VOs, as starting points for the construction of universal knowledge about volcano behavior and volcanology in a broad sense, have often developed their own techniques for data collection, used their own measurement units, cartographic projection systems, languages, lexicon/glossary, hardware standards, and so on. This situation somehow constitutes an obstacle each time the volcanological community wants to combine its efforts to better understand how volcanoes work, forecast volcanic events, or improve volcanic hazard assessments and reduce volcanic risk. This comment emphasizes the need to design and implement an interoperable environment to enable all the players to communicate and exchange information inside the volcanological community. This aim is by far most relevant to the VOs and other institutions providing data on active volcanoes. For the VOs especially, it is important to preserve the continuous feedback between the monitoring and research activities, which are the basis of the current cutting-edge volcano monitoring systems. Therefore, the interoperable environment has to keep the procedures usually adopted in the VOs working in the ways they have always done (to maintain continuity with the past), while at the same time, promoting major breakthroughs in the use of knowledge of the wider volcanological community.

To this aim, the overall architecture of a European e-RI should foresee several distributed data centers (DCs) and a coordination center (CC) to allow connection with the geographically distributed, national/regional DCs (Figure 17.6).

In order to facilitate good coordination of the distributed DC, it is useful to propose a summary classification of such structures in two different types:

- Site oriented (geographical RIs): These are represented by DCs directly connected to a volcanic area and thus to a multidisciplinary monitoring system. They are geographically identified and defined and usually correspond to VOs.
- Thematic oriented (nongeographical RIs): These centers are not involved in local monitoring but have infrastructures that produce data and services dealing with volcano monitoring and scientific research, for instance, the Space Agency databases. Furthermore, the activities of these DCs can provide complementary services to the Earth observation, catalogs, laboratories, and so on.

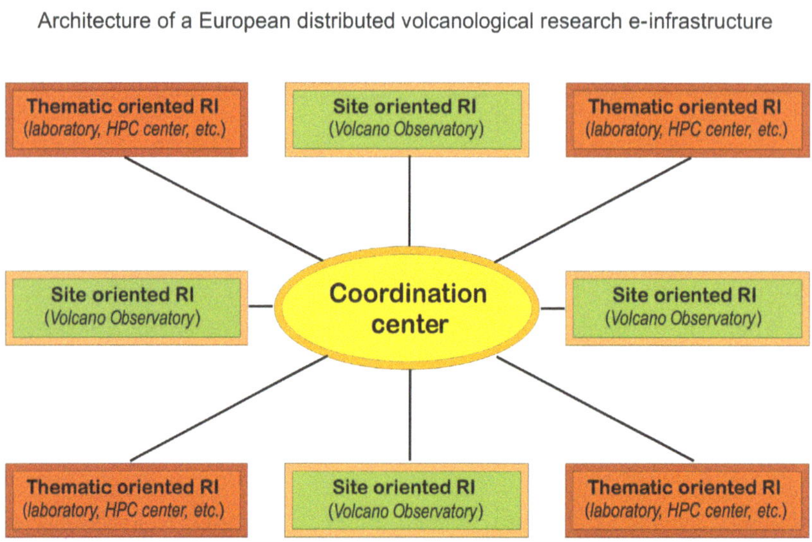

FIGURE 17.6 Overview of a possible architecture of a European distributed volcanological research e-infrastructure. Definitions and characteristics of the data centers (either thematic oriented or site oriented) and the coordination center are described in the text.

The interoperability of the CC will enable indexing the national DC data and provide an online instrument for searching the constantly updated metadata.

A few examples of the possible functions/services of the CC include

- To coordinate the standardization process of data formats at medium and long term, to make them more efficiently indexed and processed by software.
- To coordinate optimized local and remote processing and sharing of specialized software.
- To promote the homogeneity and comparability of data sets through the coordination of the purchasing of sensor and instruments.
- To foster and to disseminate "best practice," common standards and protocols for data collection.
- To coordinate and design a shared concept of relational database management system, possibly by following and implementing existing databases for volcanoes (e.g., WOVOdat for metadata, ORFEUS for seismic data, GSAC for geodetic data, etc.)
- Helping to cooperate with the local DCs in designing and implementing (hardware, software), in the preparation of the anticipated cloud, grid, or Farm architectures, which are critical for scientific networking.

The CC will eventually facilitate the setup of connections between the DCs, which in turn will lead to a real data and services distribution system.

The CC could have computational functions in order to distribute the computation workload through the DCs. Users and stakeholders should find, in a unique access point, all the information (i.e., metadata) well organized and always up-to-date, on what the national DCs are storing locally. According to the adopted policies, users may or may not be able to download the data or access the computation services, and so on.

The policy of the management of the information, in short the "data policy," should fix the functioning rules of the distributed European RI in volcanology. The fundamental rationale of the data policy for volcano observations is twofold. One is the adoption of the data access policy of the Group on Earth Observations (GEO; http://www.earthobservations.org), which aims to "provide a shared, easily accessible, timely, sustained stream of comprehensive data of documented quality, as well as metadata and information products, for informed decision-making by the year 2015"; a goal of vital importance to hazard managers at the national, as well as the European level during volcanic crises with cross-border implications. The other rationale is to provide the European research community with quality-checked volcanological data through open-access procedures to foster and sustain cutting-edge research on volcanoes and volcanological processes.

Furthermore the data policy will adhere to the EU INSPIRE directive (Infrastructure for Spatial Information in Europe; http://inspire.jrc.ec.europa.eu). In particular, the policy will rely on the INSPIRE aim that, "It must be possible to combine seamlessly spatial information from different sources across Europe and share it between many users and applications."

In general, the data, metadata, and products should be open and their access would be granted according to clear predefined rules. In some circumstances, for example, during volcanic crisis, it may be necessary to restrict some types of use of data, and even in some cases the access to data, metadata, and products.

Of course it is far from the aims of this chapter to detail the rules, which should be agreed among all the stakeholders, but it seems important to focus on a few basic concepts that must be defined in the "data policy":

- What Type/Level/Category of data, metadata, and products are access-free and what are restricted or delayed.
- What kind of use and reuse is allowed (e.g., only scientific use, commercial use, etc.).
- What Type/Level/Category of data, metadata, and products are subject to a time-limited embargo, how long the embargo is, and clear justification given why any such limitations are needed. Examples of Type/Level/Category of data that can be subject to an embargo are campaign data or the data collected during volcanic crises.
- Proper acknowledgment to the creators/owners of the data (e.g., by using permanent identifiers).

17.5 CURRENT INITIATIVES IN THE INTEGRATION OF EUROPEAN VOLCANO RIS

The following two chapters are aimed at presenting a concrete way to implement the concepts envisaged for the European RIs of volcano observations. Indeed, the ideas and comments discussed above could appear either trivial (e.g., in the case of the positive feedback between research and monitoring) or too visionary (e.g., in the case of a common data policy at European level). Here we would like to show that strategy exists to factually implement such an integrated RI in the framework of European initiatives aimed at fostering the strengthening of scientific integration in Europe.

Similar considerations about the need to federate the existing RIs involved in the volcano observation activities were made years ago also in the field of Solid Earth in Europe. These needs brought about the proposal of a long-term integration plan aimed at creating one sustainable, permanent, and distributed infrastructure, which is at the core of the European Plate Observing System (EPOS) project. Now EPOS is about to finish the preparatory phase (PP), during which the wide consortium participating in the project (about 170 institutions, belonging to 23 European countries) has been establishing the governance structure, legal framework, economic model, data policy, and technical and strategic issues for the European platform for fostering future research in solid Earth science. EPOS project is included on the road map of ESFRI, which is the strategic instrument to develop the scientific integration of Europe. At the end of the PP, EPOS will propose to enter into the construction and operational phase that is envisaged to last until 2040.

The overall scientific aim of EPOS is to promote innovative approaches for a better understanding of the physical processes controlling earthquakes, volcanic eruptions, unrest episodes, and tsunamis as well as those driving tectonics and Earth surface dynamics at the European scale. The key objectives of the EPOS project are

- Integrating existing RIs in Europe for solid Earth science (data infrastructures and facilities).
- Providing access to data and data products (adopting a common data taxonomy) through shared data policies.
- Developing e-RIs for integrated use of data, products, and services to users.
- Contributing to a community building for Earth science.
- Promoting cooperation with complementary projects in geosciences (GEO, ICT and environmental sciences).
- Providing services to different stakeholders.

According to its scientific aim, the communities involved in the activities of EPOS-PP are those of seismologists, volcanologists, geodesists, geologists, geophysicists, and experimental rocks scientists. The space agencies are also involved in the activities of EPOS-PP to define objectives and strategies in common with the solid Earth communities in current and upcoming missions for the Earth observations.

In this framework, the volcanological community is involved in the EPOS-PP project through the activities of Working Group 2 ("volcano observations"), which contributes to defining the guidelines for the implementation of the integrated services of EPOS by considering both the currently existing RIs in the different European countries and the requirements for an integrated RI discussed above.

17.6 POSSIBLE IMPLEMENTATION AND FUTURE EVOLUTIONS

Starting from the above discussions, the envisaged integrated volcano observation RI should create a coordinated infrastructure allowing the networking activities among European VOs and VRIs, the access to observations, laboratories, and service activities, and the fostering of joint research on technologies for volcano monitoring.

The following network activities are aimed at fostering cooperation among the volcanological scientific community, in a broad sense, at optimizing the use of the observation systems and at promoting a common level of knowledge and best practices on the European volcanoes. These are the premises for considering the whole VO and VRI as a "unique" distributed RI (operational interoperability).

- To define common standards, protocols, and interoperability of data among the European VOs and laboratories. In this framework, one of the aims of the network will be to complement the EPOS activities to define agreed-upon criteria for standards, quality checks, data distribution, and so on. This networking activity will also have a positive impact on the data policy, which remains quite heterogeneous at the European level so far. Indeed, the volcano observation RI will be the ideal forum to find enforceable European data policies aimed at promoting retrieval and systematic open access to volcanological data.
- To define guidelines for good practices and cooperation; to define "common languages or criteria," at the European level; to manage both ordinary and emergency phases. These guidelines can be useful to establish a "minimum" standard for monitoring active volcanoes (for instance, where volcanoes are poorly or not monitored) and thus to optimize the efforts of volcano monitoring at the European level. To share good practices on unrests and emergencies is a qualified objective for the integrated RI in volcanology (e.g., common criteria to define the levels of hazards). In this framework, the setup of a task force for rapid intervention will be a concrete action to assist local VOs in case of a large eruption with transboundary effects, or for long-lived volcanic crises that are time

consuming, or for European assistance to developing countries whose volcanoes are only poorly, or not monitored.
- To provide scientific support to civil protection authorities for hazard assessments and mitigation, preparedness and awareness of volcanic risks, and dissemination of knowledge. In addition to the role of VOs to support civil protection authorities in risk assessment and management of volcanic hazards at local scale, the European network can contribute to these authorities by defining the intervention plans at the European scale. The impact of recent Icelandic eruptions on the European air traffic emphasizes the necessity of such an action. The external communication involves the role of VOs in the local/national communities. They are usually deeply involved in the volcanic risk decision chains. The European network can help in establishing/implementing common protocols in case of hazard valuation, actions during volcanic crisis, and so on.

To facilitate the access to the observations, laboratories, and services, the RIs have to implement an e-infrastructure that integrates DCs through a CC. The e-infrastructure should foresee services that allow the discovering, sharing, analysis, and modeling of information related to the volcanoes. Examples of such services should be that aimed at distributing parameters on seismic activity (e.g., earthquakes locations, magnitude, etc., or seismic tremor amplitude) or reports on current volcanic activity, in case of cooperation with international initiatives having similar aims (e.g., Smithsonian volcanism program, GVM, etc.). Other services could be based on the massive use of high-performance computing resources to allow the modeling of volcanic processes (e.g., the modeling of either petrologic processes or dynamics of sources of ground deformations). Among the trans-national or trans-infrastructure access activity, the integrated RI should also encourage the sharing of skilled personnel between laboratories/observatories.

Although the volcano monitoring is currently performed by using cutting-edge technologies, the continuous evolution of both the techniques and the knowledge of volcanic processes requires continuous upgrading of the monitoring systems, the implementation of new instruments/methods, and the development of advanced modeling approaches to interpret the volcanic processes. The European RI in volcano observations will promote joint research activities, aimed at strengthening the above-mentioned network activities, for implementing:

- Off-line and/or real-time multidisciplinary data analysis,
- High-performance instrumentation and tests of volcano monitoring networks, components, subsystems, materials, techniques, and dedicated software,
- Adoption of innovative software solutions for making new user communities benefit from databases and computing services,

- Physical modeling of volcanic processes to improve the configuration of the observational systems,
- Technological innovations, in cooperation with the European industrial sector.

Although it is hard to define a detailed road map to implement the European integrated RI in volcano observations, a few key points can be considered from the current situation of volcanological-related activities at European level.

Notwithstanding the national initiatives aimed at implementing/strengthening the national DCs—which, however, are quite heterogeneous and often suffer from the current economic crisis—a strong boost comes at the European level from the two recently started European Supersites projects, FUTUREVOLC and MED-SUV, whose aims include the implementation of e-infrastructure for data sharing and services for two specific volcanic areas (Etna, Vesuvius/Campi Flegrei and Icelandic volcanoes). These e-infrastructures can be considered as "intermediate data centers" between the geographic/nongeographic DC and the European integrated RI (Figure 17.7), and can facilitate the implementation of the eventual services. Similar initiatives are needed from national or international agencies for the other volcanic areas (Greece, Azores, Canaries, and French volcanoes) to help them reach the same level of maturity.

The implementation of the e-infrastructures is a fundamental step toward building the integrated infrastructure in volcanology, which foresees services

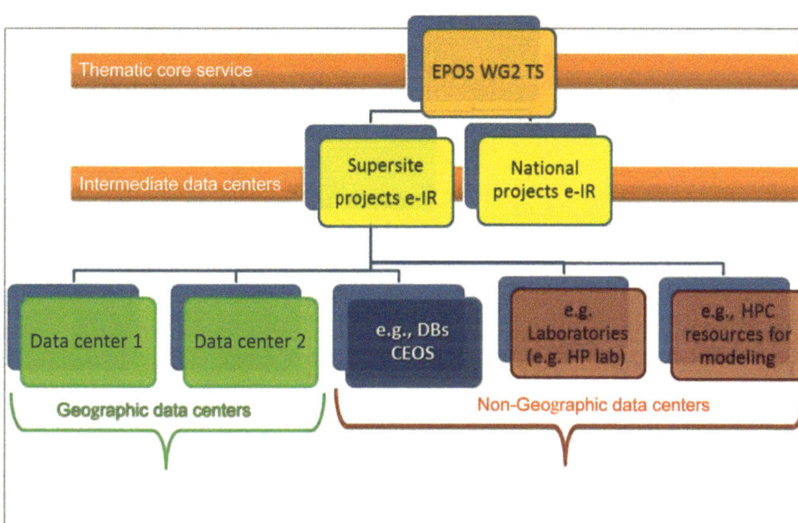

FIGURE 17.7 The three-layer architecture of the e-infrastructure, linking the geographic/nongeographic and intermediate data centers (DCs) and the thematic core service (TCs). EPOS, European Plate Observing System; WG2, Working Group 2.

aimed at fostering research activities for improving volcano monitoring, as well as initiatives for supporting the operational activities of VOs (e.g., task forces).

The effort in implementing this type of RI is not trivial. A rough estimation of the costs attains about 7–8 million euros, which could be spent in a time period lasting not less than 5 years. This implies the implementation of coordinated, long-lasting activities that the current PP of EPOS endorses. However, it should be acknowledged that the European volcanological community is strongly inclined to coordinate its activities. This community foresees the implementation of a European volcanological infrastructure by the end of the next decade.

17.7 CONCLUDING REMARKS

The state of the art of volcanological RIs in Europe is currently characterized by several single-sited RIs (VOs, volcanological research centers, laboratories, etc.), commonly located close to the active volcanic areas. Despite the cutting-edge technologies used in monitoring many of the European volcanoes and the high-quality research carried out by the European volcanological community, some limitations occur in the optimal use of the RIs due to heterogeneities in the collection and sharing of data as well as in the organization of the monitoring and research activities among different European VOs and VRIs.

To overcome these drawbacks, an integrated RI focused on the volcano observations is needed. This integrated RI should be based on the concept of the sharing and interoperability among the databases, services, monitoring networks, laboratories, best practices, data policies, and so on. Its implementation can be foreseen along a road map that includes and optimizes the effort of the current international projects, such as the Supersites projects, and national initiatives aimed at implementing integrated e-infrastructures for data sharing at different levels.

Such an integrated RI will have a noteworthy impact on both science and society. Indeed, since monitoring active volcanoes is the preferred way to gather experimental data and improve general background knowledge in volcanology, it is clear that the integrated RI will constitute a remarkable step forward in volcanology and natural hazards. The standardization and interoperability of the databases, as well as the implementation of e-infrastructure and definition of a common data policy, will not only a European impact but also a worldwide impact on volcano research and eruption forecasting. In turn, improved knowledge of the volcanic processes will allow the VOs and VRIs to implement monitoring systems through joint research activities and, therefore, reinforce the virtuous feedback between fundamental and applied research on volcanoes.

The socioeconomic impact of the integrated infrastructure on volcano observations will be twofold. First, improved monitoring of European volcanoes, advanced knowledge of volcanic processes, and strengthened cooperation with risk management agencies will enhance our shared capability to face volcanic hazards at local, regional, and continental scales (e.g., planning of actions during emergencies, etc.).

Second, the need to continuously upgrade the technological quality and dimension of monitoring systems represents an excellent opportunity for growing expertise and marketing for the European industrial sector. In this framework, the possibility of using the European volcanoes as natural laboratories to design, implement, and test new devices in very harsh or sometimes extreme environmental conditions (e.g., high temperature and/or oxidation and/or pressure conditions, seafloor environments, etc.) is a powerful opportunity for European industry (e.g., electronics, nanotechnologies, geothermal energy, etc.).

ACKNOWLEDGMENTS

This chapter summarizes many ideas and discussions of Working Group 2 "Volcano Observations" of EPOS-PP, so the authors give their sincere thanks to the WG2 participants: Agrinier, P., Allard, P., Bean, C., Blanco, M.J., Briole, P., De Natale, G., Dingwell, D., Federico, C., Gaspar, J.L., Gottsmann, J., Gudmundsson, M.T., Höskuldsson, A., Jakobsdóttir, S., Kyriakopoulos, K., Lars, O., López, C., Loughlin, S., Macedonio, G., Makropoulos, K., Martí, J., Martini, M., Matzka, J., Neri, A., Papale, P., Patané, D., Pedersen, H., Pompilio, M., Privitera, E., Reitano, D., Roland, R., Sigmundsson, F., Tait, S., Thomas, W., and Vougioukalakis, G.

REFERENCES

Aiuppa, A., et al., 2010. Patterns in the recent 2007–2008 activity of Mount Etna volcano investigated by integrated geophysical and geochemical observations. Geochem. Geophys. Geosyst. 11, Q09008. http://dx.doi.org/10.1029/2010GC003168.

Aloisi, M., Bonaccorso, A., Gambino, S., Mattia, M., Puglisi, G., 2003. Etna 2002 eruption imaged from continuous tilt and GPS data. Geophys. Res. Lett. 30 (23), 2214. http://dx.doi.org/10.1029/2003GL018896.

Árnadóttir, Th., Geirsson, H., Jiang, W., 2008. Crustal deformation in Iceland: plate spreading and earthquake deformation. Jökull 58, 59–74.

Boudon, G., Le Friant, A., Komorowski, J.C., Deplus, C., Semet, M.P., 2007. Volcano flank instability in the Lesser Antilles arc: diversity of scale, processes, and temporal recurrence. J. Geophys. Res. 112 (B8), 23.

Calvari, S., Salerno, G.G., Spampinato, L., Gouhier, M., La Spina, A., Pecora, E., Harris, A.J.L., Labazuy, P., Biale, E., Boschi, E., 2011. An unloading foam model to constrain Etna's 11–13 January 2011 lava fountaining episode. J. Geophys. Res. 116, B11207. http://dx.doi.org/10.1029/2011JB008407.

Chiodini, G., Vilardo, G., Augusti, V., Granieri, D., Caliro, S., Minopoli, C., Terranova, C., 2007. Thermal monitoring of hydrothermal activity by permanent infrared automatic stations: results obtained at Solfatara di Pozzuoli, Campi Flegrei (Italy). J. Geophys. Res. 112, B12206. http://dx.doi.org/10.1029/2007JB005140.

Corsaro, R.A., Di Renzo, V., Distefano, S., Miraglia, L., Civetta, L., 2011. Relationship between petrologic processes in the plumbing system of Mt. Etna and the dynamics of the eastern flank from 1995 to 2005. J. Volcanol. Geotherm. Res. 251 (2013), 75–89. http://dx.doi.org/10.1016/j.jvolgeores.2012.02.010.

D'Auria, L., Giudicepietro, F., Martini, M., Lanari, R., 2012. The 4D imaging of the source of ground deformation at Campi Flegrei caldera (southern Italy). J. Geophys. Res. 117, B08209. http://dx.doi.org/10.1029/2012JB009181.

Druitt, T.H., Edwards, L., Mellors, R.M., Pyle, D.M., Sparks, R.S.J., Lanphere, M., Davies, M., Barreiro, B., 1999. Santorini volcano. Geol. Soc. London, Memoirs 19, 165.

Einarsson, P., 2008. Plate boundaries, rifts and transforms in Iceland. Jökull 58, 35–58.

Gaspar, J.L., Queiroz, G., Ferreira, T., Amaral, P., Viveiros, F., Marques, R., Silva, C., Wallenstein, N., April 12, 2011. Geological Hazards and Monitoring at the Azores (Portugal). Earthzine. http://www.earthzine.org/2011/04/12/geological-hazards-and-monitoring-at-the-azores-portugal/.

Giudicepietro, F., Orazi, M., Scarpato, G., Peluso, R., D'Auria, L., Ricciolino, P., Lo Bascio, D., Esposito, A.M., Borriello, G., Capello, M., Caputo, A., Buonocunto, C., De Cesare, W., Vilardo, G., Martini, M., 2010. Seismological monitoring of Mount Vesuvius (Italy): more than a century of observations. Seismol. Res. Lett. 81 (4), 625–634. http://dx.doi.org/10.1785/gssrl.81.4.625.

Goete, J.W., 1816. Italienische Reise – Sizilien.

Greco, F., Currenti, G., D'Agostino, G., Germak, A., Napoli, R., Pistorio, A., Del Negro, C., 2012. Combining relative and absolute gravity measurements to enhance volcano monitoring. Bull. Volcanol. 74, 1745–1756. http://dx.doi.org/10.1007/s00445-012-0630-0.

Komorowski, J.-C., Legendre, Y., Caron, B., Boudon, G., 2008. Reconstruction and analysis of sub-Plinian tephra dispersal during the 1530 A.D. Soufrière (Guadeloupe) eruption: implications for scenario definition and hazards assessment. J. Volcanol. Geotherm. Res. 178 (3), 474–490. http://dx.doi.org/10.1016/j.jvolgeores.2007.11.022.

La Spina, A., Burton, M.R., Harig, R., Mure, F., Rusch, P., Jordan, M., Caltabiano, T., 2013. New insights into volcanic processes at Stromboli from Cerberus, a remote-controlled open-path FTIR scanner system. J. Volcanol. Geotherm. Res. 249, 66–76 (FTIR permanent station at Stromboli).

Lacroix, A., 1904. La Montagne Pelée et ses Eruptions. Masson et Cie, Paris, 662 pp.

Lénat, J.F., Bachèlery, P., Merle, O., 2012. Anatomy of Piton de la Fournaise volcano (La Réunion, Indian Ocean). Bull. Volcanol. 74, 1945–1961. http://dx.doi.org/10.1007/s00445-012-0640-y.

Massin, F., Ferrazzini, V., Bachèlery, P., Nercessian, A., Duputel, Z., Staudacher, T., 2011. Structures and evolution of the plumbing system of Piton de la Fournaise volcano inferred from clustering of 2007 eruptive cycle seismicity. J. Volcanol. Geotherm. Res. 202 (1), 96–106. http://dx.doi.org/10.1016/j.jvolgeores.2011.01.008.

Murè, F., Larocca, G., Spampinato, L., Caltabiano, T., Salerno, G.G., Montalto, P., Scuderi, L., 2013. Installazione di un radiometro nell'area sommitale del vulcano Etna. Rapporti tecnici INGV, n. 259.

Napoli, R., Currenti, G., Del Negro, C., Greco, F., Scandura, D., 2008. Volcanomagnetic evidence of the magmatic intrusion on 13th May 2008 Etna eruption. Geophys. Res. Lett. 35, L22301. http://dx.doi.org/10.1029/2008GL035350.

Oppenheimer, C., 2011. The Campanian eruption and human revolution in palaeolithic Europe. In: Oppenheimer, C. (Ed.), Eruptions that Shook the World. Cambridge University Press, pp. 208–210.

Patanè, D., Cocina, o, Falsaperla, S., Privitera, E., Spampinato, S., 2004. Mt. Etna volcano: a seismological framework. In: Bonaccorso, A., Calvari, S., Coltelli, M., Del Negro, C., Falsaperla, S. (Eds.), Mt. Etna: Volcano Laboratory, Geophysical Monograph, vol. 143. American Geophysical Union, pp. 147–165.

Peltier, A., Bachèlery, P., Staudacher, T., 2009. Magma transport and storage at Piton de La Fournaise (La Réunion) between 1972 and 2007: a review of geophysical and geochemical data. J. Volcanol. Geotherm. Res. 184, 93–108. http://dx.doi.org/10.1016/j.jvolgeores.2008.12.008.

Roult, G., Peltier, A., Taisne, B., Staudacher, Th., Ferrazzini, V., Di Muro, A., OVPF Team, 2012. A new comprehensive classification of the Piton de la Fournaise activity spanning the 1985–2010 period. Search and analysis of short-term precursors from a broad-band seismological station. J. Volcanol. Geotherm. Res. 241, 78–104.

Saccorotti, G., Iguchi, M., Aiuppa A. In situ Volcano Monitoring: Present and Future (this volume, chapter 7).

Salerno, G.G., Burton, M.R., Oppenheimer, C., Caltabiano, T., Randazzo, D., Bruno, N., Longo, V., 2009. Three-years of SO_2 flux measurements of Mt. Etna using an automated UV scanner array: comparison with conventional traverses and uncertainties in flux retrieval. J. Volcanol. Geotherm. Res. 183, 76–83.

Sella, G.F., Dixon, T.H., Mao, A., 2002. REVEL: a model for recent plate velocities from space geodecy. J. Geophys. Res. 197. http://dx.doi.org/10.1029/2000JB000033. ETG 11-1-31.

Schmidt, A. Volcanic Gas and Aerosol Hazards from a Future Laki-Type Eruption in Iceland (this volume, chapter 15).

Spallanzani, L., 1792. Viaggi alle Due Sicilie e in alcune parti dell'Appennino. Pavia, Comini.

Stefánsson, R., Gudmundsson, G.B., Halldórsson, P., 2008. Tjörnes fracture zone. New and old seismic evidences for the link between the North Iceland rift zone and the Mid-Atlantic ridge. Tectonophysics 447, 117–126.

Thordarson, T., Höskuldsson, Á., 2008. Postglacial volcanism in Iceland. Jökull 58, 197–228.

Thordarson, T., Larsen, G., 2007. Volcanism in Iceland in historic time: Volcano types, eruption styles and eruptive history. J. Geodyn. 43, 118–152.

Chapter 18

Integrated Monitoring of Japanese Volcanoes

Toshitsugu Fujii[1,2] and Hitoshi Yamasato[3]
[1]*Crisis & Environment Management Policy Institute, Shinjuku-ku, Tokyo, Japan,* [2]*Earthquake Research Institute, The University of Tokyo, Yayoi, Bunkyo-ku, Tokyo, Japan,* [3]*Penn State University, Geosciences Dept., University Park PA 16802, U.S.*

ABSTRACT

In Japan, 47 among 110 active volcanoes are monitored by the Japan Meteorological Agency on a 24-h basis, and several universities develop volcanological studies with their own volcano observatories at some of the recently active volcanoes. Some other research institutes also study several volcanoes for their own research purposes. The primary mission of the Japanese Coordinating Committee for the Prediction of Volcanic Eruptions is to evaluate the activities of volcanic unrest by integrating various data from various agencies; this system has worked fairly well for most of the recent volcano crises. Especially integration of seismicity and crustal deformation data from various institutions has worked promisingly to predict the onset of vulcanian explosions in some volcanoes recently activated. The reason why this kind of interagency cooperative system has worked so far is because volcanic activity in Japan has been fairly low for almost a hundred years. A new centralized system to monitor, analyze, and issue adequate volcanic information is desired to prepare for the accelerated activity of Japanese volcanoes in the near future.

18.1 INTRODUCTION

In order to minimize loss of life and property resulting from volcanic eruptions, monitoring active volcanoes is inevitable. A primary goal of volcanic monitoring is forecasting and predicting eruptions by detecting abnormal phenomena. Many eruption-prone countries have a single responsible institute for volcano monitoring and issuing volcano information such as PHIVOLCS of the Philippines, INGV of Italy, and USGS of the United States. Japanese system for monitoring volcanoes is, however, different from other countries. The Japan Meteorological Agency (JMA), an affiliate agency of the Ministry of Land, Infrastructure, Transport and Tourism (MLIT) is the responsible

Volcanic Hazards, Risks, and Disasters. http://dx.doi.org/10.1016/B978-0-12-396453-3.00018-6

organization for monitoring volcanoes and issuing alerts and warnings; however, the monitoring system of JMA is not sufficient enough to cover the whole active volcanoes and it has to obtain many volcanic observation data from other institutes belonging to different ministries in order to judge the condition of a given volcano. Some volcanoes with the observatories of national universities are equipped with observation instruments with more variety and large number than those of JMA. Due to this situation, the Coordinating Committee for Prediction of Volcanic Eruptions (CCPVE), which is a private advisory organization under the Director General of the JMA consisting of representatives of university observatories and other research institutes, plays important roles for evaluating volcanic unrest based on obtained monitoring data. Therefore, the role of CCPVE would also be discussed.

18.2 TARGET VOLCANOES FOR MONITORING

There are 110 active volcanoes in Japan (Figure 18.1). CCPVE listed up 66 volcanoes first as "active volcanoes," which had erupted within about 1,000 years.

The National Catalog of the Active Volcanoes in Japan edited by CCPVE in 1975 contained 77 volcanoes, including the northern territories. In this stage, volcanoes having historical eruption records and/or fumarolic activity were counted.

When the second edition of the National Catalog of the Active Volcanoes in Japan was published in 1991, the definition was modified to cover volcanoes that had erupted within approximately the past 2,000 years, and 83 volcanoes were designated as active volcanoes. As three volcanoes were newly assessed in 1996, the number of active volcanoes increased to 86.

In 2003, CCPVE modified the definition of active volcano to any volcano which has erupted within approximately the last 10,000 years, or which currently shows fumarolic activity according to the global standard. In this definition, 21 volcanoes were added furthermore; totally 108 active volcanoes in Japan. Based on the newest geological data, three additional volcanoes in Hokkaido were recognized to have erupted within the past 10,000 years, CCPVE concluded in 2011 that two of them were designated as new, independent active volcanoes. The other volcano is considered as a part of the volcano that was already designated as an active volcano. This brought the total number from 108 to 110.

In 2007, the CCPVE has selected 47 volcanoes that are expected to erupt within the next 100 years. The criteria of selection were mainly that the volcano experienced unrest within these 100 years; however, some volcanoes that had not experienced any unrest for several hundred years but might cause severe influence on the community nearby on the occasion of eruption were also included. Mt Fuji is an example. These 47 volcanoes were recommended as "target volcanoes" for JMA to monitor on a 24-h basis.

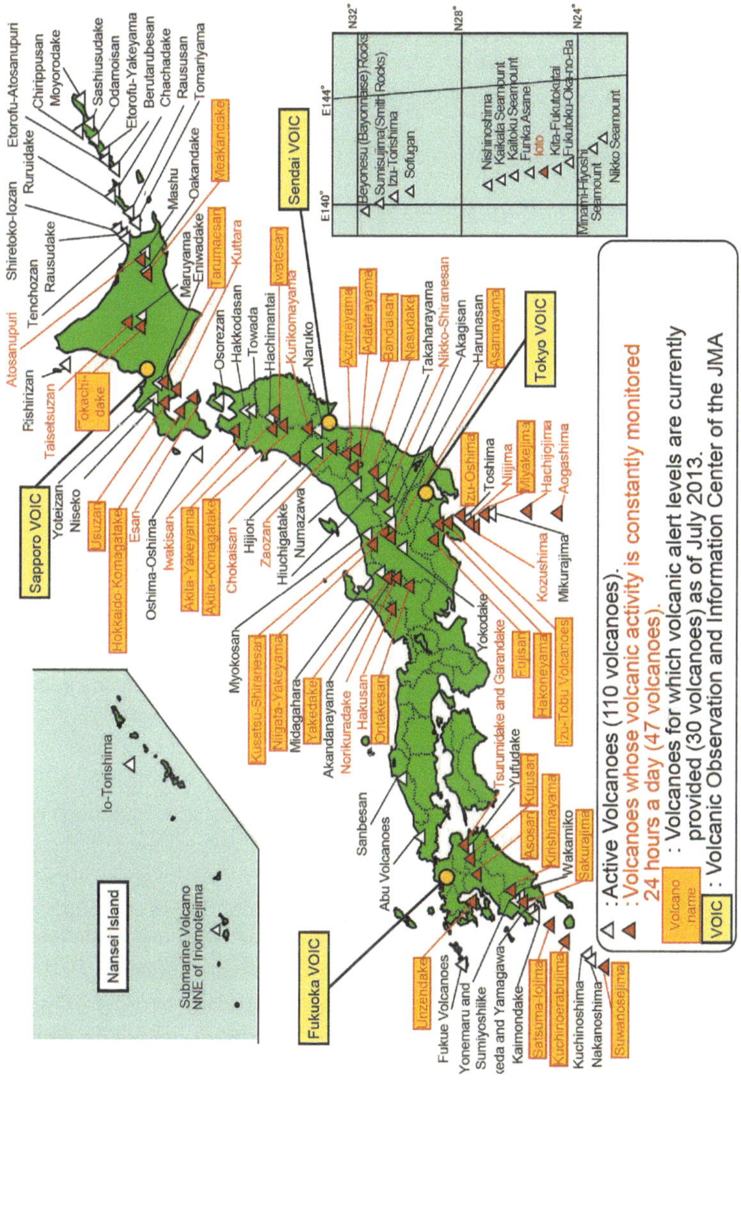

FIGURE 18.1 Active volcanoes in Japan and volcano monitoring system of the Japan Meteorological Agency (JMA).

18.3 MONITORING VOLCANOES

JMA is responsible for monitoring volcanic activity and issuing volcanic information in Japan. Another affiliate agency of MLIT, the Japan Coast Guard (JCG) monitors volcanoes under sea and remote islands, while the Geospatial Information Authority of Japan (GSI) is in charge of monitoring ground deformation using a nationwide ground deformation observation network. In addition, MLIT and prefectural erosion control departments monitor lahars at 29 volcanoes as part of their erosion control responsibilities.

18.3.1 Monitoring on Land

JMA used to monitor the main active volcanoes utilizing meteorological observatories and weather stations that are located in close proximity to the volcanoes. A mechanical seismograph was first deployed at the Kagoshima weather station, 5 km west to Sakurajima volcano, in 1888 which can be marked as the beginning of constant onsite seismic observation near an active volcano. The first volcano observatory in Japan was set up on Asama volcano in 1911 in a team effort between the Imperial Earthquake Investigation Committee of the Ministry of Education and the Nagano Weather Station.

In the 1960s, JMA began constant monitoring of 17 volcanoes using high-sensitivity seismographs deployed at meteorological observatories and weather stations on or near the volcanoes. Some of the other volcanoes were examined periodically by campaign observation.

In 2002, the Volcano Observation and Information Centers (VOICs) were established in Sapporo, Sendai, Tokyo, and Fukuoka, respectively in order to monitor volcanoes 24 h a day (Yamasato, 2005). Thirty four volcanoes were target volcanoes for VOICs in 2002 and 47 in 2009 as CCPVE recommended these volcanoes should be monitored using borehole seismographs and tiltmeters in combination with existing equipment (Figure 18.1).

In addition to seismic observation, GPS, and ground deformation observation by tiltmeter, the VOICs provide a 24-h centralized surveillance including visual observation with high-sensitivity cameras and/or cameras installed on the walls of craters and infrasonic observation. This telemetry observation goes well together with dispatching the campaign observation team regularly and temporarily. They check the changes in geothermal temperature, volcanic gas composition, deformation with GPS, and geomagnetic total intensity.

In case of unrest, JMA staff is dispatched to the volcano in order to inspect the monitoring and observation condition and to consult about the counter-measures with local organizations.

Resident Offices for Volcanic Disaster Mitigation were set up at Asama, Izu-Oshima, Miyake-jima, and Aso when meteorological observatories and weather stations close to volcanoes were abolished in 2009. These offices pursue hazard mitigation work in collaboration with local governments, while also taking charge of the campaign observation teams (Yamasato et al., 2013).

18.3.2 Geodetic Monitoring

GSI continuously monitors the Earth's crustal movement with the accuracy of millimeter analyzing data from GEONET (a Global Navigation Satellite System (GNSS)-based continuous monitoring system), GPS-based control points, and a remote GNSS-based volcano deformation monitoring system (REGMOS). With this system, GSI analyzes the process of ground buildup to an eruption and movements of magma with a certain degree of temporal and spatial accuracy (Imakiire, 2013). The GNSS network of JMA is also utilized to analyze the integrated deformation.

Izu-Oshima volcano is continuously monitored by automatic geodetic measuring system and ground deformation measurements (leveling, GNSS, and gravity) are periodically carried out by GSI in other 15 key volcanoes. GSI has also monitored geomagnetically on Fuji Volcano since December 2001.

When volcanic activity is elevated, synthetic aperture radar (SAR) interferometric analyses with satellites or aircraft are performed to detect surface deformation due to, for example, contraction of volcanic edifice or lava accumulation within the crater.

18.3.3 Monitoring of Marine and Submarine Volcanoes

Monitoring of submarine volcanoes is taken charge by JCG; that is, volcanoes in the Southern and Ryukyu areas are periodically monitored aerially or partly observed by ships (Figure 18.2; Ito, 2013). The aerial monitoring includes visual observation (discolored water) and aeromagnetic and temperature measurements. JCG also collects basic data on volcanic islands and submarine

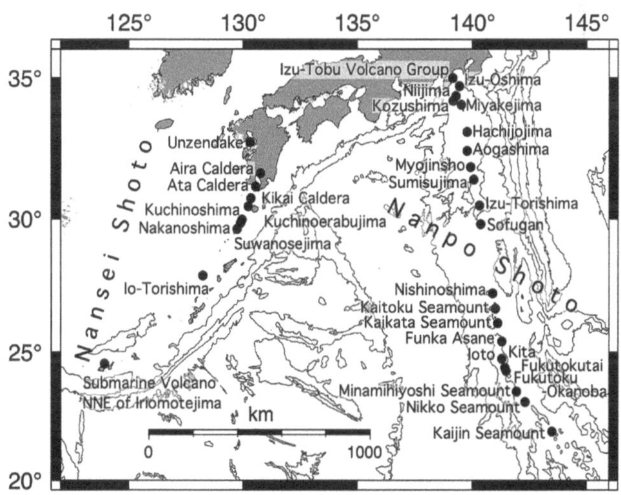

FIGURE 18.2 Volcanic islands and submarine volcanoes to be monitored by the Japan Coast Guard (Ito, 2013).

volcanoes, including topography, geomagnetic and gravitational anomalies. A self-navigated survey boat is sometimes used to survey topography with echo sounders and to sample discolored water.

18.4 OBSERVATIONAL RESEARCH BY THE NATIONAL UNIVERSITIES AND OTHER RESEARCH INSTITUTES

The national universities and the National Research Institute for Earth Science and Disaster Prevention (NIED) have their own volcanic observation networks for research purposes. Most of the seismic data obtained by the national universities and NIED are shared by JMA that utilizes them for volcano monitoring. Although the National Institute of Advanced Industrial Science and Technology (AIST) does not maintain observatory, it regularly monitors and studies volcanic gases from several volcanoes with fumarolic activities. Obtained data by the universities and other research institutes are reported to CCPVE to evaluate the activities of volcanoes.

18.4.1 Volcano Observation by Universities

The history of geophysical study of volcanic activities in Japan can be traced back to 1911, when F. Omori of the Imperial University of Tokyo monitored earthquakes at the first Japanese volcano observatory in Asama. Systematic studies on volcanoes by university researchers started in 1928, when K. Sassa of Kyoto Imperial University investigated the relationship between volcanic tremor and eruption at the Aso Volcanological Laboratory affiliated to the Faculty of Science.

In 1933, T. Minakami started geophysical observations at Yunodaira Observatory of Asama volcano, which was the predecessor of the Asama Volcano Observatory of the Earthquake Research Institute of University of Tokyo (ERI). During the eruption of Showa-shinzan of Usu volcano from 1943 to 1945, Minakami and his colleagues of ERI observed the process of the formation of a new dome caused by dacite magma intrusion, monitoring earthquakes and conducting leveling at the slope of the mountain. This initiative study on Showa-shinzan probably formed the later style of volcano observational studies in Japan, which were developed on the basis of geophysical monitoring by university researchers. This style is quite different from observational studies in other countries, such as those by USGS and by national volcanic observatories in Italy, which involve geophysical and geochemical observations and geological surveys.

Since geophysical observation on volcanoes covers various methods in long terms, national universities have established their own observatories on their allotted volcanoes. Before the formulation of the National Program for the Prediction of Volcanic Eruption (NPPVE) in 1974, the University of Tokyo established the Izu-Oshima Geo-electromagnetic Observatory in 1959 and the Kirishima Volcano Observatory in 1964 in addition to the Asama Volcano

Observatory; the Disaster Prevention Research Institute of Kyoto University constructed the Sakurajima Volcano Observatory in 1960; and the Faculty of Science of Kyushu University established the Shimabara Institute of Volcanology and Balneology in 1962, which was reorganized as the Shimabara Volcano Observatory in 1971.

The Geodesy Council submitted a recommendation report, "Promotion of a National Program for the Prediction of Volcanic Eruptions," to the Minister of Education and relevant ministers in 1973. This led to a NPPVE, which was implemented from 1974 onward as successive 5-year programs in seven times, drawn and recommended by the Geodesy Council. In 2009, it was combined with the National Program for the Earthquake Prediction into the National Program for the Observational Study for the Prediction of Volcanic Eruptions and Earthquake.

The first program aimed to construct monitoring systems appropriate to the characteristics of each volcano and to collect monitoring data needed for volcano studies and its eruption prediction. Therefore, the mainstays of the plan included strengthening volcano monitoring systems, constructing observatories, organizing campaign observation teams, promoting research to develop prediction methods, establishing a coordinating committee (CCPVE), and developing human resources.

CCPVE was established as a private advisory organization of the Director General of JMA with the office in JMA. Members of the committee included university professors, executive officers of the Ministry of Education, National Land Agency, Science and Technology Agency, and other relevant authorities.

When NPPVE was initiated, volcano observatories affiliated to national universities numbered five, namely Sakurajima Volcano Observatory (Disaster Prevention Research Institute, Kyoto University), Aso Volcanological Laboratory (Faculty of Science, Kyoto University), Shimabara Volcano Observatory (Faculty of Science, Kyushu University), Kirishima Volcano Observatory, and Asama Volcano Observatory (ERI, University of Tokyo). Later, the following observatories were added according to the program: Usu Volcano Observatory at Faculty of Science, Hokkaido University; Observation Center for Prediction of Earthquakes and Volcanic Eruption at Faculty of Science, Tohoku University; Laboratory for Earthquake Chemistry at Faculty of Science, University of Tokyo; Izu-Oshima Volcano Observatory at ERI, University of Tokyo; Kusatsu-Shirane Volcano Observatory at Tokyo Institute of Technology; and Research Center for Seismology and Volcanology at Faculty of Science, Nagoya University. With other observatories for earthquakes and volcanoes newly installed at Hirosaki University and Kagoshima University, the number of observatories reached 15 presently.

In the instrumentation point of view, although emphasis was laid on the fortification and telemeterization of seismic observation on the early stage of the Program, later years have seen strengthening of multifarious instrumentations, including those for ground deformation, electromagnetic, geochemical, and gravity observations. As a result, observation station became denser and wider in distribution. The GPS came into widespread use in large-area ground deformation observation on and around volcanoes in the late 1990s.

With recognition of the importance of geochemical and electromagnetic observations for understanding volcanic activity, equipment for geochemistry and temperature measurement of volcanic gas and crater lake water have been installed at some volcanoes including Sakurajima and Kusatsu-Shirane.

18.4.2 Volcano Observation by the NIED

Based on the NPPVE, NIED extended the Volcano Observation Network on Ioto in the early 1980s; Izu-Oshima in the late 1980s; and Miyake-jima, Fuji, and Nasudake in the 1990s without building observatories on-site.

After the national universities were reorganized as independent agencies in 2004, the Ministry of Education abandoned its policy to strengthen the university volcano observatories and decided to enhance volcano observation research stations of the NIED, provided that the obtained data would be shared among researchers of the universities. Deployment plans for the volcano observation research stations were carefully considered by the Volcano Subcommittee of the Geodesy Division of the Science and Technology Council in 2008. In accordance with this policy, NIED constructed volcano observation stations at eight sites on a total of five volcanoes (Aso, Usu, Iwate, Asama, and Kirishima) between 2009 and 2010—and one station at Mt Kusatsu-Shirane in 2011. Volcanoes covered by NIED results in 11 (Tanada and Ukawa, 2013).

The original data are shared by JMA, and the integrated data are processed continuously 24 h a day for automatic hypocenter detection and automatic estimation of pressure source based on the automatic anomaly detection of ground deformation data. The processed results are periodically reported to CCPVE for evaluating volcanic activity.

All these observational data—continuous waveform images, average short-period seismograph amplitude changes at 1-min intervals, and tiltmeter change charts—can be viewed online at the Visualization System for Volcanic Activity Web site (http://vivaweb2.bosai.go.jp/viva/v_index.html). Moreover, in January 2013, earthquake and tiltmeter data became accessible on the Volcanic Observation Network Web site (http://www.vnet.bosai.go.jp/).

NIED also analyzes ground deformation using SAR or InSAR, and the results are reported to CCPVE.

18.4.3 AIST's Research on Volcanology

AIST developed a method to date the young volcanic rocks (e.g., Ishizuka et al., 2003) and has applied it to various volcanoes. These results are indispensable for production of geological maps and improve the accuracy of the geological maps and systematic compile of the eruption data.

AIST observes the gases from active volcanoes to monitor volcanic gas discharge activity and analyze the degassing process of magma and temporal changes in hydrothermal systems. As the conventional means are analyzing volcanic gases sampled directly from fumarole, targets of the study have been limited to the area of active fumarole. Recent development of apparatus and method for observation enabled to analyze gas components within volcanic plume, and AIST started analyses of the gases from various volcanoes (Shinohara, 2013).

18.5 INTEGRATED MONITORING OF VOLCANOES IN JAPAN

18.5.1 Integration of Data from Different Monitoring Instruments

In Sakurajima volcano, the strain extension and upward tilting of the crater side were detected immediately before the occurrence of explosive eruptions (Kamo and Ishihara, 1989), implying magma intrusion into the lower part of the conduit or the reservoir at depths of 2–6 km (Ishihara, 1990). Swarms of shallow low-frequency earthquakes are also observed before eruptions; therefore, this integrated monitoring of seismicity and ground deformation was considered to be a powerful tool to predict explosive eruption. This finding was possible in Sakurajima volcano, where precise strain- and tiltmeters of 250 m long were installed within a tunnel. In other volcanoes, such a sophisticated observational tunnel was not available. Similar integrated monitoring technique utilizing the borehole-type tiltmeter was developed by JMA in the Asama 2004 eruption.

During the eruption, which began on September 1 and ended on November 23, 2004, several explosive eruptions occurred. Just before the explosive eruptions, a borehole-type tiltmeter on the northern flank of Asama volcano, 2.5 km north from the crater showed upward tilt of the crater side. At the same time, seismicity at the shallow part beneath the crater increased. Just after the explosive event, seismicity decreased rapidly and ground deformation returned to the same level of the background. The coincidence of increasing seismicity and upward tilt was observed in most of the explosive event during the 2004 eruption. These were observed several hours to several dozen hours before every event. After the 4-year quiescent interval, JMA realized the increase of seismicity with upward tilt of crater side at the same volcano on February 1, 2009 (Figure 18.3); therefore, JMA issued a warning that an eruption could occur within a day. Then an eruption occurred 13 h later, on February 2. As after the eruptions in 2004, JMA and the University of Tokyo enhanced geodetic observation, the precursory signals were detected by many tiltmeters and broadband seismographs.

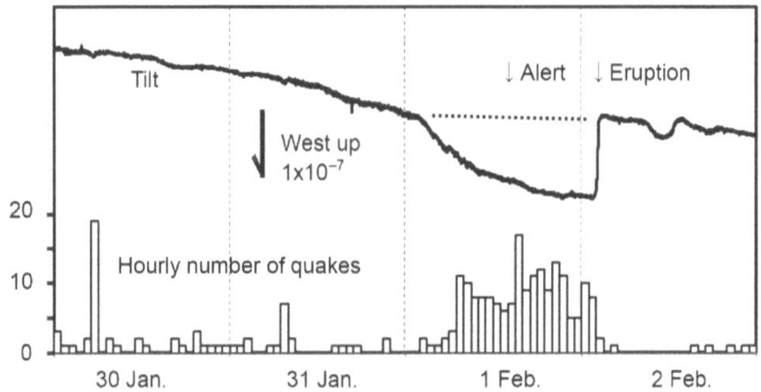

FIGURE 18.3 Tilt record and seismicity during the eruption event of Asama volcano on February 2, 2009.

Similar phenomena were observed also in the 2011 eruptions at Shinmoedake, Kirishima in southern Kyushu. Vulcanian explosions took place during February to September following the formation of lava lake within a summit crater just after the subplinian eruptions on January 26–27 in 2011 (e.g., Nakada et al., 2013). Small tilt change and low-frequency earthquake swarm were detected just before each explosion. As Shinmoedake had been activated with several phreatic explosions since 2008, JMA and NIED had installed several tiltmeters around Shinmoedake. However, just before the subplinian eruption in January, only very small tilt change could be detected even by the tiltmeter close to the crater; therefore, JMA could not issue an alert before the subplinian eruption (Kato and Yamasato, 2013).

18.5.2 Integration of Data from Different Institutions

In 47 volcanoes monitored by JMA on a 24-h basis, the instruments deployed may not cover the whole area and whole variation of measurements. Various instruments may be deployed by different institutions in multiple sites independently in some volcanoes. Although some are designed for academic use by universities and research institutes rather than for ordinary monitoring, the data from such instruments are also utilized for monitoring purposes.

The distribution of instruments and monitoring sites on Fuji is shown in Figure 18.4. The monitoring sites of JMA does not cover the whole area, and the instruments of GSI, University of Tokyo, NIED, and other organizations including Yamanashi prefecture are distributed in supplementary locations. Besides them, MLIT and prefectural erosion control departments monitor Fuji volcano using visual cameras, some of which are telemetered to JMA and are used for volcanological monitoring. Most of the seismic data from the different institutions are shared by JMA under the contract between each institution and JMA. However, some of the other data from different instruments such as ground deformation are not shared in real time. Therefore, the role of CCPVE is important in sharing the data of various

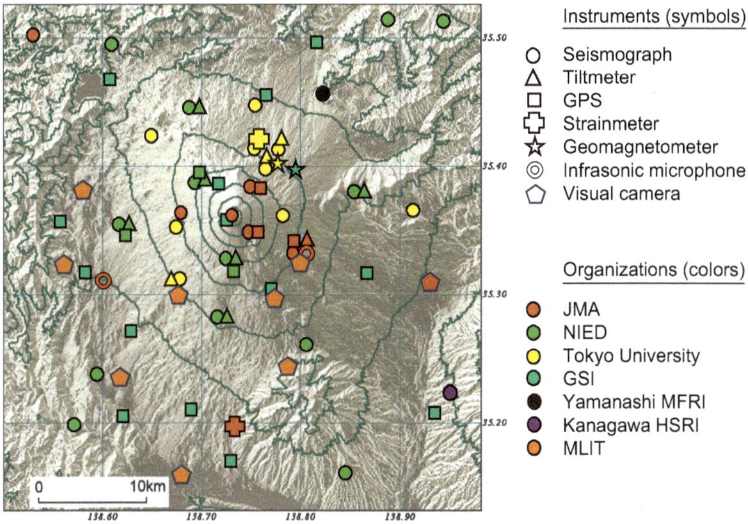

FIGURE 18.4 Monitoring network at Fuji volcano. Various monitoring instruments of multiple institutes are deployed, and most of telemetered seismic data are utilized by Japan Meteorological Agency (JMA) for the integrated monitoring. NIED, National Research Institute for Earth Science and Disaster Prevention; GSI, Geospatial Information Authority of Japan; MFRI, Mount Fuji Research Institute; HSRI, Hot Spring Research Institute; MLIT, Ministry of Land, Infrastructure, Transport and Tourism.

instruments and in assessing the current state of a given volcano because CCPVE is composed of members from the related institutions. Those data obtained by universities and other research institutes are reported in the regular or extra meetings of CCPVE for the detailed analysis of the current state of volcano activity.

18.6 ROLE OF THE CCPVE IN THE INTEGRATED MONITORING OF VOLCANOES

18.6.1 Historical View on CCPVE

NPPVE has been aimed, since its very onset, at utilizing the observed data and research achievements for the mitigation of volcanic disasters as a pragmatic need, even midway on the road toward the final goal of a practicable eruption prediction. For the purpose of comprehensive interpretation of the observed data and the pragmatic use of the achievements, CCPVE was organized in 1974 (Shimozuru, 1996).

The Committee is composed of university scientists engaged in volcanological research, scientists from research institutes such as NIED and AIST, and representatives of administrative organizations such as the Ministry of Education, Culture, Sports, Science and Technology and the Cabinet Office in charge of disaster prevention, which are also members of NPPVE. The members' term of service is 2 years, and they are commissioned by the Director General of JMA. The purposes of the committee are

1. Exchanging information on the results of monitoring and observation by related institutes and organizations, promoting researches on volcanic eruption prediction, and development of the technologies in the member institutes;
2. During volcanic eruptions, making comprehensive judgments on the phenomena and improving the quality of information about the volcano, thus contributing to disaster prevention activities; and
3. Comprehensively investigating measures for enhancing systems for volcanic eruption prediction and monitoring.

18.6.2 Integration of Interagency Monitoring Results

Volcanoes erupting frequently or those whose eruption may cause severe influence to the community are monitored by several institutes including JMA, universities, and NIED, as mentioned above. In the significant volcanoes, observation posts by multiple institutes including JMA and universities coexist and the data are shared among them to understand the nature of volcanic activity properly.

On the occasion of deployment recently, effective distribution of the monitoring posts is agreed among them under the recommendation of CCPVE; although, in the early days, installation of observation posts by different institutions was made independently without exchanging any information. Because the budget to construct the observation posts and consequently the timing of construction was controlled only within each ministry, however, adequate distribution of observation posts for a given volcano may not be attained even if recommendation of CCPVE is available for the desired distribution of instruments (Fujii, 2013).

There are three regular meetings in a year; although, the Committee may be urgently summoned during eruptions. Upon predicting a volcanic eruption, CCPVE used to announce collective opinions or the comments made by the chairperson. Today, however, its announcements mainly include the results of investigations of specific volcanoes and activity evaluation of other volcanoes in Japan. The literature references investigated by CCPVE and the committee's proceedings are published as CCPVE bulletins thrice a year. Recently, most of the references have been released almost in real time on the Web pages of JMA.

An executive board has been established to discuss the operations of the committee meetings. There are subcommittees for evaluating the activity of specific volcanoes or regions. Working groups were previously established to investigate specific topics, such as installation of activity levels for specific volcanoes and approval of new active volcanoes, though today, they serve as investigative commissions. The Usu Subcommittee was organized during the eruption of Usu volcano in 2000, and the Izu Subcommittee was formed during the eruption of Miyake-jima. The latter is the only remaining subcommittee as of 2013.

The investigative commission on volcano observation systems purposes investigating monitoring systems and integrating observational data among JMA and other related organizations. The results of the commission were used to strengthen monitoring and observation sites when JMA increased the number of volcanoes to be monitored on a 24 h basis. The investigative commission on fume research in volcanic areas prepared a database on volcanic gases, and dissolved in 2010. The database is now in trial use within JMA for further improvement.

In contingencies, the executive board or expanded board may judge an eruption, and sectional meetings and contingent CCPVE meetings are to be held when it is necessary to assess the activity of the volcano. When an eruption is expected to last for a long time, a comprehensive observation team is to be formed under CCPVE to catch changes in volcanic activity, establish new observation points, and conduct campaign observations. Because the team may need to enter the controlled areas and other dangerous zones as occasion demands, JMA is to serve as the head office and engages in negotiations with local governments, and so on.

18.7 PERSPECTIVES

Unlike the earthquake prediction issue in Japan, for which there is a government authority called the Headquarters for Earthquake Research Promotion, there is no government body for predicting volcanic eruption. CCPVE is the sole organization that evaluates and forecasts volcanic activity based on the monitored data. The comprehensive observation team is responsible for monitoring and collecting data, which are indispensable for evaluation and forecasting. Although the team members may expect to expose themselves to danger, their legal security is not covered by CCPVE but must be covered by the institutes to which they belong. This is because CCPVE is a private advisory organization of the Director General of JMA, which has neither legal responsibility nor authority over CCPVE; although, it serves as a head office of the committee.

As large-scale eruptions have not been experienced in Japan since the 1929 Hokkaido-Komagatake eruption, several large-scale eruptions may be imminent. Therefore radical strengthening of disaster prevention organizations is a pressing issue. Particularly, CCPVE should become an official organization of the government. A centralized authority such as a Volcano Agency should be ideally created to monitor, assess, and study volcanic activity. In case it is difficult to actualize it in recent trends of administrative reform, at least headquarters in charge of volcanic eruption research promotion should be immediately established. Under such headquarters, the national government should be responsible for unifying related ministries and agencies and preventing volcanic disasters.

There is another problem in the Japanese system on mitigating volcanic disaster: there are few volcanologists in JMA. JMA is supposed to be a governmental branch for civil service, and its staff members are employed among persons who passed through the national examination for government employees. As the examination includes no classification for volcanology or earth sciences except resource engineering, it is quite rare that persons trained in volcanology are hired by JMA. The staff members working in volcano monitoring are mostly trained on the job after employment. A hiring system to enroll volcanologists should be adopted by JMA. Nevertheless a new organization composed of volcanologists in various fields covering geophysics, geochemistry, and geology, and experts on volcanic hazard mitigation should be installed.

ACKNOWLEDGMENT

We greatly acknowledge Dr Setsuya Nakada for his helpful suggestions.

REFERENCES

Fujii, T., 2013. The Japanese Coordinating Committee for Prediction of Volcanic Eruptions and its Contribution to Volcanic Disaster Mitigation, vol. 380. Technical Note of NIED, 113–120.

Imakiire, T., 2013. Volcanic Disaster Measures of the Geospatial Information Authority of Japan, vol. 380. Technical Note of NIED, 129–132.

Ito, K., 2013. Observation of Volcanoes in the Sea Around Japan by the Japan Coast Guard, vol. 380. Technical Note of NIED, 133–135.

Ishihara, K., 1990. Pressure sources and induced ground deformation associated with explosive eruptions at an andesitic volcano: Sakurajima volcano, Japan. In: Ryan, M.P. (Ed.), Magma Transport and Storage. John Wiley & Sons, pp. 335–356.

Ishizuka, O., Uto, K., Yuasa, M., 2003. Volcanic History of the Back-Arc Region of the Izu-Bonin (Ogasawara) Arc, vol. 219. Geol. Soc. Spec. Pub., 187–205.

Kamo, K., Ishihara, H., 1989. A preliminary experiment on automated judgement of the stages of eruptive activity using tilt meter records at Sakurajima, Japan. In: Latter, J.H. (Ed.), Volcanic Hazards, IAVCEI Proceedings in Volcanology, vol. 1. Springer-Verlag, pp. 585–598.

Kato, K., Yamasato, H., 2013. The 2011 eruptive activity of Shinmoedake volcano, Kirishimayama, Kyushu, Japan – over view of activity and volcanic alert level of the Japan Meteorological Agency. Earth Planets Space 65, 489–504.

Nakada, S., Nagai, M., Kaneko, T., Suzuki, Y., Maeno, F., 2013. The outline of the 2011 eruption at Shinmoe-dake (Kirishima), Japan. Earth Planets Space 65, 475–488.

Shimozuru, D., 1996. Volcanic emergency management in Japan: case histories of Izu-Oshima and Unzen. In: Scarpa, Tilling (Eds.), Monitoring and Mitigation of Volcanic Hazards, pp. 787–806.

Shinohara, H., 2013. AIST's Research on Volcanology, vol. 380. Technical Note of NIED, 137–140.

Tanada, T., Ukawa, M., 2013. Volcano Disaster Mitigation Research Initiatives at National Research Institute for Earth Science and Disaster Prevention, vol. 380. Technical Note of NIED, 161–164.

Yamasato, H., 2005. Modern history of volcanic observation in Japan: especially volcano surveillance of Japan Meteorological Agency. Bull. Volcanol. Soc. Jpn. 50, S7–S18 (in Japanese with English abstract).

Yamasato, H., Funasaki, H., Takagi, Y., 2013. The Japan Meteorological Agency's Volcanic Disaster Mitigation Initiatives, vol. 380. Technical Note of NIED, 101–107.

Chapter 19

Integrating Efforts in Latin America: Asociación Latinoamericana de Volcanología (ALVO)

Hugo Delgado Granados[1], **Josè G. Viramonte**[2] **and Josè L. Palma**[3]

[1] *Departamento de Vulcanología, Instituto de Geofìsica, Universidad Nacional Autònoma de Mèxico, Mèxico,* [2] *Instituto Geonorte – INENCO, Universidad Nacional de Salta, and CONICET, Salta, Argentina,* [3] *Departamento de Ciencias de la Tierra, Universidad de Concepción, Concepción, Chile*

ABSTRACT

The Latin American region (LAR) comprises 11 countries with volcanoes characterized by a high frequency of eruptive activity. Several important eruptions have occurred within LAR in the last century with a large impact on human life, society, infrastructure, and the environment. The volcanic disasters have forced scientists and governments to find better ways to cope with eruptive phenomena. The cultural background of the LAR people as well as the usage of a common language has facilitated the communication of experiences about eruptions. It was needed, however, to establish scientific, technical, and political ties to connect volcanologists, volcanological organizations, and institutions of the region. Thus, the Asociación Latinoamericana de Volcanología was created as an association pursuing an amelioration of volcanology in the LAR for the benefit of the society. This chapter describes the integration efforts for a stronger volcanology in LAR, starting from reflections made under the SWOT (strengths, weaknesses, opportunities, threats) methodology.

19.1 VOLCANISM IN LATIN AMERICA

The Latin American region (LAR) is a very large multinational area. Eleven LAR countries have volcanoes within their territory or territorial seas, and the impact of their eruptive activity has reached wider areas outside the region (e.g., the climate impact of the eruptions of El Chichón, Mexico in 1982, Cosigüina, Nicaragua in 1835, or Huaynaputina in 1600; Robock, 2000; Oliver et al., 1996). In this region there are more than 292 active volcanoes (Siebert

Volcanic Hazards, Risks, and Disasters. http://dx.doi.org/10.1016/B978-0-12-396453-3.00019-8

and Simkin, 2002; Smithsonian Institution, 2013). The total population of the LAR with volcanoes is ~295 million inhabitants. About 30 percent of this population lives within 30 km from active volcanoes and, thus, exhibit different levels of exposure to volcanic hazards.

The LAR is also characterized by a high frequency of eruptive activity along the region (Figure 19.1). The Smithsonian database (2013) shows 1,265 eruptions in the Holocene with VEI equal or greater than 2 in Latin American countries. Since 1850, the same record of eruptions shows a nearly constant rate, with an average of nearly 4.5 eruptions per year (Figure 19.1). The LAR has experienced many eruptions of wide ranges of magnitude and impacts along its history, including many VEI 4–6 eruptions that have caused considerable destruction and disruption to society (Table 19.1).

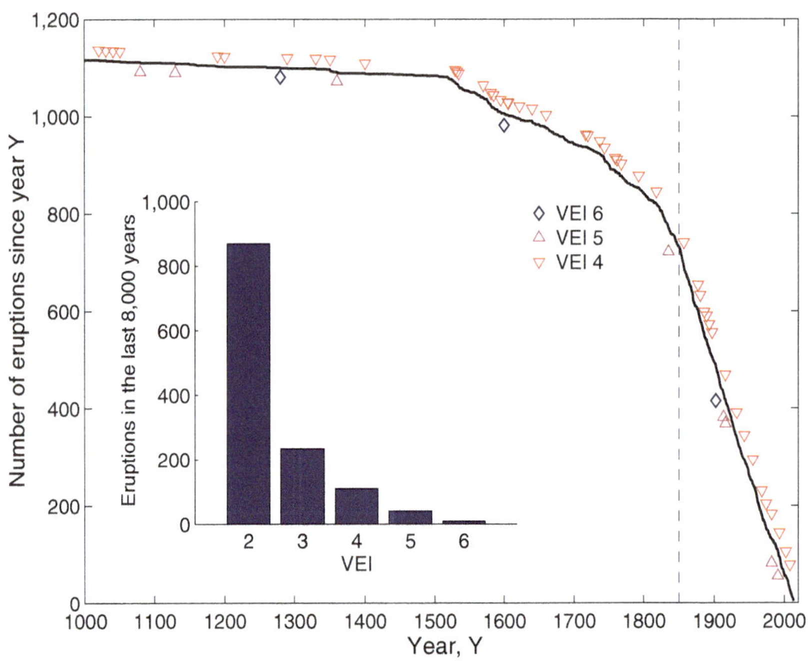

FIGURE 19.1 Number of recorded volcanic eruptions in Latin America with VEI equal or greater than 2 (database obtained from the Smithsonian Institution in October 2013). By discarding eruptions with VEI < 2 the data set is believed to be representative of the volcanic activity that actually occurred, compared to a data set that could include many VEI-1 activity reports that can be difficult to confirm. A total of 1265 eruptions have occurred within the last 8000 years (only 97 eruptions are uncertain). The main plot shows the number of eruptions between the years 1000 and 2013. The blue dashed line marks the year 1850, since when the data show 731 eruptions following an approximately constant rate of volcanism, yielding an average of 4.46 eruptions of VEI ≥ 2 per year. Three different symbols have been used to mark the timing of eruptions with VEI 6, 5, and 4. The inset plot shows the number of eruptions categorized by VEI: 870, 235, 110, 41, and 9 eruptions for VEI 2–6, respectively.

TABLE 19.1 Selected Significant Eruptions Occurred in the Latin American Region and their Impact

Volcano Name	Volcanic Type	Location: Lat	Location: Long	Location: Sumit	Volcano Number	Eruption Data	VEI	Dead Toll	Economic Losses
Region: North America Country: Mexico									
Popocatépetl	Strato volcano	19.02° N	98.62° W	5,426 m	341,090	1994–1997	2–3	Yes	5 people dead, impact to civil aviation
						2000	3		Impact to civil aviation
						2011 ~	3		Impact to civil aviation
Chichón	Strato volcano (dome)	17.36° N	92.23° W	1,150 m	341,120	1982	4–5	Yes	2,000 people dead
Parícutin	Cinder cone	19.19° N	102.19° W	3,860 m	341,060	1943–1952	≥3	Yes	1 person dead and agriculture damaged, 3 towns relocated
Colima	Strato volcano	19.51° N	103.62° W	3,850 m	341,040	1913	3	Yes	Ashes reached the city of Saltillo 725 km to the NNE of the volcano
						1997–2000	2–3	No	Evacuation of nearby towns 2 days before the eruption began in 1998
Region: Central America Country: Guatemala									
Santa María (Santiaguito)	Strato volcano (dome)	14.756° N	91.552° W	3,772 m	342,030	1989	3	Yes	?
						1990	3	Yes	?
						2004	3	Yes	?
						2013	1	No	?

(*Continued*)

TABLE 19.1 Selected Significant Eruptions Occurred in the Latin American Region and their Impact—cont'd

Volcano Name	Volcanic Type	Location			Volcano Number	Eruption Data	VEI	Dead Toll	Economic Losses
		Lat	Long	Sumit					
Fuego	Strato volcano	14.473° N	90.88° W	3,763 m	342,090	1987	1	Yes	?
						1999	1	Yes	?
						2000	2	No	?
						2002	2	No	?
						2012	1	No	?
						2013	1	No	?
Pacaya	Strato volcano	14.381° N	90.601° W	2,552 m	342,110	1984	2	Yes	?
						1987	3	Yes	?
						1989	3	Yes	?
						1990	1	No	?
						2004	1	No	?
						2010	3	Yes	?
Region: Central America País: El Salvador									
Izalco	Strato volcano	13.813° N	89.633° W	1,950 m	343,030				?
San Miguel	StratoVolcano	13.434° N	88.269° W	2,130 m	343,100				?
Santa Ana	Strato volcano	13.853° N	89.63° W	2,381 m	343,020	2005	3	Yes	?
						2006	1	No	?

Region: Central America Country: Nicaragua							
San Cristóbal	Volcanic complex	12.702° N 87.004° W 1,745 m	344,020	1997	1	No	?
				1999	2	No	?
				2003	1	No	?
				2004	1	No	?
				2006	2	No	?
				2007	1	No	?
				2009	2	No	?
				2010	2	No	?
				2013	1	No	?
Casita	Part of San Cristobal		344,020	1998		3000	El Porvenir and Roland Rodriguez towns destroyes by Lahars related Mitch Hurricane
Telica	Strato volcano	12.602° N 86.845° W 1,061 m	344,040	1981	1	No	?
				1982	2	No	?
				1987	1	No	?
				1994	2	No	?
				1999	2	No	?
				2012	1	No	?

(Continued)

TABLE 19.1 Selected Significant Eruptions Occurred in the Latin American Region and their Impact—cont'd

Volcano Name	Volcanic Type	Location			Volcano Number	Eruption Data	VEI	Dead Toll	Economic Losses
		Lat	Long	Sumit					
Cerro Negro	Cinder cone	12.506° N	86.702° W	728 m	344,070	1992	3	No	?
						1995	2	No	?
						1999	2	No	?
						2013	2	No	?
Masaya	Caldera (Santiago Pit)	11.984° N	86.161° W	635 m	344,100	1981	1	No	?
						1999	1	1 injuried	Several cars destroyed by volcanic bombs
						2008	1	No	?
						2010	1	No	Wil fire produced incandescend volcanic material
Concepción	Strato volcano	11.538° N	85.622° W	1,700 m	344,120	1982	2	No	?
						1984	1	No	?
						1985	2	No	?
						1988	2	No	?
						1999	1	No	?
						2005	2	No	?
						2006	2	No	?
						2011	1	No	?

Region: Central America Country: Costa Rica									
Rincon de la Vieja	Volcanic complex	10.83° N	85.324° W	1,916 m	345,020	1983	1	No	?
						1984	2	No	?
						1985	1	No	?
						1986	1	No	?
						1991	2	No	?
						1995	2	No	?
						1998	2	No	?
						2011	1	No	?
						2012	1	No	?
Arenal	Strato volcano	10.463° N	84.703° W	1,670 m	345,033	1968	2	87	Tabacon and La Fortuna towns destroyed by direct blast
						1987	2	Yes	?
						1988	2	Yes	?
						1989	2	Yes	?
						1990	2	Yes	?
						1991	2	Yes	?
						1992	2	Yes	?
						1993	2	Yes	?
						1998	2	Yes	?
						2000	2	Yes	?

(Continued)

TABLE 19.1 Selected Significant Eruptions Occurred in the Latin American Region and their Impact—cont'd

Volcano Name	Volcanic Type	Location			Volcano Number	Eruption Data	VEI	Dead Toll	Economic Losses
		Lat	Long	Sumit					
						2001	2	No	Temporal evacuation of several towns
						2007	2	No	Temporal evacuation of several towns
						2009	2	No	Temporal evacuation of several towns
						2010	2	No	Temporal evacuation of several towns
						2000	2	No	Temporal evacuation of several towns
Poas	Strato volcano	10.2° N	84.233° W	2,708 m	345,040	1980	1	No	?
						1981	1	No	?
						1987		No	?
						1988	1	No	?
						1989	2	No	?
						1991	2	No	?
						1991	1	No	?
						1992	1	No	?

						1994	2	No	?
						2009	1	No	?
						2011	1	No	?
						2013	1	No	?
Irazu	Strato volcano	9.979° N	83.852° W	3,432 m	345,060	1994	2	No	?
Turrialba	Strato volcano	10.025° N	83.767° W	3,340 m	345,070	2010	2	No	?
						2011	1	No	?
						2012	2	No	?
						2013	2	No	?
Region: South America Country: Colombia									
Nevado del Ruiz	Strato volcano	4.895° N	75.322° W	5,321 m	351,020	1984	1		?
						1985	3		?
						1991	3	25,000	Armero was totally destroyed by a lahar
						1994	1		?
						2012	3		?
Nevado de Huila	Strato volcano	2.93° N	76.03° W	5,364 m	351,050	2007	3	10	10 persons dead by lahars
						2008	2	Yes	?
						2012	3	No	?

(Continued)

TABLE 19.1 Selected Significant Eruptions Occurred in the Latin American Region and their Impact—cont'd

Volcano Name	Volcanic Type	Location			Volcano Number	Eruption Data	VEI	Dead Toll	Economic Losses
		Lat	Long	Sumit					
Galeras	Volcanic complex	1.22° N	77.37° W	4,276 m	351,080	1989	2	No	?
						1990	2	No	?
						1993	2	9	9 people dead including 6 volcanologist
						2000	1	No	?
						2002	1	No	?
						2004	3	No	?
						2005	2	No	?
						2006	2	No	?
						2007	3	No	?
						2008	3	No	?
						2009	1	No	?
						2010	1	No	?
						2012	1	No	?

Region: South America Country: Ecuador									
Guagua Pichincha	Strato volcano	0.171° S	78.598° W	4,784 m	352,020	1981	1	No	?
						1985	1	No	?
						1990	1	No	?
						1993	1	2	2 volcanologists died during a phreatic explosion
						1997	1	No	?
						1998	3	No	?
						2001	1	1	20 people evacuated
						2002	1	No	?
						2003	1	No	?
						2008	1	No	?
						2009	1	No	?
Tungurahua	Strato volcano	1.467° S	78.442° W	5,023 m	352,080	1993	1	No	?
						1999	3	1	18000 inhabitants were evacuated

(*Continued*)

TABLE 19.1 Selected Significant Eruptions Occurred in the Latin American Region and their Impact—cont'd

Volcano Name	Volcanic Type	Location			Volcano Number	Eruption Data	VEI	Dead Toll	Economic Losses
		Lat	Long	Sumit					
						2006	3	6	?
						2009	3	No	?
						2010	3	No	2 villages were evacuated
						2013	1	No	?
Reventador	Strato volcano	0.077° S	77.656° W	3,562 m	352,010	2002–2003	4	No	Reventador, a small village was evacuated
						2004	2	No	?
						2005	2	No	?
						2007	2	No	?
						2008	2	No	?

Region: South America Country: Ecuador (Galapagos Island)									
Cerro Azul	Shield volcano	0.92° S	91.408° W	1,640 m	353,060	1998	1	No	?
						2008	1	No	?
Alcedo	Shield volcano	0.43° S	41.12° W	1,130 m	353,040	1993	1	No	?
Fernandina	Shield volcano/ Caldera	0.37° S	91.55° W	1,476 m	353,010	1981	1	No	?
						1984	1	No	?
						1988	2	No	?
						1991	2	No	?
						1995	2	No	?
						2005	2	No	?
						2009	2	No	?
Marchena	Shield volcano	0.33° N	90.47° W	343 m	353,080	1991	2	No	?
Sierra Negra	Shield volcano	0.83° N	91.17° W	1,124 m	353,050	1980	3	No	?
Region: South America Country: Perú									
Ubinas	Strato volcano	16.355° S	70.903° W	5,672 m	354,020	2006	2	No	?
						2009	2	No	?
						2010	1	No	?
						2013	2	No	?
El Misti	Strato volcano	16.294° S	71.409° W	5,822 m	354,010	1985	1	No	?

(Continued)

TABLE 19.1 Selected Significant Eruptions Occurred in the Latin American Region and their Impact—cont'd

Volcano Name	Volcanic Type	Location			Volcano Number	Eruption Data	VEI	Dead Toll	Economic Losses
		Lat	Long	Sumit					
Sabancaya	Strato volcano	15.78° S	71.85° W	5,967 m	354,006	1986	1	No	?
						1987	2	No	?
						1988	1	No	?
						1990	3	No	?
						2000	2	No	?
						2003	2	No	?
Region: South America Country: Chile-Argentina									
Lascar	Strato volcano	23.37° S	67.73° W	5,592 m	355,100	1984	1	No	?
						1986	3	No	?
						1987	3	No	?
						1990	1	No	?
						1991	2	No	?
						1993	4	No	Talabre, a little town was partially destroyed and was moved to other place
						1993–1994	2	No	
						1996	2	No	?
						2000	2	No	?

						2002	2	No	?
						2005	3	No	?
						2006	3	No	?
						2013	1	No	?
Tupungatito	Stratovolcano	33.425° S	69.797° W	5660 m	357010	1980	2	No	
						1987	2	No	
Planchón-Peteroa	Stratovolcano	35.223° S	70.568° W	4107 m	357040	1762	4	No	
						1991	2	No	
						2010	2	No	
						2011	3	No	
Cerro Azul (Quizapu)	Stratovolcano	35.653° S	70.761° W	3788 m	357060	1932	5	No	
						1967	2		
Nevados de Chillán	Stratovolcano	36.863° S	71.377° W	3212 m	357070	1864	3	No	
						1973–1986	2	No	
						2003	1	No	
Copahue	Stratovolcano	37.856° S	71.183° W	2953 m	357090	2000	2	No	
						2012	2	No	
Lonquimay	Stratovolcano	38.377° S	71.58° W	2865 m	357100	1988–1990	3	1	Hundreds of cattle and horses died of osteofluorosis. Evacuation needed.
Llaima	Stratovolcano	38.892° S	71.729° W	3125	357110	1955	3		
						2003	2		
						2007	2		
						2008	3		

(Continued)

TABLE 19.1 Selected Significant Eruptions Occurred in the Latin American Region and their Impact—cont'd

Volcano Name	Volcanic Type	Location			Volcano Number	Eruption Data	VEI	Dead Toll	Economic Losses
		Lat	Long	Sumit					
Villarica	Stratovolcano	39.42° S	71.93 ° W	2847	357120	1948	3	No	
						1991	1	No	
						2004	1	No	
						2003	1	No	
Puyehue-Cordon Caulle complex	Fisure Vent	40.59°S	72.117°W		357550	1920–21	3	No	
						1960	3	No	
						2011	4	No	Air navigation seriously affected for a long time in Argentina Wild life, forest, enviroment, agriculture, livestock and touristic activities seriourly affected
Chaiten	Caldera Complex	42.833°S	72.646°W	1122	356041	2008	4	No	Chaiten Village destrtoyed by lahars - Air navigation affected in Patagonia
Hudson	Stratovolcano	45.90 ° S	72.97 ° W	1905	358057	1971–	3	No	
						1973	3	2	Mud avalanches burie two people and hundreds of animals
						1991	5	No	SO_2 cloud circles globe; aircraft encountered ash cloud over Australia; >1 km3 airfall on Argentina
						2011	2	No	

The most striking issue however, is the elevated death toll produced by few eruptions in the LAR. During the twentieth century two of the strongest and most deadly eruptions worldwide occurred in the LAR. The eruptions of Volcán Santa María in Guatemala on October 24–26, 1902 (VEI 6) killed 6,000 people; El Chichón volcano in Mexico on April 3–4, 1982 (VEI 5) killed ~2000 people, and the relatively minor eruption of Nevado del Ruiz volcano in Colombia (VEI 3) on November 13, 1985 claimed ~25,000 lives (Tilling, 1989). The proximity of the population to active volcanic centers is of great concern in Latin America. In the whole region approximately 7.8 million people live at less than 10 km from a volcano that has shown activity in the Holocene (Figure 19.2). A striking case is Central America, where more than 10 million people live within 20 km from a historically active volcano (Figure 19.2).

In the past, the large number of casualties related to eruptions in LAR was, in part, due to a lack of basic understanding of volcanic hazards and absence of volcano monitoring, a fact that has changed after the disaster of Nevado del Ruiz in 1985. Alvarado et al. (1999) argued that only 16 volcanoes had enough

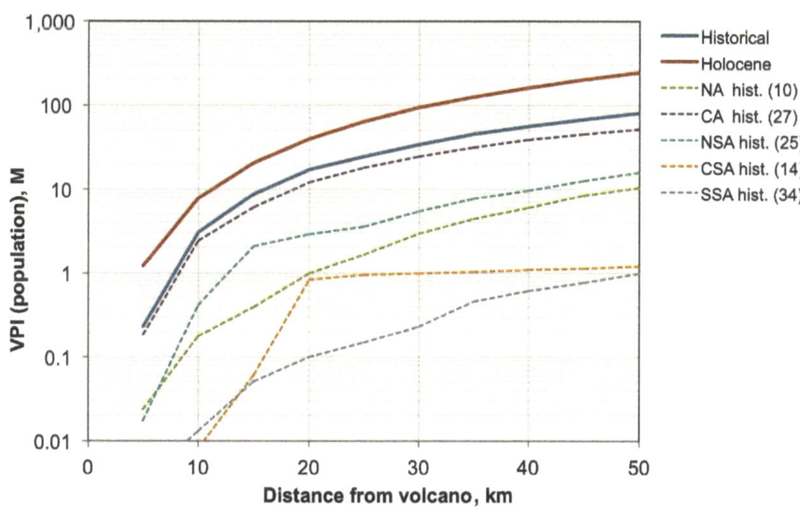

FIGURE 19.2 The Volcano Population Index (VPI) indicated for different regions in Latin America, where NA stands for North America (Mexico), CA for Central America (Guatemala, El Salvador, Nicaragua, Costa Rica, and Panamá), and NSA, CSA, and SSA refer to, respectively, Northern (Colombia and Ecuador), Central (Peru and Northern Chile), and Southern (Southern Chile) South America. Only volcanoes with historical records of activity have been considered in the calculations per region following the Smithsonian database compiled in 2008. The legend also shows, in brackets, the number of volcanoes considered in the analysis for each region. The VPI is indicated as the total population (in millions of inhabitants) living within a specified distance from a volcano in the region. VPIs for all active volcanoes in Latin America, with historical or Holocene records of activity, are included in this figure for comparison. The LandScan 2007 population data (Oak Ridge National Laboratory, LandScan 2007TM/UT-Battelle, LLC) was used to quantify the population for each VPI.

surveillance systems. By then, countries with the highest number of active volcanoes with monitoring systems were Colombia (58 percent of volcanoes instrumented) and Ecuador (50 percent). Observatories with basic monitoring systems occur in Guatemala and Mexico, followed by Costa Rica and Nicaragua. During the last two decades of the twentieth century, several new volcano observatories and research centers were established (e.g., Mexico, Guatemala, Nicaragua, Costa Rica, Colombia, Ecuador, Chile, and Argentina). Notwithstanding, at the end of the twentieth century, many active volcanoes in the LAR remained still poorly studied, with virtually little or no monitoring systems, representing a potential threat. At the end of the twentieth century only a small number of volcanoes in the LAR were relatively well studied: Colima, Popocatépetl, Santa Maria, Cerro Negro, Arenal, Irazu, Turrialba, Nevado del Ruiz, Galeras, Puracé, Cotopaxi, Guagua Pichincha, Tumgurahua, Lascar, and Villarrica. This view of Alvarado et al. (1999, 2007) has changed considerably in the last 10 years, as it is shown in the sections below.

19.2 HISTORICAL DEVELOPMENT OF THE LATIN AMERICAN ASSOCIATION OF VOLCANOLOGY

Because the LAR is a region where a single language (Spanish) is spoken, cultural and historical heritage is shared, and having so many eruptions occurring in the region, its result was not only natural and obvious but there was also a strong need to join efforts and look for international collaboration for the LAR in order to lessen the impact of volcanic phenomena. To this end, it was imperative to create an organization that would help gathering volcanologists in the LAR and enhance communication among all sorts of lines of expertise. Latin American volcanologists needed to speak to each other, establish cooperation ties, and communicate their experience, particularly those related to handling scientific information during crises. This was equally valid for scientists working at academic institutions and for geologists, geophysicists, engineers, and other technicians working at observatories or emergency offices. Nevertheless, most engineers and technicians working at volcano observatories did not attend volcanological conferences or meetings where they could learn from others' experiences. Without their work, however, little could be done to acquire signals produced remotely at volcanoes and transmitted to scientists' offices. It was also very important to create a forum for them to communicate and find information on how to handle instrumentation or get advice about new instruments or methods to solve particular problems with the hardware or software involved in transmitting data from volcanoes to their data acquisition centers. In addition, it was necessary to create the links between scientists and civil protection officials, and ensure a proper communication of scientific information to decision-makers and the public. All these needs and ideas were the driving force to constitute the Latin American Association of Volcanology.

Knowing that collaboration was strongly needed in the LAR, several efforts began in 1988 to establish a Latin American association of individuals actively working on the volcanic areas of the region (Delgado Granados et al., in press). During a meeting in Kagoshima, Japan in 1988 volcanologists from Mexico, Costa Rica, Colombia, Ecuador, and Chile met to discuss issues concerning LAR, raising for the first time the idea of gathering and connecting the Latin American volcanological community. In 1989, in Santa Fe, New Mexico, during the IAVCEI General Assembly another meeting was held with the aim of establishing a Latin American Association of Volcanology. These efforts, however, evolved very slowly mainly due to the intense work being carried out at several institutions to create new observatories and build up volcanological instrumental infrastructure in several countries of the region. In 2006, during the "*Garavolcan 2006*" meeting in Garachico, Tenerife, Spain, a plan was established to organize the Latin American volcanological community in the following years. In this organizational meeting, volcanologists from Mexico, Guatemala, El Salvador, Costa Rica, Ecuador, Chile, and Argentina were present, and agreed on a Latin American character of the association rather than Iberoamerican due to the complexities of coordinating such a wide region.

At this point, it was important to have a basic knowledge on the concerns and perspectives of the Latin American volcanological community. The "*1st Meeting of the Iberoamerican Network of Volcanology*" held in Antigua, Guatemala from February 19–22, 2008, was the opportunity to organize the Latin American volcanological community and acquire information from several countries at the same time, with the presence of volcanologists from Mexico, Guatemala, El Salvador, Nicaragua, Costa Rica, Colombia, Ecuador, Peru, Argentina, Chile, Portugal, and Spain. This meeting helped to have a first regional analysis on the nature and capacities of the region.

An important issue was the discussion about the importance of social involvement of volcanologists in Latin America for an effective volcanic risk reduction. It was critical to understand the situation and current status of efforts across Iberoamerican countries and to review how the volcanological community is pursuing the reduction of volcanic risk in the region. The analysis and evaluation of volcanic risk reduction efforts were conducted by working groups addressing different aspects of scientific and technical actions aimed at reducing volcanic risk recommended by UNESCO and the IAVCEI (i.e., IAVCEI, 1999). In order to understand and improve the conditions of Latin American volcanological science it was necessary to perform a preliminary analysis of the situation. This was achieved by a SWOT (strengths, weaknesses, opportunities, and threats) analysis carried out in 2008 at the meeting in Antigua, Guatemala. Partial results of the SWOT exercise are shown in Table 19.2.

Even though a formal organization was not established yet, preparations were made for this aim. Two meetings were further held. The first took place

TABLE 19.2 Schematic Results of the SWOT (Strengths, Weaknesses, Opportunities, and Threats) Exercise Carried Out in 2008. This Table is Focused in the Needs and Conclusions Raised With the Exercise

	Threats	Opportunities
Weaknesses	It is needed to establish national and regional, state and institutional policies, for the sustainability (economic resources, training, education, etc.), coordination and systematization of short-, medium-, and long-term studies on hazard and volcano monitoring It is needed to elaborate coordination protocols, methodologies, and information systems at national and regional levels Databases are strongly needed to establish baselines for the largest possible number of volcanoes There is a need to ensure that research and volcano-monitoring groups issue consensus views by a single spokesperson	It is needed to have in advance research, volcano monitoring, education and outreach plans and projects ready to be taken into account during volcanic crises It would be very important to reach cooperative agreements for research, volcano monitoring, and specialized human resources networking It is highly desirable to integrate and coordinate the scientific-technical sector with civil protection authorities and the community
Strengths	A safe society cannot be achieved if authorities, politicians, and the general public do not receive education about volcanic hazards There is a strong need for the implementation of international protocols to rule the behavior of scientists during volcanic crises and coordinate international technical-cooperation groups Authorities need to be aware about the need to protect the personnel and create protection systems for the personnel and monitoring equipment	Organization of forums, publication and promotion of results obtained by the research groups and networks Integration of international research groups devoted to work on projects of the region Submit proposals for research and teaching at national and international agencies for funding Incorporate volcanic hazard issues in formal education at all levels, as well as outreach activities for authorities and the general public Creation of catalogs of active volcanoes in each country and in the region to create priority lists for attention and study Creation of communication mechanisms within research and volcano-monitoring networks Review of the pertinence to separate civil protection from military Promote and update emergency plans and volcanic crisis drills

in Tenerife during the "*Cities on Volcanoes 6*" meeting from May 31–June 4, 2010. There, it was decided that the creation of the Asociacion Latinoamericana de Volcanología (ALVO) would have to take place in Manizales, Colombia, in November 6–8, 2010, prior to the "*Commemoration of the 25th Anniversary of the Nevado del Ruiz Eruption*." This decision implied that Colombia would be the place where the association would be formalized under Colombian laws. This decision had an important meaning to the volcanologists of the LAR because the date commemorated the most terrible volcanic disaster in Latin America in recent times. Many lessons have been learned from it and for that reason the region is seeking for an improvement of the monitoring systems and training people at the highest possible level so that these kind of events should not occur again. Thus, the Colombian volcanologists have emerged as one of the leading groups of the region with a lot of experience and training to deal with eruptive crises. The second meeting, prior to the creation of ALVO was held in Foz de Iguaçu, in Brazil. The details of the organization for the Colombia encounter were established during the "*Meeting of the Americas*" of the American Geophysical Union (AGU) between August 8 and 12, 2010.

During thc "*First Meeting of the Latin American Volcanological Community*," ALVO was established and registered at the Colombian authorities on November 18, 2010. ALVO was established to become the regional framework to interchange of experiences and knowledge, and to foster scientific and technical support within LAR.

19.3 ALVO FIRST STEPS

One of the main achievements of the Latin American volcanological community in recent years was the consolidation of the Latin American Volcanological Association (ALVO). This association was created with the following objectives:

1. To promote and strengthen linkages between Latin American volcanologists.
2. To promote the scientific study of volcanoes, their setting in space and time, and to make the results of such studies public.
3. To promote cooperation among scientific institutions and organizations whose objectives include fostering knowledge of volcanological disciplines.
4. To initiate and coordinate scientific research programs, especially those that rely on international cooperation.
5. To promote, advise, and support volcano-monitoring programs in ALVO country members, who ask for such activities, in a collaborative environment.
6. To support organizations and institutions responsible for volcano monitoring as well as decision-makers in case of volcanic crises in LAR countries.

7. To promote work and activities related to teaching, dissemination, and outreach of volcanology and its applications, as well as to promote or perform other activities consistent with the above purposes.
8. To establish a common reference center and gathering for the Latin American volcanological community.

ALVO's membership has increased considerably since its creation in 2010. It began with 46 members and by October 29, 2012, 152 members were registered. The current membership is more than 300 members.

An important activity developed by ALVO is the support to institutions during eruptive crises. This support is only organized when official help is requested. Under the agreement established among volcanologists and their institutions, crises such as the Telica volcano eruption in Nicaragua, in 2011, were assessed and discussions were performed as videoconferences where data were shown and discussed between local and foreign Latin American scientists. In this occasion, ALVO made the connections between the governments of Nicaragua and El Salvador in order to provide equipment and one technician to support the observations during the eruption. Another important action by ALVO was its members' participation to help and advise communities in Argentina during the eruption of Cordón Caulle (Chile) in 2011. Assessment was provided regarding secondary lahar hazards (Cordoba et al., 2011) and ash-concentration forecasts at different flight levels and thicknesses on land (Collini et al., 2012).

Regarding objectives 5 and 7, ALVO has organized several courses. Due to budget limitations, but recognizing the need for basic training in volcanology for people working at volcanoes in the region, an annual virtual course on *Basic Volcanology* has been organized at the Instituto de Geofísica of the Universidad Nacional Autónoma de México (UNAM) since 2011. The 2013 course has also been supported by IAVCEI. This course has been taught in Spanish using virtual classrooms provided by UNAM with the participation of professors from around the world. In 2011, ~50 students registered, but enrolment has increased through time (~120 registered in 2012 and 140 in 2013). In addition, a course on the use of infrared analyzers to measure CO_2 emissions from soils was organized in May 2011 in Costa Rica together with the OVSICORI-UNA (Volcanological and Seismological Observatory of Costa Rica at the Universidad Nacional). The International Course in Field Volcanology at the Central Andes under the Universidad de Salta has not stopped its activities but also has increased the support to ALVO members.

The analysis of the situation of LAR volcano science was also revised, and a long-term survey started. This study was supported by IAVCEI in order to review the state of the art of volcanology in the region. For this study, the number of active volcanoes was collected from information provided by observatories, universities, and people in charge of monitoring volcanoes along the LAR (Table 19.3). In this table, the number of volcanoes shown (257) is different if

TABLE 19.3 Number of Monitored Volcanoes per Country. This List Shows a Scheme for the LAR Indicating How Many Central Volcanoes, Monogenetic Fields or Calderas Exist and If They Are Monitored in Any Way

	Central Volcanoes			Monogenetic Fields			Calderas and/or Geothermal Fields			Total		
		Monitored			Monitored			Monitored			Monitored	
Country	Total	Number	%	Total	Number	%	Total	Number	%	Total	Number	%
Mexico	18	9	50	12	3	25	5	0	0	35	12	34
Guatemala	16	4	25	7	0	0	2	0	0	25	4	16
El Salvador	9	5	56	5	0	0	2	1	50	16	6	38
Nicaragua	13	7	54	5	1	20	1	1	100	19	9	47
Costa Rica	13	10	77	0	0	0	0	0	0	13	10	77
Panamá	2	1	50	0	0	0	1	0	0	3	1	33
Colombia	17	12	71	0	0	0	0	0	0	17	12	71
Ecuador	31	17	55	1	0	0	4	3	75	36	20	56
Peru	13	5	38	3	1	33	0	0	0	16	6	38
Bolivia	8	2	25	4	0	0	1	0	0	13	2	15
Chile	86	38	44	12	2	17	8	3	38	106	43	41
Argentina	31	5	16	7	0	0	4	1	25	42	6	14

compared with the 292 indicated by the Smithsonian Institution (2013). The main reason for this is that a different scheme has been selected for counting volcanoes. For instance, the ALVO statistics indicate how many volcanoes are central, monogenetic fields, or calderas. Smithsonian Institution (SI) includes some monogenetic volcanoes as individual volcanoes not as part of an active volcanic field. It is also important that for the time this survey was made (2010–2011) the numbers show the countries that have developed a widespread monitoring program, not taking into account the quality of the monitoring system.

In Table 19.4, a summary of how the volcanoes in LAR were monitored can be seen and the number of volcanoes with hazard maps in every country by 2011. From this table, it is evident that no LA country has developed hazard maps for all the volcanoes in its territory. However, most of the countries have made an effort to monitor active volcanoes, although the coverage is still very limited and an important effort should be made to dedicate a larger budget for these activities. Numbers in this table should be taken with caution because some volcanoes are instrumented but the table does not reflect the difficulties monitoring agencies have for taking care of their monitoring systems. Budgetary limitations forced several institutions to stop station maintenance or instrument repair, some of them donated by agencies that left instrumentation but governments have not always matched the efforts done by the international community and local scientists by providing financial support for continuous operation.

19.4 A CRITICAL VIEW INTO THE SWOT FOR THE DEVELOPMENT OF VOLCANOLOGY IN THE LATIN AMERICAN REGION

In this section an example of what ALVO is capable of doing as a united volcanological community is presented. The SWOT analysis mentioned in Section 19.2 was conducted in 2008, as a tool for understanding and decision-making. It was focused on the three actions for risk reduction and volcano monitoring: mapping of volcanic hazards, volcano monitoring, and emergency plans.

The SWOT analysis is a methodology designed to study the competitive position of an institution, taking into account its internal characteristics and surrounding conditions to determine strengths, weaknesses, opportunities, and threats. The strengths and weaknesses are internal, whereas threats and opportunities are present outside and around the entity under analysis, and thus, are external (Hill and Westbrook, 1997; Menon et al., 1999). The products of this analysis were the diagnosis and strategies about the program for reducing volcanic risk in the LAR. The analysis consists of four steps: (1) external analysis; (2) internal analysis; (3) preparation of the SWOT matrix; and (4) determination of the strategy to be used (Hill and Westbrook, 1997; Menon et al., 1999).

TABLE 19.4 Summary of Existence of Hazard Maps and Monitoring Systems for Volcanoes in the Latin American region (LAR)

Country	Holocene Volcanoes	Volcanoes with Hazard Map		Volcanoes with Seismometers		Volcanoes with Periodic Gas Sampling		Volcanoes with Periodic Spring-Water or Fumarole Sampling		Volcanoes with Deformation Monitoring		Volcanoes with Remote Sensing		Total Percentage of Instrumented and Monitored Volcanoes	
		Number	%	Number	%	Number	%	Number	%	Number	%	Number	%	Number	%
México	18	3	17	10	56	5	28	8	44	8	44	7	39	15	83
Guatemala	16	4	25	3	19	2	13	1	6	4	25	4	25	4	25
El Salvador	9	3	33	3	33	5	56	5	56	0	0	3	33	5	56
Nicaragua	13	4	31	9	69	4	31	1	8	9	69	1	8	9	69
Costa rica	13	11	85	8	62	9	69	10	77	2	15	6	46	10	77
Panamá	2	1	50	1	50	0	0	0	0	0	0	0	0	1	50
Colombia	17	14	82	12	71	6	35	6	35	9	53	3	18	13	76
Ecuador	31	11	35	19	61	6	19	7	23	9	29	16	52	28	90
Perú	13	7	54	1	8	1	8	6	46	0	0	5	38	6	46
Bolivia	8	0	0	0	0	0	0	2	25	0	0	0	0	2	25
Chile	86	9	10	40	47	3	3	10	12	29	34	33	38	44	51
Argentina	31	3	10	3	10	3	10	4	13	3	10	0	0	5	16

19.4.1 SWOT Results

During the execution of the SWOT analysis in February 2008, 23 volcanologists from 12 countries participated. A summary of the analysis results is shown in Table 19.2. It is important to mention that this view of LA volcanologists may have changed towards the year 2014.

Production of hazard maps has been one of the most important duties for the Latin American volcano observatories and other volcanological institutions throughout the past decade. The production of hazard maps has not been easy though. In Table 19.5 a list of reflections on the successes and problems encountered when performing this task is presented. Most Latin American volcanologists think that volcanic crises are a time of opportunity to test the plans conceived in advance while making hazard maps. When making hazard maps, an opportunity also exists to plan and think how to educate the officials and population, and at the same time, to look for the existence of international agreements in order to have ready, a list of potential advisors or supporters in case of crises. Most volcanologists in LAR conclude that there is capacity (scientific and technical) and enough experience to tackle the task of mapping volcanic hazards. However, the most important problem in several countries of the region has been a low budget for the task. In, countries with a large number of active volcanoes but without a national policy that establishes priorities to produce volcanic hazard maps a low financial support precludes the carrying out of research at the most inaccessible volcanoes that sometimes are the most hazardous or risky. Also, a low budget makes difficult for some institutions in the LAR to retain trained personnel working in hazard mapping. An important weakness for the LAR is the lack of a unified methodology for hazard mapping and, in some countries of the region, the lack of a single institution devoted to producing hazard maps. When two institutions or individual volcanologists produce a hazard map for the same volcano but use different methodologies, the resulting maps create confusion in the minds of the public and civil protection officials. Several Latin American volcanologists consider as an important threat the presence of foreign researchers that carry out hazard mapping during volcanic crises with no coordination with the local volcanologists.

Regarding volcanic surveillance, LAR volcanologists consider that human resources at LAR institutions have experience and training in basic monitoring, and basic monitoring systems do exist as well (Table 19.6). The existence of a large number of active, highly eruptive volcanoes is a fact that is considered a strength and opportunity for having intense training of the personnel devoted to monitoring of volcanoes in the region. This fact means not only an opportunity to acquire experience in monitoring, but also experience in handling crises before authorities, the public, and the media. Nevertheless, also the shadow of a low budget makes the institutions devoted to surveillance to have difficulties because this is an activity that requires a constant financial support for payment of salaries, maintenance, fieldwork, and

TABLE 19.5 Results of the 2008 Strengths, Weaknesses, Opportunities, and Threats about Mapping

Weaknesses	*Scarcity of financial, human, and infrastructure resources.* Latin American region (LAR) is a region where there are several volcanoes at every country and the financial resources are limited for carrying out geological research and hire trained personnel able to make hazard maps. *Lack of institutional coordination, interinstitutional and intergovernmental.* At some countries in the LAR, the making of hazard maps is not a task assigned to a particular institution so, coordination for their elaboration does not exist. *Lack of a unified methodology.* Currently, there is not a single methodology to make hazard maps, even though there is a lot of interest and models to represent volcanic processes, the making of hazard maps is still vague and ambiguous. *Lack of a scheme of priorities.* Which volcano should be mapped first? *Availability of information in an organized and centralized way.* The basic information needed to make a hazard map at some countries is not organized, not gathered, making it difficult to have geological information handy for hazard mapping. *Lack of incentives to ensure the permanency of the staff.* At several countries the professionals in charge of making these maps have no permanent positions and so, they leave the institutions before they finish or continue with the hazard mapping. *Lack of preconceived plans for research and monitoring during volcanic crisis.* Some countries lack of funding for scientific research and funds for monitoring are also limited so, the elaboration of hazard maps that might be useful during eruptive crises do not exist and as a consequence, emergency planning neither exist.	Strengths	*Experience and capacity for carrying out geological work.* *Capacity and high academic and technical level.* Several volcanologists in the LAR have a PhD or an MSc degree. Besides, several technicians with bachelor's degree level have been trained or acquired such an experience and ability that their performance is of the utmost level. *Experience in development of hazard maps.* Several institutions in the LAR have developed hazard maps for their volcanoes in the last decade. *Capacity for publication and outreach of the products.* Several institutions in the LAR have developed the ability to communicate scientific results to the public. *Existence of infrastructure (analytical laboratories, instruments, computers, software, Internet).* The existence of laboratories to process samples and get geological data useful during elaboration of hazard maps has increased in the region. *Existence of international cooperation ties.* *Management capacity.*

Continued

TABLE 19.5 Results of the 2008 Strengths, Weaknesses, Opportunities, and Threats about Mapping—cont'd

Opportunities	*Occurrence of volcanic crisis helps to rise preconceived plans.* Volcanic crises have been the main engine to accelerate the flow of financial resources in the Latin American region. *Harnessing the educational and communication system to guide the dissemination and teaching of cartography.* *Existence of international agreements and treaties.* Existence of national and international research groups and programs. Establishment of regional and international research and experts.	Threats	*Deficiency of systems to keep human resources.* At some countries, the institutions do not have appropriate systems to ensure the permanency of trained personnel. Sometimes, a person that has been trained has to leave, seeking for better job opportunities. *Conceptual issues in population, authorities, and politicians.* Incorrect concepts of hazards and risk among stakeholders and players make difficult to communicate the results and benefits of hazard mapping. *Lack of interest, knowledge and vision in population, politicians and authorities.* The value of hazard mapping is not seen unless an eruptive event occurs. Presence of external research groups during volcanic crises without coordination with local groups. Sometimes, external research groups visit a country where a volcano is about to erupt or erupting with the aim of carrying out a hazard map. In spite of the best intentions of these groups, the lack of coordination with local researchers and direct contact with the local population and media is a source for conflicts.

TABLE 19.6 Results of the 2008 Strengths, Weaknesses, Opportunities, and Threats about Surveillance

Weaknesses	*Lack of economic (for acquisition and maintenance of equipment), human (training) and infrastructure (rapid advancement of technology) resources.* *Persuasion power to convince decision-makers to invest in surveillance.* *Volcanic activity baselines are incomplete or nonexistent in some volcanoes.* *Conflicts among working groups and institutions.* *Duplication of duties.* *Insufficiency for monitoring all active volcanoes in the region.* *Instrumentation and maintenance costly or unattainable.*	Strengths	*Existence of human resources with experience and training in basic monitoring.* *Existence of basic infrastructure.* *Existence of active volcanoes with eruptive potential.* Local technicians and scientists will get experience at some point. *Existence of databases.* *Cultural and linguistic affinities.*
Opportunities	*Exchange with research groups and between working groups.* *Meeting of experts from the region.* *Existence of national and international programs for equipment acquisition.* *Creation of surveillance networks.*	Threats	*Competition among working groups during eruptive crises.* *Incorrect perception about the magnitude of the effects of volcanic activity by the population, politicians, and authorities.* *Regional and national economic crisis causing budget cuts.* *Safety of personnel and instrumentation.* Particularly, during eruptive crises it is needed to approach the volcanoes (where the instrumentation stations have been damaged and their recovery is mandatory), personnel is not secured in their physical integrity. *International programs do not take into account local needs.* Some international programs donate instrumentation that is not suitable for the volcanoes or countries for which they were donated. Or, the instrumentation needs a level of training for which the institutions do not have specialized personnel.

purchase of spare parts. Recurrent economic crises in the LAR produces, as a consequence, budget cuts in crucial monitoring activities. Some groups in LAR do not feel that they have enough power to persuade officials to devote more money to surveillance. A low financial support not only endangers the correct functioning of instrumentation but also lowers the safety measures of maintenance personnel. An important drawback in LAR is the presence of conflicts among working groups and institutions because duplication of duties makes the budget smaller and confuses the public and authorities. In some low-income countries, international organizations donate instrumentation that is the best according to the international specialists but sometimes is not the best for the volcano or the institution receiving the donation. An important issue related to the donations is the fact that new instrumentation represents additional resources for hiring or training personnel, and for maintenance and purchase of spare parts. In most cases, these additional resources never arrive at the surveillance institutions.

The existence of a legal framework of civil protection is seen as a strength in LAR for emergency planning (Table 19.7). For this task, most volcanologists in the region consider that the existence of trained personnel is a fortitude, as well as a convocation ability during volcanic crises. In most LAR countries a good communication link exists between the scientist and the civil protection authorities. Several LAR countries with volcanoes at their boundaries have the opportunity for a high level of collaboration among scientists and officials regarding emergency planning. In some countries in LAR, however, no emergency plans exist for all the active volcanoes or there is a lack of coordination between institutions. The civil protection laws are not applied in spite of their existence. In most countries of LAR no emergency drills are held and some emergency plans have no follow up or continuity. All drawbacks together may provoke failures of the civil protection system.

As can be seen in Tables 19.5–19.7, some issues can be considered a threat/weakness and at the same time, an opportunity/strength. This is because this list gathers the opinions of volcanologists from the entire LAR, and some issues are true in some countries whereas in other countries they are not an issue. However, the list shows the general concerns of the specialists of the region. For instance, "*Presence of external research groups during volcanic crises without coordination with local groups*" is considered a threat because some NGOs and scientists from foreign countries express sometimes their opinions in a way that the media and public's perception is shifted from the official, local volcanologists provoking another kind of crises. However, the presence of NGOs at the same time is considered an opportunity because of the possibilities for cooperation and financial support of activities other than monitoring, for instance, outreach in volcanology and dissemination of the hazards to the population. Perhaps, this list of strengths, weaknesses, threats, and opportunities may help scientists outside the LAR to think about what are the local needs when planning the next program to be applied in the region.

TABLE 19.7 Results of the 2008 Strengths, Weaknesses, Opportunities, and Threats about Emergency Plans

Weaknesses	*Lack of applicability of the civil protection laws.* *There are no emergency plans for all active volcanoes in the region.* *Lack of coordination between government levels and institutions.* *There are no emergency drills at most countries and volcanoes.* *Lack of continuity of plans and emergency procedures.* *Failures in the communication system between the scientific and technical sector with civil protection authorities.*	Strengths	*Existence of legislation on civil protection.* *Existence of trained personnel.* *Convocation ability during volcanic crises.* *Existence of response and emergency plans.* *System of communication between the scientific and technical sector with civil protection authorities.*
Opportunities	*Separation of civil protection from the military sector. Not at all countries of the Latin American region.* *Exchange of experience at regional level.* *Organization and training of the community.* *Regional and international reaction systems.* *Existence of NGOs.* They help a lot during crises if they are coordinated with the local institutions.	Threats	*Changes in governmental regime.* Once a government is changed, it is needed to train the new officials and decision-makers. *Loss of credibility.* *Competition between working groups.* *Volcanic crises occurring at times and situations of lack of coordination.*

19.5 FUTURE PERSPECTIVES

Since 2011, after compilation of the information shown in Tables 19.3 and 19.4, several events have occurred that have changed the monitoring level and research in LAR. The aftermath of the eruptions of Llaima and Chaiten in Chile in 2008 raised awareness of Chilean authorities in a way that they dedicated an impressive and unprecedented financial support. The eruption of Cordon Caulle in 2011 was attended under the new circumstances of official support to the agency in charge of volcano surveillance. Because the reinforcement of the monitoring system of the Chilean volcanoes has not finished, the figures shown in Table 19.4 will change strongly in the coming years.

On the other hand, volcanoes in Ecuador have also gotten the attention of national authorities and an important breath of financial support has also provoked the renewal of the monitoring systems at most active volcanoes. The results have already been noticed during the repeated eruptions of Tungurahua volcano, but will be more evident in the following years.

After the 1985 eruption of Nevado del Ruiz, Colombia started to build up an important monitoring system. Even though the official support has not been as impressive as in the case of Chile or Ecuador, Colombia is an example in LAR for a sustained financial support for monitoring activities. The eruption of November 20, 2008 of Nevado del Huila revealed the large capacity, experience, and commitment of Colombian volcanologists who have not only observed and followed the activity of this volcano, providing advice to the Colombian authorities, but also have gotten in a very close relationship with the people living in the volcano surroundings. The success of this policy was tested during the occurrence of the 2007–2008 lahars from the Nevado del Huila. Lahars with volumes of up to $350-400 \times 10^6$ m^3 significantly affected seven municipalities in southwestern Colombia (Pulgarín et al., 2011). During the 2007 lahars no victims were reported, whereas 10 casualties were reported in 2008. This contrasts strongly with the 1,100 victims produced by a large debris flow triggered after a seismogenic event in 1994.

In spite of several successful events that have been well attended by the volcanologists in Mexico, Chile, Argentina, Ecuador, and Colombia, still several matters are pending, such as establishing volcanological research systems with a long-term program of financial support to build up a stronger scientific framework that not only observes volcanoes but also studies them.

It would be highly desirable that the rest of LAR countries follow the lead taken by the above mentioned. Table 19.4 shows the state of surveillance of volcanoes in LAR; this list however, should be read with caution. For instance, Chile and Panama have ~50 percent of their volcanoes monitored but Chile has 86 active volcanoes whereas Panama has only 2. Some countries have developed a strong scientific research system (such as in Mexico where a larger effort for monitoring volcanoes is still needed), and others have not developed research systems or need more financial support for monitoring volcanoes (as in the case of Argentina), but moreover, attention from politicians and authorities is needed to strengthen institutional capabilities in most countries in LAR.

19.6 SUMMARY

The Latin American region has been the site of some of the world's largest volcanic disasters during the last quarter of the twentieth century. Eruptions such as El Chichón in 1982 in Mexico and Nevado del Ruiz in 1985 in Colombia have marked not only the region but also the volcanological history worldwide.

Notwithstanding, the current situation in LAR has been changing in terms of human and instrumental capacities as well as of communication skills and processes. For instance, after the 1985 Nevado del Ruiz eruption and catastrophe, Colombia was taken always as an example on incorrect communication among scientists, authorities, and people. Today, the Colombian volcanological community is an example of hard work, and development of methodologies and procedures to study volcanoes and anticipate their activity in time for avoiding a disaster (Pulgarín et al., 2011). They learned the hard way but the LAR volcanological community appreciates their work, experience, and willingness for sharing knowledge and experience. ALVO can be the vehicle to enhance this sharing process.

The SWOT performed in 2008 is very useful for acknowledging the concerns of the LAR volcanological community. Even though things have changed ever since the SWOT was made, several of those concerns and ideas still persist.

The study developed to revise the state of the art in LAR volcanology is still undergoing processing because the information to be synthesized is very complex. However, the preliminary numbers show differences in the view of the active volcanoes and reveal how many of the volcanoes in LAR are being monitored and how they are instrumented.

ACKNOWLEDGMENTS

The authors are indebted to Nemesio Pérez who was an initial promoter of the SWOT in 2008 with the support of the ITER Foundation and under the auspices of the Ministry of Education and Science of Spain through the Spanish Agency for International Cooperation (AECI). Also, discussions with Joan Marti, currently General Secretary of IAVCEI are also acknowledged. In particular, we thank to Gustavo Cordoba (Universidad de Nariño, Colombia), Silvana Hidalgo (Instituto Geofísico, Escuela Politécnica Nacional, Ecuador) for helping us in the compilation of LA eruption database, and especially to IAVCEI who provided the financial support for the activities to acquire information along the LAR. H. Delgado Granados was the founding president of ALVO, J. G. Viramonte is the current president, and J. L. Palma is the vice president at present.

REFERENCES

Alvarado, G., Acevedo, A.P., Monsalve, M.L., Espíndola, J.M., Gómez, D., Hall, M., Naranjo, J.A., Pulgarín, B., Raigosa, J., Sigarán, C., Van der Laat, R., 1999. El Desarrollo de la Vulcanología en Latinoamérica en el Último Cuarto de Siglo XX. Rev. Geofísica 51, 185–241 (México).

Alvarado, G.E., Soto, G.J., Pullinger, C.R., Escobar, R., Bonis, S., Escobar, D., Navarro, M., 2007. Volcanic activity, hazards, and monitoring. In: Bundschuh, J., Alvarado, G.E. (Eds.), Central America: Geology, Resources and Hazards, vol. 2. Taylor & Francis, Londres/London, pp. 1155–1188.

Collini, E., Osores, M.A., Folch, A., Viramonte, J.G., Villarosa, G., Salmuni, G., 2012. Volcanic ash forecast during the June 2011 Cordón Caulle eruption. Nat. Hazards 64. http://dx.doi.org/10.1007/s11069-012-0492-y (Springer).

Córdoba, G., Villarosa, G., Viramonte, J.G., Salmuni, G., 2010. Estudio de los riegos de lahares y flujos densos en Villa La Angostura, Rio Negro, Argentina con motivo de la erupción en curso del Cordón Caulle, Chile. Informe Interno Direccion de Protección Civil, Argentina.

Delgado Granados, H., Viramonte, J.G., Alvarado, G.E.,, Volcanología Latinoamericana: I. Estado de la situación regional en 2008 a partir del análisis de sus debilidades, amenazas, fortalezas y oportunidades, Revista Geológica de América Central, (volumen especial PASI), in press.

Hill, T., Westbrook, R., 1997. SWOT Analysis: it's time for a product recall. Long Range Plann. 30 (1), 46–52. http://dx.doi.org/10.1016/S0024-6301(96)00095-7.

IAVCEI (Subcommittee for Crisis Protocols), 1999. Professional conduct of scientists during volcanic crises. Bull. Volcanol. 60, 323–334.

Menon, A., et al., 1999. Antecedents and consequences of Marketing strategy making. J. Mark. 63 (2), 18–40. http://dx.doi.org/10.2307/1251943. JSTOR1251943.Robock, 2000.

Oliver, R., Vatin-Perignon, N., Goemans, P., Keller, F., 1996. The Geochemistry of Huaynaputina Volcano, Southern Peru. Third ISAG (17), France, p. 609.

Pulgarín, B., Agudelo, A., Calvache, M., Cardona, C., Santacoloma, C., Monsalve, M.L., 2011. Nevado del Huila (Colombia): 2007–2008 eruptions, lahars and crisis management. In: Britkreuz, C., Gursky, H.J. (Eds.), Geo-risk Management – a German Latin American Approach. Freiberger Forschungshefte, Geowissenchaften, C538, pp. 69–80.

Siebert, L., Simkin, T., 2002. Volcanoes of the World: An Illustrated Catalog of Holocene Volcanoes and Their Eruptions. Smithsonian Institution. Global Volcanism Program Digital Information Series, GVP-3. http://www.volcano.si.edu/gvp/world/.

Smithsonian Institution, 2013. Global Volcanism Program. www.volcano.si.edu/index.cfm.

Tilling, R.I., 1989. Volcanic hazards and their mitigation: progress and problems. Rev. Geophys., 237–269.

Index

Note: Page numbers followed by "b", "f" and "t" indicate boxes, figures and tables respectively

A

B

C

K

L

M

N

R

S

T

U

V

W

Z

FSC
www.fsc.org
MIX
Papier | Fördert
gute Waldnutzung
FSC® C083411

Kolibri 360 GmbH
Ferdinand-Jühlke-Straße 7
99095 Erfurt, Deutschland
produktsicherheit@kolibri360.de